CW00665993

CHEMICAL NEUROBIOLOGY

CHEMICAL NEUROBIOLOGY
An Introduction to Neurochemistry

H. F. Bradford

Professor of Neurobiology

The Royal College of Science,
Imperial College of Science and Technology
South Kensington, London

W. H. Freeman and Company
New York

Cover: Astroglial cells in tissue culture. They show immunofluorescence caused by staining with antibodies against glial fibrillary acidic protein tagged with a fluorescent marker. [Courtesy of Dr. G. P. Wilkin]

Library of Congress Cataloging in Publication Data

Bradford, H. F. (Henry F.)
    Chemical neurobiology.
    Includes bibliographies and index.
    1. Neurochemistry. 2. Brain chemistry. I. Title.
QP356.3.B73   1985   612'.8042   85-7037
ISBN 0-7167-1694-1

Printed in the United States of America

1 2 3 4 5 6 7 8 9 0 HD 4 3 2 1 0 8 9 8 7 6

*To Helen, Sonya, and Daniel
with thanks for their tolerance
and to my parents, Harry and Rose,
for their support over the years*

Contents

Chapter 6:   The Synaptosome: An In Vitro Model of the Synapse   *311*

Chapter 7:   Synapses: Their Development and Dynamics   *353*

Preface

Neurochemistry, in common with most other branches of biochemistry, is in a time of unprecedented growth and advance. In the past twenty years, we have seen striking progress in the application of biochemical techniques to questions of neural organization at the cell and molecular levels. The new techniques of immunology have greatly accelerated our progress in mapping neural pathways and have provided us with very sensitive methods for assaying and localizing many biochemical components of neural tissue. The development of simple radio-ligand binding methods for neuroreceptors has allowed us to study in detail the localization of neurotransmitter receptors and their interaction with drugs. Advances in electrical recording techniques have enabled us to study the ion-conductance properties of single transmembrane channels in synaptic junctions and membranes by patch clamping. Progress has been rapid at both the cellular and molecular biological levels. At the molecular level, recombinant DNA techniques have provided us with the amino acid sequences of the acetylcholine receptor's subunits, and the sequences of many important neuroactive peptides and their precursors are being established. At the cellular level, the noninvasive techniques of nuclear magnetic resonance, positron emission tomography, and other imaging processes enable us to monitor in situ the levels and metabolism of important and labile metabolites, such as glucose and ATP, thus making it possible for us to follow neurochemical events in intact living cells as they occur.

In terms of the number of research workers and the quantity of papers they are publishing, neurochemistry is now probably larger than the whole of biochemistry twenty years ago. The size and ramifications of the subject make it difficult to provide comprehensive coverage in a single textbook written by one person or even a small group of people. The topics I have selected for this monograph reflect the content of a neurochemistry course taught to third-year undergraduates at Imperial College, London. The book is designed for undergraduates in the medical sciences who are encountering the subject for the first time, and for postgraduates entering the field of neurochemistry. The references to the literature should also render the text useful to more advanced researchers working in the neurosciences. My intention has been to provide a summary of current knowledge and views in each of the areas presented. The referenced literature provides the key for detailed follow-up.

One advantage of a single-author textbook is the opportunity it provides to integrate related topics. With this in mind, I have inserted appropriate cross-references throughout the text; each points to a chapter that considers the topic from a different point of view.

It has become very important in recent times for neurochemists to gain a thorough grasp of the anatomy of the nervous system. The functional connections and interactions of each neuronal population are becoming more and more firmly established, and therefore consideration of precise neuronal pathways and their interconnections becomes central to the interpretation of data generated by neurochemical research. Neuroanatomy is not formally covered here, since in the space available such a text could be no more than a basic introduction too lacking in detail for its purposes. Instead, a list of useful texts and sources is offered for those who wish to pursue the subject further.

Some of the chapters have been much improved by the critical readings of colleagues who are expert in the various areas covered. I am indebted to Dr. Graham Wilkin, Dr. Michael Hanley, and Dr. Martin Rumsby for this service, as well as to the reviewers appointed by W. H. Freeman. Drs. Reg and Maureen Docherty, Dr. Brian Anderton, and Dr. Dan Monaghan have also assisted me on various points of information.

I also owe a great debt of gratitude to my secretary Mrs. Barbara Cowen who undertook the mammoth task of typing the manuscript and who helped me in many ways at the various stages of its preparation.

H. F. Bradford
Royal College of Science, London
August 1985

1

Neurons:
Organization for
Interaction

All nerve cells, or neurons, whether from the brain of a human or the nerve cord of a flatworm, perform the same basic tasks. Every neuron generates an electrical impulse in response to a chemical or mechanical stimulus, conducts the impulse through its elongated cell structure, and, at its terminal, translates the electrical activity into a chemical signal.

Once released by the neuron, the potently active chemical, called a neurotransmitter, crosses the synapse (a gap of 20 nm) to the membrane of the next neuron in the neuronal circuit. There it interacts with specific receptors and generates a new electrical impulse, which continues the signal. Alternatively, in a few instances, the chemical can be liberated into the general stream of extracellular fluid and subsequently join the flow of cerebrospinal fluid. In this case it travels, rather as hormones do, to interact with specific receptors positioned at relatively distant target sites, sometimes in adjacent brain regions (endocrine action). Instead the chemical may act very locally through receptors on neurons in the immediate neighborhood (paracrine action). The released substances mostly act simply as neurotransmitters, but a subcategory (which may be co-released with the neurotransmitter) can also act as modulating agents (neuromodulators; see Figure 4.1). These influence the extent to which their own, or adjacent, nerve terminals can either release, or respond to the action of, neurotransmitters.

1

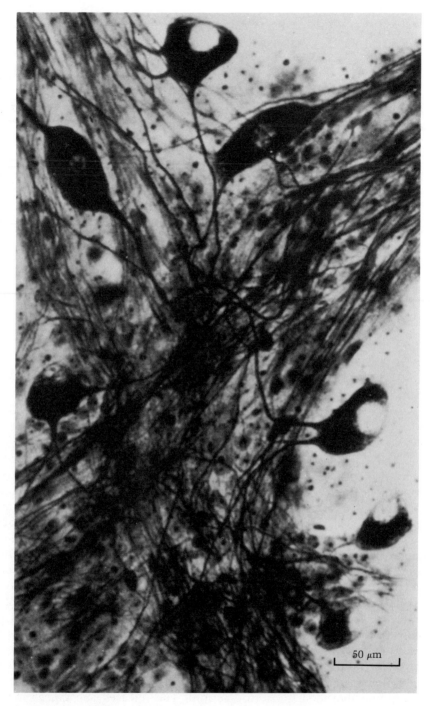

FIGURE 1.1 Neurons of the ganglion of the myenteric plexus of a cat,
stained by the silver impregnation method of Bielschowsky Gros. Bipolar
and multipolar neurons are visible. Only some of the neurons of the
ganglion are stained. Bar, 50 μm. [From ref. 146; courtesy of D. H. L.
Evans]

NEURONAL SYSTEMS

To meet the requirements of generating electrical impulses, conducting them to terminals, and liberating neuroactive substances, all neurons share the same fundamental features. Nevertheless, a wide range of neuronal cell types have evolved in the vertebrate and invertebrate nervous systems. The main cell body (also called *perikaryon* or *soma*) ranges in diameter from 10 μm basket cells of the cerebellum and ganglion neurons of similar size in the myenteric plexus (Figure 1.1) to certain 300 μm Deiter's nucleus neurons of the medulla oblongatus (Figure 1.2). The

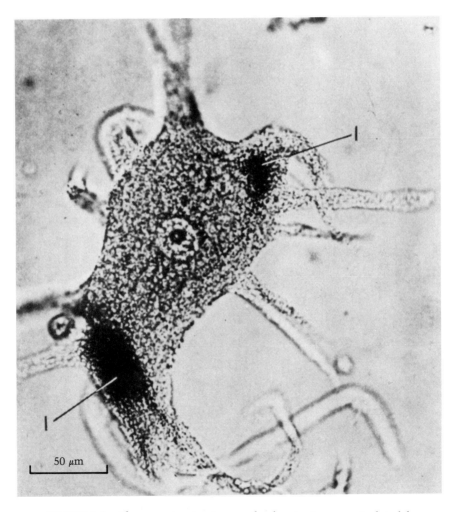

FIGURE 1.2   Phase contrast micrograph of a giant neuron isolated from the lateral vestibular nucleus (Deiter's nucleus) in the medulla oblongatus of an ox. Note the lipochrome pigment (*1*). [From ref. 3]

FIGURE 1.3 Neurons teased from
the subesophageal ganglion of a
land snail (*Helix pomatia*). (A)
Large monopolar neuron with
clearly defined nucleus and granular cytoplasm (dark-field illumination). (B) A group of monopolar
neurons showing connections to
neuropil (phase contrast). (A) bar,
25 μm; (B) bar, 200 μm.

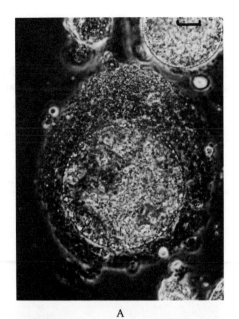

A

B

largest cell bodies occur among invertebrates. Neurons in the subesophageal ganglion of the land snail *Helix pomatia* (Figure 1.3) and the central
ganglion of the sea slug *Aplysia californica* can attain the relatively enor-
mous diameter of 300–800 μm. What advantage such large neuronal cell
bodies might confer is not known.

The ability of a neuron to respond to stimuli by generating synaptic and action potentials comes from special properties of its membrane. The total surface of the neuronal membrane, although greatly varying over the range of cell types, tends to be very large indeed. Some neurons have less than 5 percent of their surface area in the cell body itself; filamentous processes extending from the perikaryon account for the rest. The processes—dendrites at the input end of the neuron and axons at the output end—not only extend the membrane but also allow the maximum number of synaptic contacts to be made with other neurons (Figures 1.4 and 1.5).

Neurons may be classified by their processes. Neurons having many dendrites and an axon are termed *multipolar;* those having processes extending from either end are *bipolar.* The *unipolar,* or *monopolar,* neurons have a single axon, or stalk, that divides into a branching network of fibers (Figure 1.1).

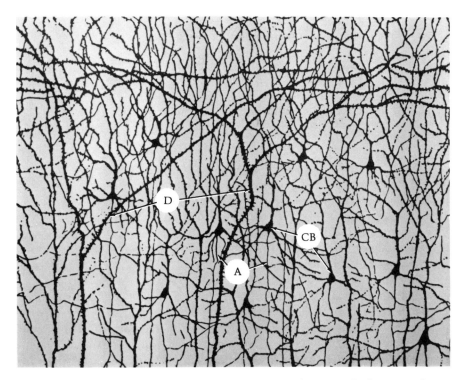

FIGURE 1.4   Golgi preparation from the visual cortex of a human infant, showing the vertical orientation of many neuronal processes. Dendrites (*D*) can be identified by the spine processes; axons (*A*) and cell bodies (*CB*) are smooth-surfaced. [From ref. 6]

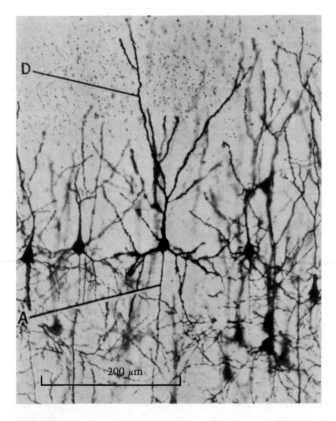

FIGURE 1.5   A section showing smaller pyramidal neurons in the outer visual cortex of a cat. Stained by the Golgi–Cox method. *A*, axon; *D*, dendrite. [From ref. 5]

Vertebrates have only one type of monopolar neuron, the sensory monopolar neuron of the dorsal root ganglion. The cell body receives an input consisting of action potentials, and it transmits this same signal as an output along its short length of axon[1] (Figures 1.6 and 1.7). It does not generate its own impulse. Monopolar neurons are much more common in the invertebrate nervous system (see Figure 1.3). There the cell body seems to function as a trophic structure, providing basic nutrients for metabolism and synthesizing protein and lipids for structural turnover, while densely branched networks of neuronal processes provide the essential signaling and impulse-generating mechanisms. The networks can even continue to generate and convey action potentials for short periods after the cell bodies are removed[2] or prevented from receiving electrical signals. Thus, the cell bodies of monopolar neurons are important in the longer term, providing sustenance and materials for structural repair and turnover, but are not critically involved in the short-term electrical activity of the cell. It is likely, however, that the continuous pas-

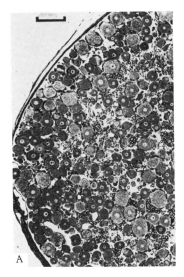
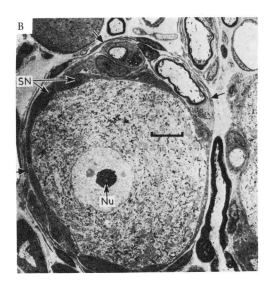

FIGURE 1.6 (A) Survey light micrograph of the cortical region of a cervical dorsal root ganglion of a rabbit. The cells are most closely packed
beneath the capsule (on left); the myelinated fibers become more numerous centrally (towards right). The wide range of perikaryal diameters, the
intermingling of cells of different sizes, and the rich capillary bed are all
evident. Plastic section, 1 μm, stained with toluidine blue. Bar, 100 μm.
[Reproduced by courtesy of J. M. Jacobs] (B) Medium-sized pale neuron
sectioned at midnucleolar level (Nu) in a survey electron micrograph of
rat cervical ganglion. Also visible are parts of the perikaryon of a small
dark neuron (top left) and of a large pale neuron (right) with an emerging
axonal process. The neurons are completely enclosed within a capsule of
satellite glial cell cytoplasm (nuclei at SN), which is more electron dense
than the neuronal cytoplasm. At the center, the ganglion cell–satellite
cell complex is partially surrounded by the processes of flattened cells
(arrows) that also enclose two of three thinly myelinated axons (top center). Bar, 10 μm. [From ref. 7]

sage of action potentials through the monopolar stalk and into the cell
body provides information about the rate at which the neuron is working
and therefore about the level at which turnover and nutrient supply are
needed, so that the perikaryon and the processes sustain their activities
interdependently.

The interdependence of the neuronal perikaryon and its fiber system
may be observed by sectioning axon bundles. Severance of the fibers in
any neuronal system causes degeneration of the region distal to the cell
body, irrespective of the point at which the section is made. Such section
is followed by outgrowth of new fibers from the proximal ends of the cut
fibers. This outgrowth is a response to the presence or absence of key

FIGURE 1.7 Scanning elec-
tron micrographs of a critical point-dried neuron
teased from a rat lumbar
dorsal root ganglion after
fixation. The stem process
of the axon emerges from
the one pole of the cell (A).
Bar, 0.5 μm. [From ref. 7]

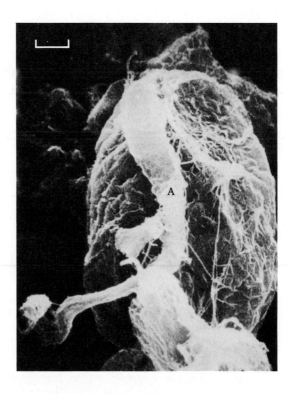

substances, such as nerve growth factor, in the stream of axonal proto-
plasm (*axoplasm*) reaching the cell body (see Chapter 7). It might also be a
response to the absence of impulse traffic in the distal region of the axons.

NEURONAL STRUCTURE

Neurons tend to have a large number of dendrites acting together to
produce either a net *excitatory* or a net *inhibitory* action. This simple
summation of inputs is one form of *neural integration,* and it determines
whether or not an action potential will be generated. If the action potential occurs, it is typically conveyed to many other neurons through the
multiple terminals of the axon.

Light microscopy reveals the typical elongated structure of neurons
of the vertebrate central nervous system to be a bushy mass of branched
dendrites at one extremity and a long, branching axon at the other (Figures 1.4, 1.5, and 1.8). In glial cells, the support cells of the nervous
system, the cell processes emanating from the perikaryon do not show
such gross asymmetry, and this helps to differentiate the two cell types.
Most neurons possess larger cell bodies than glial cells, and their nuclei
contain one or more dense, very prominent nucleoli (Figures 1.2 and 1.6).

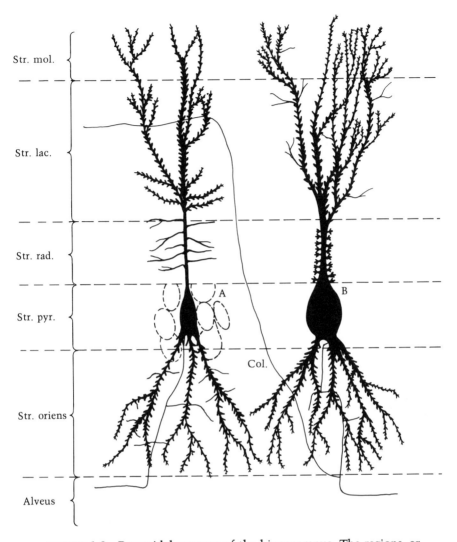

Str. mol.

Str. lac.

Str. rad.

Str. pyr.

Str. oriens

Alveus

A

B

Col.

FIGURE 1.8 Pyramidal neurons of the hippocampus. The regions, or strata (str.), of the neuron are indicated at the left of the figure. *Alveus*, white matter containing the efferent axons; *str. oriens*, region of basal dendrites; *str. pyr.*, stratum pyramidale, region of neuronal perikarya; *str. rad.*, stratum radiatum, first part of apical dendrite; *str. lac.*, stratum lacunosum, region of main branches of apical dendrite; *str. mol.*, stratum molecularis, region of final branches of apical dendrites. *A*, a small pyramidal neuron from hippocampal area CA1; note the thin lateral branches in the stratum radiatum and the absence of spines, *B*, a large pyramid from area CA3. The thin smooth axons can be seen emerging from the cell bodies. [From ref. 147]

The neuronal cytoplasm is rich in membranes making up an endoplasmic reticulum system equivalent to that found in the cells of other
tissues, particularly the liver (Figure 1.9). (Some atypical neurons have
very sparse cytoplasm, the nucleus occupying the larger portion of the
cell body. The *granule* cells are of this type; they are found in many brain
regions, including the cerebellum.) The cytoplasm also contains a large
number of free ribosomes in addition to those bound to membranes of the
rough endoplasmic reticulum. The membranes are not so densely stacked
in parallel arrays as in cells with a large secretory output, such as the
pancreatic exocrine cells, but they do show a strong tendency to organize
into parallel arrays of a special kind called *Nissl bodies.*[97] The characteristic localized membrane stacks of the Golgi apparatus can usually be
distinguished in the cytoplasm.

Many of the mitochondria in the neuronal cytoplasm are peculiarly
long, threadlike structures.[4] In culture, they can be seen by time-lapse
cinematography[10,33] to be undergoing undulating, snakelike movements
in the axoplasmic stream. Whether such long mitochondria serve some
special purpose, or possess special properties of particular value to neurons, remains unclear. They are not apparent in electron micrographs of
positively stained mitochondrial fractions prepared from brain, but they
can be seen as multiple cross-sections in positively stained tissue slices.
Negative staining often reveals their presence[4,12,143] (Figures 1.9 and 1.10;
see also Figure 6.2).

Lysosomes[13] are present in both neuronal and glial cytoplasm; as in
other tissues, they carry a large complement of acid hydrolases of latent
activity: the enzymes are detectable only when the lysosomal membrane
is ruptured[13] and are only substantially active when the assay medium is
in the pH 5 range. Most likely, the lysosomes serve the same digestive
function in neural tissue as in other tissues; in concert with similar
bodies, the peroxisomes, they reduce material ingested by endocytosis to
smaller components for use by the cells. They also serve a function in
tissue autolysis after cell damage, and they are involved in resorption,
such as takes place after nerve section and regrowth and during the turnover of cellular organelles.

Endocytosis and pinocytosis, the uptake of particles or fluid in vacuoles, are active processes in neurons of the central[14] and peripheral nervous systems, and no doubt also in glial cells. Pinocytosis is particularly
active and prominent in the growing tips (*growth cones*) of axons.[15] Observing an axon in tissue culture by time-lapse cinematography[10] (see
Figure 7.2), one gets the impression that, by pinocytosis, the advancing
axon tip can "monitor" the chemical composition of the region into
which it is about to grow.[15] In vivo, such sampling would allow chemical
guiding signals to be detected and, together with a direct interaction
between the surface membrane receptors and the chemical guiding factors, would assist in axonal pathway recognition in both development
and regeneration.

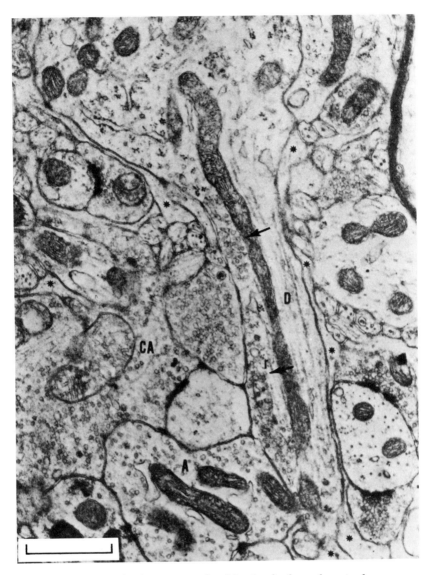

FIGURE 1.9   Branch of a primary dendrite in the lateral geniculate nucleus of a cat. A long mitochondrion is positioned both in and out of the
plane of section (lower end) and is likely to be considerably longer than
is revealed. Clusters of vesicles (*arrows*) are present in the dendrite (*D*)
together with ribosomes (*r*) and both smooth and rough endoplasmic
reticulum. Note neurotubules running longitudinally through the dendrite. Note also astrocytic sheath (*asterisks*). Adjacent to the dendrite
branch is part of a synaptic glomerulus containing a central axon (*CA*)
and the profile of a presumed nerve terminal (*A*). The dendrite forms
synapses with these axons. Bar, 1 μm. [From ref. 143]

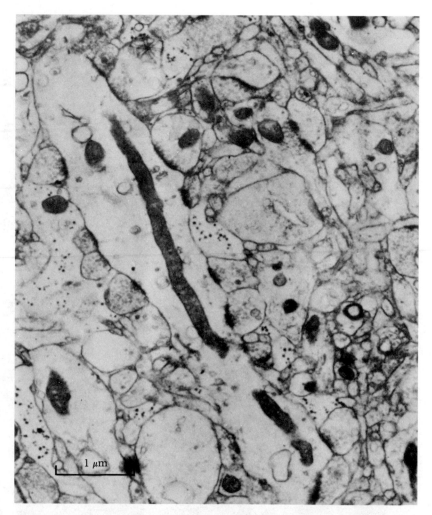

FIGURE 1.10   Dendritic trunk from the first layer of the cat cerebral cortex bearing direct synaptic contacts from axon terminals. Note long threadlike mitochondrion in multiple cross section. Bar, 1 μm. [From ref. 9]

Fibrillar Protein Systems in Neural Cells

With delicate filamentous cell processes a characteristic of neurons and glial cells alike, it is not surprising to find that in both cell types a cytoplasmic substructure confers mechanical strength to these projections. Several kinds of slender, rodlike structures made of protein provide this substructure while often serving other roles as well.

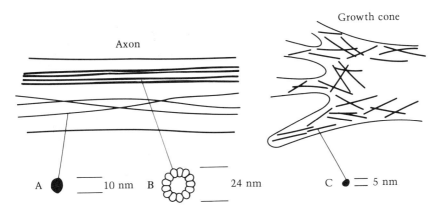

FIGURE 1.11   The three main kinds of fibrillar substructure in nerve cells. (A) neurofilament, (B) neurotubule, (C) microfilaments. [From ref. D]

Three kinds of rodlike organelles have been described in various tissues, including the nervous system. In modern terminology, they are called *microtubules, microfilaments,* and *intermediate filaments.*[16,18] Chemically distinct representatives in the nervous system include *neurotubules, neurofilaments, microfilaments,* and *glial filaments*[11] (Figure 1.11).

Neurotubules

The neurotubules found in neurons do not differ substantially in either appearance or composition from the microtubules that form the essential structure of cilia, flagella, and mitotic spindles in non-neural cells. They are tubular organelles in the range of 23–24 nm diameter with a bore of 13–14 nm and a wall thickness of 5 nm (Figures 1.9 and 1.11 to 1.14). The neurotubules run down the interior of axons, usually bearing short side-arms.

Neurotubules were first recognized when glutaraldehyde fixation of tissues was introduced, as this agent effectively preserves their structure and appearance by cross-linking them. The tubule is composed of 13 filaments of 5 nm diameter each,[18,19,L] closely stacked in parallel around a central cavity that appears to be empty (Figures 1.13 and 1.14). Each filament consists principally of a single globular protein called *tubulin* (mol wt 120,000), associated with one or two molecules of guanosine 5'-diphosphate (GDP). The tubulin of flagella is associated with other proteins called *dyneins.*[20,21] Dyneins are complexes of very large molecular weight (1–5 million) and are separated from tubulin only with difficulty. They are essentially ATPase proteins that make a transient ATP-sensitive connection with neighboring microtubules, creating a sliding action which becomes translated into a bending of the flagellum. Whether there

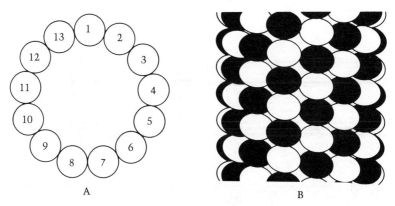

FIGURE 1.12 Schematic representation of the structure of a cytoplasmic microtubule. (A) Cross section, (B) side view. Black and white ellipsoids represent α and β subunits.

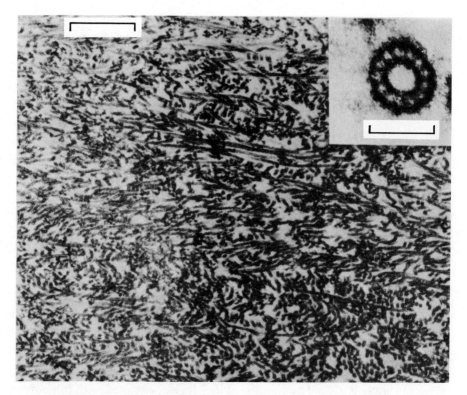

FIGURE 1.13 Microtubules from pig brain. The subunit protein (tubulin) was extracted from brain and allowed to polymerize in a test tube. The regular diameter of the microtubules, and their central hollow, are clearly seen. Inset shows a cross section of a microtubule at higher magnification, showing its 13 component protofilaments. Bar, 1 μm; inset, 25 nm. [From ref. D; courtesy of O. Behnke]

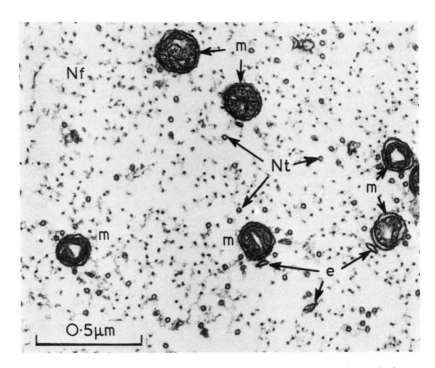

FIGURE 1.14   Electron micrograph of a transverse section through the axoplasm of a large myelinated nerve fiber: *m*, mitochondria; *e*, smooth endoplasmic reticulum; *Nf*, neurofilaments; *Nt*, neurotubules. [From ref. 38]

are "dynein equilvalent" ATPase proteins in microtubules of the nervous system is a matter of controversy.[C] Certain microtubule-associated proteins (maps) are strong candidates for this function. They do not appear to have ATPase activity, but they do electrophoretically migrate with flagella dyneins.

Tubulin has been isolated and assayed with the use of radiolabeled colchicine, a plant alkaloid that disrupts microtubular structure by binding firmly to tubulin and causing its depolymerization, thereby preventing cell division at metaphase, the stage of spindle formation. Tubulins from different tissue sources appear to be very similar molecules.[23] They can be resolved by gel electrophoresis in the presence of urea into two protein subunits of similar molecular weight but slightly different structure: α-tubulin (mol wt 53,000) and β-tubulin (mol wt 55,000). Thus, the basic structural unit of native tubulin is a dimer consisting of one β-tubulin linked to one α-tubulin subunit. α-Tubulin is slightly larger, containing an extra six amino acid residues. Native tubulin polymerizes when warmed from 0 to 37°C in the presence of guanosine triphosphate (GTP). The result is a mass of microtubules that can grow to a length of 10 μm in 30 minutes (Figure 1.13). The process is reversible. It is prevented by colchicine and vinblastine.

Neurotubules are well designed to confer mechanical rigidity and strength to the filamentous processes of neurons and glial cells. They may also be involved in dynamic functions such as axoplasmic transport and the fluidity of cell membranes.[L] The involvement of neurotubules with the mechanisms of axoplasmic transport is not clearly established. It rests on the finding that colchicine, a disrupter of neurotubules,[24,25] also blocks axoplasmic flow. Similar inhibiting action is achieved with low temperatures[C] or with vinblastine and vincristine, *Vinca* alkaloids that precipitate microtubules into paracrystalline arrays.[26] Precisely how neurotubules are involved in axoplasmic flow remains unclear, but there is evidence that calcium, a calcium-binding protein (probably calmodulin), and a Ca-Mg-ATPase are also involved.[28,C] Researchers have made several proposals:

1 The flow is created by neurotubules moving relative to one another.[C]

2 A contractile system is involved in which the ATPase of a "dynein-equivalent" protein creates cross-bridges between neighboring neurotubules, and these cross-bridges act as vectors to move substances or particles attached to the neurotubules.[22,27,183]

3 Particle movements are caused by a sliding of microtubules one along another, as occurs in flagella movement.[22]

One model built on the concept of the sliding microtubules involves energized transport of short segments called *transport filaments* from cell body to terminal by a sliding action along the surface of microtubules, by means of such ATP-activated cross-bridges, rather like a train moving down a track. The transport filaments are conceived as carrying along cell particles or proteins that attach to their surface.[L,2] Axoplasmic flow can be slowed by low concentrations of sodium vanadate and by other blockers of dynein ATPase activity, evidence in favor of a mechanism of this type.[28,29,C]

In many axonal systems, synaptic vesicles are seen in close and very regular association with neurotubules. The association is particularly well observed in the spinal cord of the lamprey, where cross-bridges between synaptic vesicle and neurotubule are readily visualized[30,31] (Figure 1.15). By time-lapse cinematography, mitochondria and other cytoplasmic inclusions have been seen to move with high velocities for short time-periods, making jerky progress (saltatory movement) along fixed pathways, which suggests that they are in close association with organized cytoplasmic substructures such as neurotubules.[10,32,33]

Both neurotubules and neurofilaments may be involved in the processes controlling membrane protein mobility and topography in many cell types.[34,36,73,C] The actions of agents that disrupt neurotubules or neurofilaments point to such a connection. Colchicine and cytochalasin B (which causes neurofilament aggregation and microfilament disruption) completely block the "capping" phenomenon, immunoglobulin aggregation at lymphocyte poles[80] (Figure 1.16). There is evidence that

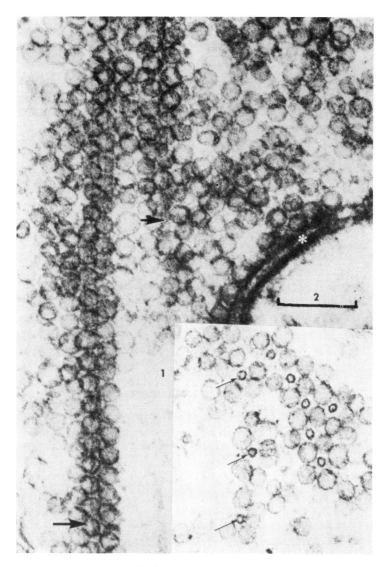

FIGURE 1.15 Neurotubules (*arrows*) in the axons of a lamprey, in close association with synaptic vesicles. Inset shows a cross section of the neurotubule system; cross-bridges are visible between neurotubules and vesicles. Bar, 200 nm. [From ref. 31]

certain transmembrane proteins of lymphocytes are "anchored" to a feltwork of microtubules positioned at the inner membrane surface. Application of nonsaturating quantities of a glycoprotein cross-linking agent such as concanavalin A prevents the capping and patching of surface antigens, probably by mooring the freely moving membrane glycopro-

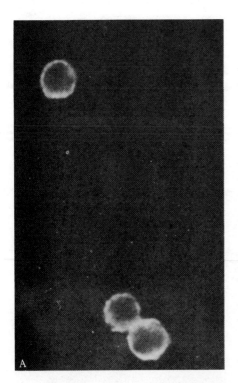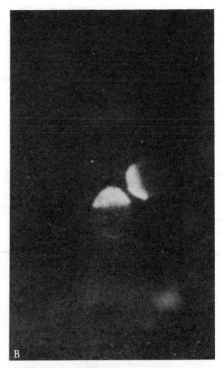

FIGURE 1.16   The capping phenomenon in lymphocytes. Patterns of immunofluorescence in mouse spleen cells incubated with fluorescence-labeled rabbit antimouse immunoglobulin in veronal buffered saline for 30 minutes. (A) At 0°C, a ring pattern is seen; (B) at room temperature, a cap pattern. The cells were washed and examined at room temperature in the presence of 3×10^{-3} M sodium azide. Note that the antimouse immunoglobulin distribution changes from (A) to (B) from generalized to clustered at the cap, (a pole) of the cell attributable to migration through the lateral plane of the cell membrane. [From ref. 80]

teins to other glycoproteins that traverse the membrane and become anchored to the submembrane microtubule and microfilament system.[34,35] Colchicine reverses the effect of concanavalin A, presumably by breaking up the extended microtubular anchoring system.[37] The actinlike contractile capability of a submembranous microfilament system might also serve to regulate receptor display and the topographic distribution of proteins in the membranes of lymphocytes[39] or fibroblasts[34] and, by extrapolation, in neurons as well (see Figure 1.17).

Neurofilaments

In the nervous system, the category of fibrous cytoplasmic organelle termed intermediate filaments is represented by the neurofilaments.[L]

These structures are distinct from neurotubules and glial filaments[11] both chemically and morphologically. They are thought to provide the basis for selective neuronal staining by methods using silver, such as the *Golgi* and the *Bielschovsky* methods (see Figures 1.1 and 1.5). Neurofilaments are 9–10 nm in diameter (see Figure 1.11) and occur throughout the neuron in axon, cell body, and dendrite. They run the length of most axons, reaching 200 μm in cultured neurons. The greatest densities of this organelle are found in axons of large diameter (Figure 1.14). The giant axons of invertebrates such as the squid (*Loligo pealli*) and the marine fanworm (*Myxicola infundibulum*) are a rich source for their preparation. In the axons of *Myxicola*, the intertwining of neurofilaments causes the whole cytoplasm to become matted into a ropelike structure that can be seized with a pair of forceps and drawn out as a long thread.

The protein structure of neurofilaments is not so clearly established as other fibrous proteins structures, but neither tubulin nor actin appear to be components. The protein subunits found in purified neurofilament preparations from various sources (brain, peripheral nerve, invertebrate axons, etc.) have been somewhat variable in molecular size, not only because of inherent differences but also because of proteolytic breakdown during isolation, at least to some extent. They appear to be particularly labile structures. Their disruption is greatly enhanced by calcium;[40] calcium ion influx during cell manipulation or disruption probably leads to breakdown of the fundamental subunits of the neurofilaments by calcium-activated proteases. In rat peripheral nerve and in brain from several mammalian species, the protein subunits of the neurofilaments turn out to be triplets of mol wt 68,000, 150,000, and 200,000.[41,42] The marine fanworm and the squid have neurofilaments with only two protein components, the large and the small.[42] The commonly observed and very abundant 50,000 to 54,000 mol wt component, often[42] reported as the major subunit of mammalian central nervous system neurofilaments, is probably derived from these larger proteins by proteolysis; indeed, studies of the events following section of peripheral nerve support this contention. During the period when neurofilaments are distintegrating morphologically, one can observe a large increase in the component of mol wt 53,000 to 73,000. When the desheathed nerves are studied by gel electrophoresis,[43] one can see a concomitant selective decrease in the three larger subunits of the mol wt 60,000 to 200,000 range.

Morphologically, neurofilaments are readily distinguished from neurotubules. In cross section they are much smaller (Figure 1.14), they often appear to be tetragonal rather than circular, and they have a much smaller central canal. They tend to course through the axoplasm in a helical route (Figure 1.11). Treatment with colchicine or one of many other agents leads to massive accumulation of neurofilaments in neurons. Large accumulations of this kind are seen in certain disease states. Neurofibrillary tangles are characteristic inclusions in the brain of patients with *senile dementia*[44] (see Chapter 9 and Figure 9.8); but their significance remains obscure.

Neurofilaments may be involved in the mechanisms of axoplasmic transport, or they may be primarily structural in function since they undoubtedly confer additional tensile strength to long processes. Their other possible functions are uncertain. Whether they interact with other fiber systems to provide anchorage or are more actively involved in contractile systems remains unexplored. Morphologically similar intermediate filaments occur in glial cells. The glial filaments, however, are very different biochemically, consisting of one protein subunit of mol wt 50,000,[11] which implies that they have a function not specifically linked to those of neuronal intermediate filaments.[45]

Microfilaments

Because of their small diameter range (5–6 nm), another class of fibrous organelles has been termed *microfilaments* (Figure 1.11). These structures appear to be composed of actin[46,47,L] and are capable of interacting with myosin in a fashion immediately suggesting that they form part of a contractile mechanism and are therefore involved in movement. This myosin–microfilament interaction leads to the formation of *decorated filaments*, which morphologically show a repeated arrowhead pattern characteristic of the contractile myosin–actin interactions of muscle. The portion of myosin that interacts with the microfilament to yield this pattern is the globular head region (heavy meromyosin).[48–50]

Microfilaments are more randomly oriented in neural cell processes than are the other fibrous protein systems. In tissue cultures, they are concentrated in the growing axonal tips (see Figure 1.11; see also Figure 7.2). The large, sheetlike cytoplasmic membranes of these advancing axonal extremities can be seen by time-lapse cinematography[10] to be undergoing large contractile waving or undulating motions as if sampling the chemical environment. Large pulsatile movements of axons can also be seen by this method of observation. Organized contractile systems involving actin filaments and myosinlike proteins almost certainly underlie such movements. Possibly offering direct evidence of the involvement of microfilaments in these cell movements is the finding that growth cone movement, endocytosis, and many other cell movements are rapidly blocked by low concentrations (1 μM) of cytochalasin B, a fungal metabolite that appears to interfere with the formation of microfilaments by preventing the polymerization of actin.[51,52,53]

AXOPLASMIC FLOW

Continuous traffic of organelles and macromolecules along the axonal processes of neurons can be readily demonstrated by constricting the axonal walls by ligature or crushing and observing the accumulation of material on both the proximal and distal sides of the constriction.[54,55] Or the progress of labeled material may be followed as it moves along the axon. Radioactive labels, fluorescent labels, and electron-dense labels (e.g., horseradish peroxidase) have all been used for this purpose.

The rates of flow in mature neurons fall into two categories: a slow rate of 1–3 mm per day that has been known for 40 years, and a more recently discovered and much faster rate, appropriate to a translocating system, operating over a wide range from 50 to 500 mm per day. Investigators continually challenge this demarcation into two systems as being too rigid; many have reported intermediate rates.[56] It seems likely that a whole spectrum of translocating systems occurs in axons, and that these differ in rate according to the neuronal types being studied. For instance the very same methods of measurement give the narrow range from 390 to 420 mm per day for the fast component of motor and sensory sciatic nerve fibers in many types of vertebrates, and it seems likely, therefore, that very similar mechanisms are operating in these fibers in a wide range of vertebrate species.[57,58]

Axoplasmic transport occurs in two directions, from perikaryon to nerve terminals (anterograde flow) and vice versa (retrograde flow). Time-lapse cinematography[10,33] shows transport in both directions simultaneously, at both fast and slow rates.

The slower processes probably involve bulk movement of axoplasm, whereas the faster and more organized processes transport individual organelles or substances along well-defined tracts within the axoplasm.[10] Mitochondria, synaptic vesicles, and many granular bodies such as lysosomes are readily observed in vitro moving through the axoplasm or through the cytoplasm of the cell body by diffusion or by streaming along organized tracts.[33] Similar movements must occur in neuronal dendrites and in glial cell processes.

One can map the anatomical projections of nerve fibers by following the progress of substances from one center to another by fast axoplasmic transport. Labeled adenosine[59] or labeled proteins[60,61] have been used to follow centrifugal transport down the axons from perikaryon to terminals. Horseradish peroxidase,[62] having heme groups that show up as electron-dense spots in electron micrographs, is a very effective marker; one maps the neural pathways by injecting it into neural tissue, from which it is taken up by pinocytosis into the nerve terminal, and then following its retrograde transport from the nerve terminal to the perikaryon.

The functions of axonal transport of materials from nerve terminal to perikaryon are likely to be at least twofold. First, there has to be a retrograde stream of organelles because the lysosomes in the terminal regions are not capable of dismembering, digesting, and catabolizing all the organelles and biochemicals continuously arriving in the anterograde axoplasmic stream. Second, information on the continuity, or state of cellular health, of the axon and its terminals must be continuously communicated to the cell body. Such communication is clearly occurring: after a nerve section at any point, regrowth occurs on the proximal edge of the cut. What these informational biochemicals are is not clear.

Nerve growth factor moves retrogradely along the axon[63,64] and it may exert a retrograde trophic influence on adrenergic neurons, particularly those of the sympathetic ganglia where the effect has been observed; this continuous flow might be informing the cell body that normal continuity is being maintained at the axonal extremities.

The flow of essential nutrients and of prefabricated proteins and organelles in the orthograde direction would supplement those taken up directly from the extracellular space by the distal or terminal regions or manufactured locally.[C] The local autonomy of the nerve terminals with respect to protein, lipid, and neurotransmitter biosynthesis—and probably with respect to the biogenesis of vesicles and mitochondria—is much greater than was thought a decade ago, though by far the largest proportion of macromolecular biosynthesis still appears to occur within the perikaryon itself.[65,66]

NEURONAL MEMBRANE STRUCTURE

During the 1970s the concept of the membrane changed radically from that of a lipid core sandwiched between two semirigid layers of partially denatured protein to that of a much more mobile and flexible system.[E–I] As elaborated from the work and ideas of S. Jonathan Singer, Garth Nicolson, and Dennis Chapman, the membrane consists of a lipid core into which are incorporated hydrophobic globular proteins (integral proteins). The core, a liquid-crystalline array of lipids, appears to be sufficiently fluid to allow the lateral diffusion of some of the integral protein components[34,36,67,68,73] (Figures 1.16 and 1.17). The introduction of *freeze-etching*, in which ultrastructure is seen in relief rather than in stained section, has made the visualization of membrane proteins a standard procedure.[69,70] In this technique, a section of tissue is quick-frozen and fractured with a sharp blade, causing the membrane to split open along its center region (Figure 1.17). Pits and mounds now visible on its fractured inner surface correspond to 8–9 nm globular proteins (Figure 1.18). Many of these integral membrane proteins are globular glycoproteins carrying a branching tree of carbohydrate residues that project from the outer surface of the cell. These glycoproteins are thought to be functionally important molecules; among other things, they probably serve as receptors for neurotransmitters and hormones (Figure 1.17).

Membrane Proteins

Few nerve cell membranes are available with the purity and high yield of red blood cell or lymphocyte membranes, and their protein and lipid content have not been analyzed and characterized in comparable detail. But nerve cells are likely to have a similar membrane organization and similar proteins and lipids. Let us, therefore, review the structural organi-

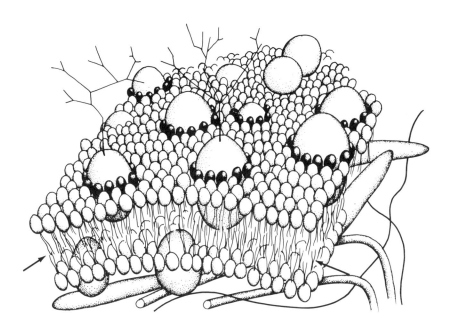

FIGURE 1.17   The fluid mosaic model of cell membranes, schematic three-dimensional and cross-sectional view. The solid bodies with stippled surfaces represent the globular integral proteins; the balls represent the polar heads of the phospholipids and the thin wavy lines their fatty acid chains. The surfaces of the integral proteins that are exposed to the aqueous solvent contain the ionic and highly polar residues of the proteins, whereas the portions of the proteins embedded in the membrane contain the nonionic and mostly hydrophobic protein residues. The bushlike structures represent carbohydrate residues of an integral glycoprotein. At long range, the integral proteins are randomly distributed in the plane of the membrane. At short range, some may form specific aggregates as shown. Some of the integral proteins may span the entire thickness of the membrane; others are only partly embedded. The bulk of the lipid is arranged as an interrupted bilayer. Certain phospholipids (dark heads) specifically associate with certain proteins forming a lipid annulus which may modify protein conformation and function. The arrows indicate the plane of cleavage in freeze-etch experiments. Neurotubules and microfilaments are seen interacting with integral proteins and peripheral proteins. [From ref. 150]

zation of lymphocyte and red blood cell membranes, with appropriate reference to nerve cell equivalents where these are known.

Certain of the integral proteins of the red blood cell membrane have been described in full structural detail. One is *glycophorin A*,[71,72] a glycoprotein of mol wt 50,000. It consists of 200 amino acid residues with 16 carbohydrate residues attached at the N-terminal region, which appears to extend well beyond the outer membrane surface. A stretch of hydrophobic amino acids 23 residues long appears to span the 10 nm thick membrane as a hydrophobic α-helix, and another domain, the C-terminal region, appears to project into the cell interior. As much as 60 percent of the molecular weight consists of the carbohydrate moieties that project outwards from the cell surface.

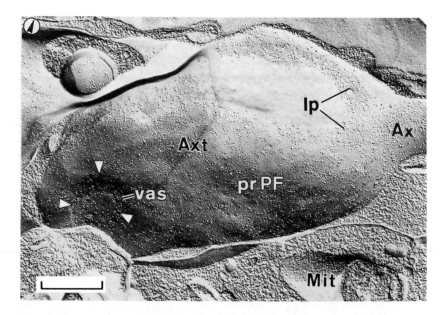

FIGURE 1.18   Nerve ending in rat spinal cord, visualized in freeze-fracture. Axon (*Ax*) with terminal (*Axt*) is impinging upon the perikaryon of a neuron. The terminal bag is seen *en face*, with the inner membrane leaflet, the protoplasmic face (*pr PF*), exposed. The active zone (*triangles*) of the presynaptic region is identified by a slight indentation of the plasmalemma and numerous discrete vesicle attachment sites (*vas*). *Mit*, mitochondria. Note the presence of numerous globular profiles of integral proteins (IP) in the axon terminal membrane. Bar, 0.5 μm. [From ref. 134]

Glycophorin is just one of several glycoproteins constituting the *type 1 integral* (or intrinsic) proteins of the red blood cell membrane. These are transmembrane glycoproteins that function as ion pumps and ion channels. Other integral glycoproteins, *type 2*, do not span the membrane but appear to act as surface antigens; they may be receptors for viruses.

One of the best characterized type 1 integral glycoproteins in the nervous system is the acetylcholine receptor. Its complete amino acid sequence has been established (see Figures 4.9 and 4.10). It is concentrated in the postsynaptic membrane of cholinergic synapses at muscle and nerve junctions, and under some conditions may occur in extrasynaptic membranes.

The voltage-sensitive sodium channel is another important transmembrane glycoprotein in neurons. Ionic currents through these channels provide the basis for action potentials and for some other forms of electrical signaling between neurons. The sodium channel glycoprotein has now been isolated from mammalian brain, neuroblastoma cells, and eel electroplax, and details of its molecular structure are emerging. The sodium channels isolated from rat brain seem to consist of a large protein subunit (mol wt 270,000) and two smaller subunits (mol wt 37,000 and 39,000). A similar protein subunit pattern is found in other neural preparations.[74,75]

A third functionally important transmembrane protein is the sodium pump (Na^+K^+-ATPase protein), responsible for producing the ionic gradients across neural cell membranes that allow the passage of the nerve impulse. This protein is present in all cells and appears to have the same molecular structure in both neural and non-neural cells.[36,77] It contains a transmembrane catalytic unit called the alpha subunit (mol wt about 100,000) and an associated glycoprotein called the beta subunit (mol wt about 45,000). The alpha subunit provides binding sites for sodium ions and ATP at the inner, cytoplasmic surface of the membrane and binding sites for potassium ions and ouabain (a pump inhibitor) at the outer surface of the membrane. The function of the glycoprotein, the beta subunit, is undetermined.

Rhodopsin is a neural transmembrane glycoprotein of the vertebrate retina. It occurs in the disk membrane of the rod-shaped outer segments of photoreceptor cells and is directly involved in photoreception. A great deal is known of its structure and organization.[78] It has a molecular weight of 36,000 and functions as a gated Ca^{2+} channel in the photoreceptor membrane.

The other category of membrane protein that has been well studied in nonneural cells consists of *peripheral proteins*. These are linked by ionic binding to the outer or inner membrane surface, from which they can be readily "salted off." As much as 30 percent of the proteins of the red blood cell membrane are of this peripheral type, and they are extracted with salt solutions. Several peripheral proteins have been characterized in detail, and *spectrin*[71,79] can be taken as an example. It appears as two bands in sodium dodecyl sulfate gel electrophoresis, one band of mol wt 240,000 (the α subunit) and the other of 220,000 (the β subunit). The two proteins are closely associated in the membrane, as shown by cross-linking with chemical reagents. Spectrin forms a scaffolding on the inner face of the membrane, giving the red cell its special disklike rigid shape[79] and regulating this shape, which changes markedly as the red blood cell passes through small blood vessels. Spectrin probably has contractile properties: it cross-reacts with myosin antisera and seems to interact with the microfilaments consisting of actin that have been detected across the cell surface. Nonfibrous monomeric actin also forms part of the spectrin scaffolding system.[79] Changes in the configuration of this system on the inner face would lead to a change in the position of the associated integral transmembrane proteins. It is the β-subunit of spectrin that attaches the protein to the inner membrane surface. It interacts with another inner peripheral membrane protein, ankyrin, which is, in turn, bound to the membrane via the cytoplasmic domain of a major transmembrane integral membrane protein called band III, which is an anion channel in the red blood cell membrane. This prevents the lateral diffusion of band III proteins in the plane of the membrane.[179] Spectrin has been detected in neural cells (e.g., cerebellar neurons) where it is restricted in location to neuronal cell bodies, probably by linking to ankyrinlike peripheral membrane proteins which have been shown to be present,[180,181,182] but which are perhaps restricted in location to the cell body. It is likely that a spectrin meshwork is serving a cytoskeletal function on the inner surface of the membranes of neuronal cell bodies, perhaps imparting local rigidity to the membrane system, or restricting the lateral diffusion of functionally important integral membrane proteins, such as neurotransmitter receptors (see Figure 4.11).

Fodrin is a major axonal protein of neurons and is related to spectrin. It is also a heterodimer, and consists of the α-subunit of spectrin together with a γ-subunit (mol wt 235,000) that is specific to brain tissue.[180,181] Fodrin is very abundant in brain, comprising 3 percent of the total protein. It is found in both cell bodies and axons, where it probably attaches to ankyrinlike peripheral proteins or to integral membrane proteins via the γ-subunit, and so forms meshworks equivalent to those produced from spectrin.[180] Unlike spectrin it readily undergoes axonal transport. This protein could play an active role in the elongation and stabilizing of the axonal membrane, as well as serve to immobilize certain integral membrane proteins.

Little information is available on peripheral proteins in neural cells, but there is evidence for the existence of neuronal surface membrane proteins that provide attachment points for actin filaments.[L]

That globular integral proteins move laterally in the plane of the membrane of many cell types, including neural cells, is now well-established;[73] rates of 1–2 μm per second have been calculated for the migration of surface immunoglobulins of lymphocytes, induced by interaction with polyvalent antibody.[80] This lateral movement can be visualized by conjugating fluoroscein to the antigen[80] (see Figure 1.16). Plant lectins are also used for this purpose. They are proteins that combine with specific carbohydrate residues of glycoproteins and cause cell agglutination. The lectin concanavalin A combines with mannose and glucose residues, and when "tagged" with fluorescein or ferritin (an electron-dense metalloprotein), it enables the position of glycoproteins in the membrane to be visualized by fluorescence microscopy or by electron microscopy. When ferritin-tagged concanavalin A was bound to the external surface of nerve terminals, the glycoproteins with which it interacted appeared to be distributed throughout the membrane surface.[81] When anticoncanavalin A antibody was applied after interaction with tagged concanavalin A, the glycoproteins were seen to be clustered in groups (Figure 1.19). The clustering has been attributed to cross-linking of several concanavalin molecules by the polyvalent antibody. For both patterns of distribution to be possible, the glycoprotein must be able to move in the plane of the membrane. The concanavalin-ferritin conjugates did not move into the synaptic cleft, suggesting that the presynaptic and postsynaptic plasma membranes are not as fluid as the membranes of other regions.[81] Alternatively, cleft material filling the 20 nm synaptic gap (see Figure 6.19) may restrict entry of the conjugates. Isolated synaptic complexes carrying portions of both pre- and postsynaptic membrane do interact with concanavalin-ferritin conjugates, but only at the exposed surface of the postsynaptic membrane where the presynaptic membrane has become detached (see Figure 6.19). Thus, the glycoproteins present in this membrane are fully capable of interacting with lectins. It is very likely, therefore, that plasma membranes of neurons and glial cells are organized in much the same way as those of non-neural cells. This would include the presence of both peripheral and integral proteins, incorporated into a bimolecular liquid-crystalline lipid array composed of phospholipids and cholesterol. The integral proteins may include many functionally important units such as

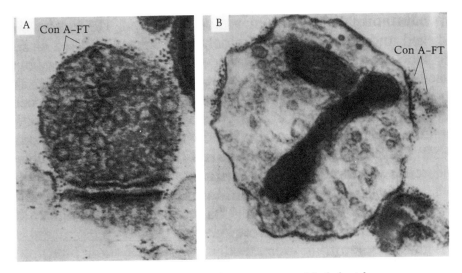

FIGURE 1.19 Fixed and unfixed synaptosomes labeled with a concanavalin A–ferritin complex (Con A-FT). (A) Synaptosome pre-fixed with glutaraldehyde. Con A-FT is dispersed over the entire surface of the synaptosome, except in the synaptic cleft. The postsynaptic membrane is broken, but the exposed internal aspect of the membrane and the attached dense material along the cleft have remained unlabeled. (B) Unfixed synaptosome, incubated with Con A-FT and anti-FT antiserum. Con A-FT is collected in discrete patches on both the presynaptic and postsynaptic membranes. [From ref. 81]

transmitter receptors, transport enzyme systems, and gated ion channels through the membrane.

Membrane Lipids

In the red blood cell, there appears to be an asymmetry in the distribution of phospholipids in the membrane, with phosphatidylcholine (lecithin) and sphingomyelin (a lecithin analogue) localized in the outer lipid leaflet and phosphatidylethanolamine and phosphatidylserine occurring mainly in the inner leaflet. Exchange between the two halves of the leaflet ("flip-flop") occurs very rarely, judging from nuclear magnetic resonance measurements (half-time 80 days).[34] No clear functional value is attributed to this asymmetric distribution, although it cannot be without significance that lecithin and sphingomyelin are electrically neutral over the wide range of pH 2 upwards whereas phosphatidylserine is negatively charged from pH 2 to pH 9 and phosphatidylethanolamine is negatively charged from pH 7.4 upwards. Whether an equivalent phospholipid arrangement exists in neural membranes has still to be established.[88] There are indications that a similar arrangement exists in myelin.[88,89]

The fatty acyl chains of the membrane phospholipids are greatly restricted in the range of chain length and the degree of saturation. Palmitic acid ($C_{16:0}$), palmitoleic ($C_{16:1}$), stearic ($C_{18:0}$), palmitoleic and oleic ($C_{18:1}$), linoleic ($C_{18:2}$), linolenic ($C_{18:3}$), and arachidonic acid ($C_{20:4}$) are all present, with C_{16} and C_{18} saturated fatty acids or those containing one double bond predominating. This particular composition appears to be of cardinal importance for forming and stabilizing the single phase of the mixed micellar structures occurring in membranes. Maintenance and control of the appropriate degree of fluidity also depend on compositional parameters of this kind.[83,1] The unsaturated fatty acid is invariably in position 2 of the glycerol moiety; the precise significance of this arrangement remains unknown. Phospholipids containing both saturated and unsaturated fatty acids appear to confer special stability on membrane micelles and prevent the phase separations that otherwise occur.

Cholesterol is the most abundant single molecular species of cell membranes. It is essentially a very rigid, planar, hydrophobic structure, tending to adsorb the hydrophobic polymethylene lipid chains onto its surface thereby reducing their molecular motion, and hence the extent of membrane fluidity. The relative content and local distribution of cholesterol will therefore allow control of membrane fluidity by controlling phospholipid movement. For instance, there is twice as much cholesterol in the outer leaflet of the membrane than in the inner leaflet,[84] suggesting that the inner leaflet has a much lower viscosity. Some membranes, however, appear to contain far less cholesterol than others[84] (Tables 1.1 and 1.2), and other lipids must serve this purpose. Mitochondrial membranes contain relatively little cholesterol, but unusual phospholipids

TABLE 1.1 Compact myelin from various animal species compared with plasma membrane of calf brain oligodendrocytes

	Composition in myelin			Composition in oligodendrocyte plasma membrane
	Human	Ox	Rat	Calf
	(% of dry wt)			(% of dry wt)
Component				
Protein	21.3	22.3	36.0	54.2
Lipid	78.7	77.7	64.0	42.5
Ganglioside	0.5	0.5	0.5	0.5
	(mol % of total lipid)			(mol % of total lipid)
Compound				
Cholesterol	40.9	44.4	42.7	36.4
Cerebroside[a]	15.6	17.3	14.1	10.4
Cerebroside sulfate[a]	4.1	2.5	2.8	2.0
Phosphatidylcholine	10.9	8.2	10.5	25.4
Ethanolamine phospholipids	13.6	15.1	18.0	7.3
Serine phospholipids	5.1	4.9	6.2	5.1
Sphingomyelin	4.7	5.1	3.4	5.4
Other lipids[b]	5.1	2.5	2.3	7.7

Note: From ref. 87; data are collected from various sources.

[a] Compare with glycolipids in Table 1.2.

[b] Includes phosphatidylinositol.

such as cardiolipin (diphosphatidyl glycerol) are present.[85] Few data equivalent to that shown in Table 1.1 exist for nerve cells.

THE ORGANIZATION OF MYELIN MEMBRANES

The white matter of the nervous system consists of axons wrapped in myelin, a thick sheath of fatty membrane (see Figures 2.8 to 2.10). This axonal covering electrically insulates the nerve fiber and allows a considerable increase in the rate of nerve impulse conduction. Current concepts of the molecular organization of the myelin membranes have benefited considerably from the advances in the understanding of other membrane systems.[88,89] Although myelin is morphologically an extension of the plasma membrane of the myelin-forming cells, biochemically its composition is very different from that of typical eukaryotic plasma cell membranes.

TABLE 1.2 Lipid composition of membranes from mammalian cells

Lipid class	Plasma membranes	Lysosomal membranes	Nuclear membranes	Endoplasmic reticulum membranes	Mitochondrial membranes	Golgi membranes
Phosphatidyl choline	18.5	23.0	44.0	48.0	37.5	24.5
Sphingomyelin	12.0	23.0	3.0	5.0	0	6.5
Phosphatidyl ethanolamine	11.5	12.5	16.5	19.0	28.5	9.0
Phosphatidyl serine	7.0 ⎤	6.0	3.5	4.0	0	2.5
Phosphatidyl inositol	3.0 ⎦		6.0	7.5	2.5	5.0
Lysophosphatidyl choline	2.5	0.0	1.0	1.5	0	3.0
Diphosphatidyl glycerol	0.0	5.0[a]	1.0	0.0	14.0	0.0
Other phospholipids	2.5[b]	–	–	–	–	3.5[c]
Cholesterol	19.5	14.0	10.0	5.5	2.5	7.5
Cholesterol esters	2.5	8.0	1.0	1.0 ⎤		4.5
Free fatty acids	6.0	–	9.0	3.5	9.0	4.5
Triglycerides	7.0	2.5	4.0	5.0 ⎦		18.0
Other neutral lipids	8.0 ⎤	6.0[d]	1.5	–	6.0	16.0
Glycolipids	⎦	–	–	–	0	–

Note: Values are expressed as weight percentages of total lipids and are averaged data; all entries have been rounded off to 0.5 percent. Dash means not mentioned or not present. Data are from various sources quotes in ref. 85.

[a] "Solvent front" (cardiolipid plus phosphatidic acid).
[b] Includes phosphatidic acid and lysocompounds other than lysophosphatidyl choline.
[c] Lysophosphatidyl ethanolamine.
[d] Mono- and diglycerides.

The Myelination Process

It is now clear that the wrapping of extensions of glial membranes around axons leads to their myelination, but the details of the process remain obscure. In the central nervous system, *oligodendroglia* serve this purpose; in the peripheral nervous system, *Schwann cells* perform the equivalent task. The essential sequence of events leading to myelination in the peripheral nervous system during embryonic development is shown in Figure 1.20. A bundle of Schwann cells and axons, enclosed together by a common membrane, begin to multiply. One axon becomes isolated from the others in its bundle and lies in a single furrow, or trough, on the surface of a Schwann cell.[86] If the axon fiber is longer than about 3 μm, it somehow signals to the Schwann cell to begin myelination. (Since some fibers with similar apposition to Schwann cells do not become myelinated, this signal probably emanates from the axon itself, although nothing is known of its nature.) The membranes that form the walls of the furrow come together to form a long sheet of double membrane, or *mesaxon*, which first becomes loosely bound around the axon and subsequently tightens up, or contracts (undergoes compaction) to form the mature myelin sheath.

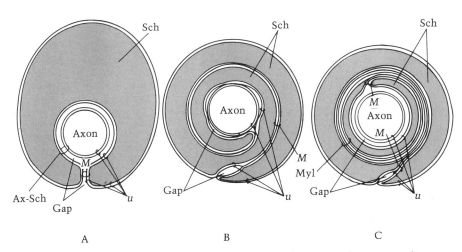

A   B   C

FIGURE 1.20 Diagrams illustrating the development of nerve myelin (*Myl*). (A) Earliest stage. A small axon is enveloped by a relatively large Schwann cell (*Sch*). Note the positions of the axon-Schwann membranes (*ax-Sch*) and of the membranes (*u*) that combine to make a mesaxon (*M*). At this stage, a gap is present between the unit membranes. (B) Intermediate stage. The membranes of the mesaxon and, to some extent, the axon-Schwann membranes, come together, with a closure of most of the gap. (C) Layers of compact myelin (*Myl*) by contact of the cytoplasmic surfaces of the mesaxon loops. [From ref. 151]

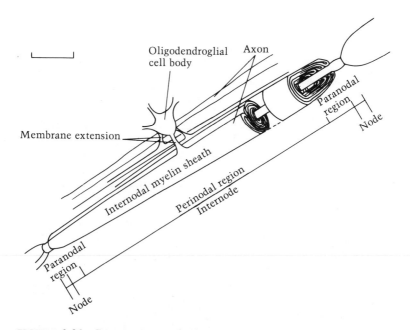

FIGURE 1.21 Perspective scale drawing of the interrelationships between the oligodendroglial cell, its processes, and the myelin sheath. On average there are some 20 to 40 processes per oligodendroglial cell, each up to 12 μm long. The length of the mature myelin sheath is usually at least ten times greater than the diameter of the oligodendroglial cell body. Bar, 10 μm. [From ref. 87]

In the central nervous system, the oligodendroglia have multiple membrane extensions, up to 50 in the optic nerve. Each myelinates a single stretch on a different axon (internodal myelin sheath), and many axons are seved by a single oligodendroglial cell (Figure 1.21). This massive output of plasma membrane to form the wrappings for local nerve fibers can require the cells to produce three times their own weight in myelin every day[89,90] (Figure 1.22).

Myelin Composition

The composition of myelin is very different from that of glial cell membranes, and the membranous outgrowths (extensions) provide a bridge of mixed composition. Myelin is characteristically richer in lipid than in protein and has an unusual lipid content (see Table 1.1). It is enriched in cholesterol, cerebrosides, and ethanolamine phospholipids. The plasmalogen form of the latter predominates, comprising as much as 80 percent of this type of phospholipid.[91] In plasma membranes the major representative is the more usual diglyceride form, and any special functions contributed by the plasmalogen in myelin remains unclear. The same must

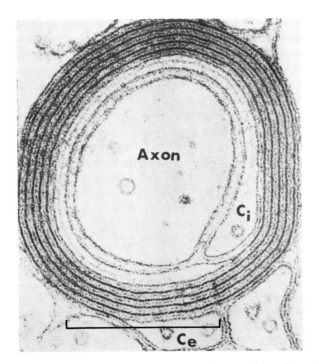

FIGURE 1.22   Transverse section of a small myelinated axon from kitten spinal cord, showing the myelin-related cytoplasm situated both internal (Ci) and external (Ce) to the compacted myelin layer. Bar, 0.25 μM. [From ref. 148]

be said for the cerebrosides. These glycosphingolipids contain galactose, sphingosine, and long (C_{24}) fatty acids that are often hydroxylated. The galactose residues appear to be located at the membrane surface, since a large proportion of them are accessible to galactose oxidase[92] and cerebroside antisera readily aggregate myelin,[93] showing that the galactose residues are accessible at the myelin surface for antibody-antigen interaction. All available evidence shows that myelin lipids are in the liquid crystalline state in situ. Thus concanavalin A receptors can diffuse laterally,[81,94] and physical techniques show that cholesterol[95] and phospholipids[96] are fully mobile. Cholesterol appears to control the extent of membrane fluidity owing to its capacity to adsorb polymethylene fatty chains onto its relatively rigid planar hydrophobic structure, and it also reduces water and ionic permeability. The slight enrichment of cholesterol in myelin, therefore, is likely to establish an intermediate degree of liquidity and reduce membrane permeability, properties appropriate to a membrane system with the primary function of electrically insulating axons. The relative enrichment in lipid over protein adds to these properties, although many of the proteins of myelin are themselves special compositional features. Two of the proteins, *basic protein* and *proteolipid protein*, are the major components of the myelin membrane, constituting as much as 80 percent of the total. Both are extractable into chloroform-methanol (2 : 1 by vol.), but the addition of salt solutions (20 percent vol.) to cause phase separation precipitates the water-soluble basic protein and leaves the proteolipid protein in solution in the chloroform-rich phase, revealing its very hydrophobic nature.

The membrane proteins remaining after chloroform-methanol extraction consist of high-molecular-weight proteins and glycoproteins. They include the *Wolfgram proteins*[98] (W1 and W2) whose existence was first proposed to account for the amino acid composition of isolated myelin, which could not be duplicated by any theoretical combination of proteolipid protein and basic protein amino acids. Wolfgram proteins have been positively identified, isolated, and sequenced, and their immunological properties have been studied; they are now regarded as important components of myelin.[90,100,101,102] Dicarboxylic acids account for about one quarter of their total amino acids and hydrophobic amino acids for another third. Their function is still unknown, but new insights may come from the recent attribution to Wolfgram proteins of the 2′,3′-cyclic nucleotidase (CNPase) activity characteristic of myelin.

Some of the other high-molecular-weight proteins are contaminants[100,103] because of the presence of axon sheath or oligodendroglial plasma membranes, which are difficult to remove entirely. It is most unlikely, however, that the myelin membrane would contain only three or four proteins.[88] Many of these other minor protein species are very likely natural components of the oligodendroglial plasma membrane that become incorporated into compact myelin during its formation, but at comparatively low concentrations owing to the overriding presence of the new and major proteolipid and basic proteins of myelin.

Special interest attaches to the glycoproteins found in this fraction,[106,107] since in other membranes they constitute the surface recognition sites for interaction with a wide range of agents such as neurotransmitters and hormones.

Myelin Proteolipid Protein

The very hydrophobic proteolipid protein is extractable from myelin of the central nervous system with chloroform-methanol, and it remains soluble in the dense chloroform-rich layer after the phase separation that follows the addition of salt solution. Proteolipid protein appears to be absent from peripheral nerve myelin, and it (or similar proteolipids) occurs only in small amounts in spinal cord. Proteolipids differ from lipoproteins in being soluble in organic solvents. The arrangement of the protein-lipid molecular complex provides an outer covering of lipid that confers solubility in lipid solvents. This arrangement ensures a close association with the lipid region of the membrane and indicates that the protein is of the integral category.

The myelin proteolipid constitutes about 50 percent of brain myelin proteins. Although its investigation has been hampered by the difficulty encountered in handling the protein once removed from the membrane, a great deal is now known about it. Its molecular weight is 23,500,[108] and it has a high proportion of the neutral and aromatic kinds of amino acid residues, which provide hydrophobic side chains. It also contains 2 percent by weight of fatty acids (palmitic and oleic) that appear to be covalently bound, clearly adding to its hydrophobicity. (These are added in the Golgi membrane system.) The proteolipid and its lipid-free apoprotein appear to be able to take up various configurations according to the local environment. Physical measurements (circular dichroism and optical rotatory dispersion) show that the apoprotein has a high degree (60 percent) of α-helical structure in organic solvents and that this is reversibly changed into the β-pleated structure when it is transferred to an aqueous system.[110]

The strongly hydrophobic nature of the apoprotein is reflected in its strong affinity for lipids, both neutral and zwitterionic. It has a particular affinity for cholesterol and binds acyl chains very tightly, reducing their movement.[111] It has been proposed that a spherical structure of 3.8 nm diameter is adopted in the myelin membrane, which would allow the proteolipid protein to span the whole width of the lipid bilayer and be exposed at each surface of the membrane. At the cytoplasmic apposition of the bilayers, it could interact directly with basic protein (see Figure 1.23). This picture is supported by the ready labeling of the protein with radio-iodine by lactoperoxidase; because of its size, this enzyme would only have access to the outer surface where the myelin layers are less tightly apposed.[114] Freeze-fracture micrographs show the presence of particles exposed at the hydrocarbon center of myelin membranes, by the plane of freeze-fracture (Figure 1.17). Similar profiles can be seen at the outer surface of phosphatidylserine liposomes[116,117] (model membrane sacs) previously incubated with the apoprotein. These profiles are probably the proteolipid positioned in the lipid layers of the liposome with its outer region projecting at the surface. Untreated liposomes showed only a smooth surface.

Myelin Basic Protein

The other major protein component of myelin, basic protein, is readily extractable from both central and peripheral nervous systems with diluted mineral acid and with salt solutions, which also shows its surface location and ionic linkage to the myelin membrane. Its high isoelectric point (above pH 10.5) indicates the preponderance of arginine and lysine residues. The molecular weight of the protein is 18,500 in bovine central

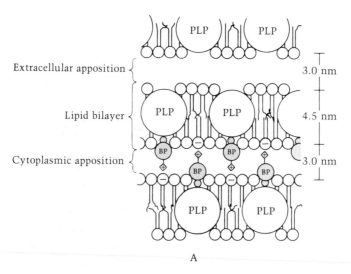

A

FIGURE 1.23   Diagrammatic representation of molecular organization in
central nerve myelin, showing the arrangement of basic protein (*BP*) and
proteolipid protein (*PLP*). (A) Cross section, (B) longitudinal section. The
basic protein is located exclusively at the cytoplasmic apposition and has
some hydrophobic sequences of polypeptide chain on the N-terminal
section of the protein penetrating the lipid surface. Ionic interactions
with lipid occur predominately on the C-terminal section of the protein.
Basic protein molecules adopt two extremes of orientation at the cytoplasmic apposition (as shown), owing to the way the repeating unit of
compact myelin is formed by compaction of two units of the oligodendrocyte plasma membrane. The proteolipid protein is buried in the lipid
bilayer with a slight exposure at the external apposition. Dimensions for
the repeating unit and the lipid phase and apposition regions are from X-
ray diffraction results. Other proteins (dark globular structures) are incorporated into the model and will include enzymes such as the 2',3'-cyclic
nucleotide 3-phosphohydrolase. Such components are distributed between the external and cytoplasmic apposition regions. Carbohydrate-
containing proteins and ganglioside (G) (with branching carbohydrate
residues attached) are exclusively located at the external apposition.
[From ref. 87]

nervous myelin; similar or lower molecular weights are found in other
species (14,000 in rat, mouse, and hamster) and in peripheral nerve myelin.[87] The N-terminal alanine in bovine preparations is blocked by acetylation. Arginine methylation and also serine or threonine phosphorylation occur as posttranscriptional modifications of the protein.

There is no clear evidence for any extensive α-helical or β-pleated
structure in the basic protein, and researchers have concluded that it
assumes a random coil configuration. Physical studies indicated that an
elongated shape (prolate ellipsoid) exists in aqueous solution, with a high
axial ratio (1.5 × 15 nm);[112] it is likely that this shape is adopted *in situ* in
the myelin membrane, where it may exist as a dimer.[113]

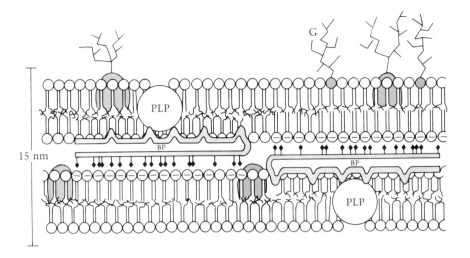

B

Unlike the proteolipid, myelin basic protein appears to be located only at the cytoplasmic surface with five N-terminal hydrophobic sequences (8–10 residues) being devoid of positive charges and therefore capable of penetrating into the lipid region. Most of the elongated molecule, however, will be positioned at the membrane surface with the C-terminal region making the ionic linkages typical of peripheral membrane proteins. This location at the inner membrane surface was deduced from the failure of many agents that label membrane proteins to gain access to the basic protein of intact myelin, and is in accord with the tight juxtaposition of the cytoplasmic surfaces in myelin lamellae (Figures 1.22 and 1.23). For instance, lactoperoxidase, which introduces radio-iodine into tryptophan residues, does label proteolipid protein but does not interact with the basic protein.[114] Dansyl chloride,[115] which should react with the multiple -NH₂ groups, is equally ineffective. These tightly apposed cytoplasmic surfaces can be separated by hypertonic salt solutions, suggesting that ionic forces predominate in their interaction. A surface carrying such a preponderance of ionic sites would provide a very appropriate environment for the basic protein with its excess of positive charges, especially as basic protein readily forms insoluble complexes with acidic lipids,[117] particularly sulfatides and phosphatidylserine.

The myelin basic protein has received special attention over the years since it was recognized as the antigen responsible for an autoimmune disease called experimental allergic encephalomyelitis, an experimental neuroallergy. This is a valuable model for the study of simple inflammatory diseases in the central nervous system,[118] of which human multiple sclerosis may be an example.

Molecular Organization of Myelin

Although there is not yet final agreement on the molecular arrangement
of the various components of mature compact myelin, models can be
constructed from their properties and their interactions when isolated,
and from studies of the extent of their labeling by various chemical
probes applied to intact myelin. The model of Rumsby and Crang (Figure
1.23) shows the integral position of the globular proteolipid protein, with
its open access at both membrane surfaces, and the peripheral location of
the basic protein at the cytoplasmic apposition, with its fine hydrophobic
fingers penetrating the lipid regions. Studies on intact myelin with probes
that interact with carbohydrate residues (e.g., galactose oxidase with tritiated borohydride) show labeling of high-molecular-weight glycoproteins.
These and similar proteins are also accessible to lactoperoxidase and
should be located at the outer membrane surface as shown in the model.
In addition, concanavalin A–ferritin complexes (Con A-FT) allow direct
visualization of the high mobility of these glycoproteins in the plane of
the membrane (Figure 1.24), another feature shared by myelin and membranes in general (cf. Figure 1.19).

Other proteins that must find a place in such a model are the few
enzymes associated with the sheath. They include the much-studied
2',3'-cyclic AMP phosphodiesterase, which has become an accepted
"marker" enzyme for myelin[119] although its function remains obscure,
and also cholesterol ester hydrolase,[120] as well as a phosphoprotein kinase
and phosphatase[121,122] capable of phosphorylating and dephosphorylating
myelin basic protein.

Current views of the molecular arrangement of the constituents of
myelin envision a membrane structure not grossly dissimilar in properties from cell membranes in general. In spite of its very high lipid content
(75 percent), the electrical resistor properties of myelin appear to be similar to those of other membranes. The key to its very effective insulating
action on the axonal membrane appears to be the high degree of compaction of the myelin layers. Such compaction excludes extracellular fluid,
and as it is only at the node region that the axonal membranes are exposed to the ionic milieu of the extracellular fluid, the ionic currents
associated with the nerve impulse have to operate from node to node,
thereby accelerating the rate of conduction of the nerve impulse (see
Figures 2.8 to 2.10).

THE FORMS AND ULTRASTRUCTURE OF SYNAPSES

The synapses through which neurons receive inputs on their dendrites
and transmit outputs at their axon terminals are highly structured cell
regions.[J,K] At a few invertebrate synapses, such as the giant synapses in
earthworms and crayfish, the nerve impulse is directly conveyed from

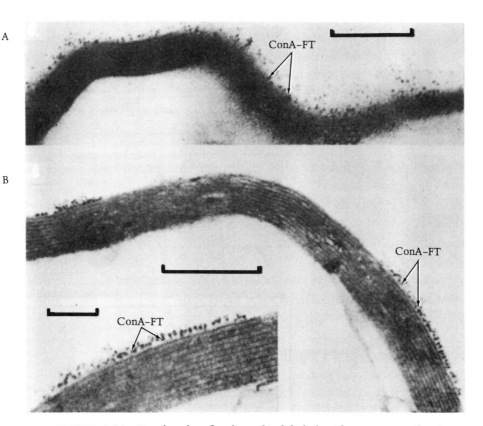

FIGURE 1.24   Fixed and unfixed myelin labeled with a concanavalin A–
ferritin complex (*Con A-FT*). (A) Myelin fragment pre-fixed with glutaraldehyde; the bound Con A-FT is uniformly and randomly dispersed
over the outer myelin surface (which in some places is cut tangentially).
Bar, 0.25 μm. (B) Unfixed myelin fragment incubated with Con A-FT and
anti-FT antiserum. Con A is collected in patches. Bar, 0.36 μm. Inset is
detail of a patch. Bar, 0.125 μm. [From ref. 81]

neuron to neuron through tightly apposed membranes forming a low
resistance junction, or *electrical synapse*. This type of transmission,
called *ephaptic*, does not require the mediation of chemical neurotransmitters and is always excitatory in action, the electrical currents being
conducted directly from cell to cell.

   In the far greater proportion of synapses, however, the communicating neurons are separated by a 20 nm gap that appears to be largely filled
with densely staining *cleft material* (see Figure 6.16), though not exclusively so. Present concepts require that the Na^+, K^+ or Cl^- ions, whose
movement across the membrane provide the synaptic potentials, are able
to diffuse freely in this space. Similarly, chemical neurotransmitters secreted by the nerve terminals should be able to move freely through the
cleft and interact with their specific receptors, which are positioned in

the postsynaptic membrane (see Chapters 4 and 7). This cleft material shows staining properties parallel to those of other components characteristic of neuron-to-neuron synaptic structures: *presynaptic dense projections* on the incoming (presynaptic) side and *postsynaptic density*, or thickening, on the target (postsynaptic) side.

Synaptic Structures

The diagram of typical synaptic components (Figure 1.25) shows the array of specialized structures found in nerve terminals and their junctions. The role of the majority of these structures remains uncertain, though the characteristic appearance of some of them allows easy recognition of synapses in electron micrographs. A typical synapse of the cerebral cortex after positive staining can be seen in Figure 6.16. Most of the specialized synaptic structures have been isolated in sufficient bulk and purity for biochemical study. The isolation of *junctional complexes* and *postsynaptic densities* are described in Chapter 6 (see Figures 6.17 to 6.20). Intact *coated vesicles* (or *complex vesicles*), with their surrounding elaborate hexagonal basket structure, which forms the coat, have also been prepared in bulk (Figure 1.25)[123] and their major macromolecular components characterized.[124,173,174] The coat itself consists essentially of protein and is readily disintegrated by proteolytic enzymes.[124] The pure coated vesicles contain one major protein, *clathrin* (mol wt 180,000), which is also widely distributed throughout the presynapse.[127] In addition to clathrin, two proteins of about 38,000 and 35,000 mol wt are probably also intrinsic to coated vesicles.

The evidence suggests that clathrin is synthesized in neuronal cell bodies on free and bound ribosomes, the nascent peptide chains then being released into the cytosol. Clathrin is carried by axoplasmic transport to the synaptic regions where it becomes associated with a range of structures, including coated vesicles and pre- and postsynaptic membranes.[175]

Clathrin is not a glycoprotein, which may be the reason for its easy association and dissociation from membranes.[129] Synaptic vesicles in general are probably formed by an endocytotic process. They can be labeled by incubating isolated nerve terminals with electron-dense markers such as horseradish peroxidase or colloidal thorium dioxide.[125] This indicates their formation by an inwardly directed pinching-off from the plasma membrane. The basket, or coat, of vesicles could be formed in the endocytotic process if clathrin made up the inner lining of the membrane,[126,127] perhaps at localized regions called "coated pits," which could pinch off to become coated vesicles (Figure 1.25).[127,177] This is known to occur in the neuromuscular junction[176] and other tissues.[177] It seems that coated vesicles can take up Ca^{2+} by an ATP-requiring mechanism comparable to the Ca^{2+} sequestering activity performed by the sarcoplasmic reticulum of muscle.[128] Thus, coated vesicles may also be involved in capturing or localizing Ca^{2+} in the nerve terminals. They are also seen in

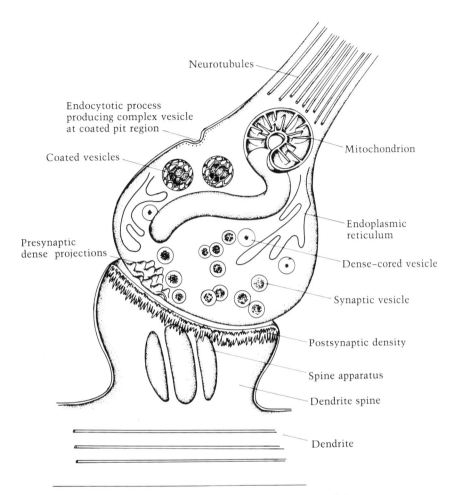

FIGURE 1.25   Diagrammatic version of an axon terminal forming a synapse on a dendrite spine. The structures shown are not to scale. Endocytosis produces complex vesicles consisting of a hexagonal basketwork of
fibers (cytonet), which form part of the inner surface of the nerve terminal membrane at regions called "coated pits".

the Golgi areas of neurons where, as in other cells (e.g., renal tubules,
oocytes), they probably participate in intracellular transport of newly
synthesized membrane and protein and in receptor-mediated endocytosis
of specific substances.[175] Coated vesicles have been described in dendrites
and postsynaptic regions; they are not limited to presynaptic nerve endings. The principal function of these interesting structures is still thought
to be the recapturing of membrane from the nerve terminal membrane by
endocytosis after the exocytotic release of neurotransmitter from synaptic vesicles (Chapter 7). This recapturing of membrane would prevent the
plasma membrane from expanding in area.[125,130] After shedding of the

coat, conventional synaptic vesicles would be formed since the "bare" vesicle seems to be indistinguishable from conventional synaptic vesicles in size and other properties. The similar staining properties of the clathrin coat and the network of presynaptic densities and projections suggest that the former may become incorporated into the latter during its life history.[130,131]

The *common synaptic vesicles* are very regular structures with a clear double membrane (see Figure 6.13). They are 50 nm in diameter and, together with the synaptic thickenings, form a very characteristic synaptic structural configuration, allowing the pre- and postsynaptic regions of the synapse to be distinguished morphologically. But even these well-known structures have been described as artifacts of fixation[132] and possibly derived from a labile endoplasmic reticulum of the type more readily seen in negatively stained terminals (Figures 1.9 and 6.2).

These vesicles have been isolated as greatly enriched fractions, judged morphologically (Figure 6.14), and they are found to be enriched in various transmitters (e.g., acetylcholine, catecholamines, serotonin) depending on the brain region of origin. A mixture of transmitter types is always present, as the vesicle preparations are derived from a mixed population of nerve terminals, each employing a different neurotransmitter. Other chemical components are also present, including ATP and soluble proteins or other polyionic substances. They can be released from the vesicles by various procedures. *Vesiculin* (mol wt 180,000) is a peptidoaminoglycan believed to bind acetylcholine within the cholinergic vesicle of the electric ray (*Torpedo*).[138] *Chromogranins* (mol wt 82,000) are proteins associated with noradrenaline in adrenergic vesicles of the splenic nerve[139] and chromaffin granules of the adrenal medulla.[140] The large (40–60 nm) *dense-cored granules* contained in the nerve terminals of the posterior pituitary contain either oxytocin or vasopressin for release to the bloodstream. These granules also contain neurophysins (mol wt 200,000), probably formed from the same mother protein as the peptide hormones themselves.[141] *Dense-cored vesicles* are larger structures (60–80 nm diameter) found in neurons and their terminals in many different regions of the nervous system. They are believed to specifically contain catecholamines or indoleamine neurotransmitters. Agents releasing these neurotransmitters from neural tissue will often remove the central density of these vesicles.[82]

The *presynaptic grid* is the regular hexagonal arrangement of presynaptic densities that can be seen in *en face* electron micrographs, both positively stained and freeze-fractured, and can be inferred from transverse sections (Figures 1.26 and 1.27). The hexagonal spacing of the dense projections of the grid suggests a role as guides for positioning the vesicle close to the terminal membrane ready to discharge their transmitter (see Figure 7.11). In this case the grid would not allow the vesicle to approach closer than 10 nm from the membrane itself, and a connecting tube would be required to convey transmitter from vesicle to cleft. *Synapto-*

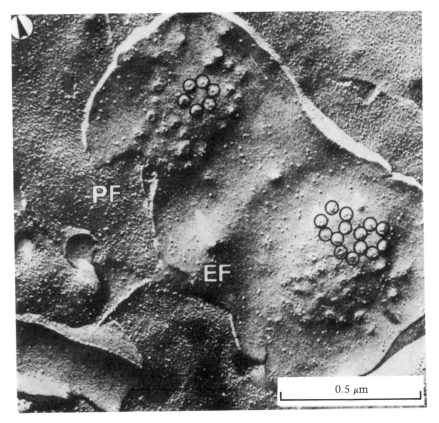

FIGURE 1.26   Presynaptic nerve terminal from rat spinal cord, showing hexagonal arrangement of vesicle attachment sites in freeze-etch replica of active zone on the presynaptic membrane. Numerous protuberances, often with craterlike openings, are encircled in order to visualize the geometry of vesicle-to-plasmalemma relationship. *PF*, cytoplasmic fracture face; *EF*, external fracture face. [From ref. 152]

pores, structures seen in micrographs of conventionally stained and freeze-etched nerve terminals,[133,134] have been proposed to serve this "channel" function.[133,134]

The usually globular organization of the substructure of postsynaptic thickenings is described in Chapter 6. A great deal remains to be discovered about the detailed organization and function of these thickenings. They are synthesized by the target cell and apparently held in position by physical continuities with the postsynaptic membrane, with cleft material, and possibly even with the presynaptic density. In the latter case, complete *transsynaptic* connections would have to exist. Which cell synthesizes such connecting structures is an interesting question. It has been

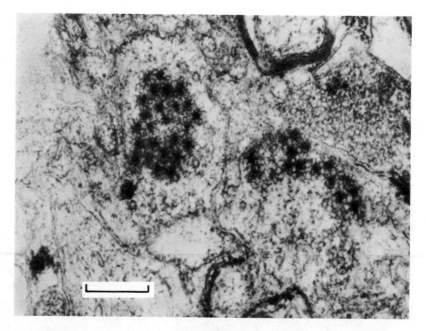

FIGURE 1.27   Synapse of spinal cord sectioned *en face*, showing regularly arranged dense projections in presynaptic processes. Bar, 0.4 μm. [From ref. 153]

proposed that strong ionic forces hold the two components together, but exposure to conditions that should separate such ionically linked structures is not very effective.[136] Electron microscope sections made in the plane of the thickening often show them as perforated structures, even complete ring structures[135] (Figure 1.28), which might account for the commonly seen double thickenings in synapses (see for example, Figure 6.16). In developmental terms, the postsynaptic thickenings can appear either before or during contact formation between axon and postsynaptic target, depending on the neural system involved,[136] and can therefore be formed in the absence of a presynaptic fiber. Postsynaptic densities, for instance, form in the Purkinje cells of mutant Weaver mice in which the presynaptic parallel fibers do not develop. These fibers are the major presynaptic input to Purkinje cells.[137]

Many catecholaminergic neurons (those employing monoamines as neurotransmitters) do not form conventional synapses of the kind described above but terminate in strings of swellings called axonal varicosities (Figure 1.29). Many varicosities do not form structured synaptic contacts with pre- and postsynaptic thickenings, but they usually contain synaptic vesicles or granules and are capable of releasing neurotransmitters into the extracellular space. A single impulse releases transmitter from each of a string of axonal varicosities (see Chapter 7).

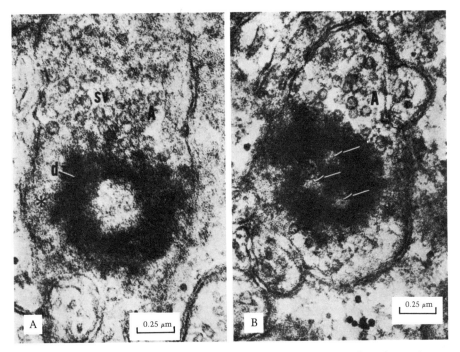

FIGURE 1.28   (A) A synapse in cat cerebral cortex viewed *en face.* The postsynaptic density (*d*) forms a ring and is surrounded by the profile of the axon terminal (*A*), which contains synaptic vesicles (*sv*). (B) A synaptic junction viewed *en face.* The postsynaptic density has three perforations (*arrows*) and is surrounded by the profile of an axon terminal (*A*). [From ref. 135]

Synaptic Systems

In the central nervous system, synapses form in a wide range of pre- and postsynaptic configurations[154,K] (Figures 1.30 and 1.31). Connections may occur from axon direct to dendrite or to dendrite spine and not only from axon terminals (*bouton terminaux*) but midaxon (*bouton de passage*). Midaxon contacts may be multiple, occurring repetitively as the incoming fiber climbs up the dendritic tree.

Another arrangement involves *serial synapses*[142] in which single dendrites, axons, or axon terminals receive or make several synaptic contacts in a small space and often in a serial arrangement, which affects their ability to respond to an invading nerve impulse.

Axoaxonic synapses occur when one terminal forms a synapse on a second terminal (Figure 1.30), modulating its synaptic effectiveness (presynaptic inhibition or facilitation). Such synaptic arrangements are seen in greatest abundance in neuronal groupings that function as relay cen-

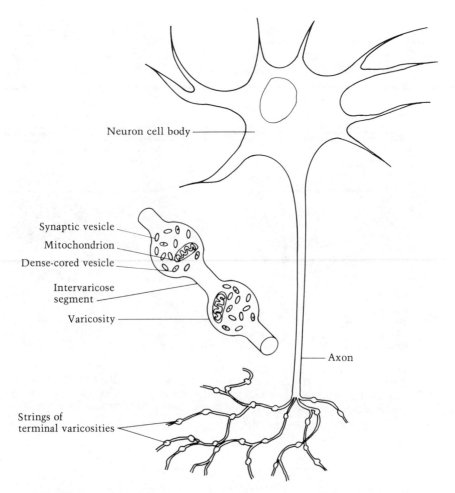

Neuron cell body

Synaptic vesicle

Mitochondrion

Dense-cored vesicle

Intervaricose
segment

Varicosity

Axon

Strings of
terminal varicosities

FIGURE 1.29   Diagram of a neuron with terminal varicosities.

ters collecting and redistributing neuronal signals[K] (e.g., the thalamus, the lateral geniculate nucleus of the visual system, and the cuneate and gracile nuclei of the upper regions of the spinal cord).

In another more common arrangement, the axon terminates on the cell body, forming an *axosomatic synapse*. Its function is usually inhibitory, as in the basket cell endings around the Purkinje cell bodies in the cerebellum. The inhibitory effect might be related to the close proximity of the terminals to the cell body, where the nerve impulse is generated when the cell membrane potential is lowered to the critical level.

Dendrodendritic synapses are synaptic systems which (Figure 1.32)

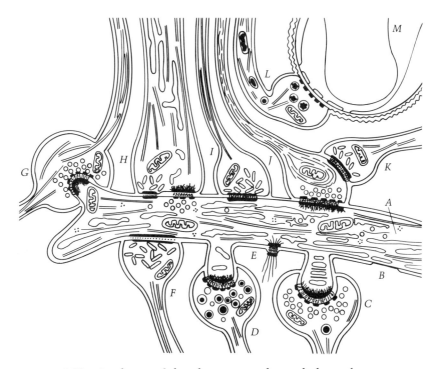

FIGURE 1.30 A scheme of the ultrastructural morphology of synapses,
showing various junctional structures grouped around a dendrite (A). The
tight junction (B) and the desmosome (E) are without synaptic significance. Excitatory synaptic boutons are shown (C, G) containing small
spherical translucent synaptic vesicles. D, a bouton with dense-cored,
catecholamine-containing vesicles; F, an inhibitory synapse containing
small flattened vesicles; H, a reciprocal synaptic structure between two
dendrite profiles, inhibitory towards dendrite A and excitatory in the
opposite direction; I, an inhibitory synapse containing large flattened
vesicles; J and K, two serial synapses: J is excitatory to the dendrite and
K is inhibitory to J; L, a neurosecretory ending adjacent to a vascular
channel (M) surrounded by a fenestrated endothelium. All the boutons in
the diagram are of the terminal type except G, which is a *bouton de
passage*. [From ref. 172; artwork by R. E. M. Moore]

have been observed in several brain regions,[143] including the lateral geniculate nucleus, the olfactory bulb, the retina, the superior colliculus, and
the dorsal horn of the spinal cord. Such synapses show typical presynaptic aggregations of synaptic vesicles and their associated synaptic thickenings, demonstrating that these structures are not restricted to axon
terminals (Figure 1.9). These dendrodendritic synapses are often reciprocal. This curious two-way organization implies the existence of a two-

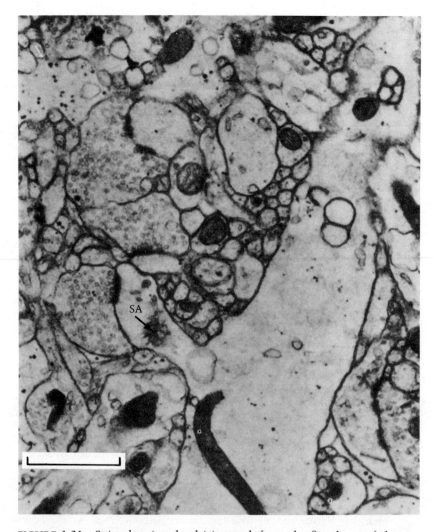

FIGURE 1.31   Spine-bearing dendritic trunk from the first layer of the cat cerebral cortex. Showing axo-dendritic synapses at dendrite spines. Note spine apparatus (SA). Bar, 1 μm. [From ref. 9]

component synapse, one region forward-directed and the other backward-directed. With one of these components being inhibitory in action, reciprocal synapses could mediate localized feedback inhibition (Figure 1.32). Dendrodendritic synapses have been seen in the olfactory bulb between granule cell dendrites and mitral cell dendrites; a feedback inhibition mechanism would explain the observed long-lasting self-inhibition of the mitral cells that follows their activation.[144]

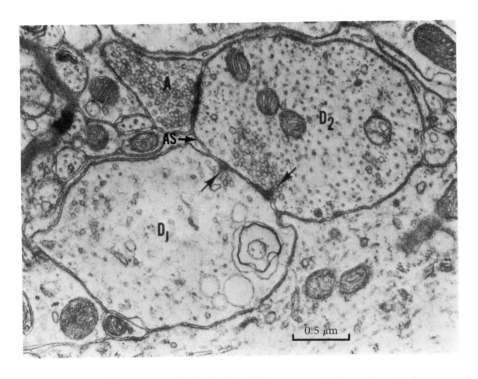

FIGURE 1.32 Reciprocal dendrodendritic synapse in lateral geniculate nucleus of cat. Arrows indicate direction of presumed chemical transmission. One dendrite (D_1) probably belongs to a Class V cell and the other (D_2) to a neuron whose axon projects to the visual cortex. Note the small axon terminal (A) synapsing with D_2 and the astroglial sheath (AS) that partially surrounds these three processes. Bar, 0.5 μm. [From ref. 143]

Complex synaptic interactions are also seen in multiple-component structures called *glomeruli*, or *synaptic islands*. Where several axons and dendrites form specific contacts in a localized region of the tissue usually encapsulated within glial cells, and they often carry both excitatory and inhibitory components. The synaptic glomerulus of the cerebellum is a well-known example of this type of synaptic complexity (see Figures 1.33, 1.34, and 1.9).

Excitatory and Inhibitory Synapses   Certain rough principles have emerged for identifying the two categories of neuron-to-neuron synapse, excitatory and inhibitory. Electrophysiological and other evidence shows the distribution of the excitatory synapses to be on the upper reaches of the dendritic tree of central neurons. These axodendritic synapses are attached either directly onto the dendrite surface or onto dendrite spines containing a characteristic *spinal apparatus* of unknown function (see

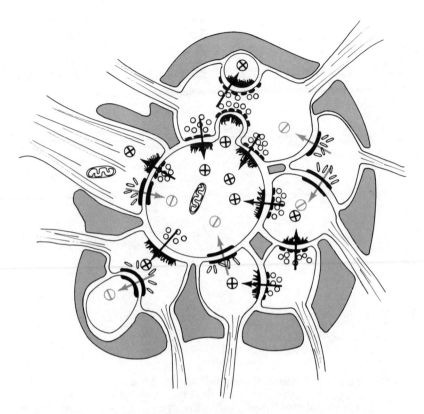

FIGURE 1.33   A synaptic glomerulus, showing various arrangements of
synapses grouped around a centrally placed terminal dendritic expansion,
seen in cross section. Both excitatory (+) and inhibitory (−) synapses are
shown; the direction of transmission is indicated by arrows. A glial capsule surrounds the whole complex. [From ref. 172; artwork by R. E. M.
Moore.]

Figures 1.25 and 1.30 to 1.31). Inhibitory synapses, in contrast, usually are
attached to initial segments of dendrites or to cell bodies (axosomatic).

   In ultrastructural terms, too, there appear to be differences. Gray[142]
has shown that in simple two-component synapses, those designated
inhibitory from their electrophysiological properties and distribution
possess a symmetrical arrangement of pre- and postsynaptic thickenings;
these have been designated Type II synapses. Excitatory synapses, on the
other hand, show an asymmetrical arrangement of these thickenings,
with the postsynaptic component a larger and more conspicuous structure. These are known as Type I synapses (Figure 1.30).

   Another common feature of inhibitory synapses is the presence of
flattened or elongated vesicles, seen after aldehyde fixation.[142,145] A wide
survey of inhibitory synapses supports this correlation.

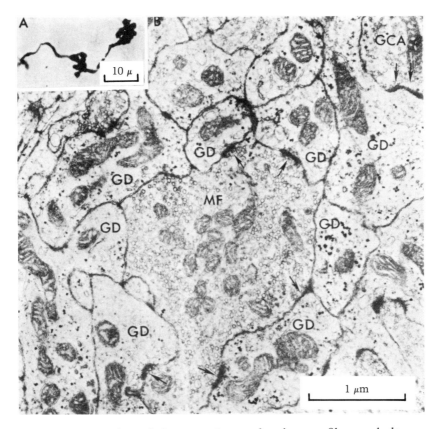

FIGURE 1.34   Light and electron micrographs of mossy fibers and ultra-structural organization of cerebellar glomerulus in the spectacled caiman (*Caiman crocodilus*) a close relative of the crocodiles. (A) Golgi stain of a mossy fiber rosette at the cerebellar granular layer. (B) Electron micro-graph of a similar area, showing the mossy fiber terminal (*MF*). The mossy fiber sac is seen to establish multiple synaptic junctions (*arrows*) with the dendrites of the granule cells (*GD*), which surround the mossy-fiber terminal. Smaller areas of synaptic contact between mossy fiber and granule cell dendrites are shown in the upper left and lower right corners of the micrograph. The upper right corner shows a synaptic relationship between a Golgi cell axon (*GCA*) and a granule cell dendrite (*two arrows*). Note the difference in size and shape between the synaptic vesicles in the mossy fiber and in the Golgi cell terminals (see Figure 1.33 for diagrammatic model of the synaptic interactions). [From ref. 155]

Synaptic contacts appear to cover a large proportion of the dendrite surface and in some regions appear to leave little or no free space. Figures 1.35 and 1.36 show the dense synaptic coverage on apical dendrites of pyramidal neurons from hippocampus (see also Figures 1.9 and 1.10). Few areas of the dendritic membrane are free of synaptic terminals.

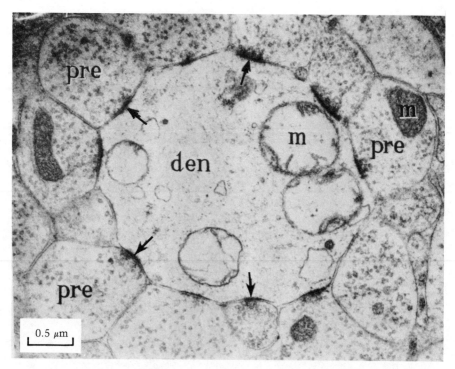

FIGURE 1.35   Section from the stratum lacunosum of a rabbit hippocampus, cut at right angles to the apical dendrites (see Figure 1.8 for location). A large dendrite (*den*) is cut transversely and its membrane is in contact with presynaptic processes over the whole of its surface. Three of these processes are labeled (*pre*). Characteristic membrane thickenings are visible, some indicated by arrows. These are limited in extent to part of the synaptic cleft. [From ref. 147]

SPECIFIC BIOCHEMICAL MARKERS
FOR NEURONS

The first indication that a particular biochemical is localized in neurons comes from its enrichment in gray matter over white. The disappearance or substantial depletion of a compound after treatment designed to destroy neurons is another indicator. Absolute specificity in the localization of substances to neurons or to glia is hard to prove. But the treatment of neural tissue sections with fluorescence-labeled antibodies to supposed glial or neuronal components employed as immunohistochemical reagents, and the analysis of isolated cell preparations and cell fractions by direct biochemical and immunochemical methods, can be used to show that certain proteins are greatly enriched and sometimes specifically located in neurons.

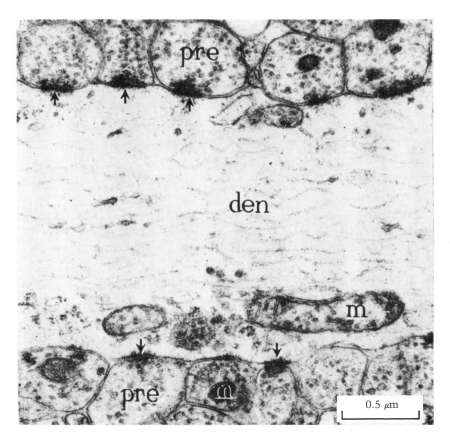

FIGURE 1.36   Section taken parallel to the apical dendrites from the region of the stratum lacunosum of rabbit hippocampus (cf. Figure 1.35). The main branch of an apical dendrite (*den*) is seen cut longitudinally; it contains dendrite tubules and mitochondria (*m*). The whole of the dendritic surface membrane is in contact with presynaptic processes, two of which are labeled (*pre*). Characteristic membrane thickenings are visible, some indicated by arrows. [From ref. 147]

Neurotransmitter-Synthesizing Enzymes

All evidence points to a specific localization of many of the key enzymes of neurotransmitter biosynthesis in specific neuronal subpopulations (Chapter 4). For example, glutamate decarboxylase, which produces γ-aminobutyrate from glutamate, appears to be specifically in the neuron and nerve terminals that release γ-aminobutyrate (called gabaergic neurons).[157] Choline acetylase,[158] which forms acetylcholine from acetyl coenzyme A (CoA) and choline, appears to be a specific marker for cholinergic neurons.

Tyrosine hydroxylase,[159] which employs molecular oxygen to form L-dihydroxyphenylalanine (L-dopa), is present only in neurons synthesizing catecholamines (i.e., noradrenaline and dopamine). Tryptophan hydroxylase[160] and dopamine β-hydroxylase[161] are biochemical markers respectively for indoleamine (5-hydroxytryptamine, serotonin) and norepinephrine-containing neurons. These conclusions were reached from immunohistochemical localization of the proteins, employing as labeling agents antisera to the purified enzymes in question.

The 14-3-2 Protein

Another biochemical substance localized in neurons is the 14-3-2 protein. Its curious name derives from its elution pattern during column chromatography. This protein has been isolated and antisera have been raised to it, which have greatly expedited studies of its distribution in neural tissues. It is greatly enriched in the nervous system—in the rat it is 200 times more abundant in the brain than elsewhere—and it was found in immunologically cross-reacting forms in the nervous systems of all mammalian species, birds, fish, and reptiles.[162,163,164,165]

The 14-3-2 protein has been quantified by complement fixation assay, quantitative immunoelectrophoresis, and radioimmunoassay.[165] It has been localized to neurons by degeneration studies and by immunohistochemistry,[166] though it is not present in all neurons.[167] Sedimentation velocity and equilibrium data indicate that the protein consists of two dimers of mol wt 39,000 (i.e., a total of 78,000), and SDS-polyacrylamide gel electrophoresis gives a monomer of mol wt 48,000.[168] It is a very acidic protein, containing a high proportion of glutamate and aspartate residues, but it differs considerably in composition from S-100 protein, an acidic protein localized to glial cells (see Chapter 2).

The 14-3-2 protein is now designated neuron-specific enolase (NSE) since it has been shown to be identical with an isoenzyme of the glycolytic enzyme enolase (2-phospho-D-glycerate-hydrolase; EC 4.2.1.11).[165] Its isolation in a modified procedure can now be monitored by following enolase activity.[165] It is a valuable biochemical marker for following the differentiation of neurons during embryonic development (e.g., in the rat or monkey cerebellum).[165] It constitutes 1.4 percent of the total soluble protein and 14 percent of the total enolase activity of adult rat brain.[165] The non-neuronal enolase isoenzyme has been shown to be localized to glial cells.[169] A third isoenzyme is a hybrid containing subunits from the other two types.

Tetanus and Cholera Toxins

The bacterial toxin that causes tetanus binds specifically to neuronal cell surfaces in a variety of peripheral and central neural cultures.[170] Specific gangliosides (GD_{1b} and GT_1) are the receptors for this agent.[171] Neurons in dorsal root ganglion cultures may be unambiguously distinguished from glia and fibroblasts by their ability to bind both tetanus toxin and antibodies to the Thy-1 antigen.[170] Such cells are further distinguished by the absence of glial markers such as glial fibrillary acidic protein. (See Chapter 2.)

The GM_1 ganglioside is the receptor for the toxin causing cholera. Cholera toxin specifically labels neurons of the cerebellum because of the unique presence of the GM_1 ganglioside on neuronal cell surfaces in this brain region.[176]

Other Neuron-specific Proteins

A variety of additional proteins are purported to be located specifically in neurons, based on subcellular localization, immunohistochemistry, and immunoelectrophoresis using antisera raised against the isolated proteins.

GP-350   A brain-specific sialoglycoprotein, GP-350, with a molecular weight of 11,600 and an isoelectric point (pI) of 2, is localized to Purkinje cells, pyramidal cells, and stellate cells in both membrane-bound and soluble forms.[178]

Synaptin   Synaptic vesicle membranes contain a glycoprotein called synaptin, which has also been detected in synaptic plasma membranes. It has a molecular weight of 45,000 and a pI of 4.2.[178]

D_1, D_2, and D_3   Three brain-specific proteins were detected by immunoelectrophoresis and found to be localized to synaptic membranes. D_1 is composed of two polypeptide chains of 50,000 and 116,000 molecular weight, D_2 is a single polypeptide of 139,000 molecular weight, and D_3 consists of three chains of molecular weights between 14,000 and 50,000.[178] The functions of these proteins is unknown but they can be used as synaptic markers. The level of D_2 parallels the rate of synapse formation and may be involved in intercellular recognition processes during synaptogenesis.

P-400   The membrane-bound protein P-400 is found only in the molecular layer of the cerebellum, where it is present in the dendrites of Purkinje cells. It has a molecular weight of 400,000 but its function is unknown.[178]

REFERENCES

General Texts and Review Articles

A   Landon, D. N., ed. (1976) *The Peripheral Nerve.* Chapman and Hall, London.
B   Watson, W. E. (1976) *Cell Biology of the Brain.* Chapman and Hall, London.
C   Thoenen, H., and G. W. Kreutzberg, eds. (1981) *The Role of Fast Transport in the Nervous System. Neurosci. Res. Program Bull.* 20 (1): 1–138.

D Bray, D. (1974) *Endeavour* 33 (120): 131–136.

E Parsons, D. S., ed. (1975) *Biological Membranes.* Oxford University Press (Clarendon Press), Oxford.

F Stryer, L. (1981) "Introduction to Biological Membranes" *Biochemistry,* 2d ed., pp. 205–231. Freeman, New York.

G Finean, J. B., R. Coleman, and R. H. Michell. (1984) *Membranes and Their Cellular Functions,* 3d ed. Blackwell, Oxford.

H Jamieson, G. A., and D. M. Robinson. (1976, 1977) *Mammalian Cell Membranes,* vols. 1 and 2. Butterworths, London and Boston.

 I Bretscher, M. S. (1973) "Membrane Structure: Some General Principles," *Science* 181: 622–629.

 J Jones, D. G. (1978) "Some Current Concepts of Synaptic Organization," *Adv. Anat. Embryol. Cell Biol.* 55 (Fasc. 4): 1–69.

K Shepherd, G. M. (1980) *The Synaptic Organization of the Brain,* 2d ed. Oxford University Press, London.

L Lasek, R. J., M. L. Shelanski, B. R. Brinkley, J. S. Condeelis, M. H. Ellisman, R. D. Goldman, T. S. Reese, F. Solomon, and E. W. Taylor. (1981) *Cytoskeletons and the Architecture of Nervous Systems. Neurosci. Res. Program Bull.* 19 (1): 1–153.

M Kandel, E. R., and J. H. Schwartz. (1981) *Principles of Neural Science.* Edward Arnold, London.

N Shepherd, G. M. (1983) *Neurobiology.* Oxford University Press, London.

O Morell, P. (1984) *Myelin,* 2d ed. Plenum Press, New York.

Additional Citations

 1 Kandel, E. R. (1976) *Cellular Basis of Behavior.* pp. 80–82. Freeman, New York.

 2 Tauc, L. (1960) *J. Physiol.* (Lond.) 152: 36P.

 3 Roots, B. I., and P. V. Johnstone. (1964) *J. Ultrastruct. Res.* 10: 350–361.

 4 Whittaker, V. P., and M. N. Sheridan. (1965) *J. Neurochem.* 12: 363–372.

 5 Sholl, D. A. (1956) *The Organization of the Cerebral Cortex.* Methuen, London.

 6 Cajal, S. R. (1928) *Degeneration and Regeneration of the Nervous System* (trans. R. May). Oxford University Press, London.

 7 Lieberman, A. R. (1976) In *The Peripheral Nerve* (ed. D. N. Landon), pp. 188–278. Chapman and Hall, London.

 8 Hosli, L., P. F. Andres, and E. Hosli. (1977) *Neurosci. Lett.* 6: 79–83.

 9 Colonnier, M. (1968) *Brain Res.* 9: 268–287.

10 Pomerat, C. M., W. J. Hendelman, C. W. Railborn, and J. F. Massey. (1967) In *The Neuron* (ed. H. Hyden), pp. 119–178. Elsevier, Amsterdam.

11 Liem, R. (1982) *J. Neurochem.* 38: 142–150.

12 Whittaker, V. P., I. A. Michaelson, and R. J. A. Kirkland. (1964) *Biochem. J.* 90: 293–303.

13 Koenig, H., D. Gaines, T. McDonald, R. Gray, and J. Scott. (1969) *J. Neurochem.* 11: 729–743.

14 Waxman, S. G., and G. D. Pappas. (1969) *Brain Res.* 14: 240–244.

15 Cowan, M. (1979) *Sci. Am.* 241 (3): 112–133.

16 Gaskin, F., and M. L. Shelanski. (1976) *Essays Biochem.* 12: 115–146.

17 Bray, D., and D. Gilbert. (1981) *Ann. Rev. Neurosci.* 4: 505–523.

18 Ringo, P. L. (1968) *J. Ultrastruct. Res.* 17: 266–277.

19 Synder, J. A., and J. R. MacIntosh. (1976) *Ann. Rev. Biochem.* 45: 699–720.

20 Huang, B., G. Piperno, and D. J. L. Luck. (1979) *J. Biol. Chem.* 254: 3091–3099.

21 Tang, W. J., C. W. Bell, W. S. Sale, and I. R. Gibbons. (1982) *J. Biol. Chem.* 257: 508–515.

22 Johnson, K. A., and J. S. Wall. (1983) *J. Cell Biol.* 96: 669–678.

23 Olmsted, J. B., and G. G. Borisy. (1973) *Ann. Rev. Biochem.* 42: 507–540.

24 Karlson, J. O., and J. Sjostrand. (1969) *Brain Res.* 13: 617–619.

25 Norstrom, A., H. A. Hausson, and J. Sjostrand. (1971) *Z. Zellforsch. Mikrosk. Anat.* 113: 271–293.

26 Hammond, G. R., and R. S. Smith. (1977) *Brain Res.* 128: 227–242.

27 Hepler, P. K., J. R. McIntosh, and S. Cleland. (1970) *J. Cell Biol.* 45: 438–444.

28 Ochs, S. (1974) In *The Peripheral Nervous System* (ed. J. I. Hubbard), pp. 47–72. Plenum Press, New York.

29 Adams, R. J. (1982) *Nature* (Lond.) 297: 327–329.

30 Smith, D. S. (1971) *Philos. Trans. R. Soc.* (Lond.) *Sec. B.* 261: 363–370.

31 Smith, D. S., U. Järlfors, and R. Beránek. (1970) *J. Cell Biol.* 46: 199–219.

32 Burdwood, W. O. (1975) *J. Cell Biol.* 27: 115A.

33 Forman, D. S., A. L. Padjen, and G. R. Siggins. (1977) *Brain Res.* 136: 197–213.

34 Bretscher, M. S., and M. C. Raff. (1975) *Nature* 258: 43–49.

35 Heath, J. P. (1983) *Nature* (Lond.) 302: 532–534.

36 Edelman, G. M. (1976) *Science* 192: 218–226.

37 Edelman, G. M., I. Yahara, and J. L. Wang. (1973) *Proc. Natl. Acad. Sci.* (USA) 70: 1442–1446.

38 Landon, D. N., and S. Hall. (1976) In *The Peripheral Nerve* (ed. D. N. Landon), pp. 1–105. Chapman and Hall, London.

39 Berlin, R. D. (1975) *Ann. N.Y. Acad. Sci.* 253: 445–454.

40 Gilbert, D. S., B. J. Newby, and B. H. Anderton. (1975) *Nature* (Lond.) 256: 586–589.

41 Schlaepfer, W. W., and L. A. Freeman. (1978) *J. Cell Biol.* 78: 652–662.

42 Shelanski, M. L., and R. K. H. Liem. (1979) *J. Neurochem.* 33: 5–13.

43 Schlaepfer, W. W., and S. Micko. (1978) *J. Cell Biol.* 78: 369–378.

44 Nikaido, T., J. Austin, L. Truef, and R. Rhinehart. (1972) *Arch. Neurol.* 27: 549–554.

45 Yen, S. H., D. Dahl, M. Schachner, and M. L. Shelanski. (1976) *Proc. Natl. Acad. Sci.* (USA) 73: 529–533.

46 Adelman, M. R., and E. W. Taylor. (1969) *Biochemistry* 8: 4964–4977.

47 Pollard, T. D., E. Shelton, R. R. Weihing, and E. D. Korn. (1970) *J. Mol. Biol.* 50: 91–97.

48 Nachmias, V. T., H. E. Huxley, and D. Kessler. (1970) *J. Mol. Biol.* 50: 83–90.

49 Spooner, B. S., J. F. Ash, J. T. Wrenn, R. B. Frater, and N. K. Wessels. (1973) *Tissue Cell* 5: 37–46.

50 Ishikawa, H., R. Bischoff, and H. Holtzer. (1969) *J. Cell Biol.* 43: 312–328.

51 Brown, S. S., and J. A. Spudich. (1979) *J. Cell Biol.* 83: 657–662.

52 Hartwig, J. H., and T. P. Stossel. (1976) *J. Cell Biol.* 71: 295–303.

53 Weihing, R. R. (1976) *J. Cell Biol.* 71: 303–307.
54 Zellanà, J. (1968) *Z. Zellforsch. Mikrosk. Anat.* 92: 186–196.
55 Kapeller, K., and D. Mayor. (1969) *Proc. R. Soc.* (Lond.), *Sec. B.* 172: 53–63.
56 Bradley, W. G., D. Murchison, and M. J. Day. (1971) *Brain Res.* 35: 185–197.
57 Ochs, S. (1972) *J. Neurobiol.* 2: 331–345.
58 Ochs, S. (1972) *Ann. N.Y. Acad. Sci.* 193: 43–57.
59 Schubert, P., and G. W. Kreutzberg. (1975) *Brain Res.* 85: 317–319.
60 Karlsson, J. D., and J. Sjostrand. (1972) *Brain Res.* 37: 975–982.
61 Cowan, W. M., D. I. Gottlieb, A. E. Jendrikson, J. L. Proce, and T. A. Woolsey. (1972) *Brain Res.* 37: 21–51.
62 Heuser, J., and T. S. Reese. (1972) *Anat. Rec.* 172: 329–330.
63 Hendry, I. A., K. Stöckel, H. Thoenen, and L. L. Iversen. (1974) *Brain Res.* 68: 103–121.
64 Iversen, L. L., K. Stöckel, and H. Thoenen. (1975) *Brain Res.* 88: 37–43.
65 Droz, B. (1973) *Brain Res.* 62: 383–394.
66 Droz, B., H. L. Koenig, and L. di Gianberardino. (1973) *Brain Res.* 60: 93–127.
67 Singer, S. J. (1975) In *Cell Membranes* (eds. G. Weisman and R. Claiborne), pp. 35–44. H. P. Publishing, New York.
68 Nicholson, G. L. (1977) *Biochim. Biophys. Acta* 457: 57–108.
69 Pinto de Silva, P., and D. Branton. (1970) *J. Cell Biol.* 45: 598–605.
70 Tillack, T. W., and V. T. Marchesi. (1970) *J. Cell Biol.* 45: 649–653.
71 Marchesi, V. T., H. Furthmayr, and M. Tomita. (1976) *Ann. Rev. Biochem.* 45: 667–698.
72 Marchesi, S. L. (1975) *Proc. Natl. Acad. Sci.* (USA) 72: 2964–2968.
73 Cherry, R. J. (1979) *Biochim. Biophys. Acta* 559: 289–320.
74 Catterall, W. A. (1982) *Trends Neurosci.* 5: 303–306.
75 Catterall, W. A. (1980) *Ann. Rev. Pharmacol. Toxicol.* 20: 15–43.
76 Skou, J. C., and J. G. Norby, eds. (1979) *Na+K+-ATPase: Structure and Kinetics.* Academic Press, London.
77 Robinson, J. D. (1983) In *Handbook of Neurochemistry*, 2d ed., vol. 4 (ed. A. Lajtha), pp. 173–193. Plenum Press, New York.
78 Hubbellward-Bownds, M. D. (1979) *Ann. Rev. Neurosci.* 2: 17–168.
79 Ralston, G. B. (1978) *Trends Biochem.* 3: 195–198.
80 Taylor, R. B., P. H. Duffus, M. C. Raff, and S. de Petris. (1971) *Nature (New Biol.)* 233: 225–229.
81 Matus, A., F. de Petris, and M. L. Raff. (1973) *Nature (New Biol.)* 294: 278–280.
82 Langercrantz, H. (1976) *Neuroscience* 1: 81–92.
83 Chapman, D. (1976) In *Mammalian Cell Membranes* (ed. G. A. Jamieson and D. M. Robinson), pp. 97–137. Butterworths, London and Boston.
84 Emmelot, P. (1976) In *Mammalian Cell Membranes* vol. 2 (ed. G. A. Jamieson and D. M. Robinson), pp. 1–54. Butterworths, London and Boston.
85 Van Hoeven, R. S., and P. Emmelot. (1972) *J. Membrane Biol.* 9: 105–111.
86 Webster, H. de F., J. R. Martin, and M. F. O'Connell. (1973) *Dev. Biol.* 32: 401–411.
87 Rumsby, M. G. (1978) *Biochem. Soc. Trans.* 6: 448–462.
88 Rumsby, M. G., and A. J. Crang. (1978) *Cell Surf. Rev.* 4: 1247–1362.
89 Crang, A. J., and M. G. Rumsby. (1978) *Adv. Exp. Med. Biol.* 100: 235–248.

90 W. T. Norton, and S. E. Poduslo. (1973) *J. Neurochem.* 21: 759–773.

91 Fishman, M. A., H. C. Agrawal, A. Alexander, J. Gotterman, R. E. Martensen, and R. F. Mitchell. (1975) *J. Neurochem.* 24: 689–694.

92 Linington, C., and M. G. Rumsby. (1978) *Adv. Exp. Med. Biol.* 100: 263–273.

93 Oxberry, J. M., and N. A. Gregson. (1978) *Brain Res.* 78: 303–313.

94 Matthieu, J. M., A. Daniel, R. H. Quarles, and R. O. Brady. (1974) *Brain Res.* 81: 348–353.

95 Rawlins, E. A. (1973) *J. Cell Biol.* 58: 42–53.

96 Dawson, R. M. C., and R. M. Gould. (1976) *Adv. Exp. Med. Biol.* 72: 95–113.

97 Porter, K. R., and M. A. Bonneville. (1963) *Introduction to Fine Structure of Cells and Tissues.* Harry Kimpton, London.

98 Wolfgram, F., and K. Kotorii. (1968) *J. Neurochem.* 15: 1281–1290.

99 Norton, W. T. (1975) In *The Nervous System*, vol. 1 (ed. D. B. Tower), pp. 467–481. Raven Press, New York.

100 Norton, W. T., and W. Cammer. (1984) In *Myelin*, 2d ed. (ed. P. Morris), pp. 147–195. Plenum Press, New York.

101 Nussbaum, J. L., J. P. Delannoy, and P. Mandel. (1977) *J. Neurochem.* 28: 183–191.

102 Roussel, G., J. P. Delannoy, J. L. Nussbaum, and P. Mandel. (1977) *Neuroscience* 2: 307–313.

103 Drummond, R. J., and G. Dean. (1980) *J. Neurochem.* 35: 1155–1165.

104 Sprinkle, T. J., M. R. Wells, F. A. Garver, and D. B. Smith. (1980) *J. Neurochem.* 35: 1200–1208.

105 Zanetta, J. P., P. Sarlieve, A. Reever, G. Vincendon, and G. Gombos. (1977) *J. Neurochem.* 29: 355–358.

106 Quarles, R. H., J. L. Everly, and R. O. Brady. (1973) *J. Neurochem.* 21: 1177–1191.

107 Poduslo, J. F., R. H. Quarles, and R. O. Brady. (1976) *J. Biol. Chem.* 251: 153–158.

108 Nussbaum, J. L., J. F. Ronayrenc, J. Jolles, P. Jolles, and P. Mandel. (1974) *FEBS Lett.* 23: 285–297.

109 Rothman, J. E. (1981) *Science* 213: 1212–1219.

110 Folch-Pi, J., and P. J. Stoffyn. (1972) *Ann. N.Y. Acad. Sci.* 195: 86–107.

111 Papahadjopoulos, D., W. J. Vail, and M. Moscarello. (1975) *J. Membrane Biol.* 22: 143–164.

112 Epand, R. M., M. A. Moscarello, B. Zierenberg, and W. J. Vail. (1974) *Biochemistry* 13: 1264–1267.

113 Harris, R., and J. B. C. Findlay. (1983) *Biochim. Biophys. Acta* 732: 75–82.

114 Poduslo, J. F., and P. E. Bruan. (1975) *J. Biol. Chem.* 250: 1099–1105.

115 Crang, A. J., and M. G. Rumsby. (1977) *Biochem. Soc. Trans.* 5: 110–112.

116 Vail, W. J., D. Papahadjopoulos, and M. A. Moscarello. (1974) *Biochim. Biophys. Acta* 345: 463–467.

117 London, Y., and F. G. A. Vosseberg. (1973) *Biochim. Biophys. Acta* 478: 478–490.

118 Mackay, I. R., P. R. Carnegie, and A. S. Coates. (1973) *Clin. Exp. Immunol.* 15: 471–482.

119 Braun, P. E., and R. L. Barchi. (1972) *Brain Res.* 40: 437–444.

120 Eto, Y., and K. Suzuki. (1972) *J. Neurochem.* 19: 117–121.

121 Miyamoto, E. (1975) *J. Neurochem.* 24: 503–512.

122 Miyamoto, E., and S. Kakinchi. (1975) *Biochim. Biophys. Acta* 384: 458–465.

123 Kanaseki, T., and K. Kadota. (1969) *J. Cell Biol.* 42: 202–220.

124 Pearse, B. M. F. (1975) *J. Mol. Biol.* 97: 93–98.

125 Fried, R. C., and M. P. Blaustein. (1978) *J. Cell Biol.* 78: 685–700.

126 Gray, E. G., and R. A. Willis. (1970) *Brain Res.* 24: 149–168.

127 Pearse, B. M. F., and M. S. Bretscher. (1981) *Annu. Rev. Biochem.* 50: 85–101.

128 Blitz, A. L., R. E. Fine, and P. A. Toselli. (1977) *J. Cell Biol.* 75: 135–147.

129 Bretscher, M. S. (1973) *Science* 181: 622–629.

130 Jones, D. G. (1978) *Adv. Anat. Embryol. Cell Biol.* 55: (4) 1–69.

131 Jones, D. G., and H. F. Bradford. (1971) *Tissue Cell* 3: 177–190.

132 Gray, E. G. (1976) *Prog. Brain Res.* 45: 207–236.

133 Akert, K., H. Moor, K. Pfenninger, and C. Sandri. (1969) *Prog. Brain Res.* 31: 223–240.

134 Sandri, C., J. M. Van Buren, and K. Akert. (1977) *Membrane Morphology of the Nervous System.* Elsevier, Amsterdam (Prog. Brain Res. 46).

135 Peters, A., and I. R. Kaiserman-Abramof. (1969) *Z. Zellforsch. Mikrosk. Anat.* 100: 487–506.

136 Cotman, C. W., and G. A. Banker. (1974) In *Reviews of Neuroscience*, vol. 1. (eds. S. Ehrenpreis and I. J. Kopin), pp. 2–61. Raven Press, New York.

137 Hirano, A., and H. M. Dembitzer. (1973) *J. Cell Biol.* 56: 478–487.

138 Whittaker, V. P., and H. Stadler. (1980) In *Proteins of the Nervous System* 2d ed. (eds. R. A. Bradshaw and D. M. Schneider), pp. 231–255. Raven Press, New York.

139 Banks, P., and K. B. Helle. (1971) *Proc. R. Soc.* (Lond.), *Sec. B.* 261: 305–310.

140 Kirshner, N., and A. G. Kirshner. (1971) *Proc. R. Soc.* (Lond.), *Sec. B.* 261: 279–289.

141 Pickering, B. T. (1978) *Essays Biochem.* 14: 45–81.

142 Gray, E. G. (1969) *Prog. Brain Res.* 31: 141–155.

143 Famiglietti, E. V. (1970) *Brain Res.* 20: 181–191.

144 Rall, W., G. M. Shepherd, T. S. Reese, and M. W. Brightman. (1966) *Exp. Neurol.* 14: 44–56.

145 Uchizono, K. (1965) *Nature* (Lond.), 207: 642–643.

146 Gabella, G. (1976) *Structure of the Autonomic Nervous System.* Chapman and Hall, London.

147 Hamlyn, L. H. (1963) *J. Anat.* 97: 189–201.

148 Bunge, R. P. (1970) In *The Neurosciences: Second Study Program* (ed. F. O. Schmitt), pp. 782–797. Rockefeller University Press, New York.

149 Singer, S. J., and G. L. Nicolson. (1972) *Science* 175: 720–731.

150 Hayes-Griffith, O., and P. C. Jost. (1978) In *Molecular Specialization and Symmetry in Membrane Function* (eds. A. K. Solomon and M. Karnovsky), pp. 30–60. Harvard University Press, Cambridge, Mass.

151 Robertson, J. D. (1960) *Prog. Biophys.* 10: 344–418.

152 Pfenninger, K., K. Akert, H. Moor, and C. Sandri. (1972) *J. Neurocytol.* 1: 129–149.

153 Gray, E. G. (1963) *J. Anat.* 97: 101–106.

154 Bellairs, R., and E. G. Gray. (1974) *Essays on the Nervous System*, pp. 155–178. Oxford University Press (Clarendon Press), Oxford.

155 Llinas, R. (1970) In *The Neurosciences: Second Study Program* (ed. F. O. Schmitt), pp. 409–426. Rockefeller University Press, New York.

156 Szentágothai, J. (1970) In *The Neurosciences: Second Study Program* (ed. F. O. Schmitt), pp. 427–443. Rockefeller University Press, New York.

157 McLaughlin, B. J., J. G. Wood, K. Saito, E. Roberts, and J.-Y. Wu. (1974) *Brain Res.* 85: 377–391.

158 Kimura, H., P. L. McGeer, J. H. Peng, and E. G. McGeer. (1981) *J. Comp. Neurol.* 200: 151–201.

159 Pickel, V. M., S. C. Beckley, K. K. Sumal, T. H. Joh, D. J. Reis, and R. J. Miller. (1981) *Acta Histochem. Suppl. Band 24* (S): 97–105.

160 Joh, T. H., T. Shikimi, V. Pickel, and D. J. Reis. (1975) *Fed. Proc.* 34: 3281.

161 Cimarusti, D. L., K. Saito, J. E. Vaughan, P. Barber, E. Roberts, and P. E. Thomas. (1979) *Brain Res.* 162: 55–67.

162 Zomzeley-Neurath, C., and A. Keller. (1977) *Neurochem. Res.* 2: 353–377.

163 Zomzeley-Neurath, C. E. (1983) In *Handbook of Neurochemistry* 2d ed., vol. 4 (ed. A. Lajtha), pp. 403–433. Plenum Press, New York.

164 Bock, E. (1978) *J. Neurochem.* 30: 7–14.

165 Zomzeley-Neurath, C., and W. Walker. (1980) In *Proteins of the Nervous System* 2d ed. (eds. R. A. Bradshaw and D. M. Schneider), pp. 1–57. Raven Press, New York.

166 Pickel, V., D. J. Reis, P. J. Marangos, and C. Zomzeley-Neurath. (1976) *Brain Res.* 105: 184–187.

167 Cimino, M., B. K. Hartman, and B. W. Moore. (1977) *Proc. Int. Soc. Neurochem.* 6: 304.

168 Marangos, P. J., C. Zomzeley-Neurath, D. C. M. Luk, and C. York (1975) *J. Biol. Chem.* 250: 1884–1891.

169 Schmechel, D., P. J. Marangos, M. Brightman, and F. K. Goodwin. (1978) *Science* 199: 313–315.

170 Fields, K. L., J. P. Brockes, R. Mirsley, and L. M. B. Wendon. (1978) *Cell* 14: 43–51.

171 Schachner, M., and M. Willinger. (1979) *Prog. Brain Res.* 51: 23–44.

172 Williams, P. L., and R. Warwick. (1975) *Functional Neuroanatomy of Man.* Churchill Livingstone, Edinburgh.

173 Pearse, B. M. F. (1978) *J. Mol. Biol.* 126: 803–812.

174 Garbern, J. Y., and J.-Y. Wu. (1981) *J. Neurochem* 36: 602–612.

175 Lin, C.-T., J. Y. Garbern, and J.-Y. Wu. (1982) *J. Histochem. Cytochem.* 30: 853–863.

176 Heuser, J., and T. Reese. (1981) *J. Cell Biol.* 88: 564–580.

177 Pastan, I. H., and M. C. Willingham. (1981) *Science* 214: 504–509.

178 Bock, E. (1978) *J. Neurochem.* 30: 7–14.

179 Branton, D. (1982) *Harvey Lect.* 77: 23–42.

180 Anderson, D. J. (1984) *Trends Neurosci.* 7: 355–357.

181 Lazarides, E., W. J. Nelson, and T. Kasamatsu. (1984) *Cell* 36: 269–278.

182 Moon, R. T., and E. Lazarides. (1984) *J. Cell Biol.* 98: 1899–1904.

183 Ochs, S., and R. A. Jersild. (1984) *Neurochem. Res.* 9: 823–836.

2
Glial Cells: Mechanical and Functional Supporting Cells of the Nervous System

Although glial cells are twice as numerous in neural systems as the neurons themselves, a full description of the functions they are known to serve would provide only a short statement. The extent to which these cells partner the neurons in providing the short-term or long-term output of the brain is still uncertain. A large volume of information on their morphology at the light microscopic and electron microscopic level has been gleaned mainly from studies employing postmortem fixed tissue, and this is now being extended to functional studies with living glial cells in tissue culture.

GLIAL SYSTEMS

A definition of glial cells is essentially derived by a process of exclusion. They may be defined as those cells in neural tissue that are not neurons, blood cells, capillary epithelial cells, or any other cells with a clear alternative function. Nevertheless, glial cells also possess one or two biochemical constituents by which they can be distinguished from other cells.

The origins of glial cells in the literature of neuroanatomy extend as far back as Rudolph Virchow in 1846. Their history has been fully documented by Glees[A] and by Kuffler and Nicholls.[B]

The current picture of neuroglial function includes: (1) mechanical support of neurons, (2) production of the myelin sheath, (3) rapid uptake and thus inactivation of chemical neurotransmitters released by neurons, (4) formation of scar tissue after brain injury, (5) removal of local tissue debris after cell death, (6) provision of a filter system between blood and neurons, (7) control of the composition of extracellular fluid (e.g., K^+ levels). Added to these might be activities whose existence is much less certain, such as provision of a guiding pathway for migrating neurons and their axonal processes during the embryonic development of the nervous system. Glial cells also apparently serve neurons by cooperating in certain metabolic activities and by absorbing the potassium released by active neurons.

Types of Glial Cell

Various subgroups of glial cells are identified on the basis of their size, shape, and location. The great family of glial cells are divided into *macroglia*, the larger cells, and *microglia*, the smaller cells. The macroglia include *radial glia, astroglia, oligodendroglia* and *peripheral glia*.

The largest glial cells are the astrocytes (Figures 2.1 to 2.5). Their perikarya (18–20 μm in diameter) possess large, pale-staining nuclei, glycogen granules, and in some cases prominent glial filaments (*gliofilaments*). Their long and narrow or membranous cell processes provide the characteristic star shape implied in their name. These slender structures and the swellings at their extremities (*end feet*) envelop the surface of blood capillaries (Figure 2.2B), leaving only small gaps for the diffusion of substances from the capillary endothelia through the basal lamina and into the restricted extracellular spaces of the brain. The end feet, special features of astrocytes, are also found apposed to neuronal membranes and to the pial membranes which cover the outer brain surface (Figure 2.6).

Two distinct types of astrocytes have been recognized, the *fibrous astrocytes* and the *protoplasmic astrocytes*. The latter occur mainly in gray matter dispersed among neuronal cell bodies; they provide a large proportion of the cytoplasmic volume of gray matter. Unlike fibrous astrocytes, they are not full of filamentous structures, and their processes tend to be laminar or membranous. In contrast, fibrous astrocytes occur mainly in white matter, positioned between the bundles of myelinated axons. Their characteristic fibrous cell processes as well as their perikarya are packed with gliofilaments whose major protein constituent, called glial fibrillary acidic protein, has been isolated. Antibodies to this protein, labeled with fluorescent markers, enable us to clearly visualize the fibrous astrocytes (Figure 2.3).[2]

The *oligodendroglia* (Figure 2.5C) are another major glial subgroup. They are clearly distinguished from astrocytes by their small (3–5 μm diameter) cell bodies, in which the nucleus occupies a large proportion of the volume. Oligodendroglia show well-developed endoplasmic reticu-

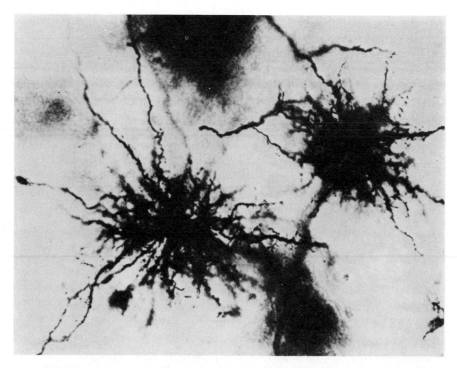

FIGURE 2.1   Fibrous astrocytes from the dentate nucleus in the cerebellum of the monkey. The cell bodies, 18 to 20 μm in diameter, emit numerous radially oriented processes extending about 45 μm. A few processes reach up to 200 μm in length. [From ref. 1]

A

B

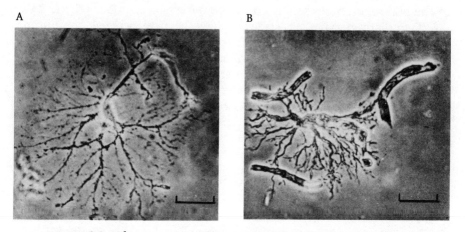

FIGURE 2.2   Phase contrast photomicrographs of astrocyte-rich fractions from rat cortex. (A) Fibrous astrocyte. (B) Protoplasmic astrocyte; note attachment of end-feet to blood capillaries. Bars, 25 μm. [From ref. 125]

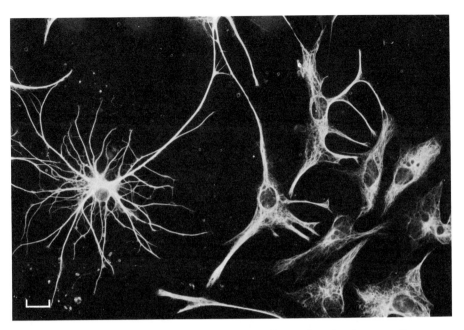

FIGURE 2.3   Astroglial cells in tissue culture, showing immunofluorescence due to staining with antibodies against glial fibrillary acidic protein (GFAP) tagged with a fluorescent marker. The cells are from 8-day-old rat cerebellum. Two categories of glial cell are present: at left is a fibrous astrocyte, from center to right are protoplasmic astrocytes. Bar, 20 μm. [Courtesy of G. P. Wilkin]

lum, Golgi membranes, and other cytoplasmic inclusions, but no gliofilaments. As indicated by their name, they have relatively few cell processes.

The oligodendroglia are found in both white matter and gray matter. Those in white matter, the *interfascicular oligodendroglia*, are responsible for the formation of the insulating myelin sheath around axons; in the white matter of developing brain, their cell processes are often seen to be continuous with the myelin sheaths they are continuously generating (see Figures 1.25 to 1.27). In gray matter, *satellite oligodendroglia* are often found in very close association with neuronal perikarya.

Microglia (Figure 2.5D) are the smallest of the neuroglia, about 2–3 μm in diameter, and are therefore readily distinguished from the macroglia. Their variable morphology, however, has made them difficult to define and study,[3,4,5] and they show a close resemblance to the connective tissue macrophages of other body regions. They are identified by their readiness to stain with silver salts, their small elongated nuclei, scant cytoplasm, and numerous small processes.[6] When exposed to stimuli that cause inflammation or other insult to the tissue, microglia are rapidly

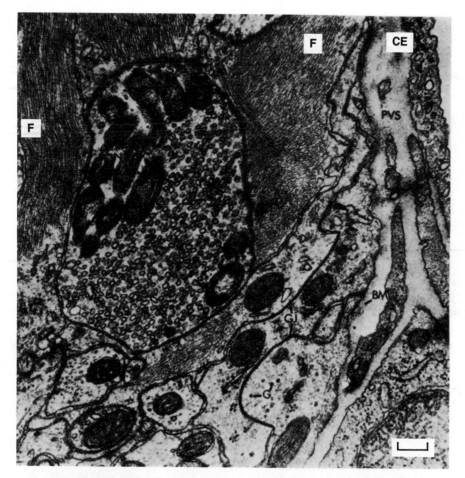

FIGURE 2.4   Fibrous astrocytic processes (F) full of glial filaments surround an axonal bouton (N) in a cat spinal cord. Perivascular astrocytes
containing glycogen (G) are joined by gap junctions (GJ), are covered by a
basement membrane (BM), and are separated from capillary endothelium
(CE) by a perivascular space (PVS) containing a pericyte. Bar, 0.1 μm.
[From ref. 37]

transformed from a passive resting state to active phagocytosis, accompanied by proliferation and migration through the tissue. It is at this stage
that their strong similarity with macrophages is most obvious.

Radial glial cells, first described at the turn of the century, have
recently reemerged as a separate group of neuroglia of considerable range
and importance. They appear during brain development only to become
transformed into other glial cell types (e.g., astrocytes) as the brain ma-
tures. They are most common and most easily recognized in submamma-
lian species. Many of the cells possess a greatly elongated, often biopolar,
shape with two or more principal cell processes extending relatively long

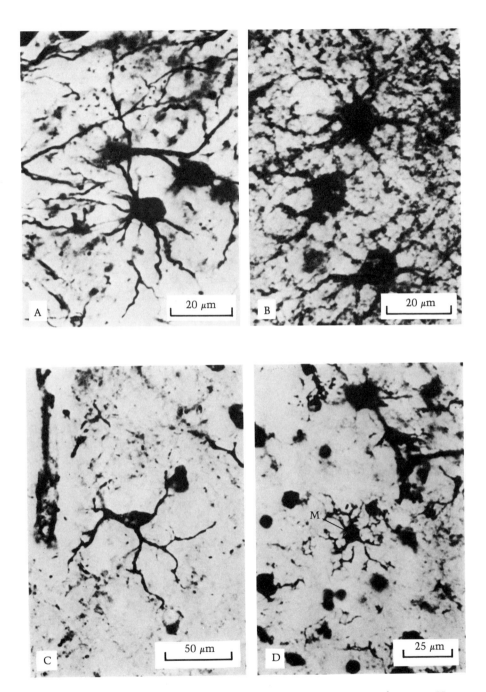

FIGURE 2.5   Types of vertebrate glia. (A) Fibrous astrocyte from rat. Hortega silver carbonate stain. (B) Human protoplasmic astrocytes. Cajal's gold sublimate stain. (C) Oligodendrocyte from rat. Weil and Davenport silver stain. (D) Microglial cell (M) from rat. Weil and Davenport silver stain. [From ref. E; (A, C, and D), K. E. Webster (original); (B), from a preparation provided by W. Blackwood]

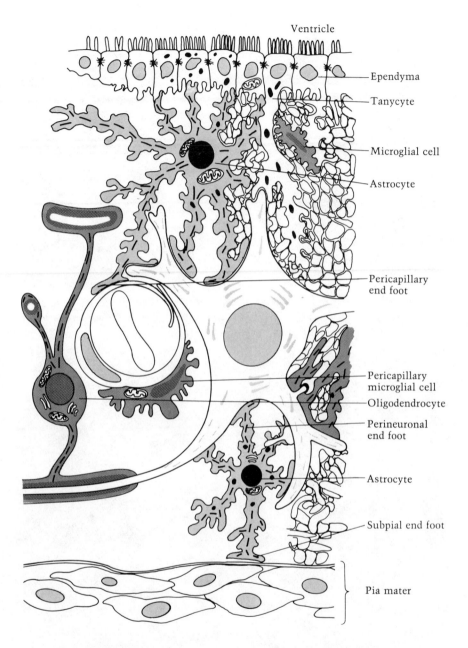

FIGURE 2.6 Types of nonneuronal cells in the central nervous system. Glial cells are shown in gray. The *ependyma* includes examples of ciliated and nonciliated cells and one *tanycyte* with a centrally directed basal process. Two *astrocytes* (top and bottom) are shown abutting a neuronal soma and dendrites; the one above also contacts a capillary, the one below expands on the *pia mater* surface. An *oligodendrocyte* (middle right) provides myelin sheaths for two axons. Also illustrated are two flattened *microglial cells*, one adjacent to a capillary (middle right) and the other within the branching network of glial and neuronal processes (neuropil) (top left) [From ref. 150; artwork by R. E. M. Moore]

distances through neural tissues and often ending on membrane surfaces or blood vessel walls. At least two types of radial glia do survive as such in the adult brain, the *Müller* cells and the *Bergmann glia.* Müller cells, found in the retina, show characteristic morphology and disposition, being greatly elongated and spanning the whole distance between the inner and outer limiting membranes of the retina. The *Bergmann glia,* of the adult mammalian cerebellum, project several elongated cell processes right across the molecular layer to the undersurface of the pial membrane, where they form conical end feet (see Figure 2.7). The cell bodies of the Bergmann glia are located in the Purkinje cell layer.

Glia of the Peripheral Nervous System

The *Schwann cells,* best known of the peripheral glia,[7] provide the myelin sheaths of peripheral nerve fibers (Figure 2.8). Extensive regions of these glia lie along the side of the myelin sheaths they generate, forming long fingers of cytoplasm containing the elongated nucleus and the usual cytoplasmic inclusions that keep the myelin sheaths serviced and alive (Figure 2.9). Unmyelinated axons of the peripheral nervous system lie in deep furrows on the surface of these peripheral glia cells (Figure 2.10). Schwann cells are elongated cells and are the major nonexcitable cells of peripheral nerve.

The cell bodies of neurons in peripheral ganglia (e.g., the dorsal root ganglia) are packed in a bed of *satellite glial cells* that cover their surface and no doubt serve functions similar to those performed by the glia of the central nervous system (see Figures 2.11 and 1.6). Several layers of the satellite cells interpose between the neurons and the inner wall of the ganglion, where they become apposed to a basal laminar membrane (see Figure 1.6). The glial or neuronal processes may be intimately associated with the plasma membrane of the partner cell, appearing sometimes to enter clefts in the surface (Figure 2.12).

At the neuromuscular junction, the outer surfaces of the nerve terminal and unmyelinated axon are ensheathed in large flat neuroglial cells called *telioglial cells* (Figure 2.13); on the muscle surface the axon and its branches occupy cavities below the telioglial cells.

Ependymal Cells

One important group of nonexcitable cells form the ependymal membrane that lines the interior of cerebral ventricles and other cavities of the brain and spinal cord. It is from the ependymal and subependymal cells that neurons and other neuroglia are ultimately derived during development. The cells forming the ependymal layer possess a variety of shapes and, no doubt, functions. Many have numerous small cytoplasmic fingers, or microvilli, projecting into the cavities and normally immersed in a stream of cerebrospinal fluid (Figures 2.6, 2.14, 2.15, and 2.16). As in the gastrointestinal tract, this device greatly increases surface area and is an

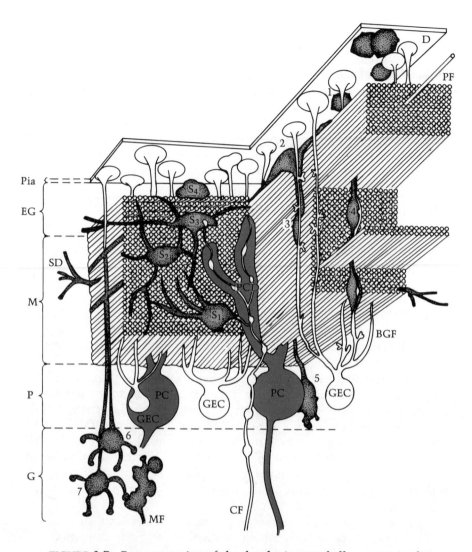

FIGURE 2.7   Reconstruction of the developing cerebellar cortex in the rhesus monkey. The thicknesses of the layers are drawn in their approximately true proportions for the 138-day monkey fetus but the diameters of the cellular elements, particularly the parallel fibers, are exaggerated in order to make the reconstruction more explicit. Postmitotic granule cells are designated by the numerals 1 through 7. *BGF*, Bergmann glial fiber; *CF*, climbing fiber; *D*, dividing external granule cell; *EG*, external granular layer; *GEC*, Golgi epithelial cell (Bergmann glia); *G*, granular layer; *M*, molecular layer; *MF*, mossy fibers; *P*, Purkinje layer; *PC*, Purkinje cell; *PCD*, Purkinje cell dendrite; *PF*, parallel fibers; S_{1-4}, stellate cells; *SD*, stellate cell dendrite. [From ref. 46]

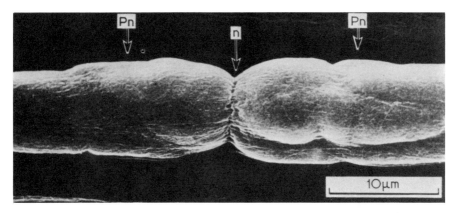

FIGURE 2.8   Scanning electron micrograph of the nodal region of a 10 μm diameter myelinated fiber from the sciatic nerve of a rat. The structure of the node (n) is obscured by the overlying sheath of fine collagen fibrils (the sheath of Plenk and Laidlaw) and the Schwann cell basal lamina; (Pn) indicates the paranodal bulbs of the internodes. [From ref. 137]

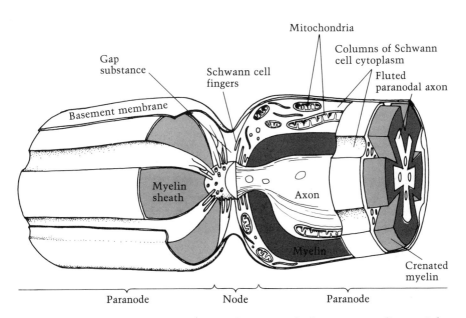

FIGURE 2.9   Structure of a node of Ranvier of a large mammalian peripheral myelinated nerve fiber. [From ref. 150]

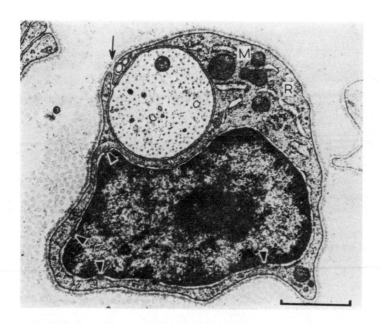

FIGURE 2.10   Schwann cell nucleus indented by unmyelinated axon of rounded cross section, from a 9-year-old child. Note nuclear pores (*arrowheads*), the mesaxon (*arrow*), mitochondrion (*M*), and granular endoplasmic reticulum (*R*) in the Schwann cell. Glutaraldehyde immersion-fixation; osmium tetroxide staining. Bar, 1 μm. [From ref. 138]

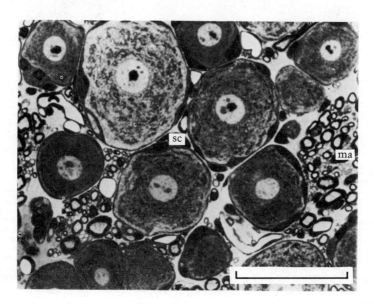

FIGURE 2.11   Neuron cell bodies surrounded by satellite glial cells (*SC*) of a rabbit dorsal root ganglion. Note myelinated axons (*ma*). Toluidine-blue-stained 1 μm plastic section. Bar, 50 μm. [From ref. 139; reproduced by courtesy of J. M. Jacobs]

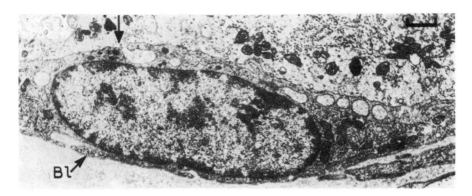

FIGURE 2.12   A satellite glial cell in close juxtaposition to a sensory neuron in a rat cervical dorsal root ganglion. A series of fingers of neuronal cytoplasm (*arrow*) can be seen penetrating the glial cytoplasm. *Bl*, basal lamina of satellite cell. Bar, 1.0 μm. [From ref. 139; reproduced by courtesy of J. M. Jacobs]

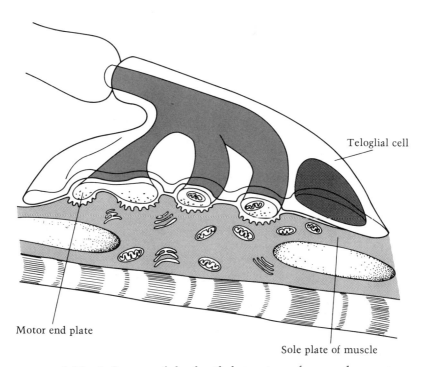

FIGURE 2.13   A diagram of the detailed structure of an *en plaque* neuromuscular junction showing ensheathing telioglial cell. Note the folding of the sarcolemma to form subsynaptic gutters, the disposition of the basal lamina, and the synaptic vesicles within the axonal terminal. [From ref. 150]

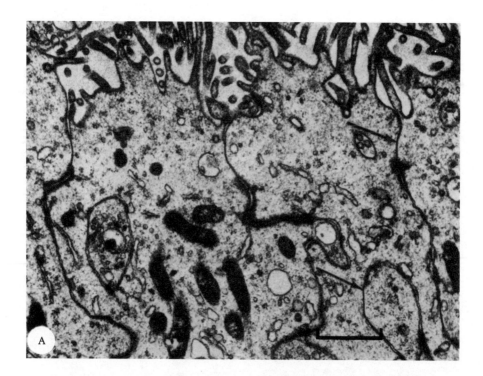

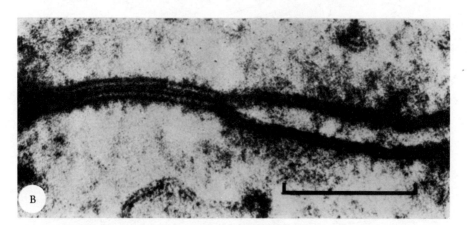

FIGURE 2.14 (A) The lining of the ventricles of the brain, formed by the folds of ependymal plasma membrane projecting into the ventricular cavities (top). The four contiguous ependymal cells are joined, at their apices and below, by tight junctions (*arrows*). Mouse brain; osmium tetroxide fixation. Bar, 1.0 μm. (B) Nerve tissue treated with uranyl acetate and dehydrated shows tight junctions; they appear as seven-layered gap junctions bisected by a median slit of constant width (left) that is continuous with the remaining interspace (center to right). Bar, 0.1 μm. [From ref. 10]

FIGURE 2.15   Scanning electron micrograph showing ependymal cells
lining the ventricle surface. Cilia (C) of the ependymal project into the
ventricle cavity. The nonciliated cells are tancytes (Ta). Supraependymal
cells (SC), a network of multipolar cells resembling neurons, are seen on
the tancyte surfaces; they may be secretory cells. Inset shows cilia at
higher magnification. Bar, 10 μm (10 μm in inset). [From ref. 158]

indication of extensive material exchange through processes of secretion
or absorption. In other regions, hairlike cilia replace microvilli and probably assist fluid flow (Figure 2.15). In mammals, the ependyma in the floor
of the third or fourth ventricle bear bulbous or rounded projections[8] (Figure 2.17).

Some cell components of the ependymal layer, the *tancytes*, send
long processes inwards to make contact with blood capillaries, neurons,
or other glial cells. *Ependymal astrocytes* and *ependymoglial cells* are
both branching forms with short side extensions (Figures 2.6 and 2.18).

From these descriptions we can see that the several classes of glial
cell merge one into the other. For example, tanycytes may well be modified radial glial cells, which themselves may be a subclass of astrocyte.

NEURON AND GLIA INTERRELATIONSHIPS

The long-standing proposal that glial cells provide a safe, nonrigid supporting and insulating matrix for neurons remains unchallenged. Astrocytes especially are suited to such a skeletal function through the

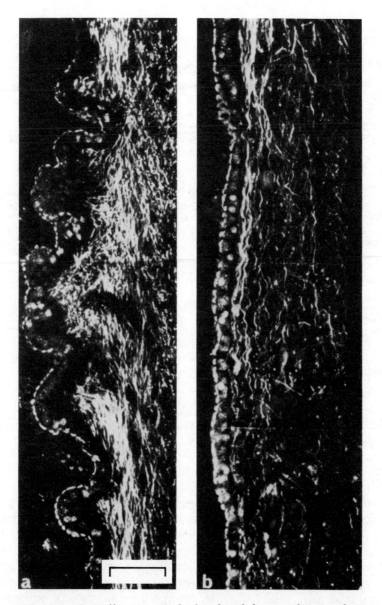

FIGURE 2.16   Differences in the height of the ependyma and structure of
the subependymal layer as seen with the fluorescence microscope in two
regions of the third ventricle of the cat. (A) Area hypothalamica dorsalis,
(B) adjoining surface of the nucleus tuberis. Stained with chrome hema-
toxylin phloxin. Bar, 40 μm. [From ref. 140]

strength contributed to their cell processes by the gliofilaments. How varied and far-reaching other relationships between neurons and glial cells may be is a fascinating question still to be explored.

The Fluid Filter System and the Blood-Brain Barrier

Neurons produce a continuous output of waste products, ions, and neuro-secretory substances, including neurotransmitters and potassium ions. They must in turn receive basic metabolic substrates and biosynthetic precursors for the manufacture of both structural and functional components, such as proteins, lipids, nucleic acids, and neurotransmitters. Activity modulators reach neurons by diffusion through the tissue or by delivery directly onto their surface from impinging nerve terminals. During electrical activity, neurons release potassium to their extracellular space, and during sustained activity potassium may reach concentrations well above normal owing to the small size of this extracellular space.

The picture, therefore, is one of exchange of materials between the neuron and its environment much as occurs in other tissues. But in order to achieve continuous well coordinated electrical activity throughout the neuronal networks, there must be rapid removal from their immediate environs of many substances that would otherwise influence neuronal activity. In the vertebrate central nervous system, the glial cells provide a densely packed filter and absorption system that almost entirely (90 percent) envelops all the blood capillaries with the closely apposed end feet of astrocytes (see Figures 2.2, 2.6, 2.18, and 2.19). The efficiency of this filter system is high because the relatively small extracellular space between the cells of the central nervous system gives a high ratio between unit cell surface area for absorption and unit volume of extracellular fluid. At the same time the ependymal cells lining the ventricular cavities of the brain must extend some influence on the composition of the fluid reaching the neurons from this source, and the concept of *ependymal secretion* is now well established, particularly for the floor of the third and fourth ventricles.[15] A feltwork of neuronal processes, nerve terminals, and also bulbous projections can be seen in these and other regions projecting past the ependymal cells and into the ventricular cavities[9] (Figure 2.17).

The densely packed glial end feet covering the surface of blood capillaries are not so tightly apposed at every point that fairly large molecules could not pass through the clefts from the bloodstream (clefts are 10–20 nm wide). This glial sheath is not, therefore, the physical basis for the long-standing concept of the blood-brain barrier.[10] Instead, the substances entering the tissue locally at blood capillaries become channeled into the narrow pathways between glial cells, where they are exposed to the battery of specific glial cell transport systems that may remove them from the extracellular fluid before they progress sufficiently far into the tissue

FIGURE 2.17   Four bulblike protrusions of nerve cells in the fourth ventricle of the rabbit, containing numerous elongated mitochondria. Bar, 5 μm [From ref. 141]

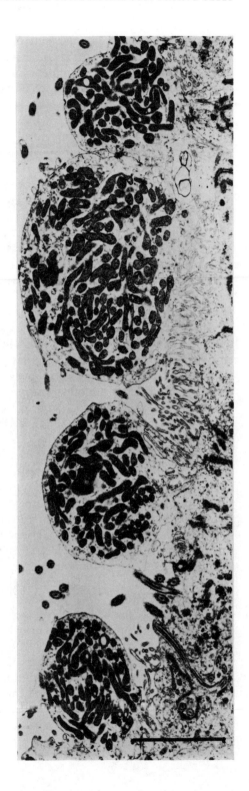

FIGURE 2.18   Tancytes in the cat infundibulum. The basal cell processes of these ependymal astrocytes are seen to cross the infundibulum and terminate with distinct end feet at the outer surface. Golgi-Bebenaite impregnation after perfusion of the animal with glutaraldehyde. Bar, 50 μm. [From ref. 140]

to encounter neurons. Substances passing into the brain by facilitated transport through the endothelial lining of blood capillaries, which is the primary blood-brain barrier, are exposed to further selection by the glial filter system. Only those compounds surviving this long and narrow route, owing to their high concentration or the absence of selective uptake systems, will reach the neurons. In this way, glial cells serve a monitoring function, determining by selective absorption the composition of the neuronal extracellular space—they do *not*, however, form a cytoplasmic barrier preventing the entry of substances at the blood-brain interface.

FIGURE 2.19   Arterioles with many glial end feet, from the white matter of the telencephalon of a 38-year-old man. The perivascular space is not always visible. Hortega method of staining. Bar, 10 μm. [From ref. D]

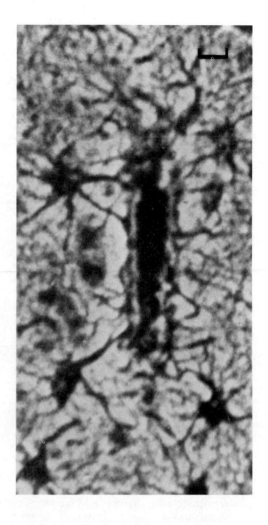

The ependymal cells, too, have sufficient space between them to allow the diffusional entry of proteins the size of microperoxidase (mol wt 18,000), horseradish peroxidase (mol wt 43,000), or albumin (mol wt, 79,000), all of which penetrate from ventricular cerebrospinal fluid to the spaces between the cells of the brain.[10,11,143] Thus, ependymal cells cannot present a barrier to the diffusion of smaller molecules in spite of the existence of tight junctions between most glial cells. The effective restriction on the entry of a wide range of substances from blood to brain appears to occur in the closely apposed endothelial cells of the blood vessel walls, which form tight junctions that overlap (Figure 2.14). These endothelial cells are also surrounded by a complete sheath of basement membrane. Large polar substances such as certain acidic dyes, inulin, horseradish peroxidase, microperoxidase, cytochrome *c*, lanthanum oxide, and many other easily detected macromolecules[12,13] are unable to enter from capillaries to brain, but they will readily enter the extracellular spaces of liver and other tissues. The first barrier, then, to entry of substances form the blood to the brain appears to be the physical one at these tight junctions that form continuous girdles of tightly apposed membranes between contiguous endothelial cells. This barrier is visualized with the aid of lanthum[10] or horseradish peroxidase (see Figure 2.20). Large molecules do not move in either direction past these tight junctions. Peroxidase introduced into the ventricles readily enters the extracellular space by passing between the ependymal cells lining the ventricles, but it cannot easily pass across the capillary wall and into the blood because of the barrier presented by the tight junctions (see Figure 2.21). In fact, the tight junctions are arranged in pairs so that small regions of extracellular space, or lacunae, are trapped between the two rings of tight junctions. It has been concluded that these junctions are unable to open and close like locks across a river because circulating peroxidase is never normally found in the interjunctional lacunae. Entry to these lacunae does occur, however, following intracarotid infusion of hypertonic solutions (e.g., mannitol[158]) that are known to make the blood-brain barrier more permeable to large and small molecules alike.[9] In general, the entry of polar substances is either greatly restricted or prevented while nonpolar or fatty substances enter much more easily. Certain key metabolic substrates disobey this rule. Glucose, for instance, shows rapid entry, via the special transport systems whose kinetic parameters have been thoroughly studied.[14] Also entering rapidly via special transport systems are a range of important precursor substances (e.g., tyrosine and tryptophan, which are required for neurotransmitter biosynthesis).

The highly selective permeability of most regions of the brain presents special problems for chemotherapy of the central nervous system, as many drugs shown to have a useful action on extracted enzymes, or tissue preparations in vitro will not penetrate the brain in vivo. In such cases, attempts are made to produce effective nonpolar analogues in order to overcome the problem. An example is the drug baclofen (Lioresal), a nonpolar version (*p*-chlorophenyl γ-aminobutyric acid), of the inhibitory neurotransmitter GABA (see Chapter 4), which does not itself pass into the brain from the blood. This drug is effective for treatment of a range of disorders, including cerebral palsy.

It would be wrong to convey the impression that all regions of the central nervous system are inaccessible to most substances in the blood. Some brain regions present a poor barrier to the influx of plasma components. Regarded as outside the blood-brain barrier are the *cirvumventricular* organs, a whole group of cerebral structures positioned around the ventricles, including the hypothalamus and median eminence, the subfornical organ, the subcommisural organ, the organum vasculosum of the lamina terminalis, the arcuate-median eminence, and the area postrema (see Figure 2.22). All of these organs are characterized by the presence of

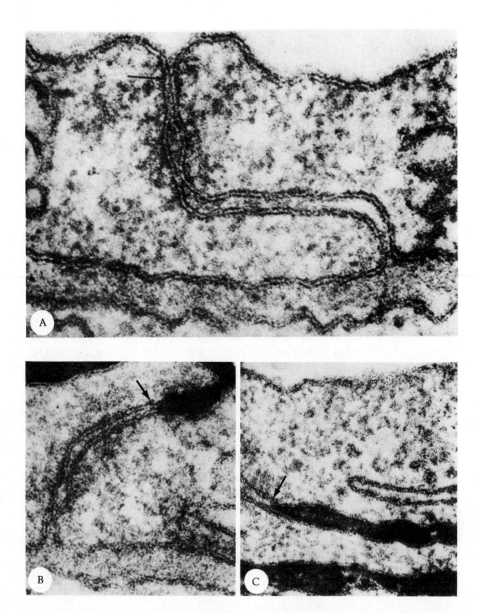

FIGURE 2.20   Tight junctions between endothelial cells in the cerebral capillaries of a mouse. The vessel lumen is at the top in each example. Aldehyde fixation. (A) A series of three punctate pentalaminar junctions delimit two compartments within the interendothelial cleft. The median lamina (second dense lamina to right of arrow) is narrower than would be expected if the two outer leaflets were merely apposed. The narrowness of this lamina suggests that the membranes are fused at those points. (B) The passage of intravascularly injected lanthanum hydroxide between endothelial cells is blocked by a tight junction (*arrow*) so that the remainder of the cleft remains free of deposit. (C) Peroxidase injected intraventricularly passes through the cleft between endothelial cells until it is stopped by a tight junction (*arrow*). [From ref. 10]

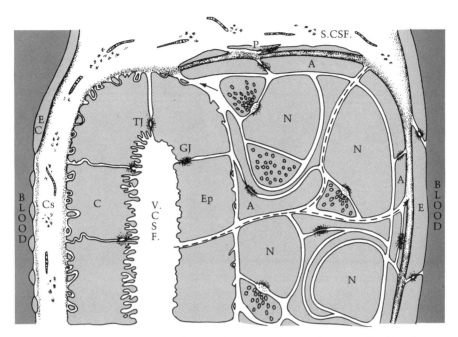

FIGURE 2.21   Diagram of sites of junctional complexes in the blood–brain–cerebrospinal fluid system. Dashed lines indicate open pathways for peroxidase injected into the ventricle. The thick arrow indicates site of a junctional leak where peroxidase from blood, having passed through the fenestrated endothelium of the choroid plexus, can pass along choroided stroma to enter the parenchyma of the brain. *A*, astrocytic process; *C*, choroid epithelium; *Cs*, choroid stroma; *E*, endothelium of choroid plexus vessel; *Ep*, ependyma; *GJ*, gap junction; N, neuron; *P*, pia mater; *S.CSF*, subarachnoid cerebrospinal fluid; *TJ*, tight junction; *V.CSF*, ventricular cerebrospinal fluid. [From ref. 10]

fenestrated blood vessels that are permeable to a variety of blood-borne substances[15,16] (Figures 2.23 and 2.24). Glutamic acid, for instance, and its potent neurotoxic congener, kainic acid, rapidly penetrate these regions when administered systemically, causing convulsions and cell death within minutes.[16]

The blood-brain barrier of the cerebral cortex develops its impermeability at a fairly late stage in ontogeny, often after birth.[G] In sheep and rats, impermeability increases five-fold over a 10–20 day period that approximately corresponds developmentally in the two species, though it occurs after birth in rats and before the midpoint of gestation in sheep.[17,18]

The Extracellular Space in Brain

After two decades of debate about the true size of the extracellular space in the gray matter of the mammalian nervous system,[G,E] the general

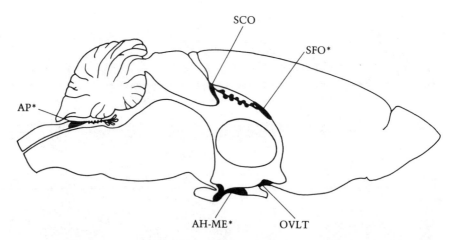

FIGURE 2.22 Diagram of a midsagittal section of rat brain showing location of the circumventricular organs. *AP*, area postrema; *SCO*, subcommissural organ; *SFO*, subfornical organ; *OVLT*, organum vasculosum of the lamina terminalis; *AH-ME*, arcuate-median eminence region of the hypothalamus. Asterisks indicate regions most vulnerable to damage by systemically administered excitotoxins, such as kainic acid. [From ref. 16]

consensus is a maximum of between 12 and 25 percent of the tissue volume depending on the region concerned, a much lower range than in liver (30 to 40 percent) and other organs. In general, the same overall conclusion applies to the invertebrate nervous system. Several complementary experimental approaches have confirmed this relatively small size-range for the extracellular space of mammalian brain, including (1) measurements of the electrical impedance of the tissue, which provide estimates of the likely volume occupied by fluid-containing conductance pathways (15 to 28 percent);[19,20,21] by the use of extracellular space markers such as [14C]sucrose or [14C]inulin, which diffuse into the extracellular spaces but are too large to enter the cells[22,23] (17 to 20 percent for four mammalian species), (3) direct measurements of the apparent sizes of the extracellular channels in electron micrographs of tissue rapidly fixed by special procedures designed to limit changes in cell sizes (and therefore also in the size of the extracellular channels) due to swelling or shrinkage.[24,25] Such studies have demonstrated how readily these channels can change their size depending on the preparation procedures. It is clear that there are narrow clefts of 15–20 nm diameter running between most cells, which could provide an extracellular space of the size range estimated by the other techniques mentioned above.

The existence of exclusively narrow extracellular channels of small

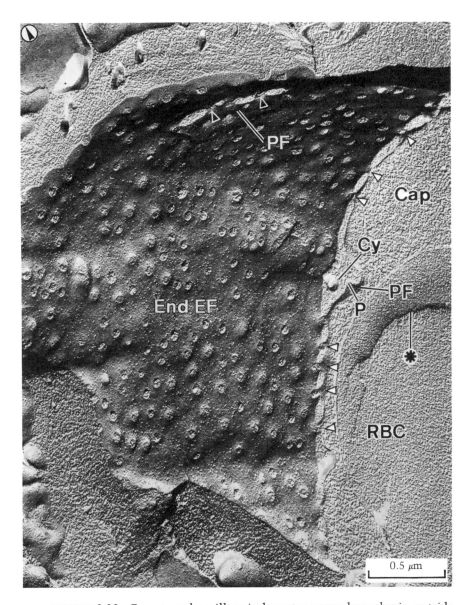

FIGURE 2.23 Fenestrated capillary in hamster neurophypophysis, outside the blood-brain barrier. The external face of the endothelial cell (*End EF*) of the capillary (*Cap*) shows numerous fenestrae in cross fracture (*triangles*). A pinocytotic vesicle (*P*) has formed in a small fold of cytoplasm (*Cy*). The majority of the field presents the endothelial external face with innumerable fenestrae in various stages of formation. Only a small step of the endothelial protoplasmic face (*PF*) is visible. The asterisk marks a random fracture step in the red blood cell (*RBC*). [From ref. 144]

FIGURE 2.24 Fenestrated capillary endothelium of the choroid plexus of the rabbit. Fenestrations are single membranes between oval endothelial cells. Black collagen fibers (cf) are present in stroma. Bar, 0.25 μm. [From ref. 145]

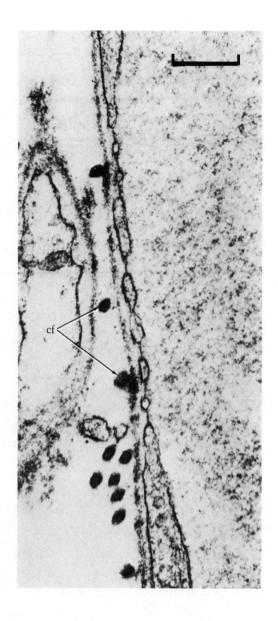

volume between the cells of the nervous system implies that small amounts of substances (e.g., sodium, potassium, neurotransmitters) liberated or accumulating in this space would rapidly reach high concentrations. Since transport systems are activated in proportion to the concentration of the substance transported, each would be quickly and efficiently absorbed into the cells. This provides an excellent transport-linked system for maintenance of the correct composition of extracellular fluids.

Glial Cell Intercommunication

Cell Junctions   Intercellular junctions are local regions of close apposition and apparent fusion between the membranes of the cells. They occur in many organs, including liver and heart, and can be isolated and studied.[19] Five broad categories exist: *gap junctions, tight junctions (zonulae occludentes), septate junctions, desmosomes (maculae adherentes),* and *synaptic junctions* (Figures 2.14, 2.20, 2.21, 2.25–2.27, and 1.30). Among glial cells, astrocytes commonly possess gap junctions that appear to allow local transference of ions, metabolites, and electrical currents.[28] They are seven-layered structures consisting of two outer components, the three-layered unit membranes (see Chapter 1) and a much reduced intervening extracellular space or gap only 2–4 nm[1,10] (Figure 2.27). Identical structures have been isolated from liver and chemically characterized;[27] they appear to be free of enzyme activity. Freeze-fracture studies show that they consist of two tightly apposed patches of homogeneous particles crowded together in polygonal arrays with center-to-center spacings of 8.5–9 nm. Each particle has a central hole of 1.5–2 nm diameter that fills with negative stain (e.g., phosphotungstic acid) and is thought to represent a central channel or low-resistance pathway through which the cells interact across the narrowed extracellular space[28] (Figures 2.25 to 2.27). A polypeptide of mol wt 25,000 to 27,000 appears to be the major protein component of isolated gap junctions, and this forms a complex together with phospholipid and cholesterol.[27,28]

In nonexcitable tissues, gap junctions probably provide a structured pathway for the transfer of ions, chemical messengers, and metabolites of molecular weight up to about 1000.[28] Excitable tissues make use of the passage of electrical currents, which can also occur across the junctions. For instance, such currents allow synchronization of myocardial cell contraction[29] and passage of nerve impulses.[31]

In the nervous system of the leech and amphibia, direct electrical coupling between two separate glial cells has been demonstrated.[32,33] Although technical difficulties have not yet allowed equivalent tests in the mammalian nervous system, it is assumed at present that the astrocytes in vivo are in similar and continuous communication through their gap junctions[34] with slow membrane potential changes being transmitted via this route. These junctions have been seen uniting astrocytes with oligodendroglia and also with neurons in cat spinal cord.[35] Electrical coupling has been recorded between glial cells and neurons in tissue culture, which suggests that neurons and glia can communicate this way in vivo.[36]

Tight Junctions   Three-layered structures consisting of two tightly apposed three-layered unit membranes but with the outer layers of each membrane apparently fused are called *tight junctions*. Such structures have been seen in the central nervous system between endothelial cells of blood capillaries, between epithelial cells of the choroid plexus (which

FIGURE 2.25   Freeze-fracture image of gap junctions between epithelial cells of a crayfish hepatopancreas. Note that the intramembrane particles are associated with the external (E) fracture face and the complementary pits with the protoplasmic (P) fracture face. Bar, 0.1 μm. [From ref. 26]

secretes cerebrospinal fluid into the ventricles), and between ependymal cells lining the ventricles[10] (Figure 2.14).

Insulating Function of Glial Cells

Astrocytes and Schwann cells often serve an insulating function by separating unmyelinated axons or nerve terminals. For instance, unmyelinated axons may lie in a trough on the Schwann cell surface, or nerve terminals at neuromuscular junctions and elsewhere may be fully encapsulated by teloglia (Figures 2.10 and 2.13). Astrocytic processes often subdivide into numerous sheetlike processes enclosing local clusters or "parcels" of nerve terminals, each isolated from the next. These sheetlike processes may form parallel stacks, which could restrict or prevent rapid lateral movement of substances and therefore guide their diffusion in the extracellular fluid from one local tissue region to another[37] (Figure 2.28).

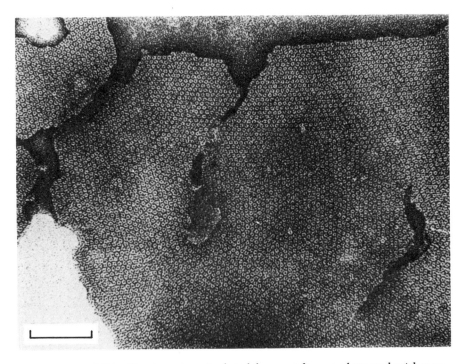

FIGURE 2.26 Gap junctions isolated from rat liver and treated with negative stain. The isolated junctions are composed of 8–9 nm particles that contain a central electron-dense 1.5–2 nm dot. This central dot is a possible location of the low-resistance polar channel for cell-to-cell communication. Bar, 10 μm. [From ref. 26]

Exchanges of Metabolites Between Neurons and Glia

Certain clearcut examples exist of metabolite exchanges between neurons and glia. Glutamine synthetase appears to be predominantly glial in location[68,40,K] which means that glutamine is synthesized in glia. A substantial proportion of glutaminase, on the other hand, appears to be neuronal.[40,I] Thus a cycle of glutamate movement between neurons and glia is likely to exist[40,43,K,L] (see Figure 4.32). Such a cycle should not be seen as exclusive because glutamine can also be used by glial cells, as judged by the properties of astrocytes in primary culture.[43] These cells possess a low-affinity uptake system for glutamine and substantial glutaminase activity.[40,43,L] Probably the flow of glutamine from glia to neurons is conditioned by such factors as rates of glutamate utilization and release.[M] Similarly, glutamate decarboxylase (GAD), which forms GABA from glutamate and is a key enzyme in the GABA-shunt pathway (an annex to the Krebs cycle in neural tissue; see Chapters 3 and 4), is found in neurons rather than glial cells. GAD enrichment in nerve terminals implies that

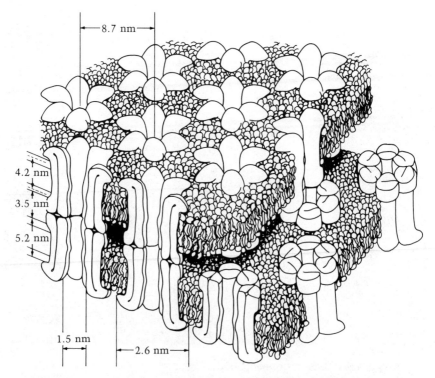

FIGURE 2.27   Models of gap junctions based on X-ray diffraction, electron microscope, and chemical data. The gap junction is depicted as a dimer of two hexamers, one associated with each membrane. The hexamers form a protein tube with an outer radius of 2.6 nm and an inner radius of 1.5 nm. Center-to-center distances and other dimensions of the gap junctions are given. An aqueous channel is conceived as extending most or all of the way through the junction formed by the hexamer, which is surrounded by the lipids of the membrane bilayer. [From ref. 30]

GABA synthesized in neurons is synaptically released and subsequently transported into glial cells, where it is further metabolized in the Krebs cycle of glial mitochondria[K,L] (see Figure 3.2).

Metabolic compartmentation of this kind exists in other forms. Glycogen granules exist largely in astrocytes rather than in neurons, and they can be mobilized as glucose following phosphorylase activation by cyclic AMP. The levels of cyclic AMP in cultured glial cells (C6 glioma cells) can be substantially elevated by exposure to beta-adrenergic agents such as norepinephrine, which also stimulates glucose release and consumption by these glioblastoma cells.[41] If these were responses typical of astrocytes, then astrocytic glycogen could provide neurons with glucose during periods of special demand, such as during activity of the

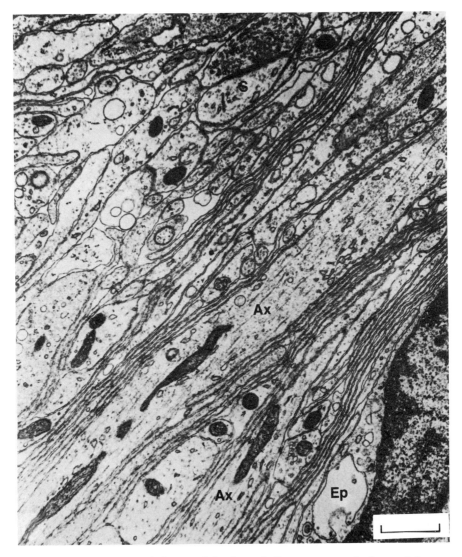

FIGURE 2.28   Glial sheets of shark medulla separating the base of the
ependyma (*Ep*) from axons (*Ax*) and from a synaptic contact (*S*). Bar, 1.0
μm. [From ref. 37]

catecholaminergic neurons or during hypoglycemia. At the same time as
they are raising cyclic AMP levels in cultured glial cells, cathechol-
amines can release this compound to the extracellular space where it
may have some trophic action on adjacent neurons.[42]
    There are likely to be many exchanges between neurons and glia that

remain to be detected. Even simple tests such as feeding radiolabeled glucose or acetate exposes metabolic differences between the two categories of cells. Glucose is preferentially taken up by neurons and acetate by peripheral glial cells,[K] and this cellular compartmentation may be visualized in autoradiographs by using isotopically labeled substrates (see Figures 3.14 and 3.15). This is not to be seen as an exclusive use of glucose by neurons, however, but rather as a quantitative difference in glucose utilization by the two cell types (see Chapter 3).

Exchanges of Macromolecules Between Neurons and Glia

Little clear evidence yet exists on the movement between neurons and glia of macromolecules such as proteins or nucleic acids, but several rather novel hypotheses have been backed with some supporting evidence.[49,53] In invertebrates, for instance, proteins may enter axons from surrounding Schwann cells, providing trophic signals, or even structural proteins to supplement those arriving in the stream of axoplasmic traffic.[44,45] Other work with labeled albumin suggests a coupled system where glial exocytosis of proteins is followed by neuronal endocytosis of the same material.[49,53,H]

Guiding Role of Glia During Neuronal Development

Another intriguing function likely to be performed by glia is assistance in ensuring the achievement of the correct connective relationships between neurons during development and nerve regeneration. Schwann cells, for instance, guide regenerating peripheral axons to their correct synaptic sites. A comparable function is served by radial glial cells, such as the Bergmann cells of the cerebellum,[46,47] which appear to guide migrating granule and other cells to their sites[48] and subsequently provide directional guidance of the elongating axons (see Figure 2.7). In vitro, for instance, neuronal processes (neurites) grow along the surface of embryonic dorsal root ganglion glia, which form up in long chains.[49] Regenerating axons in the newt spinal cord always choose to grow along preexisting grooves on the surface of glia.[52]

This physical guidance provided by glia may be supplemented, or even overshadowed, by chemical guiding stimuli or other trophic stimuli due to substances secreted by glia or present on their cell surface. Information on substances of this kind is only just beginning to be collected,[49,50,53,H,M] though some long-known agents such as nerve growth factor appear to possess the appropriate trophic influences on neurons and, appropriately, appear to be glial products.[49,53,54,H,M,N] Many of these trophic substances are macromolecules and are often proteinaceous (see Chapter 7).

Electrical Influences of Neurons on Glia

Direct electrical coupling between neurons and glia probably occurs, though it must be a rare event.[36]

The transmission of such signals across any distance, would not, however, be possible because glial cells cannot produce action potentials,[55] being devoid of active sodium-channels, and only local electrotonic (diminishing with distance) conductances through gap junctions between glia would further distribute the signal.

Indirect communication at the electrical level almost certainly does occur and may well constitute a major mode of neuron-glia communication. This form of electrical coupling involves membrane potential changes by glial cells in response to the transiently high levels of K^+ ions released to the extracellular space by neurons.

The normal level of K^+ in the extracellular fluid of brain is about 2–3 mM, lower than the 4–5 mM found in plasma.[56,57] Glial cells begin to show a displacement of resting membrane potential at K^+ concentrations, a few mM above this level. This happens because the size of this potential is dependent on the balance of K^+ ion concentration across the membranes of all cells. Glia may have special properties in this respect, behaving like true K^+ electrodes and responding with special sensitivity and specificity to changes in K^+ in their extracellular space. In fact, glial cells in general always show higher membrane potentials than their associated neurons, which probably reflects a basic difference in the permeabilities of the two cells to the various ions. Some invertebrate glial cells have a specific and high membrane permeability to K^+. The glial cells of the optic nerve of *Necturus*, the amphibian mud-puppy, are larger, and they entirely envelop the axons of this optic pathway. In the leech, even larger glial cells (*packet glia*) completely surround the neurons of the segmental ganglia. Electrodes are easily inserted into cells of such large size; their membrane potentials show a linear relationship to externally manipulated K^+ concentrations that fits well with the Nernst equation for K^+ electrodes. When external and internal K^+ concentrations are identical the membrane potential is zero, which indicates that these are the ions exclusively determining this potential[58,59] (Figure 2.29). The same does not appear to be true for the glial cells of the cerebral cortex, where other ion permeabilities make a significant contribution,[60] though here, too, the glial membrane potentials are higher than in neurons.

Electrodes positioned in the glia of the optic nerve of *Necturus* revealed that when action potentials are induced in the adjacent nerve fibers by stimulation, the glial cell membrane potential shows a synchronous depolarization. A similar response was evoked when the eyes of *Necturus* were exposed to light flashes in order to produce action potentials in the optic nerve fibers by natural stimulation of the sensory pathway[61] (Figure 2.30). Glia of the mammalian cerebral cortex respond with depolarization when cortical neurons are activated.[55,62] Potassium-sensi-

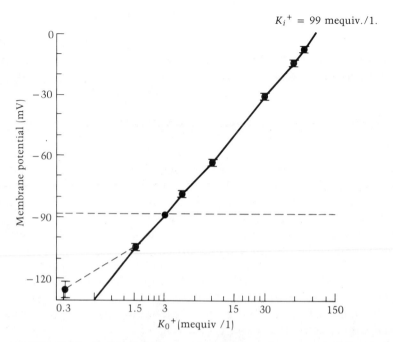

FIGURE 2.29   Relation between the glial membrane potential and the K^+ concentration in the bathing fluid (K_0) in optic nerves of the mudpuppy. Forty-two measurements were made with K^+ concentrations below or above normal. The mean of the resting potentials in Ringer's solution (3 mM K^+) was 89 mV (*dashed line*). Horizontal bars indicate ±S.D. of mean. The solid line has a slope of 59 mV for a tenfold change in K_0 according to the Nernst equation; it fits the observed points accurately between 1.5 and 75 mM. Only at 0.3 mM K^+ is there a marked deviation (dashed curve) from the curve predicted by the Nernst equation. The membrane potential is zero when the internal K^+ concentration (K_i) equals K_0. K_i^+ is therefore 99 mM. The membrane of glial cells can be used as an accurate K^+ electrode. [From ref. 59]

tive glass electrodes show that the K^+ concentrations in the extracellular space do in fact rise during neuronal activity, to levels that could cause the observed depolarization of glial cells, thus confirming that this is indeed the factor mediating the electrical coupling between the two cell types. The small size of the extracellular space in the nervous system permits even the relatively small amounts of K^+ released by neurons to rapidly reach higher concentrations than they would in tissues with a larger space.

Glial Cell Response to Injury and Inflammation

Studies with radiolabeled thymidine, which becomes incorporated into cell DNA, show that glia, unlike neurons, retain in the adult nervous

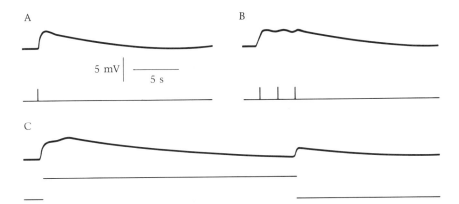

FIGURE 2.30   Effect of illumination of the eye on the membrane potential of glial cells in the optic nerve of an anesthetized mudpuppy with intact circulation. (A) Single flash of light for 0.1 second. (B) Three flashes. (C) Light stimulus maintained for 27 seconds; during such prolonged illumination the initial glial depolarization declines as the nerve discharge adapts. At the end of illumination, a burst of "off" discharges initiates a renewed glial depolarization. Lower beams monitor light. [From ref. 61]

system the ability to divide and multiply. It is the glial cells that multiply after damage to neural tissue, which leads to a local domination of tissue composition by glia (*gliosis*), which in turn leads to the formation of scar tissue. It is glial cells rather than neurons that tend to become cancerous and produce brain tumors (which are most commonly gliomas). Neuroblastomas do occur in sympathetic ganglia[N] and in the developing nervous system. The inflammation that is a consequence of injury involves the activation of resident microglia, which become phagocytic. The same transformation occurs in the reactive category of macroglial cells, including the astrocytes and oligodendroglia. These too can be seen dividing rapidly close to the site of injury,[63] producing phagocytes and also causing glial proliferation and gliosis.

Neurotransmitter Transport Systems in Glia

Neurons that synthesize and release a particular neurotransmitter usually possess a high-affinity uptake system for that neurotransmitter. Glial cells appear to have several of these systems in their membranes, rather than one.[39] These high-affinity systems characteristically respond to low concentrations of neurotransmitters in the extracellular fluid, and thus effectively and rapidly remove the potent compounds before their concentration can rise and diffuse to sites where their receptors are located.[H] For glutamate, for example, there is a very efficient high-affinity uptake to glial cells, including both separated and cultured astrocytes.[39,43] But there is some debate about the extent to which neurons, particularly perikarya, possess high-affinity uptake systems for glutamate.[39,43,O]

SPECIFIC MARKERS FOR GLIA

Histochemical localization and studies of the properties of isolated or cultured glial cells have provided information on a range of biochemical parameters that are putative or established specific constituents of glial cells. Biochemical markers of this kind are invaluable for identifying glial cells in various tissue preparations as well as providing clues about the functions of these cells.

Enzyme Markers

Carbonic Anhydrase   The enzyme carbonic anhydrase, which exists in both soluble and membrane-bound forms in the brain (EC 4.2.1.1), catalyzes the reversible combination of H_2O and CO_2 to form carbonic acid:

$$CO_2 + H_2O \rightleftharpoons H_2CO_3 \rightleftharpoons H^+ + HCO_3^-$$

This is the system primarily concerned in controlling CO_2 exchange as well as the movements of H^+ ions, bicarbonate, and other ions across cell membranes. Therefore, it also influences cell pH and water content. The enzyme contains zinc and can also hydrate aldehyde groups to form alcohols and function as an esterase. In rat brain more than half of the enzyme is particulate,[64] and judged from immunohistochemical and radioimmunoassay studies,[65,66] it appears to be principally localized in glia, particularly oligodendroglia. During development of the mammalian brain, it rises in quantity at the stage of glial proliferation, and it has also been shown to be present in primary cultures of astrocytes.[64] Both observations support its glial localization.

The presence of carbonic anhydrase in glia is indicative of the special role these cells play in controlling CO_2 exchange and pH in the extracellular spaces. The activity of the soluble form of carbonic anhydrase in astrocyte cultures can be stimulated by norepinephrine through a cyclic AMP–mediated mechanism, indicating a link to neurons in the control of the enzyme activity.[67]

Glutamine Synthetase   Immunocytochemical labeling techniques have shown that glutamine synthetase (EC 6.3.1.2) is principally a glial enzyme.[68] Its function appears to be to convert glutamate, which is being continuously absorbed from the extracellular space, into glutamine, which readily diffuses out of the glia back into the surrounding space. Glutamine, unlike glutamate, is physiologically inactive and is allowed to accumulate in the interstitial fluid to relatively high concentrations (500 μM) compared with glutamate (10 μM). At the same time that glutamine is formed in glial cells, ammonia is removed.[69] Thus, ammonia metabolism in neural tissue is also partly controlled by this enzyme.

$$\text{glutamate} + \text{ammonia} + \text{ATP} \rightarrow \text{glutamine} + \text{ADP} + P_i$$

Inhibition of glutamine synthetase by agents such as methionine sulfoximine leads to excessive accumulation of ammonia which, in turn, causes convulsions (see Chapter 9). In this way, glial cells are centrally involved in the important systems of ammonia homeostasis and the intercellular exchanges of glutamate and glutamine (Figure 4.32).

Glyercol 3-Phosphate Dehydrogenase   A side branch exists in the mainstream of glycolytic cleavage of glucose at the stages subsequent to the conversion of fructose 1,6-diphosphate into two molecules of triose phosphate. At this point one of these trioses, dihydroxyacetone phosphate, can either be converted to D-glyceraldehyde 3-phosphate (by triose phosphate isomerase) and continue in the mainstream flow or it can form a branch to L-α-glycerol phosphate through the action of the soluble enzyme glycerol 3-phosphate dehydrogenase (EC 1.1.1.8).

$$\text{dihydroxyacetone phosphate} + H^+ + \text{NADH} \rightarrow$$

$$\text{L-glycerol 3-phosphate} + \text{NAD}^+$$

This alternative pathway generates intermediates for lipid biosynthesis (phospholipids). In muscle and brain it also forms part of an efficient shuttle system for transfer of equivalents of hydrogen across the mitochondrial membrane for oxidation and energy generation. NADH and NAD^+ cannot themselves enter mitochondria. Instead, the soluble cytoplasmic enzyme and its mitochondrial counterpart, which is a different protein (a flavoprotein), shuttle glycerol phosphate across the membrane, the net result being the production of NAD^+ from NADH in the cytoplasm. The NAD^+ is required to maintain the flow of glycolysis.

The development and maintenance of the enzyme in neural tissue is controlled by glucocorticoid hormones[70,71,73] (see Table 2.1). This response appears to be specific for neural tissue. Since the inductive effect is most pronounced in separated oligondendroglia,[73] in primary cultures of astrocytes, and in C6 glioma cell lines, it is regarded as being located primarily in glial cells.[70,71,73] Immunohistochemistry has indicated localization of the enzyme in Bergmann glia and oligodendroglia of the cerebellum.[75]

TABLE 2.1   Regulation of marker enzymes in astrocytic and oligodendroglial dissociated cell cultures

	Astrocytes (nmol substrate/ min per mg protein)	Oligodendroglia (nmol substrate/ min per mg protein)
CNPase		
Control	N.D.	441 ± 11
Plus dibutyryl cyclic AMP (1 mM)	N.D.	507 ± 13
Glucose-3-phosphate dehydrogenase		
Control	4.0 ± 0.3	43.0 ± 1.8
Plus hydrocortisone (1.38 μM)	11.9 ± 0.9	688.5 ± 5.1
Lactate dehydrogenase		
Control	2022 ± 51	1147 ± 30
Plus dibutyryl cyclic AMP (1 mM)	2700 ± 132	2910 ± 112

Note: Fresh treatment-medium was added after 24 hours at 30°C. Values are mean ± SEM for three cultures of rat brain. N.D., not detectable. From ref. 73.

2',3'-Cyclic Nucleotide 3'-Phosphohydrolase (CNPase)   The enzyme CNPase (EC 3.1.4.37) hydrolyzes 2',3'-cyclic nucleotides specifically to form the corresponding 2'-phosphates. Myelin (which is made by oligodendroglia and by Schwann cells) contains a 20- to 30-fold enrichment of CNPase,[76] and bulk preparations of oligodendroglia[73,77] and Schwann cell cultures[78] also contain the enzyme, which destroys cyclic nucleotides by opening their phosphate ring. The enzyme can be enhanced twofold by treatment of these glial cell cultures with noradrenaline or dibutyryl cyclic AMP[79] (Table 2.1), and appears to be specifically glial in location.

Non-neuronal Enolase   Brain tissue contains a spectrum of enolase isoenzymes (EC 4.2.1.11) that is distinct from those found in other tissues. One isoenzyme, neuron-specific enolase (originally known as 14-3-2 protein), is strictly neuronal in location. Another isoenzyme, non-neuronal enolase, has been shown by immunocytochemical analysis to be localized in glial cells[82,153,154] This glial enzyme is similar, and probably identical, to liver enolase,[154] but in brain it is a valuable biochemical marker for glial cells.

Specific Proteins and Other Glial Markers

Despite continuing controversy concerning the specific cell location of several proteins, evidence continues to accumulate that at least some proteins are predominantly located in glia. However, some highly localized neuronal sites for these proteins have also been found. It may be that glia are the site of synthesis, with subsequent migration to neurons, or it may be that cross-reactivity of the antisera or cross-contamination of preparations employed in antisera production is providing misleading evidence for the minor neuronal content.

Among these proteins is *S-100 protein*,[80,82] so named because it is completely soluble in saturated ammonium sulfate solution at pH 7. It is a small molecule of low immunogenicity (mol wt 21,000–24,000) and is found in all vertebrate and several invertebrate nervous systems, where it is 104-fold more concentrated than in other tissues. S-100 protein appears late in development, at a time when histological maturity is well advanced. It constitutes about 1 percent by weight of the soluble proteins of brain homogenates. Its function is likely to be related to its special composition, being very acidic and readily soluble in aqueous media. This acidity is due to its high content of glutamate and aspartate residues (19 and 11 percent respectively of total residues). It characteristically migrates rapidly in gel electrophoresis. The protein appears to consist of three slightly unequal subunits of mol wt about 7000 each. It is known to be principally glial (astrocytic) in location from its higher concentration in white matter than in gray, its histochemical localization, and its production by glial cells in tissue culture.[80,81,83,84,94,108] Detection of S-100 protein in various neuronal preparations,[85,86,87] microdissected cells, or bulk-prepared neuronal perikarya (tissue cultures and synaptosomes) is probably explained by glial contamination and other artifacts.

The ability of S-100 protein to specifically bind 8 to 10 calcium ions per molecule at high-affinity sites leads it to change its conformation in a fashion that exposes its hydrophobic regions and enhances monovalent cation movements across model bilayers.[88,89] Monovalent cations at physiological concentrations inhibit this binding, the order of effectiveness being $K^+ > Na^+ > Li^+$. This indicates that one kind of function the protein may serve in glial cells is cation transport. The presence of S-100 protein as an integral constituent of the acidic proteins of nuclear chromatin suggests a role in genomic regulation in neural tissue, but there is no definitive evidence on what the precise functions of this protein are, in spite of its selective concentration in neural tissue.[81]

Another glial-specific protein is *glial fibrillary acidic protein* (GFAP), isolated from glia-enriched brain areas[82,90,91,92] and subsequently found by immunohistochemistry to be localized principally in fibrous astrocytes and radial glial cells,[91,93,94] where it resides in the prominent gliofilaments (see Figure 2.4).

The peak in production of GFAP during mouse brain development occurs at a time (10–14 days postnatal) when astroglia are differentiating.[95] The protein consists of one to seven peptide monomers of mol wt 40,000 to 55,000 each.[96,97,98]

Additional markers for glia include *rat neural antigen 1* (Ran-1), a Schwann cell–specific surface antigen in rats. Antibodies against this surface antigen were raised against a neural tumor that arose in the rat spinal cord and adjacent nerve roots.[155]

Galactocerebroside, the major glycolipid of myelin, is found specifically in oligodendroglial cells in culture. These cells constitute about 25 percent of the cells in optic nerve cultures and are identified through their lack of GFAP. They label intensely with antigalactocerebroside antibody,[156] whereas GFAP-positive cells in the same culture show no such binding.[156,157] Freshly isolated Schwann cells are galactocerebroside positive but become negative after 24 hours in culture, reflecting the differences in response of these two cell types in tissue culture.

NEURONS AND GLIA IN CELL CULTURE

Tissue culture has become a powerful and very popular approach to research problems in cellular neurobiology. The various origins of the neural cells being cultured (including their origin in normal tissue or cancer cell lines) and the various methods of culturing them has provided a rich array of techniques and preparations for studying the neurochemical and functional properties of neurons and glia.[N] The tissue culture approach must be seen as complementary to the use of other neural tissue preparations. The strengths and weaknesses of these various techniques are listed in Table 2.2.

Categories of Cell Culture

Primary Cultures Cultures started from cells or tissues taken from organisms and grown for more than 24 hours are termed *primary cultures.* When a primary culture is subcultured it becomes a *cell-line,* and consists of the numerous lineages of the cells originally present in the primary culture. Cells in primary cultures are immediate descendants of cells in situ, and their behavior and function should be close to those of the in situ cells.

Cell Culture Lines Since cell lines are serially transplanted, the more rapidly proliferating cells (e.g., glia) will be preferentially selected and will overgrow the nondividing cells (e.g., neurons). Established cell lines originate from primary cell cultures either spontaneously or after treatment with carcinogens, radiation, or viruses.[N] The cells become transformed by these treatments and acquire heritable morphological and growth properties associated with tumor cells. After much subculturing, they become stable from generation to generation. These cells are capable of indefinite cell division. Established glial cell lines from rats include RG-179 and EH-118 MG, which is astrocytic in character.[N] Single trans-

TABLE 2.2   Comparison of tissue cultures and other neural preparations

Alternatives	Viable cells	Permanently established lines	Biochemically differentiated cells	Functionally and generally homogeneous populations	Quantities sufficient for biochemical studies	Characterized with respect to specific cell type	Normally regulated intra- and intercellularly
Clonal lines	Yes	Yes	Yes	Yes	Yes	No, or uncertain	Doubtful or not known
Primary culture (explant and dissociated)	Yes	No, but could provide a starting point	Yes	Not generally, but methods are being developed	Yes/No	Potentially	Potentially in some cases
Organ culture (including reaggregation systems)	Yes	No	Yes	No	Yes/No	No	Potentially
Bulk dissociation and purification	Yes/No	No, but could provide a starting point	Yes	Possibly	Yes	Yes, ideally	No?
Hand dissection of specific cells	Yes/No	No	Yes	Possibly	Only with great difficulty	Yes, ideally	No?
Localization in situ of specific cell types using specific staining or autoradiographic techniques	Yes	No	Yes	Yes, relative to the extent that the tagged function is cell-type specific	Cytochemistry	Potentially	Yes
Brain slices	Yes/No	No	Yes	No	Yes	No	Yes?
Brain homogenates	No	No	Doubtful	No	Yes	No	No

Note: From ref. M.

formed cells that have been isolated and cultured to produce a single cell type by mitosis are called clonal cell lines (see below).

Explants or Organ Cultures   A type of primary culture that has been very popular and useful in the past is the explant, formed when small, thin fragments of embryonic or early postnatal neural tissue, and the outgrowth of cells from these preparations, survive in culture medium. A disadvantage of this approach is the necessity for the cells to migrate to the boundaries of the explant before they can be easily visualized or individually studied. However, spontaneous activity, excitability, and conduction have been demonstrated in cerebellar explants[99] and spinal cord, and cerebral cortex cultures form morphologically and functionally normal synapses[100,103] and dendritic trees.[104] Thus, much basic nervous system tissue architecture and sometimes specific organotypic organization are maintained in these explants.[N]

Dissociated Neural Cell Cultures   Another type of primary culture, the *dissociated neural cell culture*, is produced when neural embryonic tissues are dissociated by mechanical or enzymatic procedures to produce cell suspensions that are then introduced into culture medium. Dissociated cells from various brain regions and cerebellum, and from spinal cord have been successfully cultured.[N,73] The advantages of such cultures are numerous; they include the ready visualization of cells and the possibility of studying individual cell types in isolation over long periods in pure neuronal or glial cell[73] cultures. The actions of trophic factors on neuronal growth can be studied, and it is possible to investigate the processes involved in the reorganization of the cells as they begin to interact.[101,102] Thus, the formation of nerve-muscle junctions can be followed as they form in cultures of dissociated neural and muscle cells.[103,N]

Purified oligodendroglial and astrocytic cultures derived from dissociated cerebral cortex cells also show distinct and separate pharmacological responses. Two oligodendroglial cell markers, CNPase and glycerol 3-phosphate dehydrogenase, are induced by dibutyryl cyclic AMP and hydrocortisone. Dibutyryl cyclic AMP induces lactate dehydrogenase particularly effectively in oligodendroglial cultures. Cyclic AMP levels are greatly enhanced in both types of dissociated cell culture by norepinephrine, and this is blocked by the alpha-adrenergic blocker phentolamine in oligodendroglia (Tables 2.2 and 2.3). Prostaglandin E also potently stimulates cyclic AMP formation in oligodendroglia.

Clonal Cell Lines   Neuroblastoma and glioma cell lines originally obtained from single tumor cells provide special opportunities for investigating the unique differences between neurons and glia.[104,105,106] The particular advantages concern the production of large homogeneous populations of these cells (by cloning), which can then be grown under

TABLE 2.3   Regulation of cyclic AMP levels in astrocytic and oligodendroglial dissociated cell cultures

	Astrocytes (pmol cyclic AMP per mg protein)	Oligodendroglia (pmol cyclic AMP per mg protein)
Control	8.7 ± 0.2	8.9 ± 0.4
Norepinephrine (3 μM)	28.7 ± 2.9	50.1 ± 9.0
Norepinephrine plus phentolamine (3 μM)	180.6 ± 10.1	62.0 ± 4.4
Adenosine (100 μM)	44.3 ± 1.6	14.3 ± 0.7
Prostaglandin E (3 μM)	67.6 ± 8.8	187.4 ± 31.2

Note: Incubation was for 5 minutes at 30°C. Values are mean ± SEM for three rat-brain cultures. From ref. 73.

conditions that can be controlled and manipulated. These cell lines include central glial cells (e.g., C6 glioma, astrocytic cell type), peripheral glial cells (Schwann cell tumor lines, RN-2),[107] and various neuronal cell clones (e.g., C-1300). (See Figures 2.31 and 2.32.) The shape of these cells very much depends on the culture conditions employed. Cells grown in suspension cultures are mostly very rounded (20–40 μm diameter) with few if any processes. However, when grown on flat surfaces (e.g., monolayer culture) they show remarkable morphological diversity, with cell processes and neurites (neuroblastoma) readily forming and cell bodies often enlarging in size. Examples of their properties include the biosynthesis of S-100 protein and the induction of glycerol 3-phosphate dehydrogenase by hydrocortisone (cf. Table 2.1), which involve enhancement of cyclic AMP levels and occur only in glial cell lines.[108] The same is true for lactate dehydrogenase induction by noradrenaline in C6 glioma cells, which is preceded by a rise in cyclic AMP.[109] The mobilization of glial glycogen by noradrenaline is also mediated by the induction of cyclic AMP[41] and this leads to the release and metabolism of glucose. These effects result from the interaction of noradrenaline with beta receptors on the glial cell surface, since beta antagonists prevent the responses.[74] Thus, under the appropriate conditions catecholamines should be able to release glucose from the glycogen stores of glial cells in vivo. The two clonal cell lines neuroblastoma and glioma produce a wide range of different specific proteins,[109] emphasizing their essentially different nature.

The biosynthesis and properties of several macromolecules have been studied in clonal glial cell cultures. S-100,[111] mucopolysaccharides, glycolipids, and glycoproteins[78] have been investigated and found to be identical with those produced in whole brain. Such properties imply the presence and normal production of organized and properly controlled multienzyme sequences in what are essentially cancer cells in culture.[112]

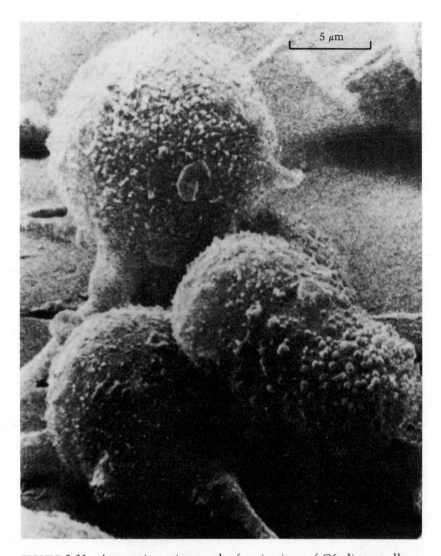

FIGURE 2.31   A scanning micrograph of projections of C6 glioma cells. Specimen was prepared by fixation in paraformaldehyde and glutaraldehyde. Critical-point drying and coating was carried out with conductive carbon followed by 40 percent palladium and 60 percent gold alloy. Note surface projections, which are 30–60 μm in diameter. Bar, 5 μm. [From ref. 146]

Microelectrode studies on neuroblastoma cells (C-1300) in culture show that their electrophysiological properties are essentially similar to those of normal neurons. This includes generation of tetrodotoxin-sensitive action potentials and sensitivity to applied acetylcholine,[104] with hyperpolarization, depolarization, and action potentials as typical re-

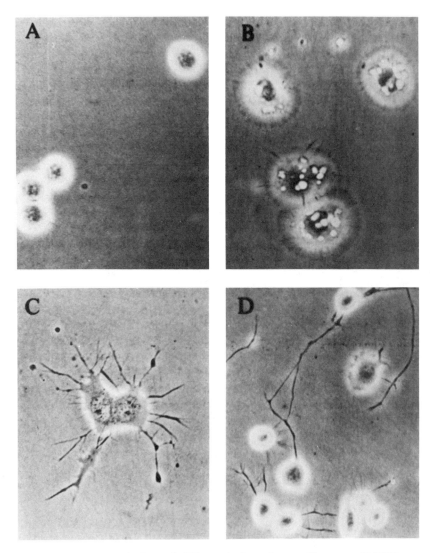

FIGURE 2.32 Morphological differentiation of neuroblastoma C-1300.
Cells were plated in serum-free medium in 60 mm tissue culture dishes
at 2×10^5 cells per plate. Phase photomicrographs were taken at various
times after plating. Time after transfer to tissue culture dish: (A) 5 minutes, (B) 15 minutes, (C) 5 hours, (D) 24 hours. [From ref. 147]

sponses[110] (Figure 2.33). Many of the enzymes for neurotransmitter synthesis are present and active,[104] but cyclic AMP appears to be an important agent in stimulating and maintaining some of these enzymes[105] in
the cells. Cyclic AMP also stimulates the outgrowth of neurites from
cultured neuroblastoma (Figure 2.34). Such triggering and control of bio-

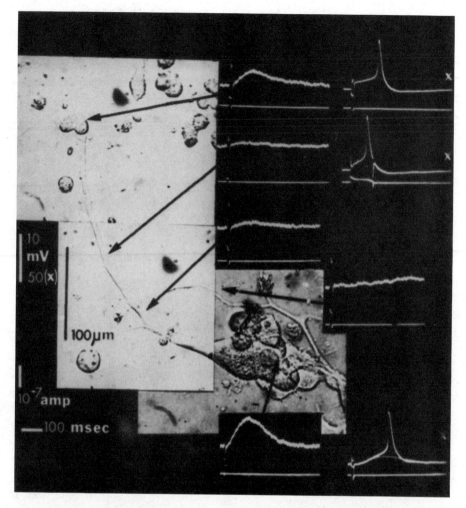

FIGURE 2.33   Distribution of acetylcholine sensitivity over the surface of
a differentiated neuroblastoma cell with one process, 3 weeks after plating. Sensitivity to acetylcholine is restricted to the cell body, where acetylcholine application evoked an action potential (lower right), and to the
neurite growth cone. Extracellular electrical stimulation of the axon or
its tip gave rise to an action potential that propagated back to the cell
body (two top right traces; points of stimulation as for acetylcholine
application). Calibration for voltage, current, and time is given at the
lower left; *50(X)* means 50 mV and applies to records marked with X.
[From ref. 148]

synthesis in cell cultures by cyclic AMP is not an unexpected phenomenon, since this important substance is formed by the enzyme adenyl
cyclase, whose activity may be modulated by various catecholamines and
hormones (see Table 2.3). These biodynamic agents are thus able to exert
their actions through the formation of cyclic AMP.[104]

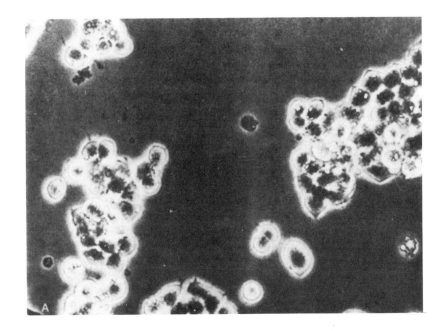

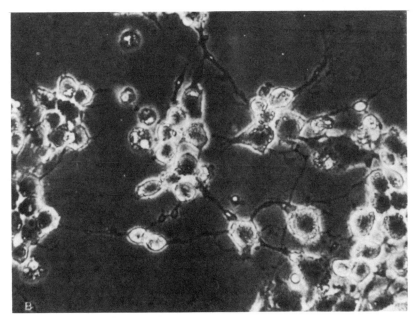

FIGURE 2.34 Expression of differentiated functions in mouse neuroblastoma mediated by dibutyryl cyclic AMP. (A) Cells in the presence of serum. (B) Cells in the presence of serum and 1 mM dibutyryl cyclic AMP for 48 hours. [From ref. 152]

BULK-PREPARED NEURONS AND GLIA

Isolated Neurons

Because as much as 90 percent of the cell surface of many neurons reside in axons and dendrites, bulk preparations of cell bodies (perikarya) of neurons must be viewed as very specialized subneuronal preparations, not necessarily displaying the properties typical of the whole neuron. Early methods of cell isolation were not able to produce enough tissue for easy biochemical analysis. For instance, large (100–300 μm diameter) neurons from lateral vestibular nucleus can be hooked and pulled out using thin (7 μm diameter) stainless steel wires,[113] and microbiochemical analysis has been applied to these cells (see Figure 1.2). Somewhat larger quantities were prepared by pushing disaggregated blocks of ox vestibular nucleus through nylon sieves of diminishing size (apertures of 300–350 μm^2 followed by 110 μm^2) in isotonic sucrose at 0–4°C. The separated neurons were then isolated by hand using small nylon loops (150–200 μm diameter) at a maximum rate of 600 neurons per hour.[114]

Much larger numbers of neuronal perikarya, shorn of most of the length of their cell processes, were prepared from mechanically or enzymatically (trypsin) disaggregated cerebral or cerebellar cortex tissue, which was subsequently separated into fractions enriched in various cell types by centrifugation into density gradients, such as those formed from ficoll (a sucrose polymer of mol wt 400,000) dissolved in isotonic sucrose[115–125] (Figures 2.35 and 2.36). The yields are small, representing only about 1–2 percent of the starting tissue content of neurons, though higher yields of neurons have been estimated on the basis of DNA content.[128] This means that the cells isolated from young adult rats may not be a representative sample of the general neuronal population present in the whole tissue. However, the perikarya of Purkinje cells, granule cells, and interneurons can be clearly distinguished in preparations from cerebellum.[123]

The purity and intactness of the neuronal perikarya preparations have been questioned, but morphological examination and measurement of biochemical markers for glial cells suggest that contamination with glia is low,[119,121] with preparations containing 70–90 percent neurons in terms of particle counts being reported.[121,125,128] These perikarya are probably in some cases damaged and of limited use for metabolic experiments, but they can be of value for studying the chemical composition of neuronal cell bodies.

Isolated Glial Cells

Bulk preparations of glial cells have been produced from the same tissue samples. They separate as fractions of lower density in the same sucrose gradients described above for preparing neuronal perikarya, though albumin gradients have also been employed[120] to prepare glia. Well-pre-

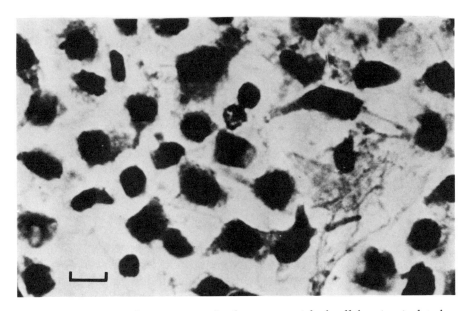

FIGURE 2.35 Photomicrograph of neuron-enriched cell fraction isolated from rabbit brain. Most of the structures present are neuronal perikarya with proximal dendrites. A few small, dark nuclei and some fibrous material are seen as contaminants of the fraction. Bar, 10 μm. [From ref: 118]

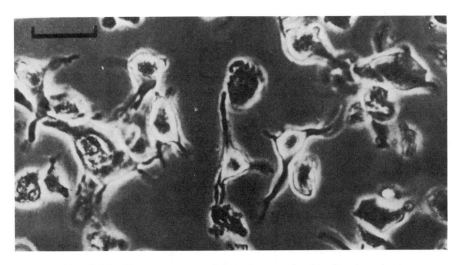

FIGURE 2.36   Phase contrast photomicrograph of isolated rat brain neuron fraction. Many cells have portions of processes attached. Note absence of small debris. Bar, 2 μm. [From ref. 149]

served astrocytes and neurons have been isolated by using gentler methods of disaggregation[121,125] (Figures 2.1 to 2.3, 2.37, and 2.38) but the yields of glial cells are at low levels, usually much less than the yield of neuronal perikarya.[125]

Glial cell preparations isolated by these density gradient procedures are usually rather intact structures with a full complement of long processes. Morphological and other estimates put their purity in the 50–60 percent range[119,120,125] (Figures 2.2, 2.37, and 2.38). The likely extent of contamination with broken nerve-fibers, dendrites, and nerve terminals is small, about 15–20 percent as judged by biochemical markers for neuronal structures.

Uses of Bulk Preparations

The various laboratories working with enriched preparations of neuronal perikarya and glial cells have applied them in different ways. Most have concentrated on compositional and dynamic studies, making comparisons between the two types of cell fraction. This has included studies of the rates of biosynthesis of proteins,[131] lipids,[127] amino acids, and many other cell components, as well as rates of respiration and energy metabolism,[115] the size and turnover of amino acid pools,[132] and the uptake and release of neurotransmitters.[126,130,131]

There is not always great consistency in the properties of these cells when prepared by different methods. Examples of differences shown by the neuronal preparations when compared with glial cells are their higher rates of protein synthesis[118,133] (but see ref. 119), their lower rates of respiration,[133] their larger amino acid pools (but see ref. 131) turning over more slowly, and their Ca^2-dependent release of neurotransmitters.[130,133,136] In addition, the neuronal perikarya may show great enrichment of lysosomal enzymes[129] and RNA.[125]

Other applications are more directly concerned with *functional* analysis of the role of neurons and glia in intact neural tissue. In this case, the cell types are separated and analyzed after inducing some specified response or activity of the intact brain. In this way biochemical responses occurring in the brain region under scrutiny can be further localized to cell type.

A reduction in protein labeling, which occurs specifically in the visual cortex of rats reared in the dark, has been localized to neuronal perikarya by this approach.[134,135] Time-course experiments strongly suggest that the reduced labeling is due to reduced synthesis of a group of rapidly labeling proteins (including tubulin) that are normally transported out of the neuronal cell bodies and into the axoplasmic stream.[133] In normal animals both intact neurons and their derived perikaryal fractions do, in fact, often show considerably higher rates of protein labeling than the glial fractions prepared simultaneously,[118,133,134,135] but some preparations show equivalent rates in the two cell types.[119]

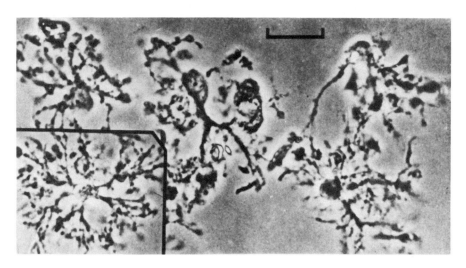

FIGURE 2.37   Astroglia isolated from rat brain. Processes are extensive, branched, and of irregular diameters. The framed area shows a single typical rat brain astroglial cell with symmetrically arranged, highly branched processes. Bar, 2 μm. [From ref. 149]

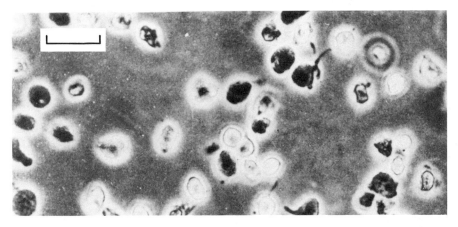

FIGURE 2.38   Oligodendroglia isolated from bovine white matter. A few cells have a short process. Nuclei are not generally discernible at this magnification. Note the absence of astroglia and small debris. Bar, 2 μm. [From ref. 149]

Another experimental use of isolated cells (e.g., cerebellar cells) has been to culture them and study such phenomena as the reappearance of cell processes, their capacity to transport neurotransmitters, and the density of neurotransmitter receptors on their surface.[128,123] When young animals are used, comparisons can be made between development of cells in vivo and in vitro after isolation.[124]

Each of the various preparations of neurons and glia—whether whole tissue organ cultures, cloned tumor cell lines, or isolated cell body preparations—has its advantages and limitations and must be regarded as providing complementary information on the properties of neurons or glial cells (Table 2.2). Those features consistently expressed in all of the preparations will provide the most convincing evidence for the specific biochemical properties of each cell type.

REFERENCES

General Texts and Review Articles

A   Glees, P. (1955) *Neuroglia Morphology and Function.* Thomas, Springfield, Ill.

B   Kuffler, S. W., and J. G. Nicholls. (1984) *From Neuron to Brain,* 2d ed. Sinauer, Sunderland, Mass.

C   Watson, W. E. (1976) "Brain As an Epithelium," *Cell Biology of Brain,* pp. 1–40. Chapman and Hall, London.

D   Windle, W. (1958) *The Biology of Neuroglia.* Thomas, Springfield, Ill.

E   Johnston, P. V., and B. I. Roots. (1972) "Considerations of Some Features of the Organization and Ultrastructure of Nervous Systems," *Nerve Membranes,* pp. 5–44. Pergamon Press, Oxford.

F   Peters, A., S. L. Palay, and H. D. F. Webster. (1976) *Fine Structure of the Nervous Systems: The Neurons and Supporting Cells.* Saunders, Philadelphia.

G   Bradbury, M. (1979) *The Concept of a Blood-Brain Barrier.* Wiley, New York.

H   Varon, S. S., and G. G. Somjen, eds. (1979) *Neuron–Glia Interactions.* Neurosciences Research Program Bulletin, vol. 17, no. 1, pp. 1–239.

I   Feldman, J., N. B. Gilula, and J. D. Pitts. (1978) *Intercellular Junctions and Synapses.* Chapman and Hall, London.

J   Schoffeniels, E., G. Franck, L. Hertz, and D. B. Tower, eds. (1978) *Dynamic Properties of Glial Cells.* Pergamon Press, Oxford.

K   Bito, L. Z., H. Davson, and J. D. Fenstermacher, eds. (1977) *The Ocular and Cerebrospinal Fluids.* Academic Press, New York.

L   Sato, G. ed. (1973) *Tissue Culture of the Nervous System.* Plenum Press, New York.

M   Fedoroff, S., and L. Hertz. (1977) *Cell, Tissue, and Organ Culture in Neurobiology.* Academic Press, New York.

N   Poduslo, S. E., and W. T. Norton. (1975) *Methods Enzymol.* 35: 561–579.

O   Balazs, R., and J. Cremer. (1973) *Metabolic Compartmentation in the Brain.* Macmillan, London.

P   Hertz, L., E. Kvamme, E. G. McGeer, and A. Schousboe, eds. (1983) *Glutamine, Glutamate, and GABA in the Central Nervous System.* Alan Liss, New York.

Q   Bradford, H. F., ed. (1982) *Neurotransmitter Interaction and Compartmentation.* Plenum Press, New York.

R   Treherne, J., ed. (1982) *Glia–Neuron Interactions.* Cambridge University Press, Cambridge (*J. Exp. Biol.* 95).

Additional Citations

1 Chan Palay, V. (1977) *Cerebellar Dentate Nucleus: Organization, Cytology, and Transmitters.* Springer-Verlag, Berlin.
2 Bignami, A., and D. Dahl. (1974) *J. Comp. Neurol.* 153: 27–38.
3 Sturrock, R. R. (1976) *J. Anat.* 122: 521–537.
4 Das, G. D. (1976) *Virchow's Archiv. (Cell Pathol.)* 20: 287–298.
5 Das, G. D. (1976) *Virchow's Archiv. (Cell Pathol.)* 20: 299–305.
6 Ling, E. A., J. A. Paterson, and A. Privat. (1973) *J. Comp. Neurol.* 149: 43–71.
7 Bunge, R. S. (1970) In *The Neurosciences: Second Study Program* (ed. F. O. Schmitt), pp. 782–797. Rockefeller University Press, New York.
8 Brightman, M. W., and S. L. Palay. (1963) *J. Cell Biol.* 19: 415–439.
9 Chan-Palay, V. (1976) *Brain Res.* 102: 103–130.
10 Brightman, M. W., and T. S. Reese. (1969) *J. Cell Biol.* 40: 648–677.
11 Reese, T. S., N. Feder, and M. W. Brightman. (1971) *J. Neuropathol.* 30: 137–138.
12 Reese, T. S., and M. J. Karnovsky. (1967) *J. Cell Biol.* 34: 207–217.
13 Milhorat, T. H., D. A. Davis, and B. J. Lloyd. (1973) *Science* 166: 1514–1516.
14 Bachelard, H. S., P. M. Daniel, E. R. Love, and O. E. Pratt. (1973) *Proc. R. Soc.* (Lond.) *Ser. B* 183: 71–82.
15 Weindl, A. (1973) In *Frontiers of Neuroendocrinology* (eds. L. Martini and W. F. Ganong), pp. 1–32. Oxford University Press, London.
16 Olney, J. W., and M. T. Price. (1978) In *Kainic Acid As a Tool in Neurobiology* (eds. E. G. McGeer, J. W. Olney, and P. O. McGeer), pp. 239–263. Raven Press, New York.
17 Ferguson, R. K., and D. M. Woodbury. (1969) *Exp. Brain Res.* 7: 181–194.
18 Evans, L. A. N., J. M. Reynolds, M. L. Reynolds, N. R. Saunders, and M. B. Segal. (1974) *J. Physiol.* (Lond.) 238: 371–386.
19 Fenstermacher, J. D., C.-L. Li, and V. A. Levin. (1970) *Exp. Neurol.* 27: 101–114.
20 Van Harreveld, A. (1966) *Brain Tissue Electrolytes.* Butterworths, London.
21 Bourke, R. S., E. S. Greenberg, and D. B. Tower. (1965) *Am. J. Physiol.* 208: 682–652.
22 Levin, V. A., J. D. Fenstermacher, and C. S. Patlack. (1970) *Am. J. Physiol.* 219: 1528–1533.
23 Kessler, J. A., J. D. Fenstermacher, and E. S. Owens. (1976) *Am. J. Physiol.* 230: 614–618.
24 Van Harreveld, A., J. Crowell, and S. K. Malhotra. (1965) *J. Cell Biol.* 25: 117–137.
25 Van Harreveld, A., and F. I. Kahtab. (1969) *J. Cell Sci.* 4: 437–453.
26 Gilula, N. B. (1978) In *Intercellular Junctions and Synapses* (eds. J. Feldman, N. B. Gilula, and J. D. Pitts), pp. 1–22. *Receptors and Recognition, Ser. B*, vol. 2, Chapman and Hall, London.
27 Goodenough, D., and W. Stoekenius. (1972) *J. Cell Biol.* 54: 646–656.
28 Bennett, M. V. L., and D. A. Goodenough. (1978) *Neurosci. Res. Prog. Bull.* 16 (3): 373–486.
29 Weidmann, S. (1966) *J. Physiol.* (Lond.) 187: 323–342.
30 Makowski, L., D. L. D. Caspar, W. C. Phillips, and D. A. Goodenough. (1977) *J. Cell Biol.* 74: 629–645.
31 Furshpan, E. J., and D. D. Potter. (1959) *J. Physiol.* (Lond.) 145: 289–325.

32 Kuffler, S. W., and D. D. Potter. (1964) *J. Neurophysiol.* 27: 290–320.
33 Kuffler, S. W., and J. G. Nicholls. (1966) *Ergeb. Physiol.* 57: 1–90.
34 Palay, S. L., and V. Chan-Palay. (1974) *Cerebellar Cortex Cytology and Organization.* Springer-Verlag, New York.
35 Morales, R., and P. Duncan. (1975) *Anat. Rec.* 182: 255–266.
36 Walker, F. D., and W. J. Held. (1969) *Science* 165: 602–603.
37 Brightman, M. W., J. J. Anders, D. Schmechel, and J. M. Rosenstein. (1978) In *Dynamic Properties of Glial Cells* (eds. E. Schoffeniels, G. Franck, L. Hertz, and D. B. Tower), pp. 21–44. Pergamon Press, Oxford.
38 Hertz, L. (1982) In *Handbook of Neurochemistry,* 2d ed., vol. 1 (ed. A. Lajtha), pp. 319–355. Plenum Press, New York.
39 Hertz, L. (1979) *Prog. Neurobiol.* 13: 277–323.
40 Patel, A. J., A. Hunt, R. D. Gordon, and R. Balazs. (1982) *Dev. Brain Res.* 4: 3–11.
41 Newburgh, R. W., and R. Rosenberg. (1972) *Proc. Natl. Acad. Sci.* (USA) 69: 1677–1680.
42 Rindler, M. J., M. M. Boshor, N. Spitzer, and M. H. Saier. (1978) *J. Biol. Chem.* 253: 5431–5436.
43 Schousboe, A., L. Hertz, G. Svenneby, and E. Kvamme. (1979) *J. Neurochem.* 32: 943–950.
44 Lasek, R. J., and M. A. Tytell. (1981) *J. Exp. Biol.* 95: 153–165.
45 Gainer, H., I. Tasaki, and R. J. Lasek. (1977) *J. Cell Biol.* 74: 524–530.
46 Rakic, P. (1973) *J. Comp. Neurol.* 147: 523–546.
47 Sotelo, C., and J. P. Changeux. (1974) *Brain Res.* 77: 484–491.
48 Rakic, P. (1979) In *The Neurosciences: Fourth Study Program* (ed. F. O. Schmitt), pp. 109–127. MIT Press, Mass.
49 Varon, S., and M. Manthorpe. (1982) *Adv. Cell Neurobiol.* 3: 35–95.
50 Arenander, A. T., and J. de Vellis. (1981) *Brain Res.* 224: 117–127.
51 Louis, J. C., B. Pettman, J. Courageot, J. F. Rumigny, P. Mandel, and M. Sensenbrenner. (1981) *Exp. Brain Res.* 42: 63–70.
52 Turner, J. E., and M. Singer. (1974) *J. Comp. Neurol.* 156: 1–18.
53 Varon, S., and R. Adler. (1981) *Adv. Cell Neurobiol.* 2: 115–163.
54 Varon, S., C. Raiborn, and P. A. Burnham. (1974) *Neurobiology* 4: 317–327.
55 Kelly, J. P., and D. C. Van Essen. (1974) *J. Physiol.* (Lond.) 238: 515–547.
56 Katzman, R., and H. M. Pappius. (1973) *Brain Electrolytes and Fluid Metabolism.* Williams and Wilkins, Baltimore.
57 Prince, D. A., H. D. Lux, and E. Neher. (1973) *Brain Res.* 50: 489–495.
58 Nicholls, J. G., and S. W. Kuffler. (1965) *J. Neurophysiol.* 20: 519–525.
59 Kuffler, S. W. (1967) *Proc. R. Soc.* (Lond.), Soc. B 168: 1–21.
60 Ransom, B. R., and S. Goldring. (1973) *J. Neurophysiol.* 36: 855–868.
61 Orkand, R. K., J. G. Nicholls, and S. W. Kuffler. (1966) *J. Neurophysiol.* 29: 788–806.
62 Ransom, B. R., and S. Goldring. (1973) *J. Neurophysiol.* 36: 869–878.
63 Skoff, R. S. (1975) *J. Comp. Neurol.* 161: 595–612.
64 Kimelberg, H. K., S. Narumi, S. Biddlecome, and R. S. Bourke. (1978) In *Dynamic Properties of Glial Cells* (ed. E. Schoffeniels, G. Franck, L. Hertz, and D. B. Tower), pp. 347–357. Pergamon Press, Oxford.
65 Delaunoy, J. P., D. Filippi, G. Laurent, and P. Mandel. (1978) *Brain Res.* 155: 201–204.

66 Ghandour, M. S., O. K. Langley, G. Vincendon, G. Gombos, D. Filippi, N. Limozin, C. Dalmasso, and G. Laurent. (1980) *Neuroscience* 5: 559–571.

67 Church, G. A., H. K. Kimelberg, and V. A. Sapirstein. (1980) *J. Neurochem.* 25: 323–328.

68 Norenberg, M. D., and A. Martinez-Hernandez. (1979) *Brain Res.* 161: 303–310.

69 Benjamin, A. M., and J. H. Quastel. (1975) *J. Neurochem.* 25: 197–206.

70 Leveille, P. J., J. F. McGinnis, D. S. Maxwell, and J. de Vellis. (1980) *Brain Res.* 196: 287–306.

71 Chaise, P., and J. P. Roscoe. (1976) *Brain Res.* 109: 423–425.

72 McCarthy, K. D., and J. de Vellis. (1979) *Life Sci.* 24: 639–650.

73 McCarthy, K. D., and J. de Vellis. (1980) *J. Cell Biol.* 85: 890–902.

74 de Vellis, R. J., B. S. McEwan, R. Cole, and D. Inglish. (1974) *J. Steroid Biochem.* 5: 392–393.

75 Fisher, M., D. A. Gapp, and L. P. Kozak. (1981) *Dev. Brain Res.* 1: 341–354.

76 Kurihara, R., J. L. Nussbaum, and S. Mandel. (1971) *Life Sci.* 10 (2): 421–429.

77 Poduslo, S. E. (1975) *J. Neurochem.* 24: 647–654.

78 Reddy, N. B., V. Askansas, and W. K. Engel. (1982) *J. Neurochem.* 39: 887–889.

79 McMorris, F. A. (1977) *Trans. Am. Soc. Neurochem.* 8: 143.

80 Moore, B. W. (1972) *Int. Rev. Neurobiol.* 15: 215–225.

81 Moore, B. W., and D. MacGregor. (1965) *J. Biol. Chem.* 240: 1647–1653.

82 Bock, E. (1978) *J. Neurochem.* 30: 7–14.

83 Møller, M., A. Ingild, and E. Bock. (1977) *Brain Res.* 140: 1–13.

84 Schubert, D., S. Heinemann, W. Carlile, H. Tarikas, B. Kimes, J. Patrick, J. Y. Steinbach, W. Culp, and B. Brandt. (1974) *Nature* (Lond.) 249: 224–227.

85 Hydén, H., and L. Rönnbäck. (1975) *Neurobiology* 5: 291–302.

86 Haglid, K., A. Hamberger, H-A. Hansson, H. Hydén, L. Persson, and L. Rönnbäck. (1974) *Nature* (Lond.) 251: 532–534.

87 Bock, E., and A. Hamberger. (1976) *Brain Res.* 112: 329–335.

88 Calissano, P., S. Alema, and P. Fasella. (1974) *Biochemistry* 13: 4553–4560.

89 Calissano, P., and A. D. Bangham. (1971) *Biochem. Biophys. Res. Commun.* 43: 504–509.

90 Eng, L. F., J. J. Vanderhaeghen, A. Bignami, and G. Gerste. (1971) *Brain Res.* 28: 351–354.

91 Bignami, A., and D. Dahl. (1977) *J. Histochem. Cytochem.* 25: 466–469.

92 Uyeda, C. T., L. F. Eng, and A. Bignami. (1972) *Brain Res.* 37: 81–89.

93 Bignami, A., D. Dahl, and D. C. Rueger. (1980) *Adv. Cell Neurobiol.* 1: 285–310.

94 Antanitus, D. S., B. H. Choi, and L. W. Lapham. (1976) *Brain Res.* 103: 613–616.

95 Jacque, C. M., O. S. Jorgensen, N. A. Baumann, and E. Bock. (1976) *J. Neurochem.* 27: 905–909.

96 Dahl, D. (1976) *Biochem. Biophys. Acta* 420: 142–154.

97 Yen, S.-H., M. Schachner, and M. L. Shelanski. (1976) *Proc. Natl. Acad. Sci.* (USA) 73: 529–533.

98 Davison, P. F., and R. N. Jones. (1981) *J. Cell Biol.* 88: 67–72.

99 Hild, W., and I. Tasaki. (1962) *J. Neurophysiol.* 25: 277–304.

100 Crain, S. M., E. R. Peterson, B. Crain, and E. J. Simon. (1977) *Brain Res.* 133: 162–166.

101 Hausman, R. E., and A. A. Moscona. (1972) *Proc. Natl. Acad. Sci.* (USA) 72: 916–920.

102 Hausman, R. E., and A. A. Moscona. (1976) *Proc. Natl. Acad. Sci.* (USA) 73: 3594–3598.

103 Nelson, P. G. (1975) *Physiol. Rev.* 55: 1–61.

104 McMorris, F. A., P. G. Nelson, and F. H. Ruddle. (1973) *Neurosci. Res. Prog. Bull.* 11 (5): 417–536.

105 Waymire, J. C., K. Gilmer-Waymire, and R. E. Boehme. (1978) *J. Neurochem.* 31: 699–706.

106 Varon, S. (1975) *Exp. Neurol.* 48: 93–100.

107 Pfeiffer, S. E., and W. Wechsler. (1972) *Proc. Natl. Acad. Sci.* (USA) 69: 2885–2889.

108 McGinnis, J. F., and J. de Vellis. (1976) *Fed. Proc.* 35: 1636.

109 De Vellis, J., and G. Brooker. (1972) *Fed. Proc.* 31: 513.

110 Peacock, J. H., and P. G. Nelson. (1973) *J. Neurobiol.* 4: 363–374.

111 Gysin, R., B. W. Moore, R. T. Proffitt, T. F. Deuel, K. Caldwell, and L. Glaser. (1980) *J. Biol. Chem.* 255: 1515–1520.

112 Labourdette, G., and P. Mandel. (1978) *Biochem. Biophys. Res. Commun.* 85: 1307–1313.

113 Hydén, H. (1960) *J. Neurochem.* (1964) 6: 57–72.

114 Roots, B. I., and P. V. Johnstone. (1964) *J. Ultrastruct. Res.* 10: 350–361.

115 Rose, S. P. R. (1967) *Biochem. J.* 102: 33–43.

116 Norton, W. T., and S. E. Poduslo. (1967) *Science* 167: 1144–1146.

117 Satake, M., S. I. Hasegawa, S. Abe, and R. Tanaka. (1968) *Brain Res.* 11: 246–250.

118 Hamberger, A., C. Blomstrand, and A. L. Lehninger. (1970) *J. Cell Biol.* 45: 221–234.

119 Sellinger, O. Z., J. M. Azcurra, D. E. Johnson, W. G. Ohlsson, and Z. Lodin. (1971) *Nature (New Biol.)* 235: 253–256.

120 Cohen, J., P. L. Woodhams, and R. Balazs. (1979) *Brain Res.* 161: 503–514.

121 Farooq, M., R. Ferszt, C. L. Moore, and W. T. Norton. (1977) *Brain Res.* 124: 69–81.

122 Chao, S. W., and M. G. Rumsby. (1977) *Brain Res.* 124: 347–351.

123 Garthwaite, J., and R. Balazs. (1981) *Adv. Cell Neurobiol.* 2: 461–489.

124 Woodhams, P. L., G. P. Wilkin, and R. Balazs. (1981) *Dev. Neurosci.* 4: 307–321.

125 Farooq, M., and W. T. Norton. (1978) *J. Neurochem.* 31: 887–894.

126 Nagata, Y., K. Mikoshuba, and Y. Tsukada. (1974) *J. Neurochem.* 22: 493–503.

127 Norton, W. T., and S. E. Poduslo. (1971) *J. Lipid Res.* 12: 84–90.

128 Sinha, A. K., S. P. R. Rose, L. Sinha, and D. Spears. (1978) *J. Neurochem.* 30: 1513–1524.

129 Sinha, A. K., and S. P. R. Rose (1972) *Brain Res.* 39: 181 196.

130 Sellström, A., and A. Hamberger. (1977) *Brain Res.* 119: 189–198.

131 Sellström, A., L-A. Sjoberg, and A. Hamberger. (1975) *J. Neurochem.* 25: 393–398.

132 Johnson, D. E., and O. Z. Sellinger. (1973) *Brain Res.* 54: 129–142.

133 Rose, S. P. R., and A. R. Sinha. (1969) *J. Neurochem.* 16: 1319–1328.

134 Rose, S. P. R. (1977) *Philos. Trans. R. Soc.* (Lond.), *Sec. B* 278: 307–318.

135 Blomstrand, C., and A. Hamberger. (1969) *J. Neurochem.* 16: 1401–1407.

136 Rose, S. P. R., and A. K. Sinha. (1974) *J. Neurochem.* 23: 1065–1076.

137 Landon, D. N., and S. Hall. (1976) In *The Peripheral Nerve* (ed. D. N. Landon) Chapman and Hall, London, pp. 1–105.

138 Ochoa, J. (1976) In *The Peripheral Nerve* (ed. D. N. Landon), pp. 106–158. Chapman and Hall, London.

139 Lieberman, A. R. (1976) In *The Peripheral Nerve* (ed. D. N. Landon), pp. 188–278. Chapman and Hall, London.

140 Flieschhauer, K. (1960) *Z. Zellforsch. Mikrosk. Anat.* 51: 467–472.

141 Leonhardt, H. (1967) *Z. Zellforsch. Mikrosk. Anat.* 79: 172–180.

142 Kessel, R. G., and R. H. Kardon. (1979) *Tissues and Organs: A Text-Atlas of Scanning Electron Microscopy.* Freeman, New York.

143 Westgaard, E. E., and M. N. Brightman. (1973) *J. Comp. Neurol.* 152: 17–44.

144 Sandri, C., J. M. Van Buren, and K. Akert. (1977) *Membrane Morphology of the Nervous System* Elsevier, Amsterdam (*Prog. Brain Res.* 49).

145 Tennyson, V. M., and G. D. Pappas. (1968) *Prog. Brain Res.* 29: 63–85.

146 Henn, F. A., D. J. Anderson, and D. G. Rustad. (1976) *Brain Res.* 101: 341–344.

147 Schubert, D., S. Humphreys, F. de Vitry, and F. Jacob. (1971) *Dev. Biol.* 25: 514–546.

148 Harris, A. J., and M. J. Dennis. (1970) *Science* 167: 1253–1255.

149 Raine, C. S., S. E. Poduslo, and W. T. Norton. (1971) *Brain Res.* 27: 11–24.

150 Williams, P. L., and R. Warwick. (1975) *Functional Neuroanatomy of Man.* Churchill Livingstone, Edinburgh, London, and New York.

151 Chan-Palay, V. (1977) *J. Comp. Neurol.* 176: 467–493.

152 Furmanski, P., D. J. Silverman, and M. Lubin. (1971) *Nature* (Lond.) 233: 413–415.

153 Schmechel, D., P. J. Marangos, A. P. Zis, M. Brightman, and F. K. Goodwin. (1978) *Science* 199: 313–315.

154 Marangos, P. J., A. M. Parma, and F. K. Goodwin. (1978) *J. Neurochem.* 31: 733–738.

155 Fields, K. L., C. Gosling, M. Megson, and P. L. Stern. (1975) *Proc. Natl. Acad. Sci.* (USA) 72: 1286–1300.

156 Raff, M. C., R. Mirsky, K. L. Fields, R. P. Lisak, S. H. Dorfman, D. H. Silberberg, N. A. Gregson, S. Liebowitz, and M. Kennedy. (1978) *Nature* (Lond.) 274: 813–816.

157 Steck, A. J., and G. Perruisseau. (1980) *J. Neurol. Sci.* 47: 135–144.

158 Newelt, E. A., J. A. Barranger, M. A. Pagel, J. M. Quirk, R. O. Brady, and E. P. Frenkel. (1984) *Neurology* 34: 1012–1019.

3
Brain Glucose and Energy Metabolism: The Linkage to Function

BRAIN ENERGY METABOLISM IN VIVO

The Vulnerable Brain

The brain is remarkable in its dependence on the blood for its immediate supply of oxygen and essential energy substrates. Any interruption in their delivery leads within seconds to unconsciousness and within minutes to irreversible changes resulting in cell death. In this sense the brain is a very vulnerable organ.

The rapidity of the lethal consequences of a short supply of essential substrates highlights the paradoxically low levels stored within the brain. For example, glycogen storage levels in liver, muscle and brain are in the approximate ratio of 100/10/1, the brain levels being 2–4 μmol/g of tissue. The considerable nutrient buffering capacity of glycogen stores in liver and muscle simply does not exist within the brain.

What the Brain Takes from the Blood and Adds to It

Most organs in the body enjoy an affinity for many substrates present in the blood. They are able to absorb and metabolize a wide range of keto acids, amino acids, and fatty acids, for example, in addition to glucose itself, and the supplies of all these substances are fully adequate to maintain normal function. The brain, in contrast,[A,B,C] must receive its few substrates through the filter of the blood-brain barrier. (See Chapter 1.)

Entry to the brain is by facilitated transport and is restricted to a relatively narrow range of substances.[52] In isolated brain-tissue slices the blood-brain barrier is circumvented and a much wider range of nutrients can be employed, which shows that the restriction is due only to the existence of the barrier and does not indicate limited metabolic potential. Thus, although the brain in vivo is an organ of high metabolic rate, it is very largely dependent on the blood to supply the essential fuels for its activities. It uses about 20 percent of the total oxygen entering the body of a recumbent human, with glucose, the primary substrate, being taken up by the brain at 15–20 μmol/g of brain per hour, a rate two to three times lower than for smaller mammals (see below). The process can be described by Michaelis-Menten kinetics with a K_m of 5–9 mM and a maximum velocity (V_{max}) of 1.6–2.8 μmol/g of brain per minute in experimental animals.[52] In the human, this involves oxygen uptake at 80–90 μmol/g of brain per hour with carbon dioxide released to venous blood at almost the same rate. These figures indicate that glucose is almost fully oxidized to CO_2, with only relatively small amounts (5–13 percent) appearing in venous blood as lactate and pyruvate.[2,A,B,C] During continuous intravenous infusion of [^{14}C]glucose into various mammals, including anesthetized dogs, monkeys, and recumbent humans, the specific radioactivity of carbon dioxide in venous blood of the saggital sinus approaches that of systemic glucose, but only after 1 hour. This result shows that all metabolic carbon dioxide is derived from glucose, but through pathways linked to large pools of oxidizable substrates.[B,C] (See Figures 3.1 and 3.2 and Tables 3.1 and 3.2.)

Such facts as these have been established during the past 40 years principally by the technique of arteriovenous (A-V) difference measurement,[A] which relies on sampling venous blood leaving the brain (usually taken from the jugular vein at its superior bulb and systemic arterial blood. Blood flow is measured by the difference in levels between the arterial blood and venous blood of a nonmetabolized substance such as inhaled nitrous oxide, krypton (^{188}Kr), or xenon (^{133}Xe). Venous blood from the jugular bulb gives the cerebral blood flow (CBF) for the entire brain. In experimental animals, venous blood sampled from the superior saggital sinus provides a value for mainly cortical flow. From these data the cerebral metabolic rate (CMR) for any given substrate (e.g., glucose, oxygen) can be calculated (i.e., CMR = CBF × A-V difference).

Clear differences have been detected between the in vivo metabolic rates of the brains of different mammalian species. Values for cerebral oxygen and glucose consumption (the cerebral metabolic rate) in the rat are twice those for the dog and often three times the values for humans[A,3,4,5] (Table 3.3). These observations demonstrate the existence of an inverse ratio between brain oxygen consumption and body weight, which is probably due to the relatively high neuron-packing density and the higher neuron-to-glia ratio in the brains of smaller animals.[14]

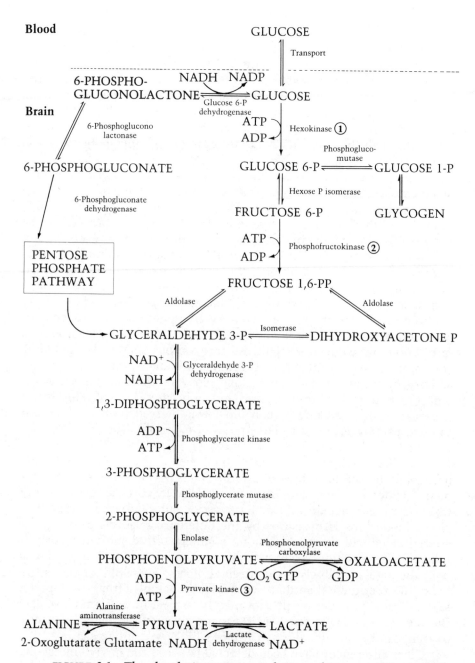

FIGURE 3.1 The glycolytic sequence in brain and its connections with related pathways. The numbers refer to control points and are explained in Table 3.1

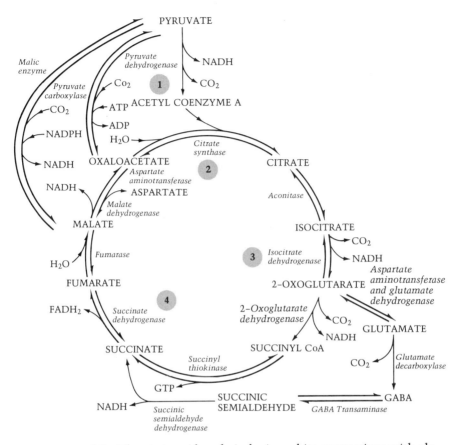

FIGURE 3.2 The citric acid cycle in brain and its connections with glutamate, aspartate, and GABA formation. The numbers refer to control points (see Table 3.1).

Glucose: The Primary Substrate

The central concept of the preeminence of glucose as the substrate indirectly supporting all the energy-requiring activities of the mammalian brain remains unchallenged in the mid-1980s.[A,C] It was the prevailing concept 50 years ago and has been modified to only a small extent since by the realization that certain other substrates may be effectively employed by the brain as energy sources if they are available in the blood at a sufficiently high level. Principal among these are ketones. In the late 1960s, it was observed that in obese humans who fasted for 38 to 41 days or in experimental animals that had greatly diminished blood glucose

TABLE 3.1   Control points in mainstream energy metabolism

Location	Enzyme	Modulator	Action
Control points in glycolysis (see Figure 3.1)			
1	Hexokinase	Glucose 6-phosphate	Product inhibition
2	Phosphofructokinase	ADP; P_i; AMP; cyclic 3,5-AMP; NH_4; fructose 6-PO_4; fructose diphosphate	Allosteric stimulation
		ATP; citrate	Allosteric inhibition
3	Pyruvate kinase	Phosphocreatine (synergistically with ATP; competitively with phosphoenolpyruvate)	Inhibition increased in the presence of ADP at high concentrations
Control points in the citric acid cycle (see Figure 3.2)			
1	Pyruvate dehydrogenase complex	Phosphorylation (inhibited by ADP)	Inhibition
		Dephosphorylation by $Mg^2 \pm Ca^2 \pm$-activated phosphatase	Stimulation
		$NADH/NAD^+$ ratio	Inhibition (high) Stimulation (low)
2	Citrate synthetase	$NADH/NAD^+$ ratio	Inhibition (high) Stimulation (low)
		Availability of oxaloacetate	Stimulation (high) Inhibition (low)
3	Isocitrate dehydrogenase	$NADH/NAD^+$ ratio	Inhibition (high) Stimulation (low)
4	Succinate dehydrogenase	ATP	Inhibition
		Reduced coenzyme Q_{10}	Inhibition

Note: Data are from refs. A, 21, and 49.

TABLE 3.2   Recovery of isotopic carbon in metabolites of [$U^{14}C$]glucose in brain tissue

Metabolite	Cortex slices			In vivo		
	% total		pool size (μmol/g tissue)	% total		pool size (μmol/g tissue)
Glutamate	7.1	(5.9)	8.18	43.5	(59.5)	11.04
Aspartate	0.75	(6.2)	4.16	9.4	(12.8)	2.72
Glutamine	1.6	(13.4)	4.58	13.5	(18.4)	4.70
GABA	1.6	(13.4)	1.11	6.0	(8.2)	1.81
Alanine	0.88	(7.2)	0.92	0.85	(1.1)	0.36
Lactate	72.0	(–)	N.M.	9.7	(–)	N.M.
CO_2	15.5	(–)	N.M.	N.M.		N.M.

Note: Experiments were of 1 hour duration at 37°C. Values in parentheses are percentage recovered only in amino acids. N.M., not measured.

TABLE 3.3 Cerebral blood flow and cerebral metabolic rate in brains of different species

Species	Difference in oxygen between arteries and veins (μmol/ml)	Cerebral blood flow rate[a] (ml/g min)	Cerebral metabolic rate[b]	
			A (μmol O_2/g min)	B (μmol glucose/g min)
Human[c]	3.02	0.45 ± 0.02	1.35 ± 0.07	0.27
Dog[d]	4.46 ± 0.45	0.63 ± 0.08	2.63 ± 0.22	0.47
Rat[d]	4.33 ± 0.22	1.08 ± 0.05	4.60 ± 0.09	0.77

Note: Values for rat and dog pertain to cerebral cortical tissue (venous samplings are from superior saggital sinus). Values for human are more representative of whole brain. Data are from refs. 2–5 and various sources quoted in ref. A.

[a] Cerebral blood flow method: human, ^{85}Kr; dog, N_2O; rat, ^{133}Xe.

[b] Obtained by multiplying the A-V difference by the cerebral blood flow.

[c] Unanesthetized.

[d] Lightly anesthetized with thiopental (dog) or nitrous oxide (rat), which have minimal effects on cerebral blood flow and metabolic rate.

levels, certain ketone bodies (β-hydroxybutyrate, acetoacetate) rose to relatively high levels in the blood. Under these circumstances the A-V difference for the ketone compounds became greatly augmented. In fact, the ketones could account for half or more of the oxygen used by the brain[6,7,8,A,52] (Table 3.4). At the same time, the A-V difference for glucose was reduced to half its normal level and was responsible for only 30 percent of the oxygen used. Normal humans fasting for only 12 to 16 hours showed a significant increase in cerebral uptake of ketone bodies, directly proportional to the ketone content of the blood.[9,11] The key enzymes involved in the use of the ketones as energy substrates are D-β-hydroxybutyrate dehydrogenase, acetoacetate-succinyl-CoA-transferase, and acetoacetylthiolase (Figure 3.3). They are localized in mitochondria and appear to be present in adult animal brains at a level adequate to respond to substantial rises in blood ketone level.[6,8,10] The blood ketone content is very high in suckling rats, in large part owing to the considerable content of these substances in rat milk. At this stage these compounds dominate the primary nutrition of the suckling rat's brain, with 25 to 76 percent of oxygen uptake being ascribed to their oxidation.

Entry of ketones to the brain is, as expected, by way of facilitated transport across the blood-brain barrier via the monocarboxylic acid carrier system which also transports lactate and pyruvate, and the efficacy of this transport appears to increase during starvation.[12,13,52]

Apart from ketone bodies, normally present only at low levels in blood and hardly used at all, glucose is the substrate that throughout life supports the energy requirements of the brain. Only glucose, mannose, or compounds that are rapidly converted to glucose elsewhere in the body

TABLE 3.4   Arterial levels and arteriovenous differences of major brain substrates in fasting humans

	Oxygen	Carbon dioxide	Glucose	B-Hydroxybutyrate	Acetoacetate	2-Amino-nitrogen
A-V difference (mmol/liter blood)	2.96 ± 0.34	1.90 ± 0.37	0.26 ± 0.02	0.34 ± 0.06	0.06 ± 0.01	0.09 ± 0.03
Arterial level (mmol/liter blood)	—	—	4.49 ± 0.41	6.67 ± 0.37	1.17 ± 0.27	3.14 ± 0.19
Oxygen substrate equivalent for oxidation (mmol oxygen/liter blood)[a]	—	—	0.87	1.53	0.24	0.42
Carbon dioxide equivalent generated[b]	—	—	0.87	1.36	0.24	0.34

Note: Volunteers fasted for 38–41 days before measurements were taken. Data from ref. 7 are the mean for three persons.
[a] Values add up to A-V difference for oxygen.
[b] Values add up to A-V difference for CO_2 (actually negative).

FIGURE 3.3 Main pathways of
ketone body utilization. CoA
equals coenzyme A.

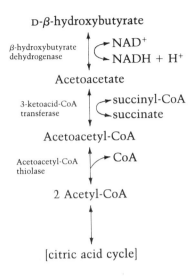

can revive patients or animals from hypoglycemic coma. None of the
ketones are effective,[9] probably because glucose itself is necessary to
prime the citric acid cycle.

The glycogen content of the brain at the normal low level of 2–4
μmol/g is able to support brain metabolism for only three minutes[A] and
therefore would not be of much value as an emergency source of glucose
beyond an extremely brief period.

The Regional Use of Glucose

Glucose Autoradiography Direct autoradiography of brain sections after feeding [2-^{14}C]glucose in vivo has been used to determine the regional
glucose utilization in rat brain over short time-intervals (5 minutes).
These brief intervals are necessary to capture and visualize the localized
uptake of isotopic glucose and accumulation of glucose 6-phosphate, before they are further metabolized and dispersed[29,C] (Figure 3.4). A method
has been devised that should allow quantitative evaluation of the complete metabolism of glucose through the use of a series of glucose molecules specifically labeled with ^{14}C at different points on the carbon
skeleton.[C]

Deoxyglucose Autoradiography Another powerful technique involving
tissue-slice autoradiography after uptake in vivo of a glucose analogue[15,16,F] has allowed both quantitative and qualitative studies of regional glucose utilization in the brain both at rest and during functional
response.

The basis of the technique is very simple. It depends on the accumulation within brain neurons and glia of radioactive 2-deoxyglucose 6-
phosphate (DG6P) after systemic administration of ^{14}C-labeled deoxyglu-

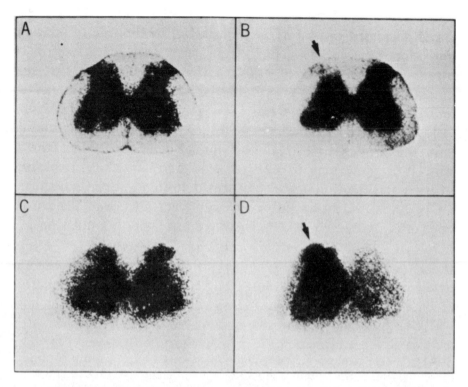

FIGURE 3.4   Functional activity of dorsal horn neurons of the spinal cord
as revealed by glucose and deoxyglucose autoradiography. (A) Autoradiography of rat lumbar spinal cord obtained by perfusion with glutaralde-
hyde-formaldehyde 45 minutes after injection of [^{14}C]glucose into the
anesthetized, unstimulated rat. (B) Autoradiograph prepared as in (A),
except that the sciatic nerve was stimulated for 45 minutes after injection of [^{14}C]glucose. The arrow points to the decreased optical density in
the dorsal horn ipsilateral to the side of stimulation. (C) Autoradiograph
of rat lumbar spinal cord obtained by sacrifice of the animal 45 minutes
after injection of [^{14}C]DG into the anesthetized, unstimulated rat. (D)
Autoradiograph prepared as in (C), except that the sciatic nerve was stimulated for 45 minutes after injection of [^{14}C]glucose. The arrow points to
the increased optical density in the dorsal horn ipsilateral to the side of
stimulation. [From ref. 57]

cose (DG). This compound is carried into the brain efficiently through the
blood-brain barrier by the glucose transport system. Once it enters cells
becomes phosphorylated by hexokinase to DG6P, its further metabolic
progress is blocked because it will not function as a substrate for any of
the usual enzymes and pathways employing glucose 6-phosphate. For
this reason, DG6P is essentially trapped within the tissue, at least long
enough for the appropriate measurements to be made (half-life 6–10
hours)[16] (Figure 3.5). The concentrations of radioactive DG6P are measured by a quantitative autoradiographic technique involving densitome-
try applied to histological sections of brain prepared after killing the

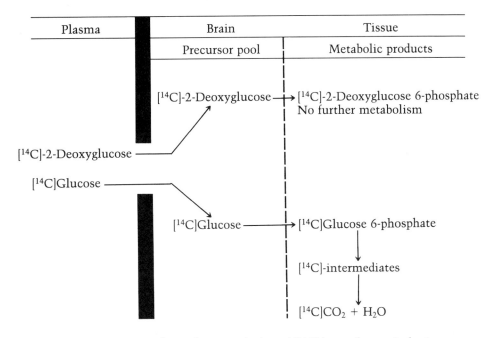

FIGURE 3.5   Uptake and accumulation of [^{14}C]deoxyglucose in brain
tissue. Glucose 6-phosphate is subsequently metabolized via the glycolytic sequence of enzymes, but deoxyglucose 6-phosphate is not further
metabolized and accumulates in the cell.

animal by decapitation and freezing the brain. The data can be presented
on color-coded maps to show the relative rates of DG6P accumulation
(i.e., glucose utilization) in the different regions represented in the autoradiographs.[16] The extent to which the technique can be relied upon for
these quantitative estimates of deoxyglucose uptake has, however, become a matter of dispute. Evidence exists[28,30] that DG6P does not all
remain trapped in the tissue but is rapidly dephosphorylated, with both
DG and DG6P subsequently removed by the blood.[61]
    Whatever the reliability of the quantitative aspects of the technique,
the qualitative results are striking. Functional response to various modal-
ities of sensory stimulation can be readily seen in terms of increased
deoxyglucose uptake. Again it is clear that exogenous glucose is the primary substrate employed to meet the increased energy demands associated with augmented neural activity. For instance, sciatic nerve stimulation in the rat for lengthy periods (e.g., 45 minutes) has been shown to
cause a clear increase in DG6P accumulation by the cells of the ipsilateral dorsal horn of the lumbar spinal cord.[57] In contrast, autoradiography
using [^{14}C]glucose as substrate and prolonged sciatic nerve stimulation
showed a decrease in accumulation of radioactivity, presumably because
of the accelerated metabolism of glucose and dispersion of its metabolites.

Application of olfactory stimuli such as amylacetate, camphor, or cheese produced a marked increase in DG6P accumulation in the olfactory bulb in the rat, and auditory stimuli caused equivalent responses in the brain regions that mediate hearing[58] (Figure 3.6).

A reduction in sensory input leads to a reduced rate of glucose utilization, as has been shown for both auditory and visual systems.[16,17] Bilateral visual deprivation in the rhesus monkey leads to a generalized reduction in silver grains in autoradiographs of the visual cortex (Figure 3.7), and monocular deprivation causes the appearance in the visual cortex of alternate light and dark vertical columns 0.3–0.5 mm wide, revealing the

FIGURE 3.6 Functional activity of the auditory system in the conscious rat as revealed by deoxyglucose autoradiography. (A) Normal conscious rat. (B) Both external auditory canals were obstructed with wax and the animal was placed in a sound-proof room. (C) Only one auditory canal was obstructed, the canal contralateral to the side of the brain with reduced autoradiographic density in auditory structures. Note the inferior colliculi, leminisci, and superior olives. [From ref. 58]

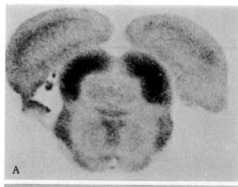

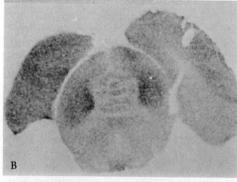

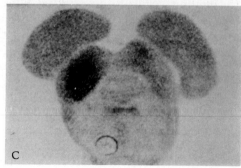

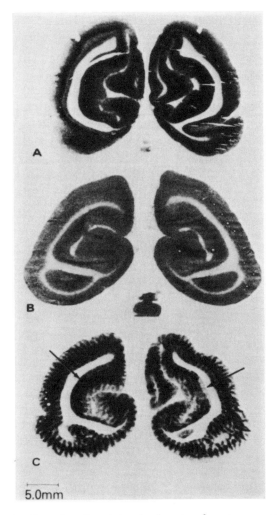

5.0mm

FIGURE 3.7 Functional activity in the visual system as revealed by deoxyglucose autoradiography. Autoradiographs of coronal brain sections from Rhesus monkeys at the level of the striate (visual) cortex. (A) Animal with normal binocular vision. Note the laminar distribution of the density; the dark band corresponds to cortical-cell layer IV. (B) Animal with bilateral visual deprivation. Note the almost uniform and reduced relative density, especially with virtual disappearance of the dark band corresponding to layer IV. (C) Animal with right eye occluded. The half-brain on the left side of the photograph represents the left hemisphere contralateral to the occluded eye. Note the alternate dark and light striations, each approximately 0.3 to 0.4 mm wide, representing the ocular dominance columns. These columns are most apparent in the dark band corresponding to layer IV but extend through the entire thickness of the cortex. The arrows point to regions of bilateral asymmetry where the ocular dominance columns are absent. These are presumably areas with normally only monocular input. The one on the left, contralateral to the occluded eye, has a continuous dark lamina corresponding to layer IV, which is completely absent on the ipsilateral side. These regions are believed to be the loci of the cortical representations of the blind spots. [From ref. 17]

presence of ocular dominance columns, segregated but interdigitated inputs from each retina. Such effects can be seen only in animals that have binocular vision and, therefore, ocular dominance columns.

Such examples illustrate the striking correlation between levels of functional activity and levels of glucose uptake and phosphorylation in the nervous system, implying that a rapid, large, and effective modulation of energy metabolism is occurring during functional response and is accompanied by equivalent increases in local cerebral blood flow. These findings reflect the constantly varying fuel requirements of the main energy-utilizing components of the nervous system, those concerned with generating electrical activity, nerve impulse traffic, and restoration of neuronal membrane potentials. Essentially, this energy utilization centers on the activity of the sodium pump, Na^+K^+-ATPase, which is responsible for the coupled inward transport of K^+ and outward transport of Na^+ to maintain membrane potentials at rest and restore their associated ionic gradients after electrical activity.[18] The long-standing and widely accepted estimate for the proportion of energy consumption due to this key process is 40 percent.[18,19]

The linkage of glucose metabolism to neuronal activity via the sodium pump has been demonstrated by the use of agents such as ouabain, which blocks the pump but not the generation of nerve impulses (spike activity), since the latter uses the energy already stored in the ionic gradients. When applied in vitro to the rat posterior pituitary gland, for instance, the uptake of glucose to the preparation was blocked but not spike activity or vasopressin release.[20] The Na^+K^+-ATPase releases ADP from ATP. This release leads to a change in the ATP-to-ADP ratio, which controls the rate of oxidative phosphorylation. A decrease in this ratio allows acceleration of the ATP-generating process. The various stages in the coupled flow of metabolic processes from glucose uptake through glycolysis to the citric acid cycle and the accompanying chemical reduction of nicotine-adenine dinucleotides (NAD) are all increased in parallel, owing largely to allosteric or direct modulation of the activity of rate-limiting enzymes of glycolysis such as hexokinase, phosphofructokinase, and pyruvate kinase. These and the many other factors controlling the flux through energy-yielding metabolic pathways from glucose to ATP production are identical with those found for other tissues;[21,K] key control steps in glycolysis and in the citric acid cycle are listed in Table 3.1.

This pacemaker role of Na^+K^+-ATPase is a special feature of the nervous system and becomes especially important in nerve terminals, dendrites, and cell regions of small diameter where the ratio of surface area to volume is relatively large because the Na^+ and K^+ fluxes across the membrane produce larger changes in ionic concentrations in the cytoplasm. The resulting higher rate of activity of the sodium pump is responsible for the elevated respiration that characterizes these smaller cell regions (see Chapter 6).

In Vivo Autoradiography   The autoradiographic technique that monitors the accumulation of radioactive DG6P in histological brain sections prepared after in vivo administration can be extended to follow changing patterns of glucose metabolism by in vivo monitoring of radiolabeled DG6P. The technique can also be applied to study regional brain glucose utilization in humans, since the harmless gamma-emitting radionucleide [2-^{18}F]fluoro-2-deoxy-D-glucose enables detection of radioactivity through the skull by its emission of positrons. It is readily phosphorylated by hexokinase to the corresponding 2-fluoro-2-deoxy-D-glucose phosphate, which accumulates in the cells in an analogous way to DG6P but at a 20 to 30 percent lower rate of formation. A transverse-section scanning device (computerized tomography) allows three-dimensional resolution of these metabolic changes by PET scanning (position emission tomagraphy) as they occur.[22,23,F,G,H]

The visualized functional responses in glucose phosphorylation are accompanied by increases in local cerebral blood flow. These changes in blood flow have been known for many years and have been measured by a variety of techniques.[A,24,25] The mode of coupling of cerebral blood flow to increased functional activity within cerebral tissues remains unknown but is thought to involve vasodilation owing to such factors as increased H^+ concentrations from augmentation of lactate and CO_2 formation,[A] raised extracellular K^+, or raised adenosine levels due to AMP degradation.[26] Control of cerebral blood flow through chemical influences directed at extracerebral sites such as the carotid and aortic chemoreceptors may also be involved.[A,25]

Metabolism During Extremes of Brain Activity: Anesthesia and Convulsion

As might be expected, parallel reductions in cerebral blood flow and cerebral oxygen utilization occur in conditions where brain functional activity is reduced, as in barbiturate anesthesia and hypothermia.[A] In thesc situations the tissue levels of the energy phosphates, ATP, ADP, AMP, and phosphocreatine (PhCr), are either increased or unchanged while levels of key glycolytic intermediates such as glucose 6-phosphate and fructose 6-phosphate increase substantially, indicating the inhibition of phosphofructokinase, the glycolytic rate-limiting enzyme (Table 3.5).[A] Citrate, too, may increase leading to inhibition of the citric acid cycle at the isocitrate dehydrogenase stage (see Table 3.1). Following such slowing in mainstream energy metabolism, there would be reduced oxidative phosphorylation and a general winding down of metabolic flow.

During continuous generalized seizure states the opposite set of conditions predominate. Cerebral blood flow, oxygen uptake, and the flux through mainstream energy pathways are all enhanced. Cerebral glucose and glycogen levels fall, accompanied by a moderate fall in levels of ATP

TABLE 3.5   Metabolic changes in the brain during extremes of cerebral activity

	Anesthesia		Seizure		
	Control	Percent change	Control	Percent change at 1 min	Percent change at 5 min
Energy phosphates[a,b] (µmol/g tissue) n = 10					
Phosphocreatine	3.63 ± 0.09	40 ↑	4.59 ± 0.06	50 ↓	45 ↓
ATP	2.95 ± 0.04	6 ↑	2.95 ± 0.03	12 ↓	7 ↑
ADP	0.33 ± 0.101	21 ↓	0.0257 ± 0.002	32 ↑	16 ↑
AMP	0.050 ± 0.006	36 ↓	0.030 ± 0.001	66 ↑	43 ↑
Glycolytic intermediates[b,c,g] (µmol/g tissue) n = 5					
Glucose-6-phosphate	0.099 ± 0.014	72 ↑		25 ↑	19 ↑
Fructose	0.0122 ± 0.0013	59 ↑		25 ↑	6 ↑
3-Phosphoglycerate	0.0903 ± 0.0122	38 ↓		83 ↑	33 ↑
Pyruvate	0.122 ± 0.005	41 ↓	0.141 ± 0.012	104 ↑	31 ↑
Citric acid cycle intermediates[b,c,g] (µmol/g tissue) n = 5					
Citrate	0.332 ± 0.012	6 ↓	0.330 ± 0.003	19 ↑	25 ↑
Succinate	0.470 ± 0.054	6 ↓	–	–	–
Oxaloacetate	0.0067 ± 0.0008	12 ↑	–	–	–
2-Oxoglutarate	–		0.159 ± 0.066	34 ↓	3 ↓
Malate	–		0.389 ± 0.014	59 ↑	93 ↑
Glucose utilization[b,e,f] (µmol/g tissue per min) n = 6					
CMR$_{gluc}$ { Sensorimotor cortex	1.2	43 ↓	0.77 ± 0.06(6)	400 ↑	260 ↑
{ Auditory cortex	1.58	48 ↓	–	–	–
Oxygen uptake[b,d] (µmol O$_2$/g brain per min)					
CMR$_{O_2}$ Thiopental 46mg/kg	1.47	54 ↓	3.4	270 ↑	262 ↑
Pentobarbital 2mg/kg	–	10 ↓	–	–	–
Cerebral blood flow[b,d] (ml/g per min)	0.54	48 ↓	0.7	971 ↑	671 ↑

Note: ↑ = increase; ↓ = decrease. Cerebral metabolic rates for oxygen (CMR$_{O_2}$) or for glucose (CMR$_{gluc}$) are calculated from Arterio-Venous difference in blood content (µmcl/ml) and the cerebral blood flow rate (CBF); i.e., CMR = CBF × A-V difference. Data are from ref. 50 and from various sources quoted in ref. A. [a] anesthesia, phenobarbital 225 mg/kg, mouse. [b] seizure (status epilepticus) bicuculline 1.2 mg/kg, rats. [c] anesthesia, barbiturate, rat. [d] anesthesia, thiopental, man. [e] anesthesia, thiopental, rat. [f] seizure, bicuculline, rat, values at onset and values at 1 hour. [g] anesthesia, values at 105 s (isoelectric EEG).

(e.g., 16 percent) and a much larger reduction in PhCr (e.g., 50 percent) and a rise in levels of ADP, AMP, and inorganic phosphate (P_i). But if maximum oxygenation of blood and tissue is ensured by mechanical ventilation, the fall in ATP is small and often insignificant. In sustained seizures, PhCr stabilizes at a reduced concentration and ATP levels remain close to normal provided hypoxia, hypoglycemia, or arterial hypotension do not occur.[50] During seizure states ATP *turnover* will be greatly enhanced, of course, but its levels are maintained at the expense of PhCr (see below).

Only during extremes of activity, such as anesthesia or epileptic seizures, can changes in energy phosphate levels be detected. The relatively small changes in cerebral glucose and oxygen utilization that accompany changed mental states such as anxiety, stress (e.g., that caused chemically by amphetamine), or sensory stimulation are not accompanied by experimentally detectable changes in the levels of energy phosphates or the levels of glycolytic or citric acid cycle intermediates even when there is a twofold change in metabolic rate.[A] This applies even when rapid freeze-clamping in vivo is used to fix tissue levels and minimize postmortem artifact. Measurement reveals the considerable increases in turnover and metabolic flux through all of these systems, but homeostasis is able to maintain pool sizes of metabolites and energy phosphates within a relatively narrow range[A] (Table 3.5). In summary, the brain's metabolic *rate* may increase but its metabolic *state* in terms of levels of energy phosphates is relatively unperturbed. The enhanced neuronal metabolic activity that must occur during these milder changes in brain function is largely inferred from the observed changes in metabolic rate, that is, from the increased glucose and oxygen uptake.

Compartmentation of energy phosphates may well allow larger local changes in their level than are measurable in whole tissue samples. Otherwise the shift from ATP and PhCr towards ADP, AMP and P_i, would appear to be insufficient to provide the signals necessary to trigger functional metabolic response. Other important factors, however, such as changes in redox state, no doubt play their part in initiating these responses.[A]

ENERGY PHOSPHATE TURNOVER

The Total Energy Charge of Neural Tissue

The sum total of the pools of ATP and PhCr in neural tissue is sometimes called the total energy charge, to indicate the level of this form of energy immediately available for the sodium pump or other transport biosynthetic activities. As described in the previous section, in vivo studies and indeed studies with brain slices in vitro indicate that the levels of ATP are partially conserved, or "spared," in the short term (seconds) during

(a)   Phosphocreatine + ADP + H$^+$ ⇌ Creatine + ATP

(b)   ATP → ADP + P$_i$ + H$^+$

(c)   2ADP → ATP + AMP

FIGURE 3.8 Interrelated reactions involving high-energy phosphates. (A) Creatine phosphokinase reaction. (B) Hydrolysis of ATP by ATPases. (C) Adenylate kinase reaction. Note involvement of protons in both (A) and (B).

periods of moderate increases in the utilization of energy stores, as occurs in muscle during contraction.[4,27] This sparing is achieved by a rapid conversion of ADP to ATP with PhCr as the phosphate donor. The enzyme responsible for the conversion is creatine phosphokinase. The hydrolysis of ATP generates hydrogen ions, which help to shift the equilibrium of the creatine phosphokinase reaction towards ATP formation (Figure 3.8). This ATP buffering capacity of PhCr is reflected in its two- to threefold higher tissue content (Table 3.6). The name *phosphogen* has been given to compounds that, like PhCr, can readily donate phosphate groups for the formation of ATP. In invertebrates, arginine phosphate serves the same function. But the entire store of high-energy phosphates in the brain (ATP plus PhCr) would only be enough to support energy demands for about 20 seconds,[4] which emphasizes the inadequacy of the total energy charge for the long-term maintenance of activity and the need for a continuous supply of ATP from oxidative phosphorylation.

An accurate and quantitative measure of the energy charge in the adenosine nucleotides (the adenylate energy charge) takes into account the concentration of all these nucleotides:[4,31]

$$\text{Adenylate energy charge} = \frac{[\text{ATP}] + 0.5[\text{ADP}]}{[\text{ATP}] + [\text{ADP}] + [\text{AMP}]}$$

This expression allows for the fact that ATP contains two high-energy bonds and ADP only one. The conversion of ADP to ATP through the agency of the enzyme adenylate kinase (EC 2.7.4.3) makes the extra bond available (Figure 3.8). The adenylate energy charge provides a convenient measure of the balance between production and utilization of energy in the form of adenine nucleotides. Of course, the PhCr content of the tissue must be added to this quantity in order to obtain the *total* energy charge.

The reactions involved in the hydrolysis of ATP and in its formation are linked (Figure 3.8). The ATPase reaction produces ADP and P$_i$, whereas creatine phosphokinase and adenylate kinase remove ADP with the formation of ATP. Protons formed in the ATPase reaction may assist in the creatine phosphokinase forward reaction (Figure 3.8). As a consequence there may be relatively small changes in the levels of ATP and ADP at the same time that P$_i$ and AMP accumulate. The latter two

TABLE 3.6 High-energy phosphates in fresh brain tissue

	PhCr	ATP	μmol/g tissue	ADP	AMP
Rapid fixation[a]			PhCr/ATP ratio		
Freeze-blowing ($n = 13$)	4.05 ± 0.07	2.45 ± 0.05	1.65	0.561 ± 0.022	0.041 ± 0.001
Microwave irradiation ($n = 9$)	1.69 ± 0.08	1.69 ± 0.02	1.00	1.32 ± 0.02	0.399 ± 0.011
Decapitation into liquid nitrogen ($n = 7$)	1.29 ± 0.15	1.54 ± 0.13	0.83	0.657 ± 0.037	0.403 ± 0.039
Whole body immersion in liquid nitrogen ($n = 10$)	2.82 ± 0.13	2.30 ± 0.06	1.22	0.415 ± 0.01	0.05 ± 0.001
[31P]nmr in situ					
Ref. 35[b]	2.9 – 3.3	1.2 – 1.4	2.3		
	PhCr/ATP ratio		P_i/ATP ratio		
Ref. 33 ($n = 10$)	2.12 ± 0.28		0.595 ± 0.055		
Ref. 32 ($n = 5$)	1.93 ± 0.12		–		

Note: Data are for rat brain.
[a] Fixation methods are from ref. 40.
[b] Guinea pig brain.

compounds allosterically stimulate phosphofructokinase activity, the main regulatory step in glycolysis (Table 3.1), therefore a rise in their level enhances the flow of mainstream energy metabolism. Thus, large changes in flux may occur without extensive fluctuations in ATP concentrations.

Monitoring ATP levels with [³¹P]nmr

By recording [³¹P]nmr signals from intact brain tissue both in vivo[32,33] and in vitro,[32] it has proved possible to estimate accurately the endogenous tissue levels of adenosine nucleotides, inorganic phosphate, and phosphocreatine (Figure 3.9). Both the normal levels of these constituents and departures from normality caused by various metabolic manipulations

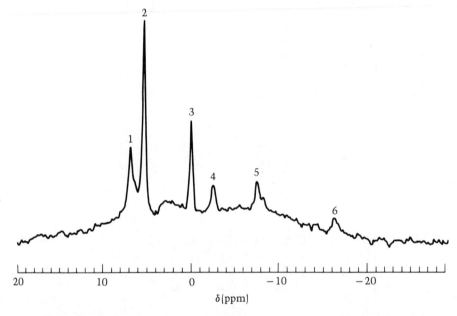

FIGURE 3.9   Nuclear magnetic resonance spectrum of cerebral slices of guinea pig brain. The tissue was superfused in Krebs-bicarbonate-buffered medium containing 10 mM glucose and gassed (O_2 and CO_2 in a ratio of 19 : 1). The spectrum presents the combined data from a block of 500 signals obtained with an interpulse interval of 8 seconds and a line-broadening of 12 Hz. A considerable proportion of the inorganic phosphate peak (65 percent) arises from the phosphate of the superfusion medium. Key: 1, sugar phosphates; 2, inorganic phosphate; 3, phosphocreatine; 4, gamma peak; 5, alpha peak; 6, beta peak. The gamma peak is due to the gamma phosphates of nucleoside triphosphates and the beta phosphates of nucleoside diphosphates. The alpha peak is due to alpha phosphates of nucleoside di- and triphosphates together with the diester phosphate of nicotinamide nucleotides. The beta peak is due to the phosphate of nucleoside triphosphate (mainly ATP). [From ref. 34]

have been reported. This relatively recent application of noninvasive nmr techniques to intact living tissues is providing new information and new insights on the dynamic aspects of energy metabolism. For instance, the best estimates of ATP, ADP, PhCr, and P_i content of brain come from the analysis of brain rapidly fixed by extremely fast freezing techniques, such as freeze-blowing; (see Table 3.6). The levels of these compounds in intact brain as measured by [³¹P]nmr spectroscopy show certain differences: Levels of ADP and P_i are lower, and PhCr and ATP higher, than has been reported for rapidly frozen brain tissue, with the PhCr: ATP ratios as high as the highest levels detected by other methods.

Changes in the relative concentrations of these compounds follow manipulations of the incubation conditions of cerebral cortex slices in vitro.[34] For instance, lowering glucose levels from 10 mM to 0.5 mM lowered sugar phosphate levels, but 0.2 mM glucose or lower were necessary before the ATP and PhCr contents of the slices were significantly reduced. Gassing with mixtures of air and CO_2 or N_2 and CO_2 in the presence of 10 mM glucose led to 30–50 percent falls in PhCr without significant change in ATP levels.[34] In addition to measuring the steady-state levels of these constituents with the use of [³¹P]nmr techniques, it has become possible to study the dynamics of their turnover.[33]

KEY SUBSTRATES AND BRAIN FUNCTION

Studies of the type outlined above show that brain tissue in vitro is remarkably resistant to relatively low glucose levels (e.g., 0.5 mM) as judged by the content of energy storage phosphates. But when blood levels of glucose decline to 2 mM, the electroencephalogram (EEG) indicates electrical failure within the brain (i.e., the EEG becomes isoelectric) and consciousness is lost. Overall neural functioning in vivo appears to be more sensitive to decreases in available glucose than is overall energy metabolism. Similar effects occur in vitro; hippocampal slices show attenuation of their synaptic potentials at glucose levels of 2 mM without any detectable energy deficit.[35]

These findings suggest that glucose participates rather directly in critical neuronal functions other than energy metabolism,[33,34] such as neurotransmitter biosynthesis. Evidence to support this proposal comes from the observation that hypoglycemia reduces the rate of production of acetylcholine,[36,37,39] presumably by restricting the supply of acetyl CoA, possibly from a metabolic compartment that preferentially feeds into acetylcholine synthesis (see Chapter 4).[39]

Oxygen, too, is critical for neurotransmitter biosynthesis as well as for oxidative phosphorylation. Molecular oxygen is involved in the hydroxylation reactions of tyrosine and tryptophan hydroxylases (see Chapter 4) and must therefore be readily available for catecholamine and serotonin synthesis. The K_m values for oxygen in these reactions are sufficiently high that any decrease in oxygen tension from normal should slow the rate of transmitter synthesis, as these are the rate-limiting enzymes in the biosynthetic pathways. Relatively mild hypoxia in rats (10 percent O_2) significantly reduces catecholamine and indoleamine biosynthesis in vivo without affecting tissue lactate levels or the adenylate energy charge.[38] Hypoxia also reduces acetylcholine synthesis rates in vivo.[39]

BRAIN ENERGY METABOLISM IN VITRO

Brain slices, synaptosomes, and isolated ganglia preparations maintained under optimal conditions in vitro have provided a great deal of basic information on neural energy metabolism and its linkage to function.[D,E]

Brain cortex slices incubated in vitro in balanced saline solution containing sufficient glucose maintain a steady oxygen uptake rate over several hours, linked to the formation of high levels of energy storage phosphates and to high tissue levels of potassium.[E] The blood-brain barrier has essentially been removed and a wide range of compounds, which now have access by simple diffusion, are metabolized by brain slices. These compounds, however, only allow maintenance of respiration and a high P:O ratio if provided in the appropriate mixtures.

It can be seen from Tables 3.2, 3.7, and 3.8 that when given alone to incubated cortex slices, substrates that are readily metabolized at various

TABLE 3.7   Respiratory rates and energy-phosphate accumulation in cerebral cortex slices incubated with various substrates

Substrate	Oxygen uptake (μmol/g tissue per hour)	PhCr	Total ATP + ADP	Inorganic phosphate
			(μmol/g tissue)	
None	38	0.37	0.56	6.7
Glucose	70	1.71	2.01	8.6
Pyruvate	78	1.18	1.8	4.6
Pyruvate and malate	–	–	–	–
Succinate	56	N.M.	1.01	10.8
2-Oxoglutarate fumarate	49	N.M.	0.68	12.3
Malate	42	1.12*	–	6.5
Oxaloacetate	59	1.26*	1.08	9.4
L-Glutamate	58	0.40*	0.90	12.1

Note: Incubation was at 37°C for 1 hour. Substrates were present alone at a concentration of 10 mM except for data marked (*), where 20–25 mM was used. N.M., not measured. Data are from various sources quoted in ref. B.

TABLE 3.8 Effect of added glucose on formation of metabolites from [^{14}C]-labeled substrates by cortex slices
Percentage of radioactivity recovered in each metabolite

Metabolite	Pyruvate (12 mM)		Acetate (1 mM)		Succinate (10 mM)		Glutamate (4 mM)	
	No glucose	Glucose	No glucose	Glucose	No glucose	Glucose	No glucose	Glucose
Glutamate	25	12	16	19	9	20	N.D.	N.D.
Aspartate	10	3	29	4	33	22	52	15
Glutamine	N.M.	1	3	35	<1	6	8	30
GABA	4	2	<3	3	4	5	3	6
Alanine	5	4	<2	1	3	4	N.D.	N.D
Lactate	39	66	N.D.	N.D.	N.D.	N.D.	N.D.	N.D.
CO$_2$	18	11	48	38	24	45	38	50
O$_2$ uptake rate (μmol/g tissue per hour)	112	142	42	67	48	70	80	108

Note: Substrates were [3-^{14}C]pyruvate, [1-^{14}C]acetate, [2–3-^{14}C]succinate, and [U-^{14}C]glutamate. Units of concentration (in mM) refer to metabolites. Incubation was for 1 hour at 37°C. N.D., not detected (too low for detection); N.M., not measured. Data are from various sources quoted in ref. B.

points on the citric acid cycle give rise to an aberrant pattern of metabolism—lowered levels of ATP and PhCr, and lowered rates of oxygen uptake—when compared with the pattern for glucose added alone. Among the features of the abnormal patterns of metabolite formation are the low rates of formation of glutamine and the exaggerated accumulation of aspartate. Both of these differences are corrected by supplementing the incubation medium with adequate glucose (5–10 mM), which, by providing sufficient glycolytic intermediates (e.g., acetyl CoA) and key citric acid cycle components, allows the cycle to turn at an adequate rate. The aspartate accumulation is due to transamination of oxaloacetate accumulating at the last stages of the cycle because operation of the cycle is retarded by the absence of acetyl CoA for condensation. As the latter becomes available, the aspartate pool size falls due to utilization of oxaloacetate. Equally, more oxidative phosphorylation occurs as NADH levels rise with increased citric acid cycle activity, and so ATP and PhCr rise to levels normally found with glucose alone. Glutamine formation also increases to normal rates; this partly reflects the restored availability of ATP, an obligatory component in the glutamine synthetase reaction. In this way, glutamine formation provides an endogenous indicator of ATP and PhCr levels.

When pyruvate plus malate or another appropriate mixture is supplied to keep the flow of metabolites through the citric acid cycle at adequate rates, the abnormality of the metabolic pattern can be reduced, and levels of ATP and PhCr approach normal (Table 3.7). But such mixtures do not support synaptic activity in hippocampal slices,[34,35] indicating the importance of other key metabolic pathways or compartments into which glucose feeds, pathways that are not directly concerned with energy metabolism.

Linkage to Excitation

The resting respiratory rates of brain slices with glucose as substrate are about half those measured for brain in vivo by arteriovenous difference. Excitation of the in vitro tissue by electrical pulses or other depolarizing treatment (e.g., exposure to veratrine alkaloids) will double respiratory rates and bring them to their in vivo levels[A,E] or even to the high rates observed during cerebral seizures. Increase in oxygen uptake is accompanied by losses of potassium content and gains in sodium, together with a diminished adenylate energy charge and PhCr content. Response patterns such as these follow as a sequel to the membrane depolarization initiated by these stimuli. There are parallel increases in ion fluxes and an associated acceleration in sodium-pump activity, augmented glucose uptake, and increased ATP production due to ADP release and other factors controlling the rate of flow of mainstream metabolism as described above (see Figure 3.8 and Table 3.1).

Lactate Formation

The details of the changes in these ionic fluxes, substrate utilization, and energy phosphate production have been studied with great facility by employing brain slices, synaptosomes (see Chapter 6), and isolated ganglia,[A,E] even though these preparations are greatly limited in their performance. Lactate, for instance, accumulates in the incubation medium and can be measured as an index of glycolytic rates (Table 3.2; Figure 3.1). The amounts of lactate formed are to some extent proportional to the volume of the incubation medium (the medium-to-tissue ratio). Relatively little lactate is detected in vivo probably because in the extracellular space it reaches a concentration at which it is readily oxidized (Table 3.2), and a greater proportion of pyruvate is oxidized via the citric acid cycle. In vitro, however, respiratory and other responses to depolarizing treatments are accompanied by increased production of lactate, indicating an acceleration of glycolytic rates.

GLUCOSE METABOLISM: THE PATHWAYS INVOLVED

Amino Acid Formation

Essentially the same pathways are involved in generating carbon dioxide, water, pyruvate, and lactate from glucose in neural tissue as in other tissues:[A,E] glycolysis, the citric acid cycle, and oxidative phosphorylation (Figures 3.1 and 3.2). These pathways have recently been reviewed in detail.[52] The hexose monophosphate shunt, which converts glucose 6-phosphate to 6-phosphonogluconate and on to ribose phosphates, is relatively inactive in brain tissue, with about 10 percent of the glucose consumed proceeding via this route.[52,A] Some activity is always necessary to provide ribose for nucleotide biosynthesis and NADP for enzyme cofactor function (e.g., isocitrate dehydrogenase).

Other distinctive features of brain glucose metabolism are the formation of γ-aminobutyric acid (GABA) and the generation of large pools of glutamate and aspartate (Table 3.9).

The GABA-Shunt Pathway

The pathway of the GABA shunt forms a short annex to the main citric acid cycle in the brain providing an alternative route (shunt) between 2-oxoglutarate and succinate (Figures 3.2 and 3.10; see also Figure 4.40). This pathway is by far most active in the brain; it is also found in the kidneys,[48] heart, and liver, but at very low levels.

TABLE 3.9   Levels of selected amino acids in brain and liver (μmol/g tissue)

	Glutamate	Aspartate	GABA	Glutamine	Alanine
Brain tissue					
Hemisphere	9.26 ± 1.09	2.40 ± 0.28	1.52 ± 0.19	2.64 ± 0.43	0.77 ± 0.10
Midbrain	6.39 ± 0.81	2.52 ± 0.13	3.06 ± 0.55	2.13 ± 0.47	0.52 ± 0.06
Cerebellum	8.42 ± 0.75	2.19 ± 0.18	1.56 ± 0.37	2.55 ± 0.58	0.88 ± 0.14
Pons medulla	4.66 ± 0.44	2.59 ± 0.21	1.32 ± 0.22	1.45 ± 0.26	0.64 ± 0.06
Whole brain	8.70	2.23	2.27	3.42	0.94
Liver tissue	4.48	0.87	0.1	<3.42	1.85

Note: Data for brain tissue are from ref. 46, for rat ($N = 10$); Data for whole brain and liver tissue are from ref. 47, for cat ($N = 10$).

In energy-generating terms, the route via the shunt is not equivalent to direct oxidation of 2-oxoglutarate in the cycle. Each of these routes produces one NADH which yields three ATP molecules during oxidative phosphorylation (Figures 3.2 and 4.40). However, substrate-level phosphorylation at the succinyl CoA synthetase step is bypassed in the shunt pathway, and no GTP and hence no ATP (via nucleoside diphosphokinase) is formed. Thus, one ATP less is generated when metabolic flow is via the GABA shunt. Since only 8–10 percent of the flow through the citric acid cycle appears to go by way of the shunt,[4] its main function is assumed to be the biosynthesis of GABA. GABA is an inhibitory neurotransmitter (see Chapter 4). It occurs in high concentration (1–3 mM) in most brain regions (see Table 3.9). The shunt itself is not universally distributed within the brain because one key enzyme, glutamate decarboxylase (GAD), is found only in those inhibitory neurons that release GABA as a neurotransmitter. Since GABA transaminase is found in most neurons and glia, it is clear that all of the enzymes of the GABA shunt will be present within GABAergic neurons. But GABA released at synapses and subsequently transported into surrounding glia can complete its transformation within the mitochondria of these cells because they contain both GABA transaminase and succinate semialdehyde dehydrogenase (Figure 3.10).

Glutamate and Aspartate Formation

The carbon skeletons of the dicarboxylic acids glutamate and aspartate are rapidly formed from glucose, as is evident from studies of the overall metabolism of [U-^{14}C]glucose (Table 3.2). The pool sizes of free glutamate and aspartate in most brain regions are high (5–10 mM, and 2–3 mM respectively) in contrast with their levels in other organs, such as the liver (Table 3.9). The precise reason for these larger pools remains uncertain, though the unusually high tissue content of 2-oxoglutarate, which

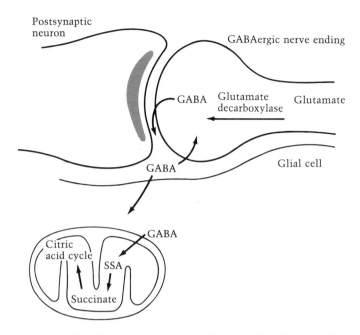

FIGURE 3.10   Compartmentation of the GABA-shunt pathway between neurons and glial cells.

also occurs, favors glutamate formation (see Chapter 4). It is likely that the high K_m for the oxidation of 2-oxoglutarate in the citric acid cycle leads to the exaggerated accumulation of this keto acid and it subsequent enhanced amination or transmination to glutamate[B,D] employing glutamate dehydrogenase or aspartate aminotransferase respectively (Figure 3.11). The latter predominates in glutamate formation in the central nervous system, although the equilibrium of the glutamate dehydrogenase reaction lies in favor of glutamate formation.[59] Active exchange transamination through the agency of the enzyme aspartate aminotransferase, which is extremely fast and universally distributed, leads to rapid conversion of glutamate to aspartate (Figure 3.11). Formation of aspartate by transamination of oxaloacetate formed in the citric acid cycle is the main route for biosynthesis of this amino acid.

Many other transamination reactions in neural tissue employ 2-oxoglutarate as the keto acid, and glutamate formation is the result. For instance, the GABA shunt not only consumes glutamate, it also produces it by transamination at the GABA transaminase stage, (see Figure 4.40). Numerous other, relatively minor pathways such as alanine transamination (Figure 3.12), and the metabolism of arginine, ornithine, and proline also give rise to glutamate.

Glutamate and aspartate are involved in many other biosynthetic activities, including protein biosynthesis. Protein synthesis occurs at

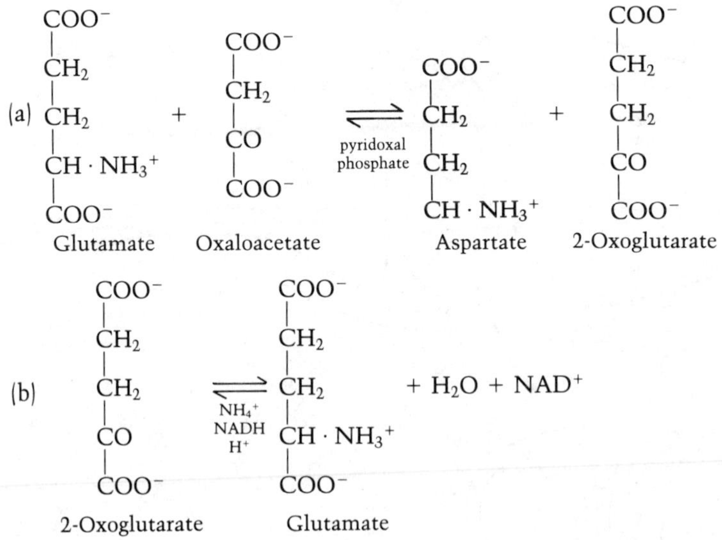

(a)

$$
\text{Glutamate} \;+\; \text{Oxaloacetate} \;\underset{\text{pyridoxal phosphate}}{\rightleftharpoons}\; \text{Aspartate} \;+\; \text{2-Oxoglutarate}
$$

Glutamate:
$$\text{COO}^- - \text{CH}_2 - \text{CH}_2 - \text{CH}\cdot\text{NH}_3^+ - \text{COO}^-$$

Oxaloacetate:
$$\text{COO}^- - \text{CH}_2 - \text{CO} - \text{COO}^-$$

Aspartate:
$$\text{COO}^- - \text{CH}_2 - \text{CH}_2 - \text{CH}\cdot\text{NH}_3^+$$

2-Oxoglutarate:
$$\text{COO}^- - \text{CH}_2 - \text{CH}_2 - \text{CO} - \text{COO}^-$$

(b)

$$
\text{2-Oxoglutarate} \;\underset{\substack{\text{NH}_4^+ \\ \text{NADH} \\ \text{H}^+}}{\rightleftharpoons}\; \text{Glutamate} \;+\; \text{H}_2\text{O} + \text{NAD}^+
$$

FIGURE 3.11 Enzymatic synthesis of glutamate and aspartate. (A) aspartate amino transferase reaction, which produces both glutamate and aspartate. (B) glutamate dehydrogenase reaction, in which the equilibrium is in favor of glutamate formation.

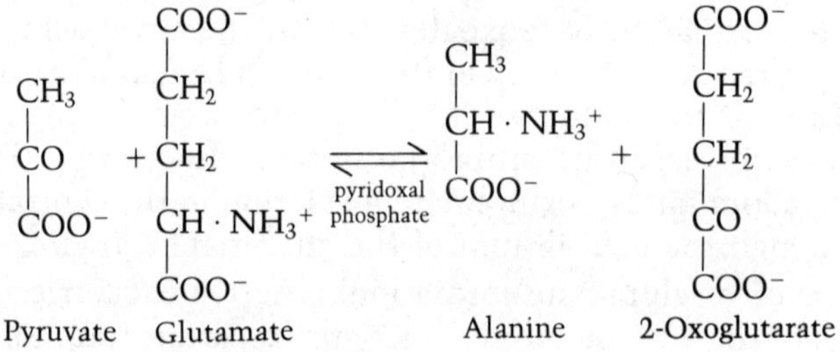

$$
\text{Pyruvate} \;+\; \text{Glutamate} \;\underset{\text{pyridoxal phosphate}}{\rightleftharpoons}\; \text{Alanine} \;+\; \text{2-Oxoglutarate}
$$

Pyruvate:
$$\text{CH}_3 - \text{CO} - \text{COO}^-$$

Glutamate:
$$\text{COO}^- - \text{CH}_2 - \text{CH}_2 - \text{CH}\cdot\text{NH}_3^+ - \text{COO}^-$$

Alanine:
$$\text{CH}_3 - \text{CH}\cdot\text{NH}_3^+ - \text{COO}^-$$

2-Oxoglutarate:
$$\text{COO}^- - \text{CH}_2 - \text{CH}_2 - \text{CO} - \text{COO}^-$$

FIGURE 3.12 Alanine aminotransferase reaction.

rapid rates in the brain, equivalent to those found in such tissues as the liver, which exports proteins. Glutamate and aspartate pools, along with those of other amino acids, actively provide and receive contributions from protein turnover. Both of these dicarboxylic amino acids are very likely to be functioning as excitatory neurotransmitters in the nervous system, though the total amounts involved in these activities can be estimated at a tenth or a hundredth of their overall tissue content (see Chapter 4).

Carbon Dioxide Fixation One important consequence of the rapid and extensive formation of dicarboxylic amino acids in the brain is the need to replenish the pools of citric acid cycle intermediates which become depleted as a result. This depletion is increased by the continuous and spontaneous decarboxylation of oxaloacetate with the formation of pyruvate and CO_2. Maintenance of the levels of citric acid cycle intermediates in the brain is due in substantial measure to the high rates of fixation of CO_2 by enzymes which form these intermediates. These are known as anaplerotic ("filling-up") reactions. Pyruvate carboxylase (EC 6.4.1.1), malic enzyme (EC 1.1.1.40), and phosphoenolpyruvate carboxykinase (EC 4.1.1.32) are all enzymes that catalyze reversible carboxylation processes that could theoretically lead to CO_2 fixation (Figures 3.1, 3.2, and 3.13). It has been calculated that during ammonia accumulation in the brain, when glutamine formation is greatly accelerated, citric acid cycle intermediates would be massively depleted within 2 or 3 minutes in the absence of anaplerotic synthesis. In fact, the pool sizes of these intermedi-

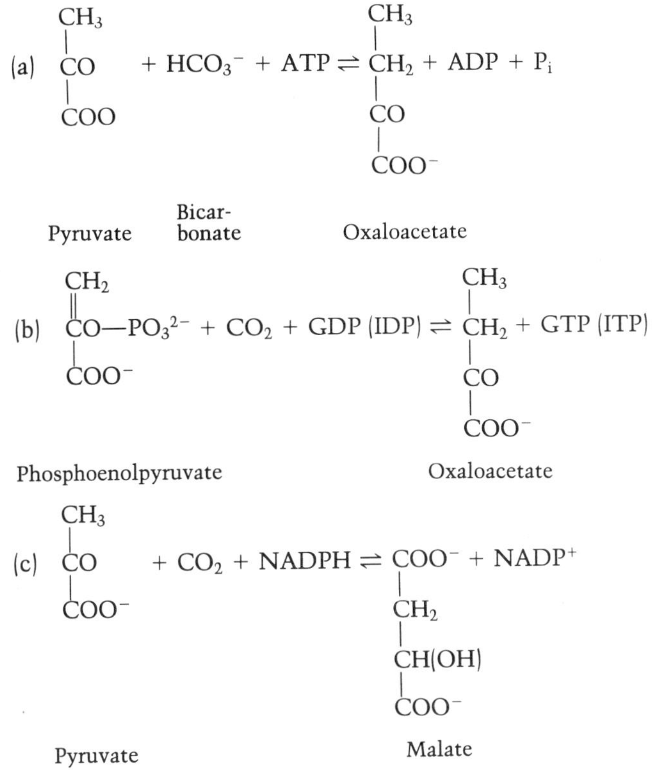

$$\text{(a)} \quad \underset{\substack{\text{Pyruvate}}}{\overset{\displaystyle \text{CH}_3}{\underset{\displaystyle \text{COO}}{|\atop \text{CO} \atop |}}} + \underset{\substack{\text{Bicar-}\\\text{bonate}}}{\text{HCO}_3^-} + \text{ATP} \rightleftharpoons \underset{\substack{\text{Oxaloacetate}}}{\overset{\displaystyle \text{CH}_3}{\underset{\displaystyle \text{COO}^-}{|\atop \text{CH}_2 \atop |\atop \text{CO} \atop |}}} + \text{ADP} + \text{P}_i$$

$$\text{(b)} \quad \underset{\substack{\text{Phosphoenolpyruvate}}}{\overset{\displaystyle \text{CH}_2}{\underset{\displaystyle \text{COO}^-}{\|\atop \text{CO}-\text{PO}_3^{2-} \atop |}}} + \text{CO}_2 + \text{GDP (IDP)} \rightleftharpoons \underset{\substack{\text{Oxaloacetate}}}{\overset{\displaystyle \text{CH}_3}{\underset{\displaystyle \text{COO}^-}{|\atop \text{CH}_2 \atop |\atop \text{CO} \atop |}}} + \text{GTP (ITP)}$$

$$\text{(c)} \quad \underset{\substack{\text{Pyruvate}}}{\overset{\displaystyle \text{CH}_3}{\underset{\displaystyle \text{COO}^-}{|\atop \text{CO} \atop |}}} + \text{CO}_2 + \text{NADPH} \rightleftharpoons \underset{\substack{\text{Malate}}}{\overset{\displaystyle \text{COO}^-}{\underset{\displaystyle \text{COO}^-}{|\atop \text{CH}_2 \atop |\atop \text{CH(OH)} \atop |}}} + \text{NADP}^+$$

FIGURE 3.13 Reversible carboxylation reactions in the brain. (A) pyruvate carboxylase, found mainly in astrocytes. (B) phosphoenolpyruvate carboxykinase. (C) malic enzyme.

ates hardly change under these conditions.[A,K] In practice, it seems that pyruvate carboxylase is responsible for by far the greatest degree of CO_2 fixation in the brain, with some 10 percent of all pyruvate used serving to replenish citric acid cycle intermediates during normal periods of metabolic flow through the cycle. Malic enzyme and phosphoenolpyruvate carboxykinase probably make negligible contributions, since their equilibrium states greatly favor decarboxylation rather than carboxylation.[A,K] In addition, numerous reversible transamination reactions can generate citric acid cycle intermediates under favorable kinetic conditions and therefore serve an anaplerotic function. An example is the alanine aminotransferase reaction (Figures 3.1 and 3.12), which can generate 2-oxoglutarate where the tissue concentration of alanine allows the reaction to proceed to the right.

Glutamine and Alanine

Two other amino acids whose carbon skeletons are rapidly supplied by glucose are glutamine and alanine. Alanine is formed by transamination of the pyruvate formed by glycolysis (Figure 3.1). The enzyme responsible is alanine aminotransferase (EC 2.6.1.2), with glutamate providing the amino group (Figure 3.12).

Glutamine is formed directly from glutamate, which, we recall, is itself rapidly formed from 2-oxoglutarate. The enzyme responsible in this case is glutamine synthetase (EC 6.3.1.2), shown to be exclusively localized to glial cells (see Figure 4.32; see also Chapters 2 and 4). Glutamine formation requires ATP and ammonia, and it occurs at rapid rates. It is found in tissues at high concentration (1–3 mM) (Tables 3.2 and 3.9) and accumulates in cerebrospinal fluid at levels of about 0.5 mM. Because its formation requires both glutamate and ammonia, it is important in the metabolism and intercellular transport of both substances and may be an important precursor of glutamate released as a neurotransmitter (see Figure 4.32). Glutamine levels in the brain rise dramatically in conditions of hyperammonemia, such as in liver and kidney failure, suggesting a function for glutamine in trapping and inactivating ammonia (see Chapter 9). Glutamine also participates in the biosynthesis of both protein and nucleotide (guanosine monophosphate, or GMP) in neural tissue and is thus widely involved in brain metabolism.

CELLULAR COMPARTMENTATION OF METABOLISM

Neuronal and Glial Substrates

A considerable body of evidence from both in vitro and in vivo studies suggests that some specialization in the use of substrate exists between neurons and glia.[A,41] Glucose, for instance, seems to be preferentially used by neurons, whereas a range of other substrates feeding directly into the citric acid cycle (e.g., acetate, pyruvate, succinate, glutamate, GABA, phenylalanine, and leucine) are mainly taken up and metabolized by glial cells. This difference has been directly demonstrated for dorsal root ganglia by autoradiography[42,43,44] (Figure 3.14 and 3.15). Other functional studies involving neuronal stimulation in vivo support this compartmentation concept.[45] That GABA is produced only in neurons and glutamine only in glial cells directly illustrates compartmentation at the level of biosynthesis (see Figure 3.10; see also Figure 4.32). Whereas such differences in enzyme localization are absolute, it is likely that at the level of energy substrate utilization the compartmentation is not absolute but reflects differential rates of uptake and utilization.

Computer Models

Computer analysis of the compartmentation of glucose and related substrates in the brain has involved models constructed from the best available data for the concentrations of intermediates in the tissue, from the constants for enzyme reactions linking these intermediates, and from rates of turnover based on the use of radiolabeled substrates. The quantitative patterns of metabolism developing with time are compared with those known to exist for a given substrate, and gross anomalies are dealt with by altering assumptions about pool size or enzyme activity until the most acceptable model is formed.[41,51,56]

For glucose and amino acid metabolism in the brain, the closest correlation between computer-derived and experimentally derived data occurs when the model allows for the existence of at least two separate pools of citric acid cycle intermediates, one containing a large pool of glutamate with relatively little conversion to glutamine and another containing a much smaller pool of glutamate that is rapidly converted to glutamine. The large compartment forms GABA, which is transaminated primarily in the small compartment. Entry of ^{14}C from [U-^{14}C]glucose appears to be first to the large compartment, with relatively slow subsequent migration of ^{14}C to the smaller compartment.[41,51,B,D] The small compartment preferentially uses a number of substrates, including amino acids and short-chain fatty acids.[41]

Although metabolic compartments do not necessarily correspond to anatomically separate cells, in this case all evidence points to the neurons as the large compartment and glial cells as the small compartments (Figure 3.10). This pattern of compartmentation develops in ontogeny in parallel with the appearance of glial cells.[41] Both neuronal and glial pools can be subdivided into subpools representing glial cell types or cell regions such as nerve endings and dendrites.[41,54]

In addition to neurons and glial cells, the cells of the vasculature provide a third cellular compartment in the central nervous system. The capillary endothelial cells comprise only a few percent of the cellular

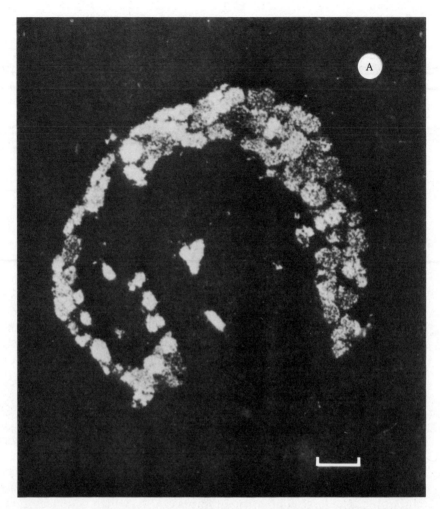

FIGURE 3.14   Preferential use of glucose by neurons of dorsal root ganglion. Autoradiography of dorsal root ganglia after incubation for 90 minutes with [¹⁴C]glucose. (A) Dark-field autoradiograph showing photomontage of a section through a whole ganglion. Radioactivity is localized to a greater extent over neuronal cell bodies than over the satellite glial cells. The sensory fibers in the interior of the ganglion are unlabeled. Exposure time, 24 days. Bar, 100 μm. (B and C) Phase-contrast and corresponding dark-field autoradiograph of the dorsal root ganglion. Silver grains are localized to a lesser extent over the satellite glial cells (*arrows*) than over the neuronal cell bodies (*N*). The sensory fibers (*SF*) are devoid of label. Exposure time, 24 days. Bar, 50 μm. [From ref. 43]

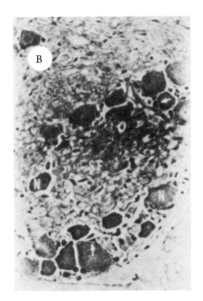
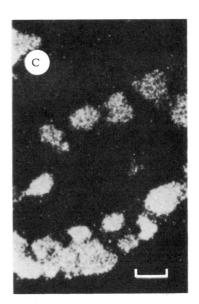

volume of the brain but contain fivefold to sevenfold the number of mitochondria seen in systemic endothelia.[53] They appear to be very active metabolically and are probably responsible for a large proportion of the fatty acid and amino acid oxidation.[52]

The glial cells contribute about half the cellular volume of the brain (Chapter 2). Astrocytes comprise a large proportion of the glia and contain concentrations of mitochondria in their relatively few processes.[55] This accords with their high respiratory rates, roughly equivalent to those of neurons.[55] Oligodendroglia contain large numbers of mitochondria but fewer glycogen storage granules than astrocytes, and much of their metabolism is geared to the synthesis and turnover of myelin in the central nervous system. They are generally regarded as being metabolically the more active glial cells, and in tissue culture show differences from astrocytes both in their metabolic properties[55] and in their ability to respond to various inducing agents with formation of cyclic-AMP and glycolytic enzymes (see Tables 2.1 and 2.3).

Thus, glial cells are likely to be specialized in their metabolism and complementary to the activity of neurons, which seem to be especially geared towards the use of glucose as substrate. For instance, there is evidence that astrocytes supply citric acid cycle intermediates to neurons and that this prevents the depletion of these metabolites, which could occur due to efflux of glutamine or release of glutamate as a neurotransmitter from glutamatergic nerve terminals. The anaplerotic enzyme pyruvate carboxylase is selectively expressed in astrocytes, and this would provide the source of carbon (Figure 3.13). There is probably a net transfer

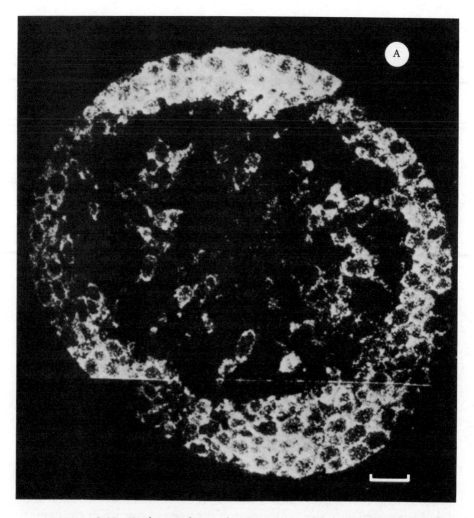

FIGURE 3.15 Preferential use of acetate by satellite glial cells in dorsal root ganglion. Autoradiography of dorsal root ganglia after incubation for 90 minutes with [14C]acetate. (A) Dark-field autoradiography showing photomontage of a section through a whole ganglion. Radioactivity is preferentially localized in the satellite glial cells surrounding the large neuronal cell bodies. The sensory fibers in the interior of the ganglion are unlabeled. Exposure time, 11 days. Bar, 100 μm. (B and C) Phase-contrast and corresponding dark-field autoradiograph of dorsal root ganglion. Silver grains are predominantly located over satellite glial cells (*arrows*); the neurons (N) and sensory fibers (SF) show little labeling. Exposure time, 5 days. Bar, 50 μm. [From ref. 43]

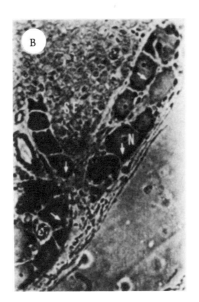 

of 2-oxoglutarate (and malate) from astrocytes to neurons, involving transport of the keto acid into nerve terminals by Na^+-dependent, high-affinity uptake processes. Such transport systems have been shown to exist in nerve-terminal-enriched preparations.[60]

In summary, glucose is still seen as a vital molecule for energy metabolism in neural tissue. The explanation for the apparent specialization of the nervous system for glucose remains uncertain, but it is likely to be linked to its multiple participation in the metabolic activities of the cell, including neurotransmitter biosynthesis, and to the supply of the complete sequence of glycolytic and citric acid cycle intermediate.

REFERENCES

General Texts and Review Articles

A Seisjö, B. K. (1978) *Brain Energy Metabolism.* Wiley, Chichester.

B Bradford, H. F. (1968) "Carbohydrate and Energy Metabolism," in *Applied Neurochemistry* (eds. A. N. Davison and J. Dobbing), pp. 222–250. Blackwell, Oxford.

C Sacks, W. (1983) "Cerebral Metabolism *in Vivo*," in *Handbook of Neurochemistry* 2d. ed., vol. 3 (ed. A. Lajtha), pp. 321–351. Plenum Press, New York.

D Balázs, R. (1970) "Carbohydrate Metabolism," in *Handbook of Neurochemistry* (ed. A. Lajtha), vol. 3, pp. 1–36. Plenum Press, New York.

E   McIlwain, H., and H. S. Bachelard. (1985) *Biochemistry and the Central Nervous System*, 5th ed. Churchill Livingstone, Edinburgh and London.

F   Plum, F., A. Gjedde, and F. E. Samson, eds. (1976) "Neuroanatomical Functional Mapping by the Radioactive 2-Deoxy-D-Glucose Method," *Neurosci. Res. Prog. Bull.* 14 (4): 459–518.

G   Smith, C. B., and L. Sokoloff. (1981) "The Energy Metabolism of the Brain," in *Molecular Neuropathology* (eds. A. N. Davison and R. H. S. Thompson), pp. 104–131. Edward Arnold, London.

H   Sokoloff, L. (1981) "The Relationship Between Function and Energy Metabolism: Its Use in the Localization of Functional Activity in the Nervous System," *Neurosci. Res. Prog. Bull.* 19 (2): 159–210.

I   Sokoloff, L. (1983) "Measurement of Local Glucose Utilization in the CNS and Its Relationship to Local Functional Activity," in *Handbook of Neurochemistry* (ed. A. Lajtha), 2d ed., pp. 225–257. Plenum Press, New York.

J   Passoneau, J. V., R. A. Hawkins, W. D. Lust, and F. A. Welsh, eds. (1980) *Cerebral Metabolism and Neural Function*. Williams and Wilkins, Baltimore.

K   Hawkins, R. A., and A. M. Mans. (1983) "Intermediary Metabolism of Carbohydrates and Other Fuels," in *Handbook of Neurochemistry*, 2d ed. (ed. A. Lajtha), pp. 259–294. Plenum Press, New York.

L   Gibson, G. E., and J. P. Blass. (1983) "Metabolism and Neurotransmission," in *Handbook of Neurochemistry*, 2d ed. (ed. A. Lajtha), pp. 633–651. Plenum Press, New York.

Additional Citations

1   Nelson, S., D. W. Schulz, J. V. Passonneau, and O. H. Lowry. (1968) *J. Neurochem.* 15: 1271–1279.

2   Himwich, W. A., and H. E. Himwich. (1946) *J. Neurophysiol.* 9: 133–136.

3   Nilsson, L., and B. K. Seisjö. (1976) *J. Neurochem.* 26: 353–359.

4   Nilsson, L. (1974) *Acta Physiol. Scand.* 92: 142–144.

5   Häggerdal, M., J. R. Harp, L. Nilsson, and B. K. Seisjö. (1975) *J. Neurochem.* 24: 311–316.

6   Sokoloff, L. (1973) *Ann. Rev. Med.* 24: 271–279.

7   Owen, O. E., A. P. Morgan, H. G. Kemp, J. M. Sullivan, M. G. Herrera, and G. F. Cahill. (1967) *J. Clin. Invest.* 46: 1589–1595.

8   Hawkins, R. A., A. L. Miller, J. E. Cremer, and R. L. Veech. (1974) *J. Neurochem.* 23: 917–923.

9   Gottstein, U., W. Müller, W. Berghoff, H. Gartner, and K. Held. (1971) *Klin. Wochenschr.* 49: 406–411.

10 Cremer, J. E., and D. F. Heath. (1974) *Biochem. J.* 142: 527–544.

11 Persson, B., G. Settergren, and G. Dahlquist. (1972) *Acta Paediatr. Scand.* 61: 273–278.

12 Gjedde, A., and C. Crone. (1975) *Am. J. Physiol.* 229: 1165–1169.

13 Moore, T. J., A. P. Lione, M. C. Sugden, and D. M. Regen. (1976) *Am. J. Physiol.* 230: 3, 619–630.

14 Tower, D. B., and O. M. Young. (1973) *J. Neurochem.* 20. 253–267.

15 Sokoloff, L., M. Reivich, C. Kennedy, M. H. Des Rosiers, C. S. Patlak, K. D. Pettigrew, O. Sakurada, and M. Shinohara. (1977) *J. Neurochem.* 28: 897–916.

16 Sokoloff, L. (1981) *Neurosci. Res. Prog. Bull.* 19: 159–210.

17 Kennedy, C., M. H. Rosiers, O. Sakurada, M. Shinohara, M. Reivich, J. W. Jehle, and L. Sokoloff. (1976) *Proc. Natl. Acad. Sci.* (USA) 73: 4230–4234.

18 Robinson, J. D. (1983) In *Handbook of Neurochemistry*, 2d ed., (ed. A. Lajtha), Plenum Press, New York, pp. 173–193.

19 Whittam, R. (1962) *Biochem. J.* 82: 205–212.

20 Mata, M., D. J. Fink, H. Gainer, C. B. Smith, L. Davidsen, H. Savaki, W. J. Schwartz, and L. Sokoloff. (1980) *J. Neurochem.* 34: 213–215.

21 Newsholme, E. A., and C. Start. (1973) *Regulation in Metabolism*. Wiley, London.

22 Phelps, M. E., J. C. Mazziota, and D. E. Kuhl. (1980) *Science* 211: 1445–1448.

23 Reivich, M., D. E. Kuhl, A. Wolf, J. Greenberg, M. Phelps, T. Ido, V. Cassella, J. Fowler, E. Hoffman, A. Alavi, P. Som, and L. Sokoloff. (1979) *Circ. Res.* 44: 127–137.

24 Olesen, J. (1971) *Brain* 94: 635–646.

25 Purves, M. J. (1972) *The Physiology of the Cerebral Circulation*. Cambridge University Press, London.

26 Rubio, R., R. M. Berne, E. L. Bockman, and R. R. Curnish. (1975) *Amer. J. Physiol.* 228: 1896–1902.

27 Bessman, S. P., and P. J. Geiger. (1981) *Science* 211: 448–452.

28 Sacks, W., S. Sacks, and A. Fleischer. (1983) *Neurochem. Res.* 8: 661–685.

29 Hawkins, R. A. (1980) In *Cerebral Metabolism in Neural Function* (eds. J. V. Passoneau, R. A. Hawkins, W. D. Lust, and F. A. Welsh), pp. 367–381. Williams and Wilkins, Baltimore.

30 Hawkins, R. A., and A. L. Miller. (1978) *Neuroscience* 3: 251–258.

31 Atkinson, D. E. (1968) *Biochemistry* 7: 4030–4034.

32 Akerman, J. J. H., T. H. Grove, G. G. Wong, D. G. Gadian, and G. K. Radda. (1980) *Nature* (Lond.) 283: 167–170.

33 Shoubridge, E. A., R. W. Briggs, and G. K. Radda. (1982) *FEBS Lett.* 140: 288–292.

34 Cox, D. W. G., P. G. Morris, J. Feeny, H. S. Bachelard. (1983) *Biochem. J.* 212: 365–370.

35 Cox, D. W. G., and H. S. Bachelard. (1982) *Brain Res.* 239: 527–534.

36 Gibson, G. E., and T. E. Duffy. (1981) *J. Neurochem.* 36: 28–33.

37 Harvey, S. A. K., R. F. G. Booth, and J. B. Clark. (1982) *Biochem. J.* 206: 433–439.

38 Davis, J. N. (1976) *J. Neurochem.* 27: 211–215.

39 Gibson, G. E., and J. P. Blass. (1983) In *Handbook of Neurochemistry* (ed. A. Lajtha), 2d ed., vol. 3, pp. 633–652. Plenum Press, New York.

40 Veech, R. L., D. Harris, D. Veloso, and E. H. Veech. (1973) *J. Neurochem.* 20: 183–188.

41 Balazs, R., and J. E. Cremer, eds. (1973) *Metabolic Compartmentation in the Brain*. Macmillan, London.

42 Schon, F., and J. S. Kelly. (1974) *Brain Res.* 66: 275–288.

43 Minchin, M. W., and P. M. Beart. (1975) *Brain Res.* 83: 437–441.

44 Duce, I. R., and P. Keen. (1983) *Neuroscience* 8: 861–866.

45 Abdul-Ghani, A-S., M. M. Boyar, J. Coutinho-Netto, and H. F. Bradford. (1980) *J. Neurochem.* 35 (1): 170–175.

46 Shaw, R. K., and J. D. Heine. (1965) *J. Neurochem.* 12: 151–155.

47 Tallan, H. H., S. Moore, and W. H. Stein. (1954) *J. Biol. Chem.* 211: 927–939.

48 Whelan, D. T., C. R. Scriver, and F. Mohynddin. (1969). *Nature* (Lond.) 224: 916–917.

49 Gutman, M., E. B. Kearney, and T. P. Singer. (1971) *Biochem. Biophys. Res. Commun.* 44: 526–532.

50 Chapman, A. G., B. S. Meldrum, and B. K. Seisjö. (1977) *J. Neurochem.* 28: 1025–1035.

51 Van den Berg, C. J., and D. Garfinkel. (1971) *Biochem. J.* 123: 211–219.

52 Hawkins, R. A., and A. M. Mans. (1983) In *Handbook of Neurochemistry*, 2d ed., vol. 3 (ed. A. Lajtha), pp. 259–294. Plenum Press, New York.

53 Oldendorf, W. H., M. E. Cornford, and W. J. Brown. (1977) *Ann. Neurol.* 1: 409–417.

54 Cremer, J. E., H. M. Teal, D. F. Heath, and J. B. Cavanagh. (1977) *J. Neurochem.* 28: 215–222.

55 Hertz, L. (1983) In *Handbook of Neurochemistry*, 2d ed., vol. 1 (ed. A. Lajtha), pp. 319–355. Plenum Press, New York.

56 Van Gelder, N. M. (1983) In *Handbook of Neurochemistry*, 2d ed., vol. 2, (ed. A. Lajtha), pp. 183–206. Plenum Press, New York.

57 Sharp, F. R. (1976) *Brain Res.* 107: 663–666.

58 Des Rosiers, M. H., C. Kennedy, C. S. Patlak, K. D. Petteigrew, L. Sokoloff, and M. Reivich. (1974) *Neurology* 24: 389 (Abstr.).

59 Patel, A. J., A. Hunt, R. D. Gordon, and R. Balazs. (1982) *Dev. Brain Res.* 4: 3–11.

60 Shank, R. P., and G. Le M. Campbell. (1983) In *Handbook of Neurochemistry* (ed. A. Lajtha), vol. 3, pp. 381–404. Plenum Press, New York.

61 Cunningham, V., and J. E. Kremer. (1985) *Trends Neurosci.* 8: 96–99.

4

Neurotransmitters: Chemical Target Seekers

The search for chemical agents that could transmit the activity of peripheral nerves onto their target organs began early in the century. However, it was not until the late 1920s and early 1930s after a long trail of research that acetylcholine was unequivocably demonstrated to be mediating the inhibitory parasympathetic influence of the vagus nerve on the frog's heart. At about the same time, it was shown to be acting at voluntary motor nerve terminals to initiate muscle contraction.[1,2] At that point, acetylcholine became the first identified neurotransmitter substance. Epinephrine, too, was an early candidate for the role of peripheral neurotransmitter, but after a long accumulation of evidence its derivative norepinephrine was finally shown in the mid-1940s to be the actual agent mediating sympathetic postganglionic transmission.[3] Dopamine, the immediate biosynthetic precursor of norepinephrine, followed a similar history, with doubt and then certainty following its progress to acceptance as a neurotransmitter in the 1950s.[4] Another neuroactive amine, serotonin (5-hydroxytryptamine), first isolated from blood in the late 1940s, was found to be present in many organs and especially concentrated in the brain.[5] Its localization and pharmacological activity showed that it was another strong contender as a neurotransmitter, both in central and peripheral nervous systems, and it was not long before its neurotransmitter function was clearly established.

In the early 1950s these four compounds, together with a few other unlikely candidates isolated from various organs including the brain (e.g., substance P, a peptide), were the full armory of agents known to be operating the chemical synapses between a neuron and its target organ, or between neurons. These few potent compounds were found to be localized to the neurons and their terminals from which they were released. They seemed to be specialized for this task and did not appear to be involved in other biochemical activities. It seemed reasonable at the time to assume that together they provided the principal means of mediating synaptic action throughout the nervous system. Indeed, it seemed a good design feature to have the "excitatory" or "inhibitory" messages mediated by a restricted range of potent substances present in small quantities and localized to nerve terminals.

It was also during the 1950s, however, that five amino acids—glutamate, aspartate, γ-aminobutyrate (GABA), glycine, and taurine—were being considered new and important contenders as neurotransmitters. They substantially departed from the ideal profile of properties outlined above. For instance, they were ubiquitous in all cells and organs, and present in high concentrations. They were also employed in a wide range of metabolic pathways and biosyntheses in both neurons and glial cells.

In the 1970s another dozen or so potent neurotransmitter candidates belonging to an entirely different chemical category were detected, isolated, and characterized. These were the neural peptides,[6] which were also fully qualified to be added to the list of compounds that could be operating chemical synapses in various regions of the animal nervous system (see Chapter 5). Like acetylcholine and the amines norepinephrine, dopamine, and serotonin, the neural peptides are more specialized in their activities than are amino acids and are present in very small quantities in localized regions of the nervous system.

More recently, a new and quite different *functional* category of substances has been identified that exerts a modifying effect on the extent of release, or on the postsynaptic actions, of conventional neurotransmitters. These *neuromodulators* are also released from neurons or nerve terminals and can act either locally or at a distance (see Table 7.1). Neuromodulators, and even secondary neurotransmitters (cotransmit-

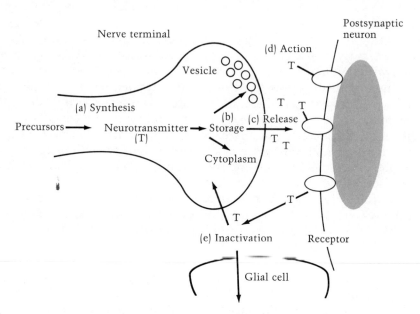

FIGURE 4.1   Principal processes occurring in the synapse.

TABLE 4.1   Established and putative neurotransmitters

System	Compound
Amino acidergic	γ-Aminobutyrate (GABA)
	Aspartate
	Glutamate
	Glycine
	Histamine
	Taurine
Cholinergic	Acetylcholine
Monoaminergic	Adrenaline
	Dopamine
	Noradrenaline
	Serotonin
	Tryptamine
Peptidergic*	Angiotensin
	Bombesin
	Carnosine
	Cholecystokinin
	Endorphins
	Luteinizing hormone releasing hormone (LHRH)
	Methione and leucine enkephalins
	Motilin
	Neuromedins (A to C)
	Neuropeptide Y
	Neurotensin
	Oxytocin
	Somatostatin
	Substance P
	Thyroid hormone releasing hormone (TRH)
	Vasoactive intestinal peptide (VIP)
	Vasopressin
Purinergic	Adenosine
	ADP
	AMP
	ATP

* Only well-established peptides are listed.

ters), which are commonly neuropeptides, may be released from the same nerve terminals together with the primary neurotransmitter.

By the mid-1980s some 30 potent compounds—water-soluble, electrically charged, and mostly amines and peptides but including such unexpected agents as adenosine and ATP—were found to be acting independently or in concert to control the communication between neurons through their points of intimate contact, the chemical synapses, and at their junctions with their various target organs (Table 4.1).

NEUROTRANSMITTER SYSTEMS

What kinds of observation first encourage the suspicion that a given substance is operating chemical synapses? Early clues may be the potency of the substance when added in small quantities to some biodynamic test system involving muscular contraction, glandular secretion, blood pressure control, or direct modulation of neuronal activity as measured by recording electrodes during small localized additions (iontophoresis). The starting point may be the purified substance itself. Other investigations start with a mixture of substances released from the nerve or organ and collected in a superfusate during induced nerve activity. The compounds are then separated and characterized, and they form a shortlist of candidates that may be serving the neurotransmitter function.

More recent strategems have been to use radiolabeled forms of the neurotransmitter candidates to investigate whether populations of specific receptors exist in the nervous system that will bind the compounds.

Neurotransmitter receptors are transmembrane proteins found at post- or presynaptic sites that very specifically bind a neurotransmitter, and many of its related drugs, at low concentrations (high-affinity binding) and convert this binding into a transmembrane ionic current or enzyme activity that is communicated to the interior of the cell. The radiolabeling of drugs known to interfere with nervous system activity may be able to reveal the existence of an unsuspected neurotransmitter system by binding to its receptors. The use of radiolabeled morphine led to the isolation of the endogenous opioid peptides. Radiolabeled benzodiazepine neuroleptic drugs are being used to detect other possible novel neurotransmitter systems by following a similar strategy.

Identification Criteria

How does one decide whether a given substance is really functioning as a neurotransmitter in the nervous system and is not, instead, causing excitation or inhibition through some artifactual membrane effect? Over the years, neuroscientists have formalized their approaches to this question and have developed the following set of requirements that must be satisfied before a substance can be considered a neurotransmitter. Where most of the requirements are met, but not all, the substance is given the status of *putative* neurotransmitter.

1 The substance is shown to be present in the neurons from which it is released.

2 Enzymes capable of synthesizing the substance are demonstrated in those same neurons.

3 Precursors and any compounds forming part of a biosynthesis route are present.

4 Calcium-dependent release of the substance to extracellular fluids is shown to occur during action potential discharge by the neuron or during local depolarization of its axon terminals.

5 The tissue local to the neuron in question possesses mechanisms for the inactivation of the proposed neurotransmitter once released.

6 The substance mimics exactly the postsynaptic action of the synaptically released neurotransmitter when added to the region of the synapse. This means that it duplicates the effect of stimulating the nerve pathway in question.

7 Pharmacological agents, such as antagonists, that interact with the synaptically released neurotransmitter also interact, in an identical manner, with the added substance.

8 Specific receptors for the substance are demonstrated in the tissue region containing the synapse.

Whether all the requirements are *rigorously* required has been much debated,[7] but when they *are* all satisfied a clear-cut, positive case emerges that the substance in question is functioning as a neurotransmitter. The eight requirements can be abbreviated as (1) Presence, (2,3) Synthesis, (4) Release, (5) Inactivation, (6,7) Identity of Action, and (8) Reception.

In historical terms, accurate chemical detection of putative neurotransmitters has usually followed some time after their detection by highly sensitive dynamic bioassays, usually a tissue preparation that includes contracting muscles of the body wall, gut, gland, blood vessel, or other structure. These preparations contain synapses and, therefore, postsynaptic receptors that are often known to be naturally operated by the substance being bioassayed. This knowledge enables the investigator to use known antagonistic drugs to ensure specificity in identifying the action of the substance concerned. An important technique today is the histochemical localization of transmitter candidates in particular neuron groups (nuclei) or nerve-fiber tracts by the formation in situ of fluorescent derivatives of neurotransmitters (e.g., catecholamines and indoleamines). Alternatively, fluorescent or radiolabeled antibodies to neurotransmitters[269,282–284] or peptides are used, or antibodies are labeled with other visualizing agents by the "sandwich" technique (Figure 4.2). This can be complemented by the use of electron-microscopic autoradiographic localization of the radiolabeled neurotransmitters themselves, which have been taken up into their specific nerve terminals by high-affinity transport after their addition to the tissue preparation in low concentrations (e.g., $0.1–5 \ \mu M$). This is based on the important concept that only those neurons releasing a given neurotransmitter have high-affinity transport systems for its uptake. Another approach is to use labeled antibodies to neurotransmitter-synthesizing enzymes since these are specifically located in the neurons that release the neurotransmitter (e.g., choline acetylase for acetylcholine, glutamate decarboxylase for GABA). And because lesion of neural pathways leads to degeneration of the nerve terminals projected by that pathway, a diminution in the amount of neurotransmitter in the group of nerve cells receiving the projection indicates the presence of this neurotransmitter in the projecting fibers. In practice, of course, a sizable proportion of the neurotransmitter candidate present in the target structure must be lost for the diminution to be detected.

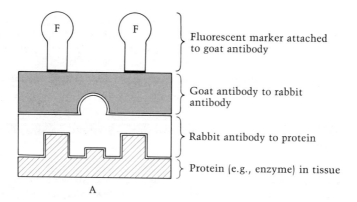

Fluorescent marker attached
to goat antibody

Goat antibody to rabbit
antibody

Rabbit antibody to protein

Protein (e.g., enzyme) in tissue

A

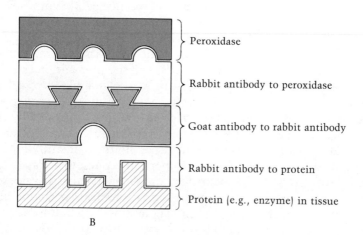

Peroxidase

Rabbit antibody to peroxidase

Goat antibody to rabbit antibody

Rabbit antibody to protein

Protein (e.g., enzyme) in tissue

B

FIGURE 4.2   Diagram showing the "sandwich" technique for histochemical labeling. (A) Immunofluorescence labeling. (B) Peroxidase labeling. Peroxidase is visualized by allowing it to produce O_2 from hydrogen peroxide in the presence of diaminobenzidine, which oxidizes to produce a dense brown reaction product. Each layer of the sandwich is applied in sequence to the tissue section.

Demonstrating the release of a proposed neurotransmitter during nerve activity, or during application of depolarizing stimuli (chemical or electrical), is most effectively performed using isolated preparations such as ganglia, brain slices, or synaptosomes (see Chapter 6), though complementary evidence obtained by washing local exposed regions of cerebral cortex of intact whole brain or placing concentric-barreled "push-pull" cannulas in deeper brain structures greatly adds to the validity of the data obtained with in vitro preparations. Sometimes it is possible to stimulate either the in vivo or the in vitro preparations by activation of neural pathways synapsing with the region being superfused. In both cases, demonstrations of a dependence on Ca^{2+} and a selectivity in the pattern of release are firm requirements, since degeneration of a preparation, reduced energy-metabolism, or membrane damage leading to increased permeability can result in nonspecific release of substances stored in high concentration in the tissue.

Simple biochemical methods for demonstrating specific receptor-ligand binding are available, and demonstration of the existence of a saturable, and therefore finite, population of specific receptors for the substance in question completes the picture of its synaptic interaction. In addition, autoradiography employing the appropriate tritiated ligand at high specific radioactivity can provide valuable maps of receptor localization.[210]

Inactivation may be either by rapid enzymatic breakdown of the substance or by its rapid removal from the synapse or extracellular space and its return to the cell interior by the very active "high-affinity" (i.e., $Km = 1-20 \ \mu M$) uptake transport systems specifically found in those neurons releasing the substance and also in the local glial cells. This latter mechanism appears to be the method of inactivation preferred by most transmitter systems. In fact, only acetylcholine and the new family of peptide neurotransmitters appear to be functionally inactivated by enzymatic conversion in the synapse (or other local regions of the extracellular space) after their release.

Second Messengers and Ionophores

In all cases, once neurotransmitters are released into the synapse they bind to their specific receptors on the postsynaptic target cell. There they exert a brief and decisive action, communicating as either an inhibitory or excitatory message. This signal is usually transmitted across the membrane as an "electrical message" in the form of a membrane potential change occurring as a result of activation of selective ion "gates" or "channels" (ionophores), which apparently either exist in some as yet undefined close physical association with the receptor or are formed by the receptor subunits themselves. Sometimes a "chemical message" is sent into the interior of the postsynaptic cell instead of an electrical potential change. In this case an enzyme system is usually activated after the neurotransmitter binds to its receptor, and the product of this enzyme activity is the message. (See Figure 4.3.) So far, we know of two such systems in which cyclic nucleotides are involved, adenylate cyclase and guanylate cyclase, which produce cyclic 3',5'-AMP and cyclic 3',5'-GMP as the *second messengers* respectively. Each of these cyclic nucleotides activates specific protein kinases by causing their self-phosphorylation (autophosphorylation). These activated enzymes then phosphorylate other proteins on the inner face of the postsynaptic cell. The proteins, once changed in this way, most likely alter membrane permeability to specific ions and create a "delayed" electrical signal. They may also affect the rate of ion transport through specific ion pumps (electrogenic pumps) and thereby cause a change in local membrane potential. The actions of

FIGURE 4.3   Scheme showing the proposed coupling between neurotransmitters or hormones and adenylate cyclase via guanine nucleotide binding proteins. Key: T, neurotransmitter or hormone; R, receptor protein; N, guanine nucleotide binding protein; C, catalytic subunit of adenyl cyclase. Subscript s indicates a stimulatory action; subscript i indicates an inhibitory action. The coupling between the three membrane components is shown as reversible. The binding of GTP by N enhances the interaction, which can result in either stimulation or inhibition of adenylate cyclase, as shown, depending on the category of N involved. GTP is removed by its hydrolysis.

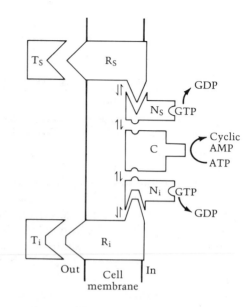

these second messengers appears to be terminated by their enzymatic destruction. Thus, phosphodiesterases open their cyclic phosphoric acid residues, rendering them inactive. A phosphatase is believed to rapidly remove the phosphate residue from the protein, thus terminating the change in membrane property initiated via the second messenger. The phosphatase dephosphorylates, and thereby inactivates, the protein kinase.

Synaptic transmission at synapses involving second messengers of this type is much slower, taking, perhaps, 100 milliseconds or longer compared with the few milliseconds at synapses where ion channels are activated directly. The precise sequence of events involved in these second-messenger systems remains largely speculative, but considerable evidence has now accumulated that supports the overall concept depicted in Figure 4.4. Specific protein substrates for cyclic nucleotide-mediated phosphorylation have been isolated or detected[178,209,244,245] (called proteins Ia, Ib, and II for the cyclic AMP system), and they appear to be localized to neurons, where they are associated with postsynaptic structures.

The two cyclic nucleotide systems show significant differences. For instance, cyclic 3',5'-AMP is found ubiquitously throughout the nervous system, whereas cyclic 3',5'-GMP shows marked regional localization (e.g., cerebellum, corpus striatum, neocortex, superior cervical ganglion) and is present at one or two orders of magnitude lower concentrations than its adenosine counterpart.[179] Cerebellar Purkinje cells are particularly enriched in components of the cyclic 3',5'-GMP system, including the cyclic nucleotide, the cyclase enzyme, and the specific protein kinase (EC 2.7.1.38).[180,181] Another difference is the localization of the cyclase

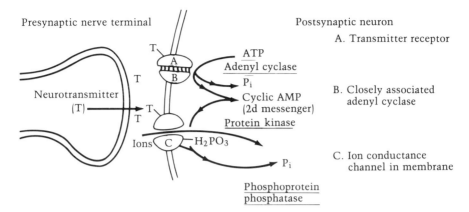

Presynaptic nerve terminal

Postsynaptic neuron

A. Transmitter receptor

Neurotransmitter (T)

B. Closely associated adenyl cyclase

ATP
Adenyl cyclase
$\overline{P_i}$
Cyclic AMP (2d messenger)
Protein kinase

Ions C H_2PO_3

C. Ion conductance channel in membrane

P_i

Phosphoprotein phosphatase

FIGURE 4.4   The second-messenger concept. Neurotransmitter receptor (A) interacts with neurotransmitter (T) and initiates cyclic nucleotide formation via a cyclase enzyme (B). The cyclic nucleotide is the second messenger. It phosphorylates a membrane protein via a protein kinase. This protein is closely linked to an ion conductance channel (C) leading to a membrane potential change (synaptic potential). This effect is reversed by a phosphoprotein phosphatase, which dephosphorylates the protein kinase and substrate protein, and by a phosphodiesterase which inactivates the cyclic nucleotide. Coupling between A and B may involve guanylnucleotide binding protein (see Figure 4.4).

enzymes. Adenylate cyclase is found almost exclusively in the plasma membrane of the postsynaptic cells, whereas guanylate cyclase is found both in membranes and in soluble cytoplasm of this region.[180,182] Neuroglia have also been reported to contain guanylate cyclase.[182] Whereas cell-free membrane preparations readily show adenylate cyclase responses to neurotransmitters and hormones, this is not so for guanyl cyclase, which also requires Ca^{2+} for its response in various intact tissues.[178] For these and other reasons, it may be that the cyclic 3',5'-GMP second-messenger system is not only more localized but is also qualitatively different in organization from its adenosine counterpart. The Ca^{2+} influx due to neurotransmitter or hormone action could instead be the primary event, which then leads to stimulation of guanylate cyclase. Levels of intracellular unbound Ca^{2+} may be influenced, in turn, by cyclic 3',5'-GMP. A slowing of sequestration of free Ca^{2+} would prolong a physiological response.[178]

A great deal is known about the specific protein kinase (EC 2.7.1.37) activated by cyclic AMP. In most tissues it consists of two isoenzymic forms, each containing two regulatory subunits and one catalytic unit.[183,244,245] These two forms differ with respect to physical properties, activation by cyclic AMP, autophosphorylation, and immunoreactivity.

The actions of these cyclic nucleotide second messengers seem manifold. In addition to regulating protein phosphorylation, changing membrane permeability to specific ions, and controlling intracellular levels of

Ca^{2+}, they can also influence rates of neurotransmitter synthesis, effect genetic expression at the transcriptional level, regulate microtubule function, and accelerate the carbohydrate and lipid metabolism that supply the necessary cellular energy for the physiological responses.[178] Thus, these cyclic nucleotide systems can influence biochemical events that form the basis of longer-term trophic responses induced in target tissues by neural activity.[244,245]

Calcium itself may be viewed as a second messenger in the presynaptic or postsynaptic region. It can accelerate protein phosphorylation through the action of Ca^{2+}-dependent protein kinases and it can regulate cyclic AMP metabolism. Moreover, in many tissues Ca^{2+} and cyclic AMP are able to interact at the level of protein phosphorylation, since each activates distinct protein kinases which can then phosphorylate the same substrate protein. In addition, Ca^{2+} modifies the activity of the cyclic-3',5'-AMP-dependent protein kinase.[178]

Another likely second-messenger system whose detailed mechanisms are yet to be defined involves the inositol phospholipids (phosphatidylinositol and polyphosphoinositides). This concept derives from the observation that stimulation of inositol lipid metabolism is a universal cellular correlate of the activation of receptors whose principal effect is to elevate cytosolic Ca^{2+} levels in the target cells. A phosphodiesterase (phospholipase C) initiates the inositol lipid breakdown with release of inositol phosphates. This is closely associated with a rise in intracellular cytosolic Ca^{2+} levels,[184–188] which can then trigger a diverse range of responses. It has not been unequivocally determined whether hydrolysis of inositol phospholipids precedes rather than results from Ca^{2+} mobilization; the problem is complicated by the wide variation in Ca^{2+} sensitivity of the inositol lipid response, depending on the tissue examined.[184] In the brain, different receptor systems mediating the response (e.g., norepinephrine, serotonin, histamine, carbechol) have quantitatively very different Ca^{2+} requirements, and rigid views of the cause or effect relationships involved are probably inappropriate at present.[186]

Enhanced, receptor-mediated breakdown and turnover of inositol lipids with release of soluble inositol phosphates, including inositol trisphosphate, appears to constitute another second messenger system that probably operates through the agency of Ca^{2+} mobilization in a wide variety of tissues.[265] Entry of inositol trisphosphate (but not other inositol phosphates) into permeabilized cells (e.g., pancreatic acinar cells or hepatocytes) leads to mobilization of intracellular Ca^{2+} stores, probably largely from endoplasmic reticulum.[189,190] It will also release Ca^{2+} from endoplasmic reticulum isolated from hepatocytes.[189] But the time course of formation of inositol trisphosphate and its mobilization of Ca^{2+} stores, about 1 to 3 seconds,[190] appears to be far too slow to implicate it in presynaptic mechanisms concerned with neurotransmitter release.[191] These are estimated to take only 200 microseconds or less between Ca^{2+} entry and first detectable neurotransmitter release (e.g., at the squid giant synapse; see Chapter 7).

THE CHOLINERGIC NEUROTRANSMITTER SYSTEM

All neural pathways using acetylcholine as their chemical transmitter are called *cholinergic*. Acetylcholine was the first chemical neurotransmitter isolated and characterized in both structure and function, and many basic features of neurotransmission systems were first investigated using cholinergic systems—particularly that most famous and widespread of synapses, the vertebrate neuromuscular junction.

Synthesis and Storage of Acetylcholine

Simple, one-step synthesis of acetylcholine from its two components acetate and choline is achieved by the important enzyme choline acetyl transferase (ChAT; EC 2.3.1.6). The choline is acetylated by acetyl CoA (Figure 4.5), mostly in nerve terminals rather than in other neuronal regions. The enzyme ChAT appears to be synthesized in the cell body and transported by axoplasmic flow (see Chapter 1) to the terminals, where it becomes active. A specific inhibitor of the enzyme is 4-napthylvinyl pyridine.

Choline is a ubiquitous substance. Primarily made in the liver, it is transported to other organs in the blood. It is a component of several common phospholipids (e.g., lecithin, sphingomyelin) found in large quantities in the cell membranes of all tissues, including nerve terminal membranes. Free choline is specifically taken up into cholinergic nerve terminals by a "high-affinity" (K_m = 0.4–4.0 μM) pump. The choline is present in the extracellular space as a result of external breakdown of previously released acetylcholine and as a result of its supply and release during the synthesis and breakdown of phospholipids. In addition, a "low-affinity" (K_m = 40–100 μM) choline uptake system appears to be present in all neurons and even some glia. The acetate moiety is derived from glucose via pyruvate and the mitochondrial pyruvate dehydrogenase complex that generates acetyl CoA.[c] Deficiencies of glucose or other cofactors essential for acetyl CoA formation (e.g., thiamine pyrophosphate) can limit acetylcholine production with consequent deficiencies in cholinergic transmission.[8] Similar effects result from restricted availability of choline in blood and cerebrospinal fluid, since choline entry to

$$CH_3 \cdot CO \cdot S\ CoA + OH \cdot CH_2 \cdot CH_2 \cdot N^+(CH_3)_3 \xrightarrow{\text{ChAT}}$$
Acetyl coenzyme A   Choline

$$CH_3 \cdot CO \cdot O \cdot CH_2 \cdot CH_2 \cdot N^+(CH_3)_3 + CoASH$$
Acetylcholine   acetyl coenzyme A

FIGURE 4.5 Enzymatic synthesis of acetylcholine: the choline acetylase reaction (ChAT).

the nerve terminal appears to be rate-limiting in acetylcholine synthesis.[9,C] Choline uptake is specifically inhibited by the drug hemicholinium-3.

Choline acetyl transferase has been isolated in a highly purified form from many mammalian brain regions and also from squid head ganglia.[C,10,11] It is a globular protein with a molecular weight of 67,000 in humans[12] and 89,000 in ox[13] (808 amino acid residues), and it carries a considerable net positive charge. Antisera raised to purified ChAT preparations have been used with striking success for histochemical localization of cholinergic neurons.[11,192] In order to render them visible, the ChAT antibodies are labeled with a fluorescent marker or with peroxidase, by the "sandwich" technique (Figure 4.2).

Organization and Control of Acetylcholine Synthesis ChAT can be isolated either in free solution or partially membrane-bound, depending on the species and the ionic strength of the buffers employed. The positive charge of the enzyme allows it to bind to the negatively charged surface of membranes. In the intact brain the enzyme probably exists in the terminal, predominantly in free solution in the cytosol but with some binding to mitochondria, synaptic vesicles, and the inner surface of the plasma membrane. Acetylcholine itself is found both in synaptic vesicles and in the cytosol. After the release of acetylcholine during nerve activity, new or "replacement" acetylcholine is synthesized and the rate of this synthesis is controlled by the availability of the substrates choline and acetyl CoA at the site of synthesis.[C] For this reason, the high-affinity uptake of choline can control acetylcholine synthesis. The fact that the high-affinity choline pump is inhibited by excess acetylcholine and accelerated by low levels of acetylcholine adds to its suitability as a control point for acetylcholine synthesis.[9]

Further control may be exercised by an inhibitory action of acetylcholine on ChAT activity, and through an action of Ca^{2+} in accelerating release of acetyl CoA from mitochondria following its entry during neural activity.

Acetylcholine Storage Both synaptoplasm and synaptic vesicles contain acetylcholine. Vesicle-bound acetylcholine (e.g., 60 percent) is not accessible to attack by acetylcholinesterase when the terminals are ruptured and their content is released, whereas the transmitter in the soluble fraction is readily attacked by this enzyme. Whether synthesis occurs in the synaptoplasm before transfer to vesicles, or whether ChAT bound to the vesicles fills these directly, is not clear. What does seem to be a consistent finding is that "newly synthesized"—that is, the most recently synthesized—acetylcholine is always the first to be released during nerve activity. This means that a "last in, first out" arrangement in the storage of acetylcholine[C] must be taken into account when models of transmitter release mechanisms are constructed (Chapter 7). Another feature of cholinergic synaptic vesicles isolated from the electric ray *Torpedo marmorata* is their content of ATP and a negatively charged mucopolysaccharide called *vesiculin,* a glycosaminoglycan with a molecular weight of 10,000. It has been characterized and is postulated to serve as a counter-ion for the positively charged acetylcholine and ATP, which exist in free solution at a total concentration of about 1 M.[15,267]

Release of Acetylcholine

A wide range of depolarizing agents—substances causing displacement of cell membrane potentials—induce acetylcholine release from a variety of neural preparations by mechanisms that require the presence of Ca^{2+} (Table 4.2). The preparations include isolated ganglia, brain slices, and synaptosomes incubated in vitro and stimulated with electrical pulses, high K^+ levels,[163] and veratridine, an agent that opens active Na^+ channels. Release from the cerebral cortex can be induced by depolarizing agents and by activation of cholinergic synapses by stimulating normal physiological neural pathways.[16] Acetylcholine release is usually followed by onset of its resynthesis in the tissue to replenish stores, thus demonstrating the control exerted over the synthetic machinery, probably by the choline high-affinity uptake system.[C] Botulinum toxin, a very potent poison produced by the bacterium *Clostridium botulinum,* appears to specifically prevent acetylcholine release.[14]

Inactivation of Released Acetylcholine Once released into the synaptic cleft, acetylcholine binds briefly to its postsynaptic receptors before being

TABLE 4.2   Acetylcholine in synaptosome beds and incubation medium after electrical and potassium stimulation

	Acetylcholine content (nmol/100 mg protein)	
	Synaptosome bed	Incubation medium
Calcium medium		
Control	60.9 ± 1.04 (23)	69.8 ± 2.57 (16)
Potassium stimulation	46.2 ± 2.02 (11)	101.0 ± 4.96 (10)
Electrical stimulation	52.8 ± 2.05 (15)	83.7 ± 5.02 (10)
Calcium-free medium (+EGTA)		
Control	64.2 ± 2.45 (9)	27.5 ± 4.72 (6)
Potassium stimulation	55.8 ± 3.0 (7)	34.9 ± 2.49 (7)
Electrical stimulation	57.7 ± 3.0 (5)	29.6 ± 5.52 (4)

Note: Synaptosome beds were incubated for 40 minutes at 37°C in Krebs-bicarbonate medium (calcium medium) or in Krebs-bicarbonate medium containing no calcium and 0.5 mM EGTA, as indicated. Potassium stimulation was by elevation of medium K^+ by 50 mM 10 minutes before the end of incubation and electrical stimulation by application of electrical pulses for the last 10 minutes of incubation. Values are means ± SEM for the number of experiments indicated in parentheses. Data are from ref. 163.

$$CH_3 \cdot CO \cdot CH_2 \cdot CH_2N^+(CH_3)_3 + H_2O \xrightarrow{\text{AChE}}$$
Acetylcholine

$$CH_3 \cdot COOH + OH \cdot CH_2 \cdot CH_2N^+(CH_3)_3$$
Acetic acid   Choline

FIGURE 4.6   Enzymatic hydrolysis of acetylcholine: the acetylcholines-terase reaction (AChE).

destroyed by the enzyme acetylcholinesterase (AChE), which is concen-trated in the cleft[F] (Figure 4.6).

In modern terminology, *cholinesterases* are enzymes that hydrolyze a wide range of choline esters (e.g., butyryl and propionyl choline). They are found in serum, erythrocytes, placenta, and elsewhere. Acetylcholin-esterase (EC 3.1.1.7), a globular glycoprotein existing in multiple forms, is present in vertebrate nerves, muscle, and erythrocytes and is specific for acetylcholine itself. It is synthesized in the cell body and distributed throughout the neuron by axoplasmic flow. Most of the detailed informa-tion about its form and structure has come from studies of the isolated enzyme after extraction from its richest known sources, the electric or-gans of electric eel (*Electrophorus electricus*) and the electric ray, which consist of stacks of many thousands of modified neuromuscular junc-tions.[17–23] Here, solubilization of the enzyme with high-ionic-strength buffers produces several globular forms, the simplest of which are mono-mers, dimers, and tetramers of catalytic subunits (each monomer of mol wt 80,000). The dimers are linked by a single disulfide bridge, and two dimers form tetramers, probably binding via van der Waals forces. The tetramers are usually joined to three-stranded "tails" by disulfide bond-ing. This 11S tetramer is probably the basic unit of the enzyme (mol wt 330,000). The 50-nm-long tail consists of a collagenlike triple helix made up of peptides of mol wt 40,000–45,000 covalently linked to each other and to the catalytic subunits by disulfide bonds. The high Stokes radius of this molecule, obtained from gel filtration, indicates its gross asymmetry. This asymmetry can be reduced by treatment with collagenase, which removes the tail and at the same time increases the molecule's sedimen-tation rate. The ready association of bundles of these molecules by their tails, due to ionic linkages along their length, produces larger forms such as that shown in Figures 4.7 and 4.8, which contains three units and therefore 12 catalytic subunits each. Ionic interaction via the tails is likely to be the mode of attachment to the collagenous extracellular basal laminar membrane of the neuromuscular junction and electric organ overlaying the sarcolemma itself. This bunch-of-grapes form, with its head and tail, appears to be one of the main structures occurring in the synapses of electric eels, rays, and other vertebrates in spite of the wide evolutionary separation of these groups. Various "tailless" globular forms are also present, but the asymmetric form appears to have special func-tional significance, as its level in the synapse can be modulated according to the degree of neural activity.[19] The precise biological reasons for the

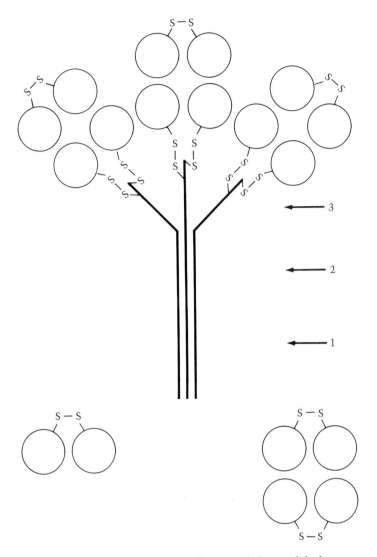

FIGURE 4.7 Structure of the collagen-tailed acetylcholinesterase molecule obtained from *Electrophorus* electric organ. Shown are the disulfide
bonds that link the individual catalytic subunits as dimers and that link
the dimers to the tail strands. Arrows indicate the points of cleavage by
collagenase: position 1 is sensitive to collagenase at 20°C, and positions 2
and 3 at 37°C. Tailless dimers and tetramer are also shown. [From ref. 19]

existence of all these forms of the same enzyme are not clear, but their
arrangement in different kinds of synapses—neuromuscular junctions
and ganglionic or central cholinergic synapses—provide versatility in the
control and efficacy of interaction of acetylcholine with its receptors in
the small space of the synaptic cleft.

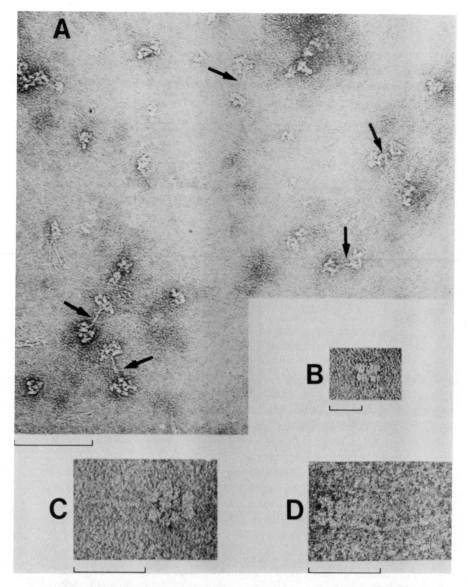

FIGURE 4.8   Electron micrographs obtained by negative staining of various molecular forms of acetylcholinesterase from eel electric organ.
(A) 14S and 18S species, characteristic of highly purified preparations prepared directly from fresh tissue extracts. The field also contains multi-subunit "heads" connected to "tails," and separated "tails". Bar, 0.05 μm. (B) Single molecule of 11S species (tetramer). Bar, 0.01 μm. (C) Single molecule of 8S species. Bar, 0.025 μm. (D) Single molecule of 18S species. Bar, 0.025 μm. [From refs. 17, 20, 23]

The histochemical detection of AChE has provided valuable information on its localization. The now-famous technique of Koelle and Friedenwald[24,25] employs acetyl- or butylthiocholine as substrates for the tissue-bound enzyme, and the cholinesterase and acetylcholinesterases are distinguished by using specific inhibitors for each enzyme type (Table 4.3). The common product of the reaction, thiocholine, is visualized by

TABLE 4.3   Some drugs interacting with cholinergic systems

Agonists (muscarinic)	Acetylcholine Muscarine Carbechol Methacholine Bethanechol Pilocarpine Arecoline Oxotremorine
Agonists (nicotinic)	Acetylcholine [a]Nicotine Carbechol Arecoline [a]Tetramethylammonium (TMA) [a]Phenyltrimethylammonium (PTMA) [a]Dimethylphenylpiperazine (DMPP) Suberyldicholine
Antagonists (muscarinic)	Atropine Scopolamine (hyoscine) [b]Benztropine Quinuclidinylbromide Pirenzipine
Antagonists (nicotinic)	D-Tubocurarine Succinylcholine (depolarizing, desensitizing) Decamethonium (depolarizing, desensitizing) Gallamine [a]Pempidine [a]Mecamylamine [a]Hexamethonium [a]Pentolinium Pancuronium α-Bungarotoxin
Releasing agent	[c]Black widow spider venom
Inhibitor of release	Botulinum toxin
Specific binding agents	α-Bungarotoxin Propylbenzilylcholine mustard Quinuclidinyl benzilate (QNB)
Pump inhibitor	Hemicholinium-3 (HC-3), triethylcholine
Synthesis inhibitor	4-Naphthylvinylpyridine
Cholinesterase inhibitors	Diisopropylphosphofluoridate (DFP) Neostigmine Physostigmine (eserine)

[a] Known principally for their actions at peripheral ganglia.
[b] Also used as a dopamine uptake blocker.
[c] Not specific to cholinergic systems.

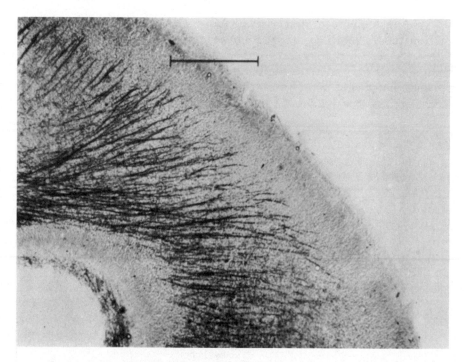

FIGURE 4.9   Acetylcholinesterase-containing fibers from the lenticular nucleus invading the lateral wall of primitive forebrain. Note absence of staining in cortical lamina. Cat embryo at about 40 days. Bar, 0.5 mm. [From ref. 26]

precipitation with lead or copper salts. It is now known that AChE is not *exclusively* associated with cholinergic neurons—for example, it is present on dopaminergic cells of the substantia nigra[219]—but it did provide the first maps of cholinergic systems and their various groupings, fiber tracts, and junctions (Figure 4.9). (Much of the mapping has since been confirmed using ChAT as the specific marker; see below).

Presynaptic Acetylcholine Receptors: Effect on Release   Receptors positioned presynaptically on the terminals of the incoming fibers and able to modulate neurotransmitter release have recently been discovered for many transmitter systems (see Chapter 7), including presynaptic cholinergic receptors in both peripheral and central nervous systems. Such receptors may exist on cholinergic nerve terminals or on the nerve terminals of other transmitter systems, for example, on terminals containing norepinephrine or dopamine. Cholinergic receptors are usually detected through the modulating influences of muscarinic or nicotinic agonists or antagonists (agents affecting cholinergic receptor subclasses; see below) on the extent of transmitter release induced by nerve stimulation or by chemical depolarizing agents. They are also detected by the inhibitory effect of cholinergic antagonists on the direct actions of acetylcholine itself in modulating its own release. Presynaptic cholinergic receptors controlling norepinephrine release have been demonstrated in heart, vascular tissue, spleen, lung, and other viscera.[67,68] In fact, most postganglionic sympathetic neurons are endowed with two presynaptic receptor systems activated by cholinergic drugs. Nicotinic agonists depolarize nerve endings and *evoke* a Ca^{2+}-dependent release of norepinephrine; muscarinic antagonists *inhibit* this release as well as the release induced by depolarizing agents and nerve stimulation (see Chapter 7).

Cholinergic presynaptic receptors controlling catecholamine release also occur in the central nervous system.[68,70,87,88] They have been demonstrated in brain slices, caudate nucleus, cerebellum, and elsewhere. The presynaptic location of such receptors is either demonstrated indirectly through their loss after sectioning of the fibers of the neural pathway proposed to bear them or more directly by the actions of drugs on transmitter release from isolated nerve terminals.

Pharmacology of Cholinergic Synapses

From the earliest days of research on synapses operated by acetylcholine, it was recognized that subcategories exist. They were distinguished at the time by their different responses to two available drugs, nicotine, a substituted pyrrolidine compound extracted from the tobacco plant (*Nicotiana tabacum*), and muscarine, a tetrahydrofuran-substituted tertiary amine extracted from the fly agaric mushroom (*Amanita muscaria*). Both are of low molecular weight, 162 and 174 respectively. Each agent *stimulates* different categories of cholinergic function, termed *nicotinic* and *muscarinic*, which are mediated by different receptor subcategories. The nicotinic type, characterized by its fast response time (1–2 milliseconds), produces its effects by directly depolarizing the postsynaptic membrane by activating Na^+ channels. The muscarinic type is slower in response and appears to operate via a second messenger, at least at some sites (e.g., the superior cervical sympathetic ganglion).[178] In this case the messenger is cyclic GMP[27] rather than cyclic AMP.

Nicotinic cholinergic synapses operate in vertebrate neuromuscular junctions, certain ganglia, central synapses, and in the electric organs of electric fish. Muscarinic synapses are found in smooth muscle, cardiac muscle, ganglia, and many central brain regions. Within the brain and central nervous system, the muscarinic type outnumbers the nicotinic type by a factor between 10 and 100. Because of differences in organization of the postsynaptic receptor systems in the two types, they are blocked as well as activated by different chemical structures. Atropine, quinuclidinyl benzilate, and a number of other drugs (see Table 4.3) specifically block muscarinic cholinergic synapses; *d*-tubocurarine (curare), hexamethonium, and α-bungarotoxin act on nicotinic receptors. These agents are very useful in distinguishing the two types of acetylcholine-operated neural inputs. General cholinergic agonists acting on both sub-categories of cholinergic receptors include carbechol, bethanechol, areco-line, pilocarpine, methacholine, and oxotremorine, although each usually show a muscarinic or nicotinic bias in its actions. (Table 4.3; Figures 8.4 and 8.5).

Postsynaptic Acetylcholine Receptors

A great deal of progress in understanding the detailed biophysical and structural aspects of the mode of action of acetylcholine in its synapses has come from the successful isolation of its postsynaptic receptors. The nicotinic acetylcholine receptor has been isolated from the electric or-gans of electric fish and from the vertebrate neuromuscular junction. The procedures employed have pioneered the way for the isolation and char-acterization of the receptor proteins of other neurotransmitter systems.

The isolation procedure has mainly involved purification from deter-gent extracts of electric organs or mammalian muscle, using affinity chromatography with either cholinergic ligands or purified snake-venom neurotoxins as the affinity agents.[37] One such agent is α-bungarotoxin, the α-neurotoxin from the venom of the elapid snake *Bungarus multi-cinctus*. Another is α-cobrotoxin from cobra venom. The muscle tissue or electric-organ tissue is first exposed to radiolabeled (^{125}I or ^{14}C acetylated) α-bungarotoxin before being solubilized with detergents.[28,29,266]

The receptor prepared from electric organs by slightly different meth-ods has an overall molecular weight in the region of 250,000,[28,29] correct-ing for bound detergent giving rise to variability. A range of subunits of differing molecular weights can be isolated from these preparations. Pro-teolytic attack of the various subunit proteins during isolation has led to controversy about the precise size of the subunits constituting the recep-tor in situ. However, with maximal inhibition of proteases, four different subunits with apparent molecular weights of 40,000, 50,000, 60,000, and 65,000 have been isolated and termed the alpha, beta, gamma, and delta subunits respectively. They are present in the receptor complex in the ratio $2:1:1:1$. Each subunit bears a carbohydrate residue and shows striking amino-acid-sequence homologies with the other subunits. The receptor from mammalian muscle appears to contain subunits of similar size and in the same ratio as exist in the electroplax[28,29,30,42] of *Torpedo*.

Recently, progress has been made on establishing the amino acid sequence of the subunits of *Torpedo* acetylcholine receptors by gene-cloning techniques following isolation of the receptor messenger RNA. This is the first application of recombinant DNA technology to a receptor protein. The sequence of a 24-residue signal peptide of the type required for insertion of the protein into the membrane and the full sequences of all of the subunits have been established.[33,34,35] Following microinjection of the receptor mRNA preparation from *Torpedo* into *Xenopus* oocytes, the receptor protein molecules appear to be formed and inserted into the oocyte membrane to produce a functional ion channel, as evidenced by the appearance of responses to topically applied acetylcholine.[32] These experiments indicate that the oocyte can provide the special posttranslational processing required for the assembly of the multisubunit membrane-bound receptor. In *Torpedo* the receptor occurs in the light form $(\alpha_2\beta\gamma\delta)$ or as a heavy-form dimer resulting from the association of two light forms by $\delta-\delta$ disulfide bridge formation. There is an associated structural membrane protein of mol wt 43,000 (Figure 4.11).

Electron micrographs of purified receptor populations show a hollow cylindrical structure with walls composed of five globular structures (Figure 4.10). Current knowledge has allowed the construction of fairly detailed hypothetical models of the *Torpedo* receptor (Figure 4.11). In these models the receptor, about 9.0 nm in diameter, typically consists of the four different subunits in the sequence $\alpha\gamma\alpha\beta\delta$,[221] grouped around a central channel 1.5–2.5 nm in diameter that transverses the membrane. The receptor is about 11 nm long, and tryptic digestion studies indicate that it projects some 5.0 nm above the membrane on the synaptic face and 2 nm above the surface of the cytoplasmic face.[29] The central channel is called the *ion conductance modulator* (ICM), the theoretical pathway through which specific ions are shunted following neurotransmitter binding. This ICM appears to be formed by amphipathic chains contributed by each of the four subunits $(\alpha, \beta, \gamma, \text{and } \delta)$ constituting the basic receptor structure,[222] at least in *Torpedo*.[29] The receptor is clearly a complex glycoprotein structure whose hydrophobic surface allows it to penetrate right across the postsynaptic membrane.

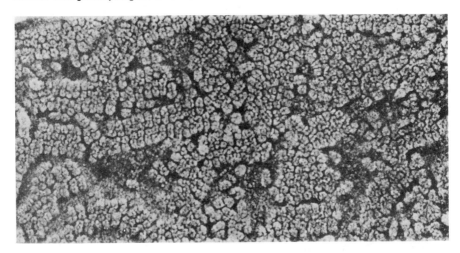

FIGURE 4.10 Acetylcholine-activated channels densely packed in the postsynaptic membrane of a cell in the electric organ of *Torpedo*. This electron micrograph shows the platinum-plated replica of a membrane that has been frozen and etched. The size of the platinum particles limits the resolution to features larger than about 2 nm. [From ref. 176; courtesy of J. Heuser and S. R. Salpeter]

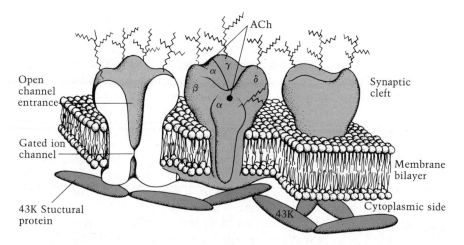

FIGURE 4.11   Three-dimensional models of the nicotinic acetylcholine receptor from *Torpedo californica* depicted as integral membrane proteins. The subunit assignments around the central channel are tentative. The sites on the two α subunits that bind acetylcholine, other cholinergic ligands, and snake-venom α-neurotoxins are shown as dark patches. The proposed shape of the central channel can be seen in the vertical section. Also shown is the membrane structural protein of mol wt 43,000 often found in association with the receptor in *Torpedo*. [From ref. 226]

Although the four subunits possess a large degree of sequence homology, they are not equivalent. They probably evolved separately to perform discrete functions, such as receptor activation, inactivation, and desensitization.[29,224] Each alpha subunit, for example, has a high-affinity binding site for acetylcholine, δ-tubocurarine, other cholinergic ligands, and snake venom α-toxins. These sites are not equivalent, however, as shown by differential binding of affinity labels or reducing agents. This is probably due to a difference in their glycosylation.[224] Other classes of effectors act primarily elsewhere on the receptor. They include the noncompetitive inhibitors (a diverse collection of local anesthetics), phencyclidine, chlorpromazine, and other agents.[223] A unique high-affinity site for noncompetitive inhibitors appears to involve binding to several subunits. It is located within the central depression of the receptor, where the distances to all five polypeptide chains are at a minimum. Multiple low-affinity sites for noncompetitive inhibitors are located at the lipid–protein interface.[224] These sites and the acetylcholine-binding sites interact with each other in an allosteric fashion.

When acetylcholine binds to the alpha subunit, a change in the conformation of the proteins leads to noncooperative binding of another acetylcholine molecule to the second alpha subunit and thence to an opening of the associated ion-channel (the ICM) for a brief period to allow the flow of a specific ion (e.g., Na^+) down its diffusion gradient. The binding of two acetylcholine molecules and the conformational change in both subunits is essential before channel opening can occur.[225] These rapid protein conformational changes have been studied by following the pattern of fluorescence changes accompanying them[29,224] during the application of compounds that enhance or block the physiological response to acetylcholine. The postsynaptic membranes in the *Torpedo* electric organ appear to consist almost entirely of closely packed receptors (see Figure 4.10). Such preparations have been used to follow the conformational changes that occur during acetylcholine binding and the subsequent activation of the ion conductance modulator. This has been achieved with the use of fluorescence spectroscopy. Agents like quinacrine (a fluorescent local anesthetic),[31] quinacrine mustard, alkyl guanidines, and even the intrinsic fluorescence of the receptor itself (owing to its tryptophan content)[31] have been employed for this purpose.[221,223,224] Also used is *histrionicatoxin,* an alkaloid from the skin of the Colombian frog *Dendrobates histrionicus;* it shows high-affinity binding ($K_D = 0.4$ μM), and at low concentrations (1 μM) it blocks Na^+ ion conductance (i.e., prevents depolarization) in the fish electric organ.[31,224]

How local anesthetics, histrionicatoxin, and alkyl guanidines exert their block on the physiological response to acetylcholine remains uncertain. They probably work either by attaching to the cholinergic ligand binding sites on the receptor subunits and physically blocking the channel or by modifying the properties of the membrane and enhancing the rate of desensitization.

It has proved possible to reconstitute solubilized and purified acetylcholine receptors in lipid vesicle preparations that respond to cholinergic agonists (e.g., carbamyl choline) by showing increased ^{22}Na permeability. This response can be blocked by cholinergic antagonists and by monoclonal antibodies to the receptor.[226] Studies with reconstituted receptors from *Torpedo* have shown ion flux rates remarkably close to those previously determined for *Torpedo* membrane preparations and those calculated from single-channel conductance measurements.[29] Such valuable preparations will allow more detailed studies of the molecular mechanisms involved in the ion gating process of the acetylcholine receptor.

Charged phospholipids (phosphatidylinositide, phosphatidylserine, and phosphatidic acid) may also serve a functional role in the receptor complex in the membrane. This question and other details of its molecular organization are among the important problems that may be solved during the 1980s.

Location of Neural Pathways Employing Acetylcholine

Detecting the presence of cholinergic pathways and determining their routes can be achieved by mapping for one or more of the various specific components of the cholinergic system (e.g., ChAT, AChE, and acetylcholine receptors; see Figure 4.9) and testing for sensitivity to iontophoresed acetylcholine, or by looking for a diminution in acetylcholine content in tissue following fiber transection. Of course one must carefully distinguish between the cell bodies and the often more distant nerve terminals of cholinergic neurons. The presence of receptors and a sensitivity to applied acetylcholine usually indicates a cholinergic input (i.e., projection) from cell bodies possibly located in other regions of the nervous system.[284]

These various approaches have shown that acetylcholine-secreting neurons provide many excitatory pathways throughout the central and peripheral systems (and also some inhibitory pathways).[G] The large-sized *mononeurons* of the ventral horn of the spinal cord drive the voluntary musculature of the vertebrate body via the nicotinic cholinergic neuromuscular junction. Within the brain itself, many cholinergic synapses have been localized by iontophoresis to the deeper layers of the cerebral cortex, Brodman layers IV, V, and VI, with some variation from region to region.[G]

There are a number of small intrinsic bipolar cholinergic interneurons in the cerebral cortex, but most of the cortical cholinergic innervation is extrinsic, originating in the subcortical cholinergic neurons of the basal forebrain (substantia innominata, nucleus basalis) and in the reticular formation[192–194] (Figure 4.12). The telencephalon has a continuous

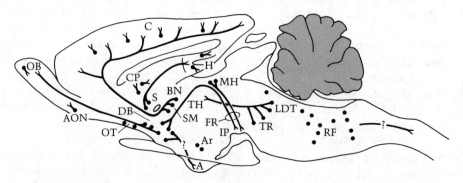

FIGURE 4.12   Schematic representation of the principal cholinergic pathways in rat brain as seen in saggital section. Key: *A*, amygdala; *AON*, anterior olfactory nucleus; *Ar*, arcuate nucleus; *BN*, nucleus basalis; *C*, cerebral cortex; *CP*, caudate putamen; *DB*, nucleus of the diagonal band; *FR*, fasiculus retroflexus; *H*, hippocampus; *IP*, nucleus interpeduncularis; *LDT*, lateral dorsal tegumental nucleus; *MH*, medial habenula; *OB*, olfactory bulb; *OT*, olfactory tubercle; *RF*, hindbrain reticular formation; *S*, septum; *SM*, stria medullaris; *TH*, thalamus; *TR*, tegumental reticular system. For clarity, classical motor and autonomic preganglionic neurons are not shown. [From ref. 192]

stream of cholinergic neurons occupying the head of the caudate puta-men and continuing into the nucleus accumbens. It leads to all compo-nents of the diagonal band and medial septum. In the rodent, the globus pallidus, itself conspicuously free of cholinergic neurons, bears a rim of large cholinergic cell bodies. This region, the equivalent of the nucleus basalis of primates, provides the major ascending cholinergic projection to the neocortex. In primates the nucleus basalis and substantia innomi-nate are by far the most extensive and well-differentiated cholinergic cell group. There are all together some six major cholinergic cells groups pro-viding cholinergic tracts in the rat brain.[194] These include projections to the hippocampus from the medial septal nucleus and from the nuclei of the diagonal band, which also projects to the olfactory bulb. The thala-mus receives fibers projecting from the pontomesencephalic reticular for-mation (Figure 4.12). In general, these pathways all have their counter-parts in brains at higher phylogenetic levels of development,[192–194] such as the brains of primates.

THE CATECHOLAMINE AND INDOLEAMINE NEUROTRANSMITTER SYSTEMS

We recall that historically, close on the heels of acetylcholine came *nor-epinephrine*,[3,4] *dopamine*,[38] and *serotonin*[5] in the development of our knowledge of chemical agents operating neuronal synapses (see page 155). Various terms are used to refer to this group of neurotransmitters, includ-ing *monoamines* and *biogenic amines*. Norepinephrine, epinephrine, and dopamine are all closely related *catecholamines*, while serotonin is an *indoleamine* (Figures 4.13 and 4.15).

Synthesis and Storage of Norepinephrine, Dopamine, and Serotonin

The pathways for synthesis of norepinephrine and dopamine follow the same route from tyrosine, dopamine being in fact the immediate precur-sor[38] of norepinephrine (Figure 4.13). The enzyme dopamine β-hydroxy-lase (EC 1.14.17.1) is present only in noradrenergic neurons. Tyrosine itself is available in tissue pools and is also transported into the brain from the bloodstream across the blood-brain barrier with relative ease owing to its hydrophobic nature. It enters via the neutral amino acid transport system, which also transports tryptophan and a range of aliphatic neutral amino acids.

Within catecholaminergic neurons tyrosine is converted to 3,4-dihy-droxyphenylalanine (dopa) in a reaction catalyzed by tyrosine hydroxy-lase. This enzyme appears to be the slowest of the sequence and is there-fore probably rate-limiting in dopamine and norepinephrine synthesis.

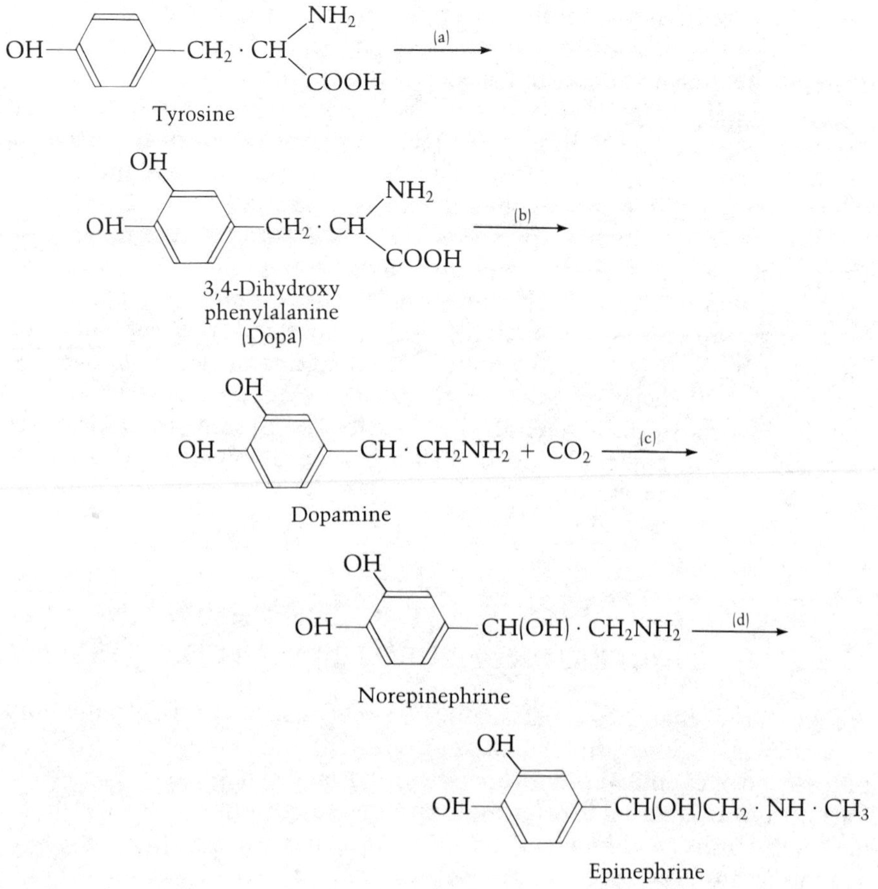

Enzymatic reaction	*Enzyme*	*Cofactors*
(a)	Tyrosine hydroxylase (EC 1.14.6.2)	Tetrahydrobiopterine Molecular O_2, Fe^{2+}, NADPH
(b)	L-Aromatic acid decarboxylase (EC 4.1.1.28)	Pyridoxal phosphate
(c)	Dopamine β-hydroxylase (EC 1.14.17.1)	[Ascorbate; Molecular O_2 Cu^{2+}
(d)	Norepinephrine *N*-methyltransferase (EC 2.1.1.28)	*S*-Adenosylmethionine

FIGURE 4.13 Enzymatic synthesis of dopamine, norepinephrine, and epinephrine.

(a) $TH\text{-}H_2 + tyrosine \rightarrow TH - H_2 \cdot tyrosine$

(b) $TH\text{-}H_2 \cdot tyrosine + \frac{1}{2}O_2 \rightarrow TH + dopa + H_2O$

(c) $Pteridine - H_2 + NADPH + H^+ \rightarrow pteridine - H_4 + NADP^+$

(d) $Pteridine - H_4 + TH \rightarrow TH - H_2 + pteridine - H_2$

FIGURE 4.14   The coupled reaction sequence for tyrosine hydroxylase and dihydropteridine reductase. Equations (a) and (b) represent tyrosine hydroxylase reactions. Equations (c) and (d) represent the coupled quininoid dihydropteridine reductase enzyme. Either NADH or NADPH may act as cofactor.

Tyrosine hydroxylase (complete name: tyrosine 3-mono-oxygenase, EC 1.14.16.2) is an interesting enzyme both because of its control over rates of transmitter synthesis and because it employs molecular oxygen, ferrous ions, and a tetrahydropteridine cofactor in the reaction[39,40,242] (Figure 4.14).

The involvement of ferrous ions was deduced from their stimulatory action when they were added during the assay of the enzyme.[239,240] Although other studies have demonstrated the presence of Fe in pure enzyme prepared from bovine adrenal medulla (0.5–0.75 mol iron per mol of enzyme),[241] there is a continuing debate on whether this metal is strictly required for tyrosine hydroxylase activity.[39,242]

The adrenal medulla contains large quantities of tyrosine hydroxylase, as does the corpus striatum of the vertebrate brain and also pheochromocytoma (catecholamine-secreting tumors). These sources have provided the tissue for the isolation, characterization, and study of the enzyme's mechanism and kinetics. It appears to have a molecular weight of about 200,000 and to consist of multiples of the smallest subunit of 60,000.[39,43] In the short term—minutes and seconds—this key enzyme is controlled by end-product (dopa, dopamine) feedback inhibition involving the pteridine cofactor. *Reduced* tyrosine hydroxylase protein is the active form of the enzyme required for the initial enzymatic reaction (Figure 4.14). This is provided continuously by the transfer of hydrogen atoms from reduced pteridine (tetrahydropteridine), which is itself produced by quinonoid dihydropteridine reductase, with NADH or NAPH as cofactor. The feedback effect of catechols and catecholamines is possible because they *oxidize* the pteridine cofactor, preventing it from generating active tyrosine hydroxylase via a reductive action.[39] Catechols also act as competitive inhibitors to the pteridine cofactor at the active site of tyrosine hydroxylase.[39,239]

But these are not the only ways in which tyrosine hydroxylase is being controlled. Another mechanism, it seems, is allosteric control through phosphorylation of the enzyme—for instance, cycle AMP–mediated phosphorylation leads to its activation. This appears to involve increased affinity of the phosphorylated enzyme for the pteridine cofactor[41,52] and a marked change in the pH dependence of its activity without change in its $V_{\max}$ for tyrosine.[39,40,41] In contrast, Ca^{2+} also activates tyrosine hydroxylase but by a mechanism involving an increased $V_{\max}$ without change in the affinities for tyrosine or pteridine cofactor.[41] Current evidence eliminates a direct action of Ca^{2+} on the enzyme and suggests that Ca^{2+}-stimulated (calmodulin-mediated) phosphorylation of the enzyme or of its effectors is the mechanism involved. In fact, this form of control has been proposed as the principal route for activation of dopaminergic nerve terminals due to depolarization with its associated Ca^{2+} influx,[41] rather than control via cyclic AMP.[39,40] In fact, predominating control mechanisms could be different in different subcategories of dopamine systems. This very important issue continues to be debated.[217]

Drugs specifically inhibiting tyrosine hydroxylase include α-methyl-p-tyrosine, a competitive inhibitor.

In summary, studies with several preparations, including the vas deferens, the pineal gland, and the adrenal glands, point to the conclusion that in the short term, tyrosine hydroxylase is controlled by multiple processes. The enhanced activity of the enzyme after nerve stimulation is due to a temporary reduction in feedback inhibition by its products and also due to an improved enzyme performance following its phosphorylation by a Ca^{2+}-stimulated or cyclic AMP–mediated protein kinase. Longer term control of tyrosine hyroxylase by hormones and drugs, over periods of days, is effected by control of the synthesis of the enzyme protein itself.[39,205,206]

Tyrosine Hydroxylase and Presynaptic Receptors A curious observation remains to be fitted into this picture: the finding that a reduction of impulse traffic in the dopaminergic neurons of the corpus striatum—for example, by lesion of the nigrostriatal tract—leads to an unexpected *increase* in striatal tyrosine hydroxylase activity. Since the effect is prevented by dopamine agonists (e.g., apomorphine), it is concluded that *presynaptic* dopaminergic receptors exist on the outer surface of terminals of dopamine neurons and that they are able to exert an inhibitory, or negative-feedback, action on the tyrosine hydroxylase contained within. This action would normally be exerted by dopamine after its release into the synaptic cleft, sharply reducing its rate of synthesis within the terminal from which it came. Therefore, this is an additional, inhibitory, control which, where dopaminergic synapses are concerned, must be balanced against the stimulatory control mechanisms outlined above.

Dopa Decarboxylase The product of tyrosine hydroxylase activity, L-dopa (Figures 4.13 and 4.14), is decarboxylated to dopamine in both dopaminergic and noradrenergic neurons by the enzyme *dopa decarboxylase* (EC 4.1.1.28), also known as aromatic L-amino acid decarboxylase, which requires pyridoxal phosphate for its activity. It is found in a range of organs in addition to brain, for example, in the kidney and liver. The enzyme from kidney has been isolated and purified and has a molecular weight of 85,000–90,000.[44,243,R] The antibodies raised to this enzyme have been used for fluorescence immunohistochemical localization of the enzyme. This technique identifies noradrenergic, dopaminergic, and serotoninergic systems, since dopa decarboxylase is present and active in all three types of neuron. Specific inhibition of the enzyme is possible through use of the drug α-methyldopa.

Dopamine β-Hydroxylase   The principal function of the enzyme dopamine β-hydroxylase is the conversion of dopamine to norepinephrine in noradrenergic neurons by hydroxylation of the ethylamine side chain, using molecular oxygen. It employs copper to effect the hydroxylation and is therefore readily inhibited by copper chelating agents. Specific inhibitors for the enzyme include disulfiram, diethyl dithiocarbamate, fusaric acid, and FLA-63, which can be used to prevent norepinephrine formation in tissues. The enzyme has been isolated and purified and appears to consist of a tetrameric glycoprotein[46] of molecular weight 75,000. Dopamine β-hydroxylase is found only in noradrenergic neurons, where it is localized to synaptic vesicles. It appears to be present in the vesicle membrane but may become incorporated in the plasma membrane of the noradrenergic nerve terminal as the result of exocytosis.

Serotonin Synthesis   A sequence of biosynthetic reactions somewhat similar to the reactions producing catecholamines operates in the central nervous system to produce serotonin, also known as 5-hydroxytryptamine, or 5-HT (see Figure 4.15). The starting substrate, tryptophan, is an essential dietary requirement, however, and cannot be synthesized *de novo*.

The hydroxylation of tryptophan to form 5-hydroxytryptophan is performed by *tryptophan hydroxylase* (EC 1.16.4), which uses the same pteridine cofactor as tyrosine hydroxylase and also uses molecular oxygen. Tryptophan hydroxylase is found principally in brain, though other nonneuronal cells like pinealocytes and mast cells contain the enzyme. It has a molecular weight of 230,000 as measured by three methods,[47] and it is almost certainly a tetramer composed of two subunits of very similar size[48,49] (mol wt 57,500 and 60,900) that are not glycoproteins.[47] Whether Fe^{2+} is a component participating in the catalytic mechanism remains uncertain, though a body of evidence suggests that it is.[49]

The K_m value for tryptophan at 50–60 μM (both in vivo and in vitro) is higher than the likely in vivo free tryptophan level of 20 μM. This implies that tryptophan availability and its rate of entry from the blood can control the rate of 5-HT synthesis. Indeed, in vivo administration of tryptophan causes large increases in the brain's 5-HT content.[50] Another rate-limiting participant in the enzyme reaction is the reduced pteridine cofactor, which occurs in neural tissues at 1–10 μM, while the K_m for this reactant, with solubilized tryptophan hydroxylase, is 31 μM.[49] Evidence

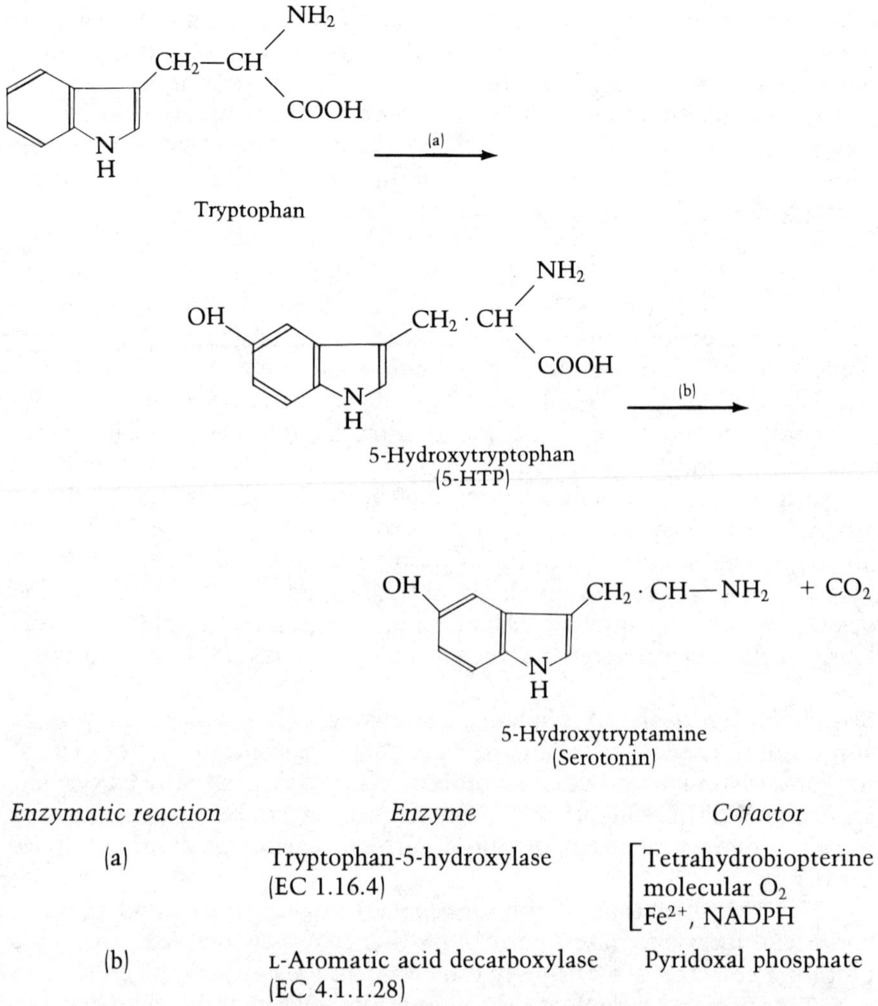

FIGURE 4.15   Enzymatic synthesis of 5-hydroxytryptamine (serotonin).

Enzymatic reaction	Enzyme	Cofactor
(a)	Tryptophan-5-hydroxylase (EC 1.16.4)	Tetrahydrobiopterine molecular O_2 Fe^{2+}, NADPH
(b)	L-Aromatic acid decarboxylase (EC 4.1.1.28)	Pyridoxal phosphate

for this as an in vivo mode of control comes from the increased tryptophan hydroxylase activity that follows intraventricular injection of tetrahydrobiopterin.[51] Allosteric control of the enzyme also seems to occur through a phosphorylating action of Ca^{2+}-dependent or cyclic-AMP-dependent kinases, as occurs for tyrosine hydroxylase.[41,49,52,211,217]

Decarboxylation of 5-Hydroxytryptophan   Current evidence strongly favors the conclusion that the next and last stage in 5-HT synthesis, 5-hydroxytryptamine decarboxylation, is catalyzed by dopa decarboxylase, the same enzyme that decarboxylates L-dopa to form dopamine and subsequently norepinephrine. The alternative is that a separate enzyme exists, 5-hydroxytryptophan decarboxylase. Evidence for the identity of the two enzymes being the same includes the observation that pure dopa decarboxylase from hog kidney also decarboxylates 5-HT to serotonin, and its antiserum precipitates the two enzymes to the same extent, indicating considerable homology in their protein structure. Antibodies to the bovine adrenal-gland enzyme allow immunohistochemical visualization of both noradrenergic and serotoninergic pathways.[45]

Storage of Norepinephrine, Dopamine, and Serotonin   In contrast with the cholinergic system, most of the biogenic amines are found and stored in so-called dense-cored, or granular, vesicles within axons and nerve terminals where biosynthesis is also localized. The terminals seem to be the main storage sites, and in the peripheral noradrenergic neurons they consist of chains of varicosities filled with vesicles and granules. Many of the vesicles are characterized by the presence of a prominent *dense-core,* which appears to be the monoamines in some kind of salt complex with carrier protein (chromogranin A), Ca^{2+}, and ATP (Figure 4.16). Agents that deplete the neurotransmitter stores (e.g., reserpine; Table 4.4), also remove the dense-core,[53] and incubation with catecholamines increases its size and density.

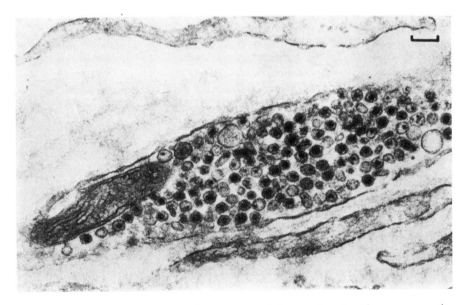

FIGURE 4.16   Varicosities from rat vas deferens showing the presence of dense-cored vesicles. Fixation was with acrylic aldehyde in sodium dichromate–osmium tetroxide. Sodium ferricyanide block staining and lead citrate section staining were applied. Bar, 0.1 μm. [From ref. 55]

TABLE 4.4   Some drugs interacting with norepinephrine systems

Drug	Presumed mechanism of action	Most prominent physiological effects
Antagonists		
Chlorpromazine*	α-Receptor antagonist	Tranquilizer
Phenoxybenzamine	α-Receptor antagonist	Antihypertensive
Propranolol	β-Receptor antagonist	–
Agonists		
Clonidine	α-Receptor agonist	Biphasic pressor response
Isoproterenol	β-Receptor agonist	Bronchodilator; hypertensive
Phenylephrine	α-Receptor agonist	Hypertensive
Releasers		
Amphetamine*	Releaser	Stimulant, euphoriant
Tyramine*	Releaser	Experimental
Storage inhibitors		
Guanethidine	Peripheral depletion and ganglionic blockade	Antihypertensive
Reserpine*	Depletion	Antihypertensive; tranquilizer
Tetrabenazine*	Depletion	Antihypertensive; tranquilizer
Pump inhibitors		
Amytriptylene	Reuptake inhibition	Antidepressant
Chlorimipramine	Reuptake inhibition	Antidepressant
Cocaine*	Reuptake inhibition; local anesthetic	Euphoriant; stimulant
Desipramine	Reuptake inhibition	Antidepressant
Imipramine	Reuptake inhibition	Antidepressant

* Also has prominent dopamine action.

Another feature of the norepinephrine storage vesicle is the presence of dopamine β-hydroxylase in its membrane. This enzyme may be used as a marker for noradrenergic vesicles.

Both large (60–120 nm diameter) and small (30–60 nm diameter) dense-cored vesicles exist with the latter predominating, the proportion of each depending on the neurons and species being investigated.[B,54] In rat heart and vas deferens, the smaller vesicles make up 95 percent of the total vesicle population. In addition to the neurotransmitter bound in these vesicles there appears to be a soluble pool, which is isolated together with the vesicles. Unfortunately, the extent to which loss from vesicles *during isolation* contributes to this pool remains uncertain. Certainly uptake from the pool to the vesicles does not seem to occur during

TABLE 4.4   (*Continued*)

Drug	Presumed mechanism of action	Most prominent physiological effects
Synthesis inhibitors		
Diethyldithiocarbamate	DBH inhibitor	–
Disulfiram	DBH inhibitor	–
FLA-63	DBH inhibitor	–
Fusaric acid	DBH inhibitor	Antihypertensive
α-Methyldopa*	Dopa decarboxylase inhibition	Antihypertensive
α-Methyl-*p*-tyrosine*	Tyrosine hydroxylase inhibition	Depressant; akinesia
Monoamine oxidase inhibitors		
Clorgyline	Type A MAO inhibition	Experimental
Iproniazid*	Broad-spectrum MAO inhibition	Antidepressant
Nialimide	Broad-spectrum MAO inhibition	Antidepressant
Tranylcypromine*	Broad-spectrum MAO inhibition	Antidepressant
COMT inhibitors		
Tropolone, pyrogallol,* rutin, quercitin	COMT inhibition	Experimental
False transmitters		
α-Methyldopamine*	Replacement of transmitter	Antihypertensive
α-Methyl-*m*-tyramine*	Replacement of transmitter	Antihypertensive
Toxin		
6-Hydroxydopamine*	Destruction of cells	Experimental
Precursors		
Dihydroxyphenylserine (DOPS)	Stimulates transmitter production	Experimental
L-Dopa*	Stimulates transmitter production	Counters parkinsonism; mild stimulant

* Also has prominent dopamine action.

isolation, since radiolabeled norepinephrine added to tissue homogenates does not appear subsequently in the vesicles.[54]

The continuous stream of dense-cored and other vesicles traveling from neuronal cell bodies to their terminal regions as part of the axoplasmic traffic can be demonstrated by nerve ligature. These particles then accumulate on the proximal side of the ligature. It appears that many of the vesicles are manufactured in the cell body and are transported empty down the axon, gradually filling with neurotransmitter on their journey.[c] A filled vesicle might contain about 10,000–15,000 molecules of neurotransmitter.[54]

Much of this information about dense-cored storage vesicles comes from studies on peripheral neurons, ganglia, and adrenal medulla, where specialized structures abound, such as the networks of terminal varicosities associated with the autonomic adrenergic nerves.

In the central nervous system, dense-cored vesicles can also be seen in dopamine- and serotonin-containing nerve terminals. It is likely that similar conclusions about vesicular organization and storage of dopamine and serotonin apply here, where the more conventional kind of neuron-to-neuron synapse predominates. Indeed, special features appear to attach to serotonin storage. Thus, a specific high-affinity "serotonin-binding protein" has been isolated from brain that is enriched in the synaptic vesicles of serotoninergic neurons, where it probably acts as a storage protein and reduces the osmotic pressure otherwise generated by the free transmitter.[164] This protein is released during synaptic activity along with serotonin and appears to become dissociated from the transmitter.

Release of Norepinephrine, Dopamine, and Serotonin

Calcium-dependent release of the neurotransmitters evoked by neural activation, or by stimuli applied directly to in vitro preparations, has been demonstrated for a wide range of preparations.[C,54,56] For example, stimulation of the splenic nerve releases norepinephrine and acetylcholine into superfusates of the spleen, and depolarizing stimuli release dopamine, serotonin, and other transmitters from slices and synaptosomes prepared from corpus striatum and other brain regions. Peripheral, sympathetically innervated tissues have been most successfully employed for release studies. Most widely used, apart from superfused whole spleen, are the heart, vas deferens, strips of vascular tissue, and iris.[56] In these organs the adrenergic neurons release the transmitter from the terminal varicosities (about 1 μm diameter). The varicosities occur at 100–300 per mm as serial swellings three to five times the diameter of the axon itself on the many terminal arborizations of the sympathetic postganglionic fibers. Many of these varicosities do not make any organized synaptic contacts, nor are they an exclusive feature of peripheral innervation. For instance, similar axonal varicosities (Figure 1.29) occur on monoaminergic neurons of the central nervous system, including neurons in the cerebral cortex and basal ganglia.[57,58] Thus, in some brain regions, monoamines are likely to be released from each of a series of varicosities as the nerve impulse passes down the axon.

Release of dopamine from the intact in situ caudate nucleus or substantia nigra can be followed by means of superfusion through a push-pull cannula whose tip is located within either brain region. One technique, employing cats, uses infusion of radiolabeled tyrosine to allow synthesis and release of radiolabeled dopamine in the correct compartment.[56,59] Lesion, or excitation, of pathways connecting with the caudate or substantia nigra allows evidence to be gathered on the neural influences controlling dopamine release.

Monitoring monoamine release in vivo has recently been improved by fixing dialysis membranes over the tips of conventional push-pull cannulae or by passing long fibers consisting of dialysis tubing right through the brain. Alternatively, monoamine release may be directly and immediately detected in vivo using electrochemical methods that employ electrodes positioned in the brain structure under study and set at a particular voltage (in vivo voltammetry).

Inactivation and Further Metabolism of Released Monoamines Facilitated reuptake to surrounding neurons and glial cells from synaptic cleft and other local regions of the extracellular space by high-affinity transport systems is the principal means of inactivating the biogenic amines.[56,62] The uptake systems require Na^+ ions for activity. They are located only in those neurons normally synthesizing and releasing biogenic amines and also in surrounding glial cells. This implies that where a biogenic aminergic neuron releases its transmitter at a synapse with, for instance, a cholinergic cell, no uptake occurs at the postsynaptic cell.

Because reuptake is such a critical stage in the inactivation of these neurotransmitters, agents potently blocking or delaying this process are often clinically valuable drugs. Tricyclic antidepressant drugs (e.g., chlorimipramine) inhibit norepinephrine and serotonin uptake. Benztropine potently inhibits dopamine uptake, the other biogenic amines being less affected (see Tables 4.4 to 4.8). These drugs enhance the interaction of the neurotransmitters with both presynaptic and postsynaptic receptors by prolonging the contact time. They are valuable in the treat-

TABLE 4.5 Typical physiological actions of adrenergic receptors

System or tissue	Action	Receptor
Cardiovascular system	Increased force of contraction	beta
Heart	Increased rate	beta
Blood vessels	Constriction	alpha
	Dilation	beta
Respiratory system	Relaxation	beta
Tracheal and bronchial smooth muscle	Relaxation	beta
Iris (radial muscle)	Pupil dilation	alpha
Uterine smooth muscle	Contraction	alpha
Spleen	Relaxation	beta
	Contraction	alpha
Bladder	Contraction	alpha
	Relaxation	beta
Skeletal muscle	Changes in twitch tension	beta
	Increased release of acetylcholine	alpha
	Increased glycogenolysis	beta
Adipose tissue	Increased lipolysis	beta

TABLE 4.6 Properties of dopamine receptors

Drug	Drug action on receptor type			
	D_1	D_2	D_3	D_4
Adenyl cyclase coupled	Yes (stimulation)	Yes (inhibitory)	No	No
Apomorphine	mixed agonist/antagonist (μM)	agonist (nM)	agonist (nM)	agonist (μM)
Bromocriptine	antagonist (μM)	agonist (nM)	N.E.*	N.E.
Butyrophenones	weak antagonist	potent antagonist	weak antagonist	antagonist (nM)
Dopamine	agonist (μM)	agonist (nM)	agonist (nM)	agonist (μM)
GTP coupled	Yes	Yes	N.E.	No
Molindone	inactive	antagonist	N.E.	N.E.
Phenothiazines	antagonist	antagonist	N.E.	N.E.
Sulpiride	inactive	antagonist	N.E.	N.E.

Note: Effective concentrations of these drugs are given in parentheses. A strong action is implied by nM and a relatively weak action by μM, determined either as potency of pharmacological action or in ligand-binding studies. Data are from refs. 65, 74, 254, and 256. N.E., not established.

ment of movement disorders (e.g., parkinsonism) or in the treatment of psychiatric illness when a diminution in specific neurotransmitter function may be desired (see Chapter 9).

High-affinity transport systems with low K_m values (1–5 μM) are well designed to serve this inactivating function: they have low capacity but turn on very quickly and can cope effectively with very small amounts of neurotransmitter. Other, less sensitive, transport systems are also present. These low-affinity systems have elevated K_m values (0.5–2 mM) but high capacity, and in concert with the high-affinity process can rapidly reduce "floods" of transmitter to low levels in the extracellular space.

Enzymatic breakdown and metabolism of the biogenic amines occurs within neuronal and glial mitochondria catalyzed by monoamine oxidase (MAO). Methylation is brought about by catechol-O-methyltransferase (COMT). (See Figures 4.17 and 4.18.) The latter enzyme,

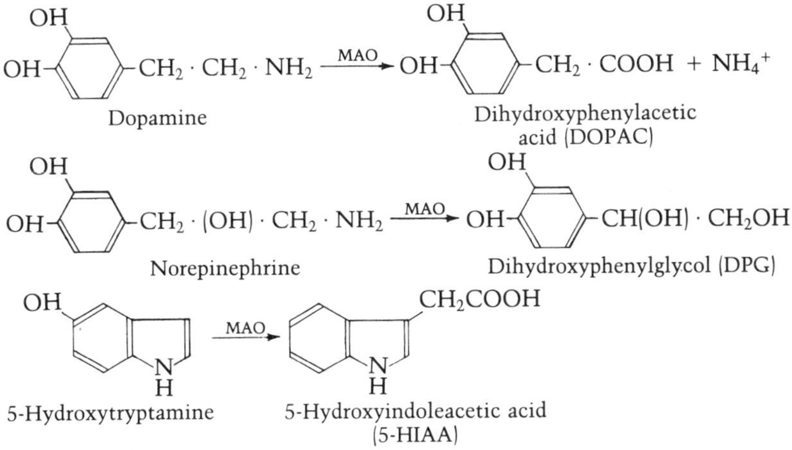

FIGURE 4.17   Enzymatic reaction of monoamine oxidase (MAO).

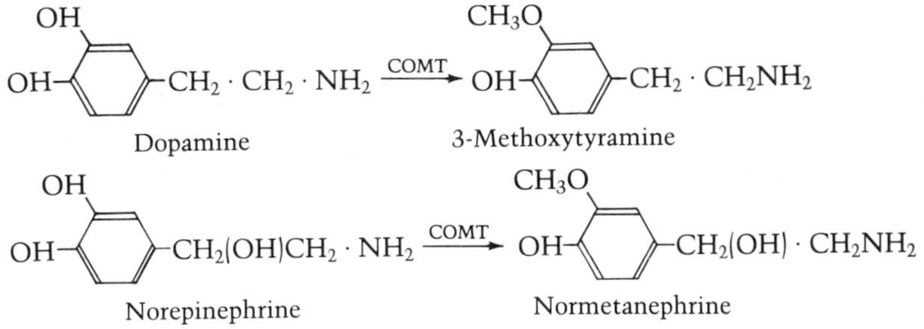

FIGURE 4.18   Enzymatic reaction of catechol-O-methyltransferase (COMT).

which only acts on substrates containing the catechol configuration, does not methylate serotonin or its metabolites. It is found principally in the cytoplasm.[Q]

Monoamine oxidase[60] (EC 1.4.3.4) is found in large quantities in glandular tissue, liver, kidney, and intestines; the brain is less enriched in this enzyme. Pig brain MAO has a molecular weight of 102,000 as judged by gel filtration.[60] The FAD cofactor has been shown to be bound covalently to the enzyme by way of a thioether linkage from the 8 position of the isoalloxazine ring to a cysteinyl residue of the protein.[60] Multiple forms of the enzyme have, however, been identified with the use of substrates and inhibitors.[60] *Monoamine oxidase A* is active towards serotonin, and tyramine and is sensitive to clorgyline inhibition. *Monoamine oxidase B* is active towards benzylamine and tyramine and is insensitive to clorgyline inhibition. In contrast, pargyline and deprinyl are potent inhibitors of the B form (see Table 4.4). Some inhibitors of MAO cause enhanced biogenic amine activity and have been used clinically in alleviating the symptoms of depressive illness, which could be caused by an imbalance in activity of biogenic amine systems (see Chapter 9). Some MAO inhibitors elicit euphoric states. Today, these drugs are not widely prescribed as antidepressants because of adverse reactions (e.g., a hypertensive effect under certain conditions). Instead, tricyclic antidepressants and benzodiazepines are the preferred antidepressants, though the potency of some of these compounds (e.g., chlorimipramine) has been attributed, at least partly, to MAO inhibition as well as to a blocking of neurotransmitter reuptake[60] (Table 4.4).

In view of the mitochondrial location of MAO, its function is likely to concern the control of levels of transmitter amines within the neuron. Free amines leaking from storage vesicles or entering the cell because of reuptake will be degraded. MAO *inhibitors* cause a rise in the level of free catecholamines, and this can in turn lead to impairment of the reuptake process.[60] The relationship between the activities of MAO and COMT in this respect remains uncertain, but methylation by the latter enzyme does not impair the activity of MAO towards these substrates.

Catechol-O-methyltransferase The enzyme catechol-O-methyltransferase COMT (EC 2.1.1.6) catalyzes the transfer of the methyl group of S-adenosylmethionine to the catecholic hydroxyl groups of norepinephrine and dopamine to produce inactive compounds;[Q] these methylated derivatives enter the bloodstream and are excreted from the body. COMT does not act on serotonin. Large amounts of the enzyme are found in liver, kidney, and brain tissue, where it no doubt serves a similar function. It is an intracellular enzyme, mostly cytoplasmic in location. It is present in both neurons and glial cells,[Q] though extraneuronal COMT is regarded as the principal site for inactivation of catecholamines released from nerve endings.[Q]

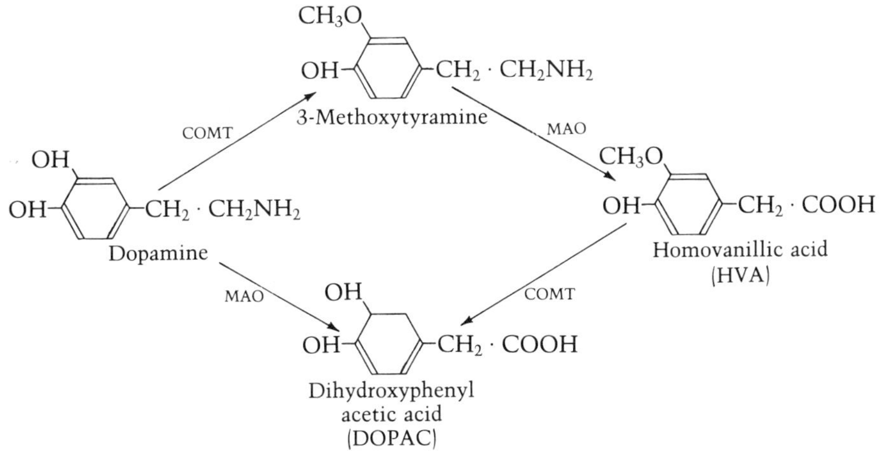

FIGURE 4.19   Interrelationships of dopamine metabolites.

COMT in the liver has a molecular weight of 24,000 and relies on -SH groups for its activity. It has an absolute requirement for a divalent cation such as Co^{2+}, Mn^{2+}, Mg^{2+}, Zn^{2+}, Fe^{2+}, or Cd^{2+}. Inhibitors of this enzyme include catechol compounds, and tropolones (which are isosteric with catechols) that compete with the substrate and chelate the divalent cation.[Q] Pyrogallol, and catechol-configurated flavenoids, such as rutin and quercitin, block O-methylation of norepinephrine and somewhat prolong the physiological effects of this amine,[61] but MAO inhibitors (e.g., iproniazid and nialamide; see Table 4.5) have a much more potent action in this respect, indicating the greater importance of both MAO and COMT acting together to control tissue levels and "spillover" of excess catecholamines in adrenergic systems. The relationship between the metabolites of dopamine and norepinephrine produced by MAO and COMT are shown in Figures 4.19 and 4.20.

5-Hydroxytryptamine Breakdown   After conversion to 5-hydroxyindole acetaldehyde by MAO, it is either dehydrogenated to 5-hydroxyindoleacetic acid (5-HIAA), which also happens to be a plant growth hormone, or it is reduced to the alcohol 5-hydroxytryptophol (5-HTOH) (Figure 4.21). Both products occur in the liver. In brain, 5-HIAA is the end product. In fact, its levels in cerebrospinal fluid are often taken as an index of the rate of serotonin turnover in the central nervous system.

Pharmacology of Biogenic Amine Systems

The vast and extremely important topic of biogenic amine pharmacology cannot be more than introduced here (see refs. A, B, H, I, R, 56, 62 and 207 for reviews). Its importance stems from the wide range of neural path-

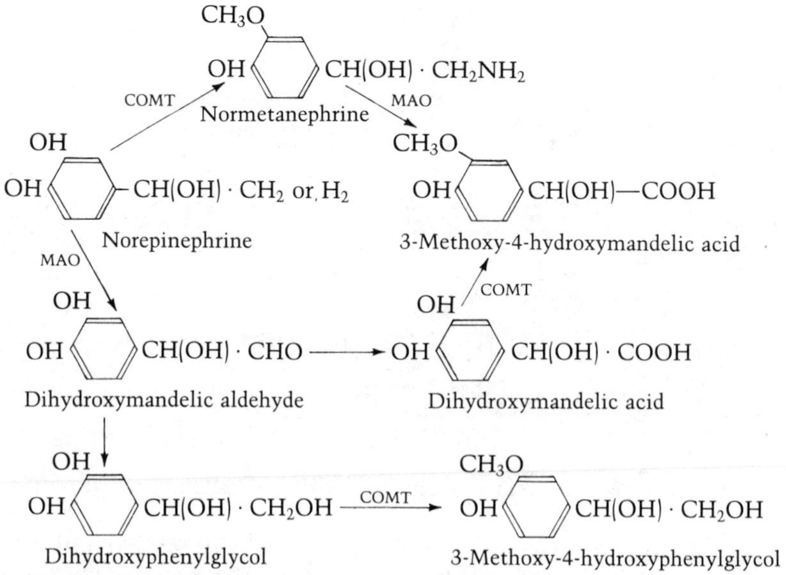

FIGURE 4.20   Interrelationships of norepinephrine metabolites.

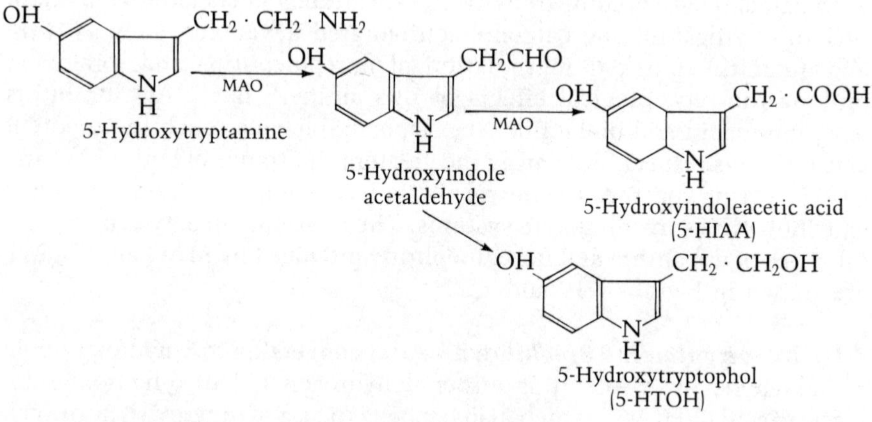

FIGURE 4.21   Catabolism of serotonin.

ways employing biogenic amines as neurotransmitters. Apart from all the sympathetic innervation of peripheral organs and vertebrate smooth muscles, a number of key processes in the central nervous system depend on the proper functioning of norepinephrine, dopamine, or serotonin neurons. For instance, all the pathways in the limbic system concerned in generating and controlling emotional states or moods, as well as neurosecretion and coordination of muscular movement via motoneuron control, employ biogenic amines as neurotransmitters. For this reason, a large proportion of clinically effective drugs for treating neurological and psychiatric disorders affect one aspect or another of the catecholamine or serotonin transmitter systems (Tables 4.4 to 4.8; see also Chapters 8 and 9).

Postsynaptic Norepinephrine Receptors

Just as subcategories of cholinergic receptors exist, so, since the early work of Dale in 1906, subcategories of norepinephrine receptors have been detected in the peripheral nervous system. They are classified as alpha and beta adrenoreceptors.

The alpha receptors are potently blocked by phentolamine, dibenamine, or phenoxybenzamine. The beta receptors are susceptible to propranolol, practolol, pronethalol, and several other compounds (see Figure 4.22). They are also distinguished by their response to adrenergic agonists. Thus alpha receptors are most sensitive to norepinephrine and least sensitive to isoprenaline; the reverse is true for beta receptors.

The two types of receptor appear to operate by quite different mechanisms. The mechanisms for alpha receptors involve the influx of Ca^{2+} followed by other transmembrane ionic fluxes such as K^+ efflux (parotid gland, liver, and adipose tissue) and sometimes by increased turnover of phosphatidylinositide phospholipids.[63,184] These mechanisms are far from being definitively described, and investigation continues in this area. On the other hand, a subclass of beta receptors (β_1) is coupled directly to adenylate cyclase, and activation of this receptor type leads to release of the second messenger, cyclic AMP, and the formation of phosphorylated proteins as described above for some muscarinic cholinergic receptors, which lead to the subsequent generation of synaptic potentials (Figures 4.3 and 4.4).

The early concept was that these two types of adrenergic receptor mediated excitatory (alpha receptor) or inhibitory (beta receptor) actions in the peripheral nervous system, but current knowledge does not support this as a firm rule, and each peripheral target organ must be considered on its merits. A list of receptors showing muscle contraction mediated by alpha receptors and muscle relaxation mediated by beta receptors is shown in Table 4.5. In the heart, however, stimulation of β_1 receptors leads to increased contraction rather than relaxation.

In terms of their clinical action, alpha blockers (e.g., phenoxybenzamine, phentolamine, and thymoxamine) are used for preventing the contraction of smooth muscle such as around blood vessels to allow vasodilation and thereby prevent excessive rises in blood pressure.

The relaxing action of beta blockers (e.g., propranolol and oxyprenolol) is employed to alleviate the ischemic pain of angina pectoris and to control cardiac arrhythmias and hypertension where they restore normal electrical activity to the heart. Agonists to beta receptors are also used clinically, such as to alleviate bronchoconstriction in asthma.

BETA BLOCKERS

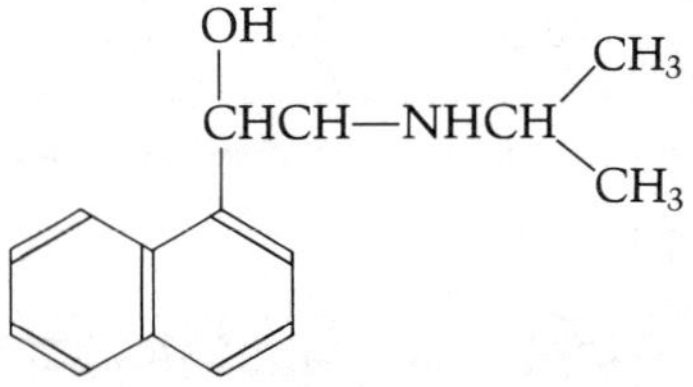

Pronethalol

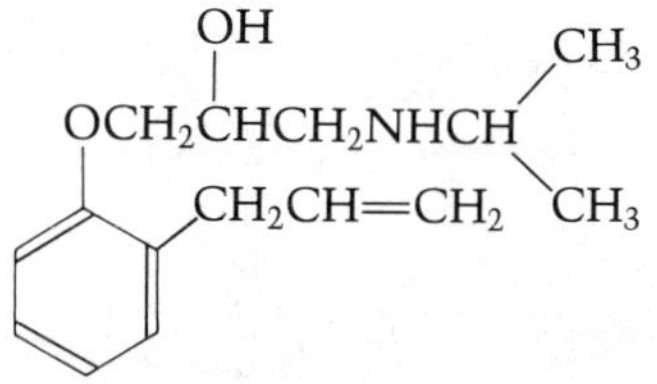

Alprenolol

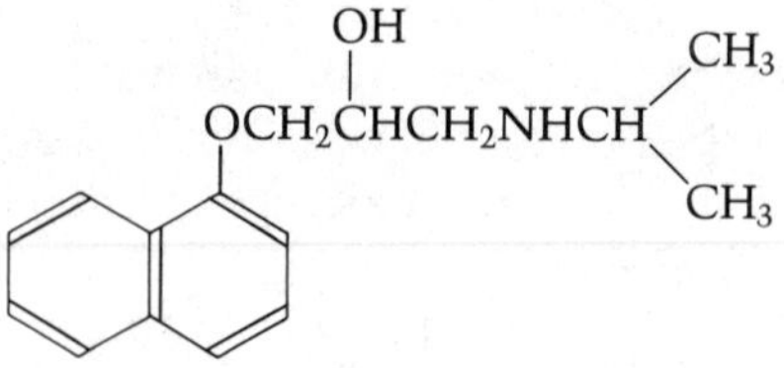

Propranolol

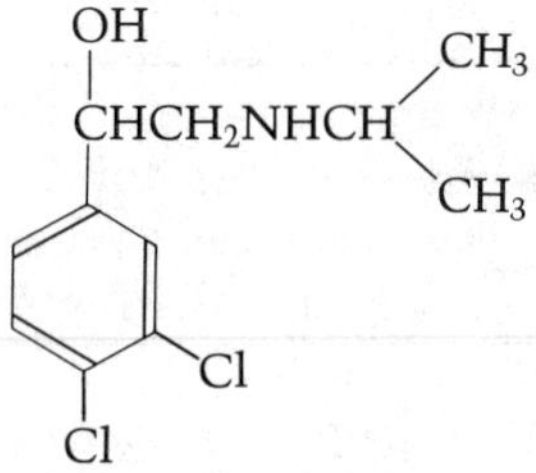

Dichloroisoprenaline (DCI)

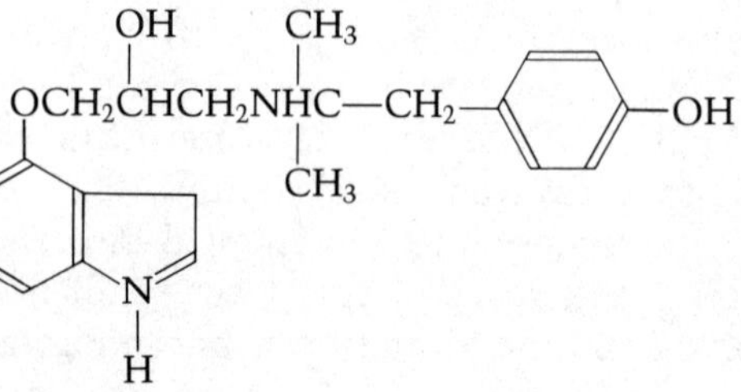

p-Hydroxybenzylpindolol

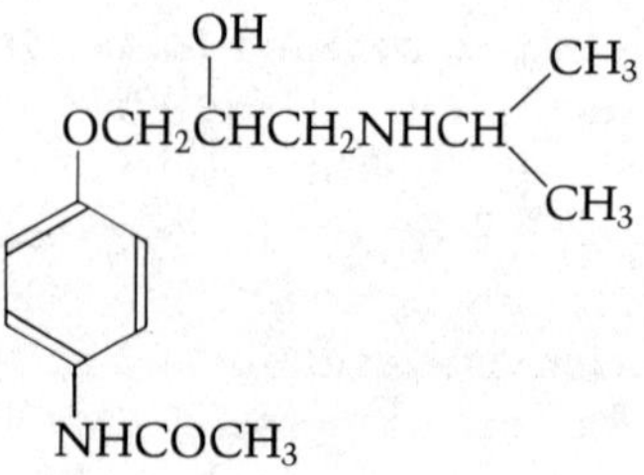

Practolol

ALPHA BLOCKERS

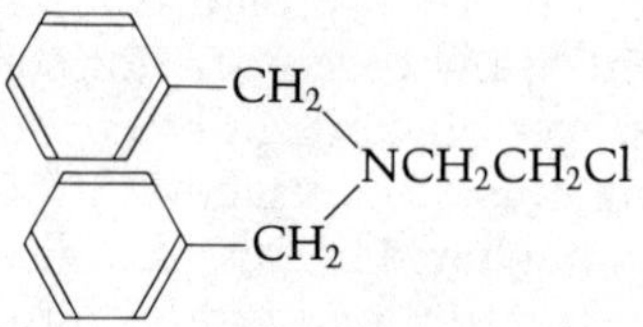

Dibenamine

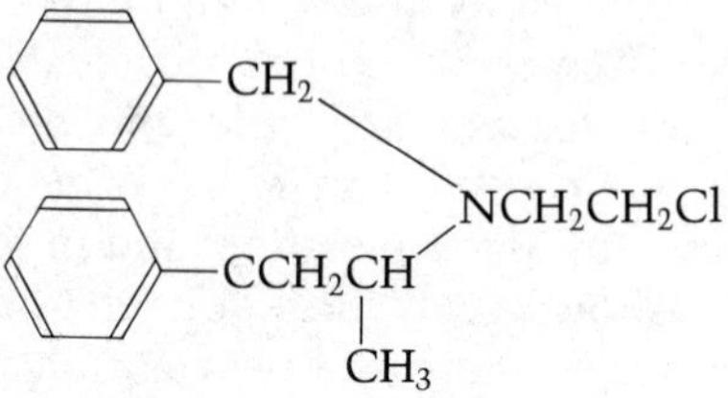

Phenoxybenzamine

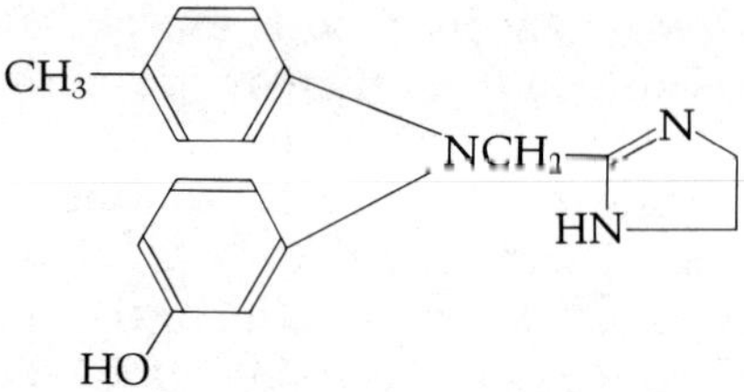

FIGURE 4.22 α Blockers and β blockers of adrenergic receptors. [From ref. 63]

Phentolamine

Guanethedine and bretylium are widely used to block noradrenergic transmission. They are effective antihypertension agents because they prevent release of transmitter in response to nerve stimulation, though their actions are not simple and precise mechanisms are not yet defined.

Dopamine Receptor Subpopulations Dopamine receptors in the central nervous system have also been classified into subcategories, D_1, D_2, D_3, and D_4, based on ligand-binding studies and responsiveness to various dopaminergic drugs (Table 4.6).[64,65,254,255,256] This is only one of several schemes of nomenclature (see refs. 254–256). The D_1 receptors appear to operate via adenylate cyclase activation and second messenger formation (cyclic AMP).[66] The D_2 receptors either couple to adenylate cyclase so as to cause its inhibition or function independently of this second messenger system. They may influence other enzyme systems, or they may act more directly on ion fluxes by regulating the opening of associated ionophores. Thus, more than one variety of D_2 receptors appear to exist.

Ligand-binding studies show that D_1 receptors preferentially bind dopamine agonists (e.g., [3H]dopamine, [3H]apomorphine, or [3H]flupenthixol). Guanine nucleotides, known to facilitate the coupling of dopamine receptors to adenylate cyclase, are able to regulate this binding (Figure 4.3). The D_2 receptors, on the other hand, are sensitive to low concentrations of both dopmaine agonists and antagonists. They preferentially bind butyrophenones (e.g., [3H]haloperidol, [3H]spiroperidol), which are antagonists, and this binding is sensitive to guanine nucleotides. The D_3 receptors bind dopamine agonists but not antagonists at low concentrations. The D_4 receptors[255] show the opposite pattern of ligand-binding affinities, displaying preference for antagonists such as [3H]haloperidol over agonists such as [3H]dopamine. They are insensitive to guanine nucleotides. It is not yet clear whether D_3 and D_4 receptor types can influence adenyl cyclase activity.[64,65,254]

Current evidence shows that all receptors D_1 to D_4 are found in the corpus striatum. The D_2 type is located on intrinsic neurons postsynaptic to dopaminergic terminals such as those of the nigrostriatal pathway. These receptors are responsible for the behavioral actions of dopamine and are cardinally involved in the inhibitory action of neuroleptics. The D_2 type is also present in the pituitary gland where it mediates the inhibition of prolactin release caused by dopamine. Intrinsic neurons of the corpus striatum also carry D_1, D_3, and D_4 types of dopamine receptor, though their precise function remains elusive. The D_4 and possibly the D_2 types appear to be located presynaptically on the terminals of neurons projecting from the cerebral cortex to the striatum (corticostriate pathway). The D_3 receptors are also located presynaptically on the terminals of the nigrostriatal pathway. These presynaptic receptors function to control dopamine *release* from these terminals (see below).

In general the D_1 receptors, which are linked to stimulation of adenyl cyclase, show poor correlation with dopaminergic functions, and their precise role remains obscure.

These various receptor locations have been deduced from changes in striatal receptor populations following lesions of the various input pathways to the corpus striatum, or destruction of its intrinsic neurons with kainic acid,[254] and they rest to some extent on surmize. There is much corroborative work to do before the picture can be finally accepted.

Dopamine terminals and receptors also occur extensively in the peripheral nervous system where they operate in sympathetic ganglia, exocrine glands, the gastrointestinal tract, and on mesenteric and renal arteries. Drugs affecting dopamine receptors are widely used as medication to control disorders of movement (chorea), tremor, hypothalamic endocrinology, and behavior. The dopamine receptor blockers include the

TABLE 4.7 Some drugs interacting with dopamine systems

Drug	Presumed mechanism of action	Most prominent physiological effects
Antagonists		
Benzodiazepines clozapine diazepam		
Butyrophenones haloperidol spiroperidol		
Diphenylbutyl-piperidines penfluoridol pimozide	Receptor blockade	Tranquilizer; antipsychotic; antinauseant
Phenothiazines chlorpromazine* fluphenazine thioridazine	Receptor blockade	Tranquilizer; antipsychotic; antinauseant
Thioxanthenes chlorprothixine flupenthixol	Receptor blockade	Tranquilizer; antipsychotic; antinauseant
Agonists		
Apomorphine	Receptor stimulation	Antiparkinsonian; emetic
Piribedil		
Bromocriptine		
Releasers		
Amphetamine	Release	Stimulant; appetite suppressant
Vesicular storage inhibitors		
Reserpine*	Depletion	Antihypertensive; tranquilizer; antipsychotic
Tetrabenazine*		
Pump inhibitors		
Amphetamine	Reuptake inhibition	Stimulant; appetite suppressant
Benztropine	Reuptake inhibition	Antiparkinsonian
Cocaine	Reuptake inhibition	Stimulant; euphoriant

* Also has prominent norepinephrine or epinephrine action, or both.

now-famous groups of tranquilizers and antipsychotic drugs: the phenothiazines, the butyrophenones, the thioxanthenes, the diphenylbutylpiperidines, and the benzodiazepines (see Table 4.7).

These agents, termed *neuroleptics,* or *anxiolytic drugs,* are effective in the treatment of depression and in the treatment of schizophrenia or other psychotic states (see Chapter 9). Their precise mode of action is still not entirely clear, though dopamine receptor blockade seems to be an important component of their action. A reduction in dopamine-mediated postsynaptic activity is probably one important consequence of this interaction, but other actions via presynaptic receptors and influences on dopamine turnover and release cannot be ruled out (see Chapter 9).

TABLE 4.7   (*Continued*)

Drug	Presumed mechanism of action	Most prominent physiological effects
Synthesis inhibitors		
Carbidopa	Dopa decarboxylase inhibition	Adjuvant for central dopa
α-Methyldopa	Dopa decarboxylase inhibition	Antihypertensive
α-Methyl-p-tyrosine*	Tyrosine hydroxylase inhibition	Depressant; akinesia
MAO inhibitors		
Clorgyline*	A-type MAO inhibition	Antidepressant
Deprenyl	B-type MAO inhibition	Antidepressant
Iproniazid;* Nialimide	Broad-spectrum MAO inhibition	Antidepressant
Tranylcypromine*	Broad-spectrum MAO inhibition	Antidepressant
COMT inhibitors		
Tropolone, pyrogallol,* rutin, quercitin	COMT inhibition	Minimal effects
False transmitters		
α-Methyldopamine*		Antihypertensive
α-Methyl-m-tyramine*	Replacement of transmitter	Mild tranquilization
Toxin		
6-Hydroxydopamine*	Destruction of cells	Experimental
Precursor		
Dopa	Stimulates transmitter production	Antiparkinsonism and mild stimulant

* Also has prominent norepinephrine or epinephrine action, or both.

Dopamine receptor *agonists* include apomorphine and bromocriptine, which are used clinically to cause vomiting and to treat neuroendocrine disorders. Bromocriptine is also used to treat parkinsonism. The therapeutic action of dopaminergic drugs in Parkinson's disease probably results from agonist action at D_2 receptors; chorea is more likely to be induced by the activation of D_1 receptors.

The similarities between catecholamine and serotonin synaptic organization leads to a considerable overlap in their pharmacology. Neuroleptics commonly affect serotonin receptors as well. Drugs acting differentially on serotonin receptors include lysergic acid diethylamide (LSD), quipazine, and 5-methoxy-*N,N*-dimethyltryptamine (5-MeODMT), all of which produce hallucination and are serotonin receptor agonists. The specific antagonists are cyproheptadine, methysergide, and 2-bromo-LSD (BOL), which are used clinically to treat migraine, diarrhea, and intestinal malabsorption symptoms. (See Table 4.8.)

Serotonin Receptor Subpopulations Similar ligand-binding studies have revealed the presence of serotonin receptor subpopulations. The subcategory S_1 preferentially binds [³H]5-HT that is, serotonin itself, whereas [³H]spiroperidol labels S_2 receptors. The latter compound is known primarily for its affinity for dopamine receptors in the corpus striatum, but in cerebral cortex it binds to serotonin receptors.

In general, serotonin agonists compete with [³H]5-HT at S_1 sites and serotonin antagonists compete with [³H]spiroperidol at S_2 sites, with potencies differing in the 10^3 range between the two subcategories of 5-HT receptor.[64] As with dopamine receptor subpopulations, guanosine nucleotide activation allows distinctions to be drawn between the two categories. The S_1 type is regulated by GTP and is linked to adenylcyclase, implying that response to their activation involves a second-messenger system. It seems likely that S_1 receptors mediate inhibition and S_2 receptors mediate excitation.[64]

Presynaptic Receptors for Monoamines Receptors located on the presynaptic terminal and activated by transmitter released either by the terminal itself (autoreceptors) or by other impinging terminals are now an established reality and play an important role in the control of synaptic activity (see Chapter 7). Essentially, their activation is thought to reduce synaptic activation by causing a reduction in transmitter release.

Peripheral epinephrine presynaptic alpha and beta receptors (α_2 and β_2) on the surface of the terminal varicosities releasing this transmitter are well studied.[67] Interaction with alpha receptor antagonists (e.g., phenoxybenzamine) enhance norepinephrine release evoked by nerve stimulation; alpha-receptor agonists reduce this release. In contrast, beta-receptor agonists (e.g., isoprenaline) enhance release during peripheral neural activation, and beta-receptor antagonists (e.g., propranolol) prevent the effect. The possibility thus exists for both enhancement and inhibition of the presynaptic release of norepinephrine by the two kinds

TABLE 4.8   Some drugs interacting with serotonin systems

Drug	Physiological effects
Agonists	
Chlorpromazine (high doses only)	Tranquilizer
LSD	Hallucinogen
5-Methoxy-*N,N*-dimethyltryptamine	Hallucinogen
Quipazine	Hallucinogen
Antagonists	
2-Bromo-LSD (BOL)	–
Cinanserin	–
Cyproheptadine	–
Methergoline	–
Methysergide	–
Storage inhibitors	
Reserpine	Tranquilizer
Tetrabenazine	Tranquilizer
Pump inhibitors	
Chlorimipramine	Antidepressant
Clomipramine	Antidepressant
Nortryptilene	Antidepressant
Synthesis inhibitors	
p-Chlorphenylalanine (PCPA)	Antisleep; mild aphrodisiac?
NSD-1034	None by itself
Monoamineoxidase inhibitors	
Clorgyline	Not precisely determined
Iproniazid	Antidepressant
Tranylcypromine	Antidepressant
Serotonin analogue/false transmitter	
Bufotenine	Hallucinogen
Toxin	
p-Chloroamphetamine (PCA)	Not precisely determined
5,6-Dihydroxytryptamine (5,6-DHT)	Not precisely determined

of receptor. The mechanisms by which control of neurotransmitter release is achieved remain obscure but may involve manipulation of Ca^{2+} availability within the terminal and an action on neurotransmitter synthesis via tyrosine hydroxylase.

Dopamine presynaptic receptors also appear to occur, both peripherally and centrally, and no doubt exert an important controlling influence on both the release and synthesis of dopamine.[67,68] A wide range of neuroleptic drugs have actions on the turnover and release of dopamine in the central nervous system,[68,69] and these actions are most likely mediated through both presynaptic and postsynaptic dopamine receptor systems.

Serotonin presynaptic receptors exist on the 5-HT–containing cell bodies of the raphe nucleus, and their activation leads to a modified depolarization response by the neurons. Much research is in progress on the existence of presynaptic serotonin receptors.

Presynaptic receptors of one transmitter category can exist on the terminals of another transmitter category. Dopamine receptors occur on peripheral noradrenergic terminals,[67,68] and cholinergic and serotoninergic receptors occur on the terminals[70,71] of central dopamine neurons. Such an arrangement allows local interaction between neurotransmitter systems (see Chapter 7).

The isolation and characterization of receptors for biogenic amines is in its infancy. The great successes in isolating the nicotonic cholinergic receptor has given impetus to the search for markers that might bind firmly enough to the biogenic amine receptors to survive the isolation procedures.[63]

Isolation of Biogenic Amine Receptor Proteins

Many attempts are in progress to isolate the integral membrane receptor proteins of the norepinephrine, dopamine, and serotonin neurotransmitter systems. The methods employed are similar to those previously found to be so successful in isolating acetylcholine receptors. These involve subfractionation of solubilized brain or other tissue by a combination of affinity chromatography and other column separation procedures. In each case a specific high-affinity ligand is used to identify the receptor system concerned. For example, [^{3}H]spiperone is used to assay solubilized or purified receptors for dopamine[236,251] or serotonin,[237] and [^{3}H]lysergic acid diethylamide can be used specifically for serotonin receptors.[273,238] Preparations purified to various extents have been reported for adrenoreceptors from a range of tissues and for dopamine and serotonin receptors from brain.

β-Adrenergic Receptor Isolation   The β-adrenergic receptor–adenylate cyclase complex that mediates the stimulatory actions of catecholamines on physiological functions in heart, smooth muscle, and many other tissues consists of at least three components. The first is the β-adrenergic receptor-binding protein containing the ligand-binding site for catecholamine agonists. The second is a nucleotide regulatory protein that binds guanine nucleotides. The third component of the system is adenylate cyclase (Figure 4.3). Both the β-adrenergic receptor protein and the nucleotide regulatory proteins have been completely purified from various sources, including erythrocytes and lung.[234,235,251] β-Adrenergic receptors can be solubilized from plasma membrane preparations by the detergent digitonin with complete retention of the ability to bind adrenergic ligands. This allowed the development of biospecific affinity chromatography procedures for the isolation of these receptors. The β-adrenergic antagonist alprenolol immobilized to Sepharose has been used for this purpose with considerable success. Frog erythrocyte β-adrenergic receptor, purified to homogeneity, consists of a single type of polypeptide, of mol wt 58,000 for the frog erythrocyte receptor and 64,000 for the lung receptor.[252] The beta receptor from turkey erythrocytes consists of two polypeptides of mol wt 40,000 and 45,000, present in the ratio of 3 or 4 to 1 respectively.[234] All of these beta-receptor preparations can be specifically and covalently labeled with the photoaffinity label [^{125}I]azidobenzylcarazolol. The full functional integrity of these purified receptors has been demonstrated by incorporating them into phospholipid vesicles, which are then fused with the plasma membranes of toad erythrocytes.[252] These erythrocytes contain the adenylcyclase enzyme and the nucleotide-binding protein, but few or no beta receptors, and thus no β-adrenergic responsiveness. After fusion with the receptor-loaded phospholipid vesicles, the cells become responsive to adrenergic agonists such as isoprenaline, and this stimulation is completely abolished by the beta-antagonist propranolol.

Solubilization of Dopamine Receptors   Dopamine D_2 receptors have been solubilized from the corpus striatum of a variety of animal species and humans, using a range of detergents[236,238,251] (e.g., digitonin, lysolecithin, or cholate), and they retain their capacity to bind ligands that are specific for dopamine receptors. [^{3}H]Spiperone has generally been the high-affinity radioligand used in these studies, but it also binds to the serotonergic receptors present in corpus striatum. This serotonergic binding component can be eliminated by adding a serotonergic selective drug, such as mianserin. The binding of [^{3}H]spiperone to these preparations shows pronounced stereospecificity for displacement by butaclamol stereoisomers. For this reason the total receptor-associated [^{3}H]spiperone binding has been defined as the difference in binding between parallel assays containing (−) or (+) butaclamol.[236,251] Only (−)-butaclamol is a clinically effective neuroleptic, and this correlates with the finding that it binds to dopamine receptors with high affinity. The solubilized receptor preparations show lower affinity for binding of specific ligands than is shown by the membrane-bound sites, and agonist binding (e.g., apomorphine) is not regulated by GTP as in membrane preparations.[236] In many other respects the solubilized dopamine D_2 receptors show quite similar pharmacological properties to the membrane-bound D_2 receptors, and efforts are now concentrated on purifying the receptor protein to homogeneity.[236,251]

Solubilization of Serotonin Receptors   Starting with rat frontal cortex, serotonin S_2 receptors have been effectively solubilized with lysolecithin and identified by [^{3}H]spiperone or [^{3}H]lysergic acid diethylamide binding.[237] They retain the high-affinity characteristics of serotonin (S_2) receptors in the original membrane. Other radioligands (e.g., [³H]ketanserin) have been shown to selectively label serotonin receptors isolated from dog, rat, and human brains.[238,253] Isolation of the serotonin S_2 receptor should soon follow these successful attempts at its solubilization.

Location of Neural Pathways Employing Biogenic Amines

The advent of the formaldehyde histofluorescence method of visualizing norepinephrine-containing and dopamine-containing neurons greatly aided the mapping of these pathways in the central nervous system.[72] The different fluorescence colors given by the aldehyde derivatives of norepinephrine and dopamine allowed these pathways to be distinguished. No stable product was produced with serotonin, however, and a modified method has been applied to localize these pathways.[73] Many sensitive and elegant techniques, used in complementary fashion, have allowed the biogenic amine systems to be mapped. These include the glyoxylic acid fluorescence techniques, immunohistochemical methods,[74] retrograde transport employing horseradish peroxidase, and axoplasmic transport of labeled proteins.[75,282,283]

Norepinephrine   Apart from the numerous pathways in the peripheral nervous system that constitute the sympathetic nervous system[B]—that is, neurons postganglionic to cholinergic sympathetic fibers arising in the central nervous system—there are *two* major central noradrenergic pathways and one minor. One major pathway arises in the locus ceruleus, A4 and A6 regions[76] consisting entirely of norepinephrine neurons. The locus ceruleus is located in the upper pons region of the medulla and projects the majority of its fibers forward (rostrally) in "bundles," giving off branches to many brain regions, including the brainstem, cerebellum, hypothalamus, amygdala, hippocampus, septum, fornix, cingulum, external capsule, and lateral and dorsal neocortex, terminating rostrally in olfactory nuclei (Figure 4.23). Some fibers travel backwards (caudally) from regions A1 and A2 into the spinal cord, cerebellum, and brain stem nuclei A1 and A2 including the cochlear nucleus.

The second noradrenergic pathway arises from neurons diffusely distributed in the brain stem of the subceruleus region (A1, A2, A5, and A7 regions).[76] They jointly give rise to the central tegmental bundle (CTB, Figure 4.23), then merge with the ventral bundle and ascend to join the median forebrain bundle (MFB). The two noradrenergic systems travel together as the ventral bundle for only a short distance until the fibers from the locus ceruleus course off dorsomedially to form the dorsal bundle. The majority of these fibers rejoin the ventral bundle by coursing ventrolaterally at the mamillary nucleus. The two bundles then ascend, together with the dopamine systems, as the median forebrain bundle (Figure 4.24).

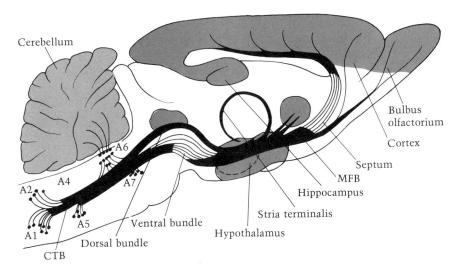

FIGURE 4.23   Norepinephrine pathways in rat brain shown in saggital section. Hatched areas represent nerve terminal fields. Pathways descending to the cerebellum and brain stem nuclei are not shown. [From ref. 77]

A third, periventricular system arises in the dorsal raphe nucleus and projects to the pretectal area, habenula, thalamus, and hypothalamus.[78]

Dopamine   Six central nervous system dopaminergic pathways have been described, in addition to the peripheral role of dopamine in sympathetic ganglia, visceral ganglia, and mesenteric and renal artery walls.

One of the central pathways is in the inner plexiform layer of the retina. Two of the others innervate the hypothalamus, namely, the incertohypothalamic system and the tuberoinfundibular system (region A12; see Figure 4.24 A and B). A major ascending pathway is the nigrostriatal system, which arises in the pars compacta of the substantia nigra[76] (region A9) and projects as the nigrostriatal bundle or tract to the corpus striatum (see Figure 4.24). In addition, some cells in the lateral substantia nigra project to the cingulum.[179] The sixth circuit is a tegmentotelencephalic system, also called the mesolimbic system. The fibers arise in the ventral tegmentum (region A8 and A10)[76] and project as part of the neostriatal tract and the median forebrain bundle to the amygdala, septum, nucleus accumbens, olfactory tubercle and frontal cortex (Figure 4.24).

Similar catecholamine pathways exist in all common laboratory animals, and homologous pathways exist in lower vertebrates.

Serotonin   Special fluorescence histochemical methods[73,85] applied to neurons full of serotonin after removal of catecholamines by neurotoxic 6-hydroxytryptamine action have revealed the serotonin pathways of the

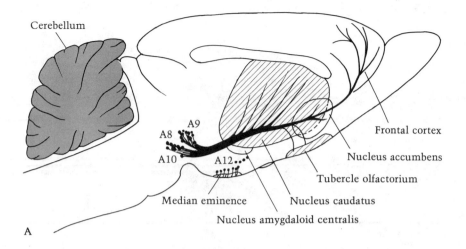

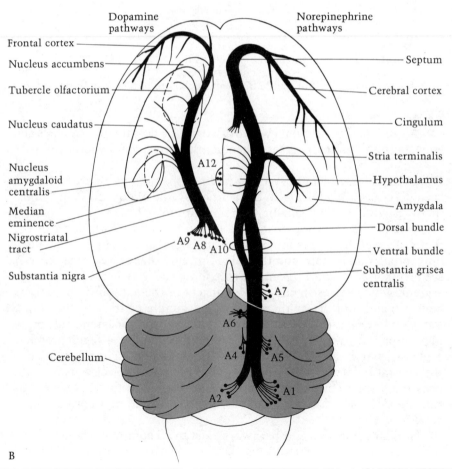

FIGURE 4.24 (A) Dopamine pathways in rat brain shown in saggital section. (B) Dopamine and norepinephrine pathways shown in horizontal section. Hatched areas represent nerve terminal fields. The numbers refer to defined groups of neurons. [From ref. 77]

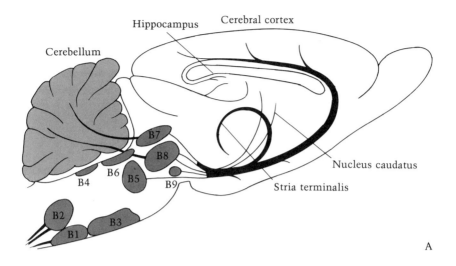

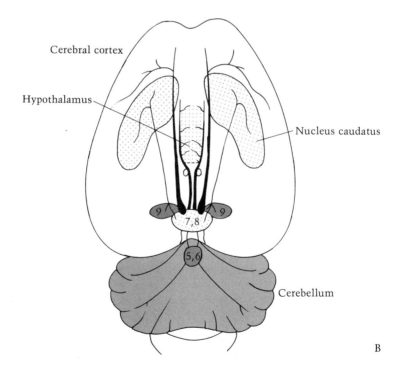

FIGURE 4.25   5-Hydroxytryptamine (serotonin) pathways in rat brain shown in saggital (A) and horizontal (B) projection. The numbers refer to defined groups of neurons. [From ref. 85]

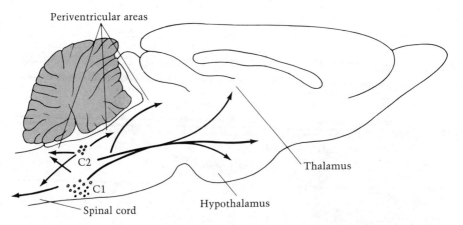

FIGURE 4.26   Schematic illustration of the epinephrine pathways in rat brain. The regions C_1 and C_2 represent the epinephrine cell-body groups in the rostral medulla oblongata. [From ref. 80]

central nervous system (Figure 4.26). The raphe nucleus, situated in the midline of the brain stem and medulla oblongata, is a diffuse nucleus consisting of nine cell groups.[73] Groups B7–9 are in the midbrain, and B1–6 are in the medulla, the pons, and the floor of the fourth ventricle. These cells produce a medial ascending pathway projecting to the hypothalamus and preoptic area. Lateral ascending tracts terminate diffusely in the corpus striatum and cerebral cortex. In addition, the cerebellum is innervated from cells in regions B5–8, and cells in the caudal medulla, B1–3, project backwards to the ventral and intermediate horns of the spinal cord.

Similar serotonin circuitry occurs in the human brain,[86] projecting from neurons grouped in a comparable way in medial regions of the brain stem, medulla, and pons-mesencephalon.

THE EPINEPHRINE
NEUROTRANSMITTER SYSTEM

Because of the relatively small quantities of the catecholamine epinephrine found in the brain compared with other monoamines and the associated technical difficulties in detecting it, not much effort was put into defining its precise role. Now, with the development of antisera to the purified form of its principal synthesizing enzyme, and the improvement of techniques for its accurate detection (e.g., high performance liquid chromatography linked to electrochemical detection), interest in this substance has rekindled (Figure 4.13).

Synthesis and Storage of Epinephrine

The main site of synthesis of epinephrine in the adrenal medulla, from which it is released directly into the blood stream. Its formation in brain slices has been demonstrated. In the brain it is particularly enriched in the brain-stem regions and in regions known to have high densities of norepinephrine-containing cells. Synthesis is by methylation of norepinephrine by the enzyme norepinephrine N-methyltransferase (EC 2.1.1.28) using S-adenosyl methionine as cofactor. This enzyme is commonly known as phenylethanolamine N-methyltransferase (PNMT). The complete sequence of enzymes (Figure 4.13 A to D) must exist in epinephrine neurons. A number of inhibitors of PNMT have been developed [e.g., 2,3,-dichloro-α-methylbenzylamine (DCMB)] that can modify epinephrine neurons pharmacologically without affecting norepinephrine or dopamine neurons in brain.

Antisera to purified PNMT from bovine adrenal gland have been used to map these epinephrine-forming neurons in rat brain.[80]

Release and Turnover of Epinephrine in Brain

Potassium depolarization, and no doubt other depolarizing treatments, releases endogenous epinephrine from brain by a calcium-dependent mechanism.[82,83] This is suppressed by clonidine and enhanced by yohimbine. Amphetamine also has a strong releasing action.[83] Endogenous release of epinephrine in vivo from the posterior hypothalamus has been monitored and found to be modulated by drugs (e.g., adrenergic β_1 and β_2 drugs) that affect dopamine and norepinephrine release in parallel.[84]

Monoamine oxidase inhibitors increase the content of all monamines, including epinephrine in brain, and alpha-adrenergic blocking drugs (e.g., yohimbine) and PNMT inhibitors (e.g., DCMB) decrease their content.[81]

The high-affinity transport of epinephrine into nerve endings and glial cells is almost certainly the principal method by which any synaptically released epinephrine is inactivated.[81] Much of this uptake may be via norepinephrine transport systems, however, and a separate epinephrine system has yet to be defined by selective transport blockers.[80,81]

Pharmacology of Epinephrine Neurons

Many of the drugs affecting the other catecholamines and the indoleamines have some action on epinephrine turnover and release, and drugs with the appropriate specificity towards epinephrine systems are still being sought.[80,81] Epinephrine will similarly stimulate cyclic AMP formation in neural tissue and has been found to be more potent in this respect than is norepinephrine in certain brain stem regions, suggesting the presence of specific epinephrine receptors linked to adenyl cyclase.[80,81]

Location of Epinephrine Neurons

Studies with antisera to PNMT show that epinephrine pathways (see Figure 4.26) project from the two brain stem nuclei (C1 and C2) located in the rostral part of the lateral reticular nucleus adjacent to the inferior olive (C1) and from the dorsomedial reticular formation just below the fourth ventricle (C$_2$) and adjacent periventricular areas. They course forward (rostrally) to the hypothalamus and thalamus, to the dorsal motor nucleus of the vagus nerve, to the nucleus of the solitary tract, and to the locus ceruleus. Some cells project backwards (caudally) to the lateral sympathetic nucleus and the central gray matter of the spinal cord.

Evidence to date, based on their anatomical projections and their physiology and pharmacology in rat brain, indicates that epinephrine neurons are likely to be involved in (1) vasomotor and respiratory control by central mechanisms, (2) thermoregulation, (3) the regulation of food and water intake and of oxytocin secretion, (4) the control of pituitary secretion. The projection to the locus ceruleus suggest that receptors for norepinephrine could control the activity of noradrenergic cell bodies in this nucleus. Indeed, epinephrine iontophoresed into the locus ceruleus inhibited the firing of these cells.[81]

Interest is especially focused on the control epinephrine neurons might exert on the central control of blood pressure, which is hinted at by the actions of a number of specific PNMT inhibitors.

THE AMINO ACID NEUROTRANSMITTER SYSTEMS

Five common amino acids have so far been seriously considered as possibly serving a neurotransmitter function. Current evidence leaves little room for doubt that glutamate and aspartate serve an excitatory role, and GABA and glycine function as inhibitory transmitters in vertebrate and invertebrate nervous systems. Taurine has inhibitory properties, but whether they are employed for some physiological purpose remains uncertain.

The outstanding difference between amino acid neurotransmitters and other neurotransmitters is their high content in neural tissue and their ubiquity and multiple involvement in mainstream biochemical processes, such as protein biosynthesis and intermediary metabolism. Their wide employment in both peripheral and central synapses in the nervous systems of many of the major animal phyla[E] indicates that they must be provided with special properties that gain them a strong competitive evolutionary selection value to overcome these disadvantages.

TWO EXCITATORY AMINO ACIDS: GLUTAMATE AND ASPARTATE

Synthesis and Storage of Glutamate and Aspartate

The amino acids glutamate and aspartate both carry a net negative charge and are close homologues in the dicarboxylic amino acid series. Glutamine and asparagine, the amidated relatives of glutamate and aspartate, carry no net charge and are physiologically inactive (see Figure 4.27). Both excitatory amino acids can be biosynthesized by transamination of their corresponding ketoacids (2-oxoglutarate and oxaloacetate, respectively), which are produced in the tricarboxylic acid cycle (Krebs cycle) whose location is in the mitochondria matrix (see Figures 3.2 and 3.11). Here glutamate is principally produced by transamination of 2-oxoglutarate by aspartate aminotransferase (EC 2.6.1.1.) and from 2-oxoglutarate and ammonium ions through the action of glutamate dehydrogenase (EC 1.4.1.3), a pyridine nucleotide–linked enzyme (Fig. 3.11). The latter favors glutamate production rather than its catabolism.[246] Although many substances feed into the citric acid cycle, radioactive glucose supplied either in vivo or in vitro rapidly gives rise to radiolabeled glutamate and aspartate greatly in excess of the likely needs of neurotransmission. Also, rapid interconversion of glutamate and aspartate occurs through the activity of the extremely rapid enzyme aspartate aminotransferase, which is found in both cytoplasm and mitochondria. Essentially, the enzyme reversibly converts 2-oxoglutarate to glutamate, employing aspartate as the donor of the amino group.

Glutamate, in particular, accumulates in larger pools in the brain than in other organs (Table 3.9), apparently because the unusually high K_m for oxidation of 2-oxoglutarate in brain tissue allows relatively large pools of this keto acid to build up and become transformed, or transaminated, to glutamate by glutamate dehydrogenase (see Chapter 3). Nerve terminals and their derived synaptosomes contain mitochondria that can generate these two amino acids from a number of substrates,

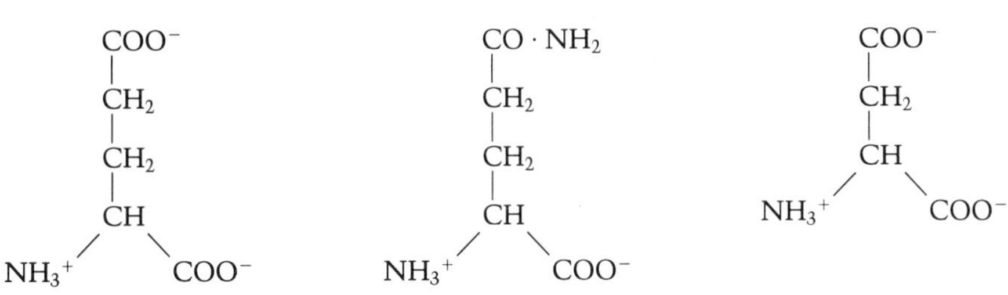

FIGURE 4.27   Structures of excitatory dicarboxylic amino acids and glutamine.

including glucose and pyruvate.[273] In addition, glutamate and aspartate are continuously formed during the turnover of tissue proteins and by the numerous transamination reactions in which they are involved. Also, rapid interconversion of 2-oxoglutarate and glutamate occurs in the malate-aspartate shuttle mechanism that serves the purpose of transferring reducing equivalents (NADH) from cytosol into mitochondria for oxidation. Because aerobic glycolysis is especially vigorous in neural tissue, the reduction of cytosolic NAD^+ to NADH occurs at a high rate and the enzymes that catalyze the reactions of the malate-aspartate shuttle are especially prevalent (Figure 4.28). Another source of glutamate is glutamine, which is readily converted to glutamate by the enzyme glutaminase (EC 3.5.1.2). This is also a mitochondrial enzyme.

With such multiple sources for these neurotransmitters, it seems very likely that compartmentation exists, partitioning transmitter production from the other metabolic activities that use or generate these compounds. The question arises, therefore, which metabolic pathways produce the glutamate and aspartate released by nerve terminals to serve a neurotransmitter function.

Two precursors that exist in liberal quantities in the brain and give rise to both glutamate and aspartate are D-glucose and L-glutamine.[215] The latter is hydrolyzed by the enzyme glutaminase, which has been isolated from brain[89] and from kidney.[90] The pig brain enzyme has a molecular weight in the range of 120,000–135,000 and appears to contain two subunits of about 64,000 each.[91] The isolated enzyme is inhibited by glutamate and activated by phosphate and a number of other anions. The enzymes isolated from kidney and brain are essentially identical. Detailed studies of the kinetics and molecular enzymology of glutaminase have been published.[91,92]

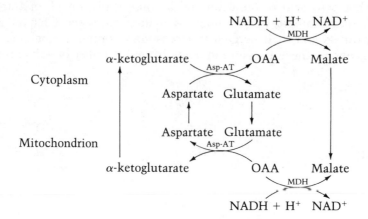

FIGURE 4.28 The malata-oxaloacetate shuttle for transport of reducing equivalents from cytoplasm to mitochondria. *Asp-AT*, aspartate aminotransferase; *MDH*, malate dehydrogenase; *OAA*, oxaloacetate.

Although both glucose and glutamine readily generate glutamate, the pools derived from glutamine are preferentially (80 percent) released from slices of hippocampus and also from incubated cortical synaptosomes during treatment with depolarizing agents.[96,97,98] This suggests that glutamine is a major precursor of neurotransmitter glutamate, and that glutaminase in nerve terminal mitochondria could be the enzyme responsible for its production. Control of glutaminase activity appears to be by product inhibition, glutamate greatly reducing the enzyme activity over the range of its likely concentrations in the terminal.[94,97] Glutaminase might be partially released from inhibition after glutamate efflux from the terminal during neural activity and allow the synthesis of replacement glutamate, a process which is seen in isolated synaptosomes and other preparations. Ammonia, too, is a moderate inhibitor and might also be involved in the control process.

Glutaminase has not yet been shown to be localized only to those neurons or nerve terminals that release glutamate, but there is evidence of its *enrichment* in glutamatergic neurons based on immunohistochemical studies with antibodies raised to glutaminase.[U,268]

In the brain, glutaminase appears to be concentrated in mitochondria, though 30–40 percent of the enzyme in the tissue is also recovered in synaptosome fractions, where it is likely to be localized to intraterminal mitochondria.[93,94] Evidence now exists that glutaminase is sited in different parts of the mitochondrion.[95] This is based on the partial inhibition of the enzyme in situ within the mitochondrion by N-ethylmaleimide and p-mercuribenzoate. This partial sensitivity seems to be due to a localization of a portion (40 percent) of the glutaminase to the inner mitochondrial matrix while the rest is attached to the outer side of the inner membrane, and possibly also to the outer membrane.[95] One consequence is that metabolites from the cytosol (e.g., phosphate and glutamate) would have access to the glutaminase located in the more superficial regions of the mitochondria, and the glutamate formed by glutamine hydrolysis would then readily enter the cytosol where it could serve a neurotransmitter function, among others. The deeper glutaminase located within the mitochondrial matrix might well provide glutamate. Glutamate is oxidized in the citric acid cycle, which is also restricted to this inner mitochondrial region.

Glutamine itself is present in cerebrospinal fluid at a concentration of approximately 0.50 mM. Its concentration in nerve terminals is very low, and a substantial inward diffusion of glutamine probably occurs in addition to its facilitated transport into terminals by a low-affinity (K_m = 0.2–2 mM) uptake systems.[274,275]

Although glutamine can also give rise to aspartate via glutamate, glucose appears to be an equivalent precursor. No special additional pathways for aspartate biosynthesis have yet been described, asparagine and asparaginase being either absent or only just detectable.

Other possible routes of production of glutamate for neurotransmitter or other purposes involve arginine, ornithine, and proline metabolism via the arginase, ornithine aminotransferase, and proline oxidase reactions in which glutamate semialdehyde is an intermediate. The rates of synthesis from these sources is slow, however, and any contributions are likely to be small.[98] A direct precursor role for 2-oxoglutarate transferred to nerve terminals from astrocytes has also been postulated.[98] This involves production of the ketoacid in astrocytes via a route involving carbon dioxide fixation by the anaplerotic enzyme pyruvate carboxylase (Figure 3.13) and high-affinity, Na^{2+}-dependent uptake of 2-oxoglutarate by nerve terminals (see Chapter 3, page 151).[98]

The ubiquitous presence and high concentration (3–10 mM) of glutamate and aspartate in neural tissues makes localization to specific neurons and within neurons a difficult task (Table 3.9). However, several studies have shown that synaptic vesicles isolated from mammalian brain contain glutamate, aspartate, and GABA at levels equivalent to or greater than those of acetylcholine and other neurotransmitters,[98–102,131] though this represents only a fractional percentage of the total present in the terminal. The majority of these compounds are recovered in the cytosolic and mitochondrial subfractions, part of which could have leaked from vesicles during their isolation. In spite of these considerations, antisera raised against a chemical complex of glutamate with glutaraldehyde have been used in immunohistochemical localization studies of glutamatergic neurons. These studies indicate that glutamate is *enriched* in neuronal systems assessed on other grounds to be glutamatergic.[U,269]

Release and Inactivation of Glutamate and Aspartate

A great deal of experimental evidence demonstrates the Ca^{2+}-dependent release of these two excitatory amino acids from both in vivo and isolated neural preparations,[100,101,276,E] the former employing superfusion cannulae.[103] In the isolated preparations, release can be evoked by electrical pulses or by depolarizing chemical stimuli such as high K^+-levels (56 mM) or veratrine (a mixture of alkaloids that activate sodium channels). For the most part the aspartate, glutamate, and GABA released from incubated synaptosomes are replaced by synthesis within the tissue, as occurs with other neurotransmitters.

A striking feature of amino acid release evoked from synaptosomes and from other in vitro preparations by depolarizing agents is its differential nature. Only the three physiologically active amino acids glutamate, aspartate, and GABA are released from cerebral cortex (Figure 4.29). Amino acid release from spinal cord synaptosomes was induced by adding K^+ ions to a final concentration of 56 mM over a range including the threshold levels of K^+ that induce depolarization (i.e., 15–20 mM). In the experiments summarized in Figure 4.30, aspartate, glutamate, GABA, and glycine were released, glycine being an inhibitory transmitter candidate at certain spinal cord interneuron synapses (see below).

In vivo release of amino acids shows the same differential pattern whether evoked by chemical agents delivered onto exposed cerebral cor-

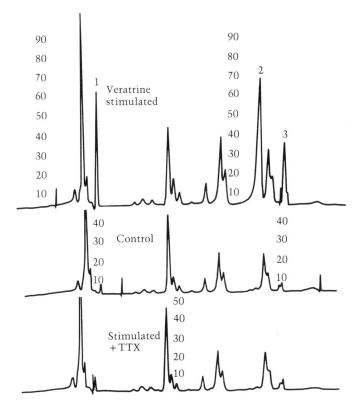

FIGURE 4.29   Chromatograms showing evoked release of physiologically active amino acids from cerebrocortical synaptosomes. The stimulus was provided by the depolarizing agent veratrine (75 μM), a mixture of alkaloids that activate voltage-sensitive sodium channels. Chromatograms show the amino acid content of incubation medium from (top trace) stimulated synaptosomes; (middle trace) unstimulated synaptosomes; (bottom trace) synaptosomes treated with veratrine in the presence of tetrodotoxin, a toxin that prevents veratrine action on voltage-dependent sodium channels. Key to amino acids: 1, GABA; 2, glutamate; 3, aspartate.

tex or by activation of synapses in the superfused area from stimulation of physiological routes[103,132] (Figure 4.31). This phenomenon provides particularly compelling evidence for a neurotransmitter role for these amino acids. Release evoked both in vivo and in vitro is blocked by morphine, a drug known to reduce or prevent neurotransmitter release.[103]

Whether amino acid neurotransmitters, including glutamate and aspartate, are synaptically released from the cytosol or from synaptic vesicles by exocytosis remains to be established. Evidence to date, based on comparisons of the specific radioactivities of amino acids released in vitro by stimulation with the radioactivities of amino acids recovered from cytosol or synaptic vesicles, suggests an origin in the cytosol.[98,101]

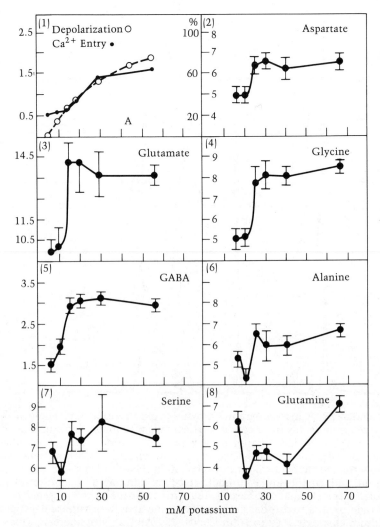

FIGURE 4.30 Release of amino acids from incubated synaptosomes isolated from spinal cord and medulla, induced by potassium at different concentrations. Values are means ± SEM for 6 to 10 experiments. Graph A shows the entry of calcium (●) and the theoretical percentage of membrane depolarization (○) induced by the potassium concentrations employed. [From ref. 100]

As with the catecholamines and serotonin,[62] inactivation of released aspartate, glutamate, GABA, and glycine is primarily by reuptake across the membranes of neural cells by high-affinity Na^+-dependent transport systems.[98,101] Once across the membrane, inactivation has been achieved, and subsequent metabolism is by different pathways. Both high-affinity $(K_m = 5-20 \ \mu M)$ and low-affinity $(K_m = 1-2 \ mM)$ transport systems exist

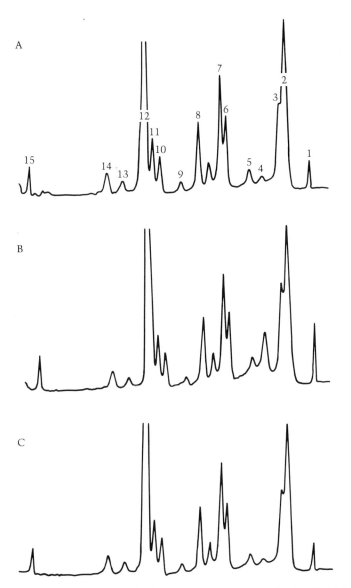

FIGURE 4.31 Typical chromatograms of superfusion fluid from sensorimotor cortex showing effects of sensory stimulation via the brachial plexus. Fractions of superfusate (8 minutes, 1 ml) were collected from sensorimotor cortex of a rat treated with γ-acetylenic GABA (100 mg/kg, IP) 45 minutes before stimulation. (A) Before stimulation; (B) during stimulation; (C) after stimulation. Key to amino acids: *1*, methionine sulfoxide plus aspartate; *2*, threonine plus glutamine; *3*, serine; *4*, glutamate; *5*, citrulline; *6*, glycine; *7*, alanine; *8*, valine (equivalent to 150 pmol and constant in each chromatogram); *9*, methionine; *10*, isoleucine; *11*, leucine; *12*, norleucine (added standard of 500 pmol); *13*, tyrosine; *14*, phenylalanine; *15*, GABA. Note the increased release of aspartate and glutamate.

for all those amino acids serving a neurotransmitter function, but only low-affinity systems occur for other amino acids,[104] once again highlighting the functional differences between these two classes of amino acids.

The cells that carry these inactivating high-affinity transport systems are those neurons that release the amino acid transmitter concerned and also the surrounding glial cells. No definite proof exists that aspartate and glutamate have separate transport systems located on different classes of neurons, and a single system is probably responsible for uptake of both dicarboxylic amino acids. In fact the nonmetabolized optical isomer D-[^{3}H]aspartate loaded into neural preparations is often used for studying glutamate uptake and release.

Glutamate taken into glial cells is either oxidized through the Krebs cycle located in the mitochondria of these cells or converted to glutamine by the enzyme glutamine synthetase, which is exclusively located in glial cells.[15] Once formed, glutamine is readily discharged from glial cells into the extracellular space, or incubation fluid, by simple diffusion; it is not avidly reconcentrated in cells by its transport systems since these are only of low affinity. Once in the cerebrospinal fluid, glutamine readily enters nerve terminals via these low-affinity systems or by diffusion. There glutaminase will convert it back to glutamate, which can again be employed for transmission purposes. Thus, a glutamate-glutamine cycle of transfer exists between neurons and glia (Figure 4.32). The existence of such a cycle is supported by the finding that isolated nerve terminals

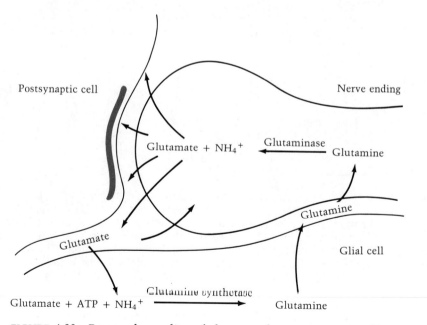

FIGURE 4.32 Proposed recycling of glutamate between neurons, glia, and extracellular space in brain tissue. Reuptake of glutamate by the nerve ending is not precluded.

(synaptosomes) contain very little of the tissue content of glutamine synthetase and a great deal of glutaminase.[93,94,105] Autoradiographic and biochemical studies also show very clearly that in dorsal root ganglia, [³H]glutamate was selectively accumulated by satellite glial cells and rapidly converted to [³H]glutamine. In contrast, [³H]glutamine preferentially entered neuronal perikarya, where it was converted in large proportions (40 percent) into [³H]glutamate.[106]

Pharmacology of Glutamate and Aspartate

Since the mid-1950s the powerful excitatory actions of glutamate and aspartate applied to brain and spinal cord neurons by iontophoresis, and the even more powerful effects of their N-alkyl derivatives, have been something of a curiosity. An extraordinarily wide range of neurons in different brain regions are excited, particularly by glutamate, though their sensitivities to this amino acid vary and neurons in some regions are quite insensitive.[E,J,K,108] Glutamate-operated synapses (or, at least, glutamate receptors) are very widely distributed and occur on neurons receiving inputs from many other transmitter systems.

It nows seem very likely that glutamate receptors exist as a population quite distinct from aspartate receptors. Evidence has come primarily from studies of the differential effects of agonists and antagonists on excitation induced by the two agents in different regions of the central nervous system and is based on the theoretical concept that aspartate, being a shorter molecule, would have difficulty fitting a receptor designed for glutamate. Glutamate, on the other hand, could interact with aspartate receptors if it adopted a "folded" configuration (Figure 4.33).

In practice, one population of receptors (NMDA receptors, aspartate type) is found to be excited by N-methyl-D-aspartate and also by L-glutamate, D-glutamate, L-homocysteate (a sulfur analogue of glutamate), ibotenate (extracted from the mushroom *Amanita muscaria*), and a number of other aspartate analogues (Figure 4.34). These receptors are seen to be activated preferentially by aspartate analogues or by glutamate analogues that are able to fold to reduce the distance between the alpha and omega carboxyl groups.

NMDA receptors are specifically blocked by Mg^{2+} at low concentrations (10 μM) and are also preferentially blocked by a group of specially synthesized longer chain mono- and diaminocarboxylic acids at low concentration. These include D-α-aminoadipate (DαAA), DD- and DL-α,ε-diaminopimelate (DAPA), and γ-D-glutamylglycine (γDGG).[108,N,O] Also potent and specific blockers are the phosphonic acid analogues (α-amino-ω-phosphonocarboxylic acids), 2-amino-5-phosphonovaleric acid (APV), and 2-amino-7-phosphonoheptanoic acid (APH). All of these compounds are conceived as interacting with the NMDA receptor in a folded configuration that reduces their effective length. Significantly, these compounds are more effective against aspartate than against glutamate-induced excitation.[261,270]

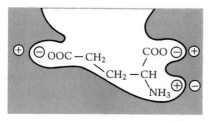

Accommodating glutamate

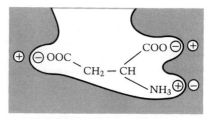

Accommodating aspartate

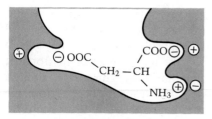

Aspartate too short
for optimal interaction

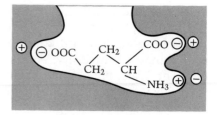

Glutamate "folds" to fit

FIGURE 4.33 Hypothetical receptors showing a preference for either glutamate or aspartate. The amino acids are shown binding to the neuronal membrane (shaded area) by way of ionic bonds between oppositely charged groups on the amino acid and membrane molecules. Aspartate is too short for optimal binding to the glutamate-preferring receptor, but glutamate can bind to the aspartate-preferring receptor by folding its molecular chain.

Two other categories or subpopulations of receptors (e.g., in dorsal horn interneurons) are preferentially excited by two compounds, quisqualate and kainate, which are glutamate analogues whose ring structure holds them in the rigid "extended" form of the glutamate configuration. They could be expected to interact preferentially with the glutamate receptor since they cannot fold to fit the aspartate receptor. Excitations produced by quisqualate, glutamate, or another agonist of this type (AMPA) are best blocked by antagonists such as glutamate diethylester (GDEE), PDA, APB, and GAMS (Figure 4.35). Kainic acid–induced excitation, however, appears to be relatively resistant to GDEE as well as to DαAA but is attenuated by 2-amino-4-phosphonobutyrate (APB). Again, NMDA, aspartate, and kainic acid–induced excitation of spinal neurons is blocked by γDGG, but this compound is less effective on excitation induced by glutamate or quisqualate.[109] Quisqualate and kainate may be exciting different subpopulations of receptors, possibly representing sites

FIGURE 4.34 N-Methyl-D-aspartic acid, related agonists, and a selection of their antagonists.

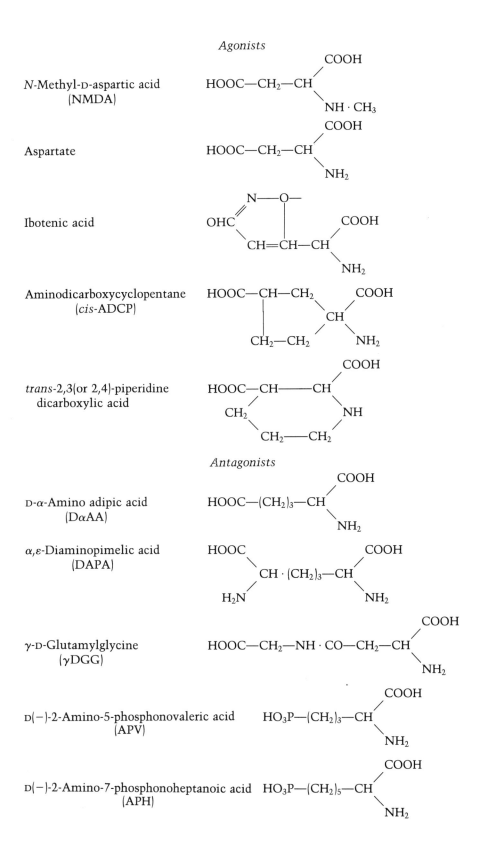

Agonists

N-Methyl-D-aspartic acid (NMDA)

$HOOC-CH_2-CH\begin{smallmatrix}COOH\\NH\cdot CH_3\end{smallmatrix}$

Aspartate

$HOOC-CH_2-CH\begin{smallmatrix}COOH\\NH_2\end{smallmatrix}$

Ibotenic acid

Aminodicarboxycyclopentane (*cis*-ADCP)

trans-2,3(or 2,4)-piperidine dicarboxylic acid

Antagonists

D-α-Amino adipic acid (DαAA)

$HOOC-(CH_2)_3-CH\begin{smallmatrix}COOH\\NH_2\end{smallmatrix}$

α,ε-Diaminopimelic acid (DAPA)

γ-D-Glutamylglycine (γDGG)

$HOOC-CH_2-NH\cdot CO-CH_2-CH\begin{smallmatrix}COOH\\NH_2\end{smallmatrix}$

D(−)-2-Amino-5-phosphonovaleric acid (APV)

$HO_3P-(CH_2)_3-CH\begin{smallmatrix}COOH\\NH_2\end{smallmatrix}$

D(−)-2-Amino-7-phosphonoheptanoic acid (APH)

$HO_3P-(CH_2)_5-CH\begin{smallmatrix}COOH\\NH_2\end{smallmatrix}$

Agonists

Quisqualic acid

$$\underset{O=C}{\overset{\overset{\displaystyle H}{N}-\overset{\overset{\displaystyle O}{\|}}{C}}{\Big|}} \quad \underset{O}{\overset{N-CH_2-CH}{\diagup}} \quad \overset{COOH}{\underset{NH_2}{}}$$

Kainic acid

$$H_3C-\overset{\overset{\displaystyle CH_2}{\|}}{C} \quad CH_2 \cdot COOH$$
$$CH-CH$$
$$CH_2 \qquad CH \cdot COOH$$
$$\underset{H}{N}$$

Glutamic acid

$$HOOC-(CH_2)_2 \cdot CH \overset{COOH}{\underset{NH_2}{\diagup}}$$

D-Amino-3-hydroxy-5-methyl-
4-isoxazolepropionic acid
(AMPA)

$$OH \qquad CH_2 \cdot CH \overset{COOH}{\underset{NH_2}{\diagup}}$$
$$\underset{N}{\underset{O}{\diagup}} \quad CH_3$$

Antagonists

Glutamate diethyl ester
(GDEE)
(not for kainic acid)

$$C_2H_5 \cdot OC-(CH_2)_2-CH \overset{COOH}{\underset{NH_2}{\diagup}}$$

D(−)-2-Amino-4-phosphonobutyrate
(APB)

$$HO_3P-(CH_2)_2-CH \overset{COOH}{\underset{NH_2}{\diagup}}$$

cis-2,3-Piperidine dicarboxylic acid
(PDA)

$$\underset{CH_2}{\overset{HOOC}{\diagdown}} \underset{CH-CH}{} \overset{COOH}{\diagup}$$
$$CH_2 \qquad NH$$
$$CH_2-CH_2$$

γ-D-Glutamyl aminomethyl sulfonate
(GAMS)

$$SO_3 \cdot CH_2 \cdot NH \cdot \underset{\underset{O}{\|}}{C} \cdot (CH_2)_2 \cdot \underset{\underset{NH_2}{|}}{CH} \cdot COOH$$

FIGURE 4.35 Quisqualic acid, kainic acid and related agonists, and a selection of their antagonists. Glutamate diethyl ester is not an antagonist for kainate.

for interaction of the "extended" and folded configurations of glutamate or perhaps indicating the presence of as yet unrecognized endogenous neurotransmitters.

In summary, evidence exists for the presence of three subpopulations of excitatory neurotransmitter receptors.[110,195] These are (1) NMDA receptors, which could be glutamate or aspartate receptors; (2) quisqualate receptors, which are likely to represent glutamate receptors; (3) kainic acid receptors, which could be identifying either a subcategory of glutamate receptors (extended configuration) or sites for an as yet unidentified endogenous neurotransmitter.[280]

Radiolabeled ligand-binding studies support this receptor classification scheme.[110] The radioligands D-[³H]APV, [³H]AMPA, and [³H]KA specifically label three different binding sites corresponding to NMDA, quisqualate, and kainate receptors, respectively. Consistent with the electrophysiological characterization of these receptors, L-glutamate has a similar high affinity for each of these sites, while L-aspartate prefers to interact with the NMDA receptor. Autoradiographic ligand-binding studies confirm that these receptors are found in synaptic fields and that they are differentially distributed throughout the CNS in an anatomically distinct manner.[212]

Studies of the effects of ions upon L-[³H]glutamate binding reveal two additional sites. In the presence of Na^+ ions, L-[³H]glutamate binding (or uptake) occurs at the Na^+-dependent L-glutamate uptake site. In the presence of Ca^{2+} and Cl^- ions, L-[³H]glutamate binding (or possibly Ca^{2+}/Cl^--dependent uptake to membrane vesicles) is greatly enhanced (fivefold) to another site, which is distinguished by its high affinity for L-APB.[110,212,257–260,282] Binding at this site is inhibited by Na^+ ions, but whether it is a true receptor is undetermined.

Evidence of a different kind has indicated an association of NMDA receptors with L-aspartate and quisqualate receptors with L-glutamate. This was provided by observations of the differential actions of the antagonist compounds APV, γDGG, GAMS, and PDA on various modes of exciting dorsal horn motoneurons in cat spinal cord (Figure 4.35).[111,261] Excitation of these cells by stimulating primary afferents entering the spinal cord activates both monosynaptic and polysynaptic pathways, which can be distinguished. The polysynaptic pathway, involving intercalated excitatory interneurons, was blocked by APV at the low concentrations that selectively inhibited excitation owing to direct application of NMDA or L-aspartate to motoneurons. In contrast, PDA, GAMS, and γDGG selectively blocked both the monosynaptic pathway and excitation due to quisqualate or L-glutamate. L-Glutamate and L-aspartate have long been favored as the candidates for neurotransmitter function at the nerve terminals of primary afferent fibers and at the terminals of excitatory interneurons respectively.

In spinal cord, depolarizations caused either by excitation of the dorsal root or by NMDA are inhibited by low concentrations of the NMDA antagonists, and in a parallel fashion, indicating that NMDA is exciting physiologically functional receptors. The smaller fraction of dorsal root–evoked synaptic excitation that is not blocked by NMDA antagonists (e.g., APV) can be reversibly depressed by γDGG or PDA, indicating that other functionally active receptors could be of the quisqualate or kainate type.

The potent agonist 2-amino-4-phosphonobutyrate blocks the main input terminals to the hippocampus (perforant path) which, on other grounds, is thought to use glutamate as a neurotransmitter.[112]

Though not yet firmly established, there is a strong possibility that neurons employing aspartate as their neurotransmitter release this compound at their terminals, where it interacts with its own specific receptor population. A similar picture may be drawn for glutamate released from its own specific neurons.

The actions of the agonist kainic acid (Figure 4.35) administered in vivo are rather dramatic. This agent, which was first extracted from the seaweed *Digenea simplex*, destroys accessible neurons that carry "glutamate" receptors by an "excitotoxic action" involving greatly extended postsynaptic excitation with its associated depolarization, excessive Na^+ influx, and increased energy demand on neurons. This appears to cause such depletion of energy reserves that the postsynaptic cell is prevented from performing the metabolic processes necessary to maintain its viability. The neurons affected are those relatively unprotected by the blood–brain barrier, such as the circumventricular organs, the median eminence,[113] and the hypothalamus (see Figure 2.22).

Glutamate itself has a similar effect in both neonatal and adult animals. Both glutamate and kainate differentially destroy dendrites and cell bodies, leaving axons and terminals to degenerate later, which supports the hypothesis that postsynaptic sites are the targets of their action.

A complex mixture of neuronal excitation and inhibitory spreading depression caused by glutamate probably plays a part in the condition called Chinese restaurant syndrome in humans, which is associated with the ingestion of large quantities of monosodium glutamate added to food for its flavor.[114] The symptoms of this syndrome include a wide range of peripheral sympathetic and central nervous system responses such as headache, hot flushes, dizziness, and nausea.

There are few uncontested specific uptake blockers for glutamate or aspartate. Compounds with such selectivity in their action would be invaluable tools in several aspects of the study of these neurotransmitters, since they should allow differential potentiation of neurons activated by glutamate or by aspartate. As with receptor agonists and antagonists, however, the close structural similarity between the two compounds, and the ability of glutamate to interact with aspartate receptors in a folded conformation, predicts considerable "crossover" in the interaction of the "active sites" of their transport systems. In fact, a single system seems to transport both molecules. Consequently, uptake blockers, too, show considerable overlap in their capacity to interfere with aspartate and glutamate uptake.

Among the compounds displaying effective and relatively specific uptake-inhibition for these two acidic amino acids, threo-3-hydroxy-L-aspartate is the most potent[115,116] ($IC_{50} = 3.2$ μM for glutamate and 4.0 μM for aspartate) with agents such as dihydrokainic acid ($IC_{50} = 176$ μM for glutamate) and its potently neurotoxic relative kainic acid ($IC_{50} = 302$ μM) as somewhat less effective agents in this respect.[117] Dimethylglutamate ($IC_{50} > 500$ μM) is in the same category[117] and has been shown to potentiate the excitatory action of glutamate iontophoresed onto neurons.

Isolation of Glutamate and Aspartate Receptors

Although there have been important advances in characterizing the pharmacological responses of amino acid receptors, little progress has been made in their isolation compared with the cholinergic and catecholaminergic receptors. This has been largely owing to the lack of availability of marker substances equivalent to α-bungarotoxin and other cholinergic agents that will remain firmly bound to the receptors during the isolation procedures. However, the specific high-affinity binding of radiolabeled glutamate and GABA to hydrophobic protein fractions isolated from crustacean muscles and from fly and locust muscles has been reported,[118,119] all of which employ both chemicals as neurotransmitters at the neuromuscular junction. In addition, a glutamate-binding glycoprotein derived from synaptic membranes has been isolated.[118,120] Such reports suggest that isolation is possible, though real progress must await the development of new marker agents.

Many studies have demonstrated the existence of specific high-affinity receptors for glutamate or aspartate on neuronal membranes.[98,121,122,123] Glutamate receptors in the cerebellum have been found to rise in number during the second and third weeks of development, when glutamatergic parallel fiber synaptogenesis is known to be occurring.[123,124]

Location of Pathways Employing Glutamate and Aspartate

Sensitivity to iontophoresed glutamate and aspartate is widespread, indicating the equally ubiquitous distribution of receptors for these excitatory compounds. Clear-cut biochemical markers for these transmitter systems are lacking, however, and the only currently available test is a search for the presence of high-affinity transport systems, though immunohistochemistry using antisera to glutamate or aspartate may soon prove their value.[269]

Both glutamate and aspartate appear to be transported by a common high-affinity carrier, and distinction between the two compounds on this basis is therefore not possible. But the two systems may be conveniently detected by the high-affinity accumulation of radiolabeled D-aspartate. This compound is not metabolized, and it accumulates at its site of uptake (Figures 4.36 and 4.37). Loss of such "high-affinity" uptake capac-

FIGURE 4.36 Schematic drawing of a cross section of the rat hippocampal formation. The systems of excitatory nerve terminals are arranged in laminae parallel to the curved cortical surface and the pyramidal (P) and granular (G) cell layers. The axons giving rise to these terminals run in lamellae perpendicular to the laminae and parallel to the plane of the section shown, but there is also a considerable longitudinal spread of axons perpendicular to the lamellae. The chief input of the region is the perforant path (PP) from the area endorhinalis to the fascia dentata and CA3. It comprises a medial (m) and lateral (l) part, exciting the middle (Mm) and outer (Mo) parts of the granular cell dendrites. This laminated projection extends to the molecular layer (M) of CA3 and is mostly ipsilateral. A separate set of cells in area entorhinalis sends PP fibers (temporoammonic fibers) bilaterally to the molecular layer of CA1. The axons of the granular cells, the mossy fibers, give rise to characteristic giant boutons exciting dendrites of cells in hilus fasciae dentatae (H, CA4) and in stratum lucidum (LU) of hippocampus CA3. The hilar cells (in area H) have axons synapsing with dendrites in the inner molecular area (Mi) of the area dentata ipsilaterally and contralaterally. Pyramidal neurons of CA3 (P) project to stratum radiatum (R, Schaffer collaterals) and stratum oriens (O) in regio superior and inferior (CA1 to CA3). These pyramidal cells have axons synapsing with dendrites in the gray area in the hippocampus and in the area dentata ipsilaterally and contralaterally. Ipsilateral axons from CA3 to the superficial part of stratum radiatum (R) of CA1 are known as Schaffer collaterals. Key: A, alveus; F, fimbria; G, stratum granulare of area dentata, H, hilus fasciae dentatae; FH, fissura hippocampi; LU, stratum lucidum (mossy fiber layer); M, stratum lacunosum-moleculare of hippocampus; Mo, Mm, and Mi, outer, middle, and inner parts of stratum moleculare of area dentata; O, P, and R, strat oriens, pyramidale, and radiatum of hippocampus; PP, perforant path from area entorhinalis with lateral (l) and medial (m) parts; short arrows indicate borders between cortical subfields. Glutamate is the putative excitatory neurotransmitter of the perfaront path (PP) on granular and pyramidal cell dendrites and also of the mossy fiber endings. [From ref. 121]

FIGURE 4.37 (A) Uptake of [³H]GABA, and (B) uptake of D-[³H]aspartate in the normal hippocampal formation of the rat. Note the striking complementarity of the distributions in (A) and (B). L-[³H]Glutamate and L-[³H]aspartate show patterns identical with (B). Surface autoradiograms of incubated slices. Key: g, stratum granulare; p, stratum pyramidale; lu, stratum lucidum; o and r, sharp transition in labeling intensity of stratum oriens and radiatum where CA1 abuts on subiculum; short arrow, limit between Mo and Mm; long arrow, limit between stratum radiatum and lacunosummoleculare of CA3. Coarse dot shows bottom of hippocampal fissure (f). See also Figure 4.36. Bar, 500 μm. [From ref. 121]

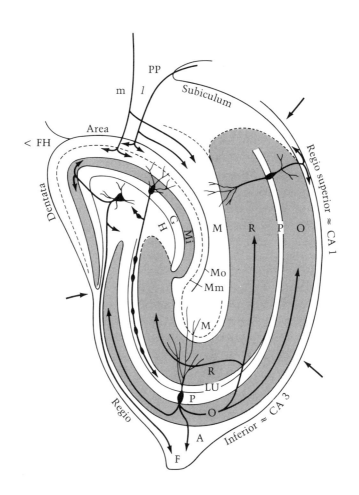

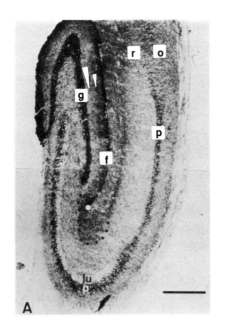

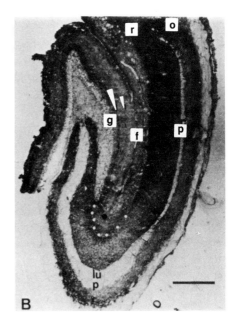

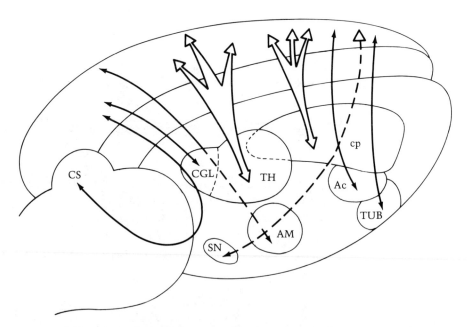

FIGURE 4.38   Glutamate-operated pathways in the brain. Proposed origin and distribution of glutamate fibers from neocortex. *Ac,* nucleus accumbens; *cp,* neostriatum; *TH,* thalamus; *CGL,* lateral geniculate body; *CS,* colliculus superior; *SN,* substantia nigra; *AM,* amygdala; *TUB,* olfactory tubercle. The projections to *Ac, TUB,* and *cp* come mainly from the frontal part of the cortex. The projection to *SN* is very small and probably also comes from the frontal part. The projection to *TH* comes from the entire cortex, but particularly from pyriform cortex. The projection to *AM* also passes through the pyriform cortex. The projections to *CGL* and *CS* come mainly from visual cortex. [From ref. 177]

ity following nerve section, coupled with a diminution in tissue level of glutamate or aspartate, indicates a neurotransmitter function for one or the other amino acid at the site concerned.

In the case of glutamate, this approach has indicated glutamateric pathways in several brain regions, including the hippocampus and corpus striatum[142,143,177] (Figures 4.36 to 4.39). For instance, sectioning of the perforant path (the major excitatory input to the hippocampus from the endorhinal area of the cerebral cortex) by lesion of the endorhinal cortex leads to a substantial and specific fall in high-affinity glutamate uptake,[141,143] as does sectioning of the corticostriate pathway, the major input from cerebral cortex to corpus striatum.

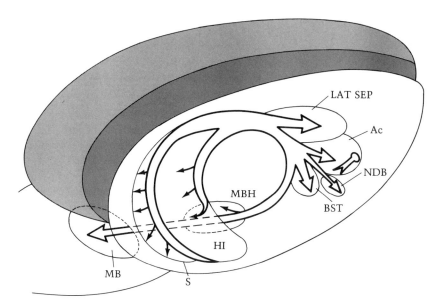

FIGURE 4.39 Glutamate-operated pathways in the brain. Proposed distribution of fornix/fimbria glutamate fibers from hippocampus-subiculum. *LAT SEP*, lateral septum; *Ac*, nucleus accumbens; *NDB*, nucleus of diagonal band; *MBH*, mediobasal hypothalamus; *MB*, mammillary body; *S*, subiculum; *HI*, hippocampus. The fibers to *LAT SEP* and *NDB* come mainly from hippocampus; the other fibers come mainly from subiculum. [From ref. 177]

TWO INHIBITORY AMINO ACIDS: GABA AND GLYCINE

GABA Synthesis

The two neutral amino carboxylic acids GABA and glycine are not closely related either metabolically or structurally in the sense that applies to glutamate and aspartate. GABA is biosynthesized by the important pyridoxal phosphate-dependent enzyme glutamic acid decarboxylase (GAD; EC 4.1.1.15).[214] This acts on glutamate, as its name implies, and removes the γ-carboxyl group as CO_2 to produce a γ-amino acid (Figure 4.40). The glutamate employed as substrate is readily produced via the citric acid cycle. Thus the carbon skeleton for GABA can be derived from mainstream metabolic traffic through the citric acid cycle and consequently rapidly labeled by substrates, such as glucose, that feed into the

FIGURE 4.40   The GABA shunt
pathway.

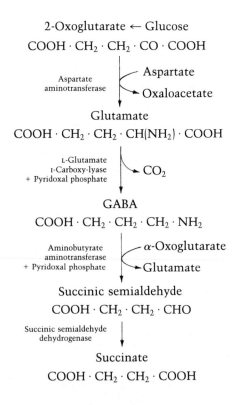

2-Oxoglutarate ← Glucose
$$COOH \cdot CH_2 \cdot CH_2 \cdot CO \cdot COOH$$

Aspartate
aminotransferase
⟨ Aspartate
↘ Oxaloacetate

Glutamate
$$COOH \cdot CH_2 \cdot CH_2 \cdot CH(NH_2) \cdot COOH$$

L-Glutamate
1-Carboxy-lyase
+ Pyridoxal phosphate
↘ CO_2

GABA
$$COOH \cdot CH_2 \cdot CH_2 \cdot CH_2 \cdot NH_2$$

Aminobutyrate
aminotransferase
+ Pyridoxal phosphate
⟨ α-Oxoglutarate
↘ Glutamate

Succinic semialdehyde
$$COOH \cdot CH_2 \cdot CH_2 \cdot CHO$$

Succinic semialdehyde
dehydrogenase

Succinate
$$COOH \cdot CH_2 \cdot CH_2 \cdot COOH$$

cycle. Glutamate decarboxylase itself appears to be localized to the nerve terminals of neurons, where it is in the soluble cytoplasmic phase, although it can be induced to bind the membranes in the presence of Ca^{2+}. The localized concentration of glutamate decarboxylase has been convincingly demonstrated by immunohistochemistry of tissue sections employing antibodies to purified GAD[125] (Figure 4.41). Its selectivity and its correlation with the presence of GABA-mediated transmission strongly indicates that only the GABA-releasing terminals contain this enzyme, for which it is therefore a chemical marker.

Until the early 1970s, two forms of GAD were though to exist at different sites in the nervous system, GAD I and GAD II.[196] GAD II was detected in some nonneural tissues (heart, kidney, liver). This classification was based on the finding that the GAD II form was stimulated by carbonyl trapping agents (e.g., amino oxyacetic acid) and by chloride ion, and the other form was inhibited by them. Investigation eventually demonstrated[197,198,199] that this was an artifact of the assay procedure, which involved collection of $^{14}CO_2$ from L-[1-^{14}C]glutamate. It was shown by purifying the radiolabeled substrate that contaminants were responsible for the increased $^{14}CO_2$ produced by the stimulating agents, due to metabolism via pathways other than those catalyzed by GAD (e.g., the citric

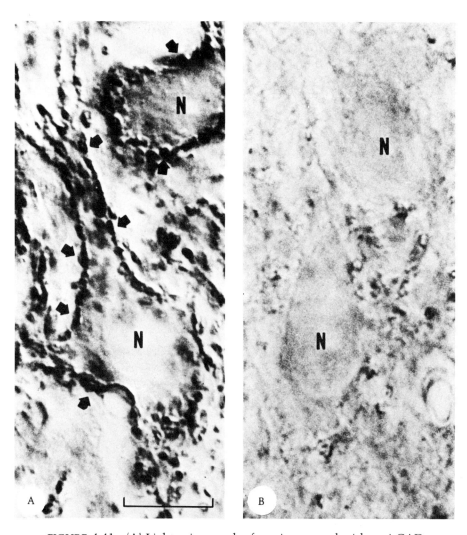

FIGURE 4.41   (A) Light micrograph of section treated with anti-GAD
serum showing dense, punctate deposits of reaction product (*arrows*) on
the soma and proximal dendrites of neurons (N) in a deep cerebellar nucleus. Nomarski optics. Rat cerebellum. (B) Light micrograph of control
section treated with nonimmune serum showing several neurons (N) in a
deep cerebellar nucleus. Localized reaction product is not seen. Nomarski
optics. Bar, 10 μm. [From ref. 125]

acid cycle). When purified substrate was employed both chloride ion and
amino oxyacetic acid substantially inhibited $^{14}CO_2$ production, and rates
of GAD activity were similar in all brain regions studied, whether measured by $^{14}CO_2$ formation from L-[1-^{14}C]glutamate or by [^{14}C]GABA production from L-[U-^{14}C]glutamate.

GAD has been isolated and purified from mouse brain and shown to have a molecular weight of 85,000.[126] It has a K_m of 0.7 mM for glutamate and 0.05 mM for pyridoxal phosphate, well below the normal in vivo concentrations of these compounds. Pyridoxal phosphate is relatively loosely bound to the apoenzyme of GAD, rendering it more vulnerable to pyridoxal antagonists.

Glycine Synthesis

The main pathway of glycine formation is via D-3-phosphoglycerate, itself derived from the glycolysis stream and hence from mainstream metabolism (Fig. 4.42). The route involves a series of organic phosphates including phosphohydroxypyruvate, phosphoserine, and serine itself, with serine levels controlling the pathway by acting as feedback inhibitor of phosphoserine phosphatase.[127] This route, with its close connections to the mainstream metabolism of glucose, probably provides the bulk source of both serine and glycine in the brain. Radioactive tracers show that a large proportion of glycine is derived from serine.[128] The parallel pathway involving the nonphosphorylated compounds D-glycerate, hydroxypyruvate, and serine may also operate.[129] Significantly, the linked enzyme D-glycerate dehydrogenase (EC 1.1.1.29) appears most enriched in those regions (e.g., medulla, spinal cord) where glycine is most concentrated and where it is known to show potent inhibitory properties.

The transamination of glyoxalate also produces glycine (Figure 4.41), and this route, too, is active in vertebrate neural tissue ("nonphosphorylation" pathway), although its relative importance remains uncertain.[130] As with glutamate and aspartate, both GABA and glycine are largely recovered in the soluble cytoplasm of neural tissue and are not especially concentrated in nerve terminals, though certain brain regions are enriched in their content, for example, GABA in substantia nigra and glycine in medulla and spinal cord (see below). But a larger proportion of total GABA than other amino acids is found particle-bound in homogenates. Whether or not these amino acids occur in functionally significant quantities in synaptic vesicles remains to be established.[131] As we have noted, glycine and GABA, like other amino acid neurotransmitters, are present in neural tissue at a content and an average concentration (2–10 mM) well in excess ($\times$ 10^3) of that typical for other categories of neurotransmitters, presumably reflecting their involvement in general metabolism (see Table 3.9).

Release and Inactivation of GABA and Glycine

Typical studies with in vitro preparations, such as cortex slices, spinal cord slices, or synaptosomes, have allowed demonstration of the Ca^{2+}-dependent release of preloaded or endogenous GABA and glycine linked to depolarization (e.g., Figures 4.30 and 6.7). Notably, glycine, though

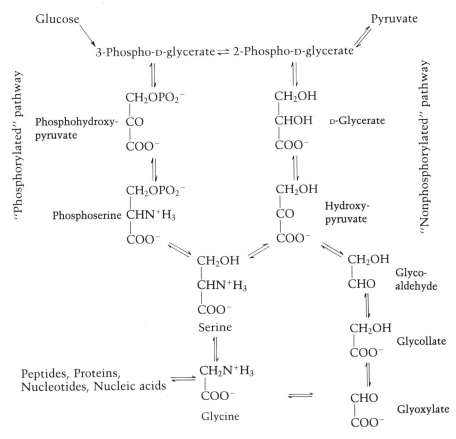

FIGURE 4.42 Biosynthetic routes for the formation of glycine in the central nervous system from either glucose or pyruvate. Both phosphorylated and nonphosphorylated pathways are shown.

present in cortical preparations, is not released from them (see Figure 6.7)[144] a feature correlating with the relative lack of potency of glycine in this region.

Studies of GABA release in vivo from cerebral cortex or substantia nigra, in response to depolarizing chemical stimuli or to neural pathway activation, also provide clear supporting evidence for a transmitter role for GABA in these brain regions.[133,143]

Again, primary functional inactivation of GABA and glycine is by transport systems displaying high affinity and low K_m values for these amino acids (see page 327). These transport systems are located specifically in the neurons releasing these compounds and in surrounding glial cells.

Once taken up, metabolic conversion or reuse follows. GABA is absorbed into glial-cell or neuronal mitochondria, where it becomes transaminated to succinic semialdehyde which, in turn, is oxidized to succinic acid and enters the citric acid cycle of the glial cell. Thus, a whole tissue sample, such as a cortex homogenate, displays the ability to synthesize GABA from glutamate and subsequently transaminate it and oxidize the product, a sequence known as the GABA shunt pathway (Figure 4.40). Analysis by histochemical and subcellular localization of the various enzymes operating this shunt shows that all but GAD, which is very largely neuronal, are located in both neuronal and in glial-cell mitochondria.[115,125] Thus, the cycle or shunt pathway involving glial cells is dependent on the uptake of GABA released by neurons (see Figure 3.10). This route is not to be seen as exclusive, since reuptake of GABA to neurons would stimulate a similar series of reactions in neuronal mitochondria. It is likely to be predominant, however, as GABA transaminase is more concentrated in glial cells.[213]

GABA-α-oxoglutarate transaminase (EC 2.6.1.19), an important enzyme in this shunt pathway, is also pyridoxal phosphate–dependent, like all transaminases, and its inhibition causes accumulation of GABA within the tissue. GABA transaminase from mouse brain has been purified to homogeneity. It has a molecular weight of 109,000 and shows a K_m for GABA of 1.1 mM.

Glycine catabolism is less well understood, but its wide involvement in the biosynthesis of many other cell constituents—such as heme, purines, glutathione, creatine peptides, proteins, nucleotides, and nucleic acids—ensures that its turnover is high.

Pharmacology of GABA and Glycine

Along with other short-chain neutral amino acids, both GABA and glycine long ago were shown to have powerful inhibitory actions when iontophoresed into the vertebrate central nervous system.[G,J] The action of glycine was more localized, being particularly potent in the spinal cord, medulla, and brain stem. Its strong action on spinal motoneurons appears to be largely mediated by small interneurons (including Renshaw cells),[272] though other glycinergic neurons may exist. The pattern of enrichment in glycine content both between brain regions and within the spinal cord correlates well both with the inhibitory potency of iontophoresed glycine and the distribution of high-affinity transport systems. GABA-mediated inhibition is found more widely throughout the vertebrate nervous system, again correlating with GABA distribution and high-affinity transport. For instance, the substantia nigra shows enrichment in these respects.

Currently, only the alkaloid strychnine is known to have a specific action on glycinergic systems. It blocks glycine action through its high-affinity binding to postsynaptic glycinergic receptors. Radiolabeled strychnine has been used to identify glycinergic synaptosomes in density gradients through its specific binding to the receptors located in the attached postsynaptic structure.[135]

The armory of drugs acting preferentially on GABA systems, in contrast, is large and includes a number of agents acting on various stages of GABA metabolism, on its uptake, and also as receptor agonists and antagonists. A selection of these drugs is presented in Table 4.9.

Since both GAD and GABA transaminase are pyridoxal phosphate–dependent, both are inhibited by antipyridoxal agents. Whether this leads to a rise in tissue or a fall, GABA level depends on the brain region, the agent, the dosage, and the period of exposure. Mostly, GAD proves to be more vulnerable, probably owing to its high K_m for the cofactor, and consequently GABA levels fall with associated onset of seizures. This greater vulnerability of GAD to antipyridoxal agents has been linked to its looser binding of pyridoxal phosphate[134] and its relatively low saturation with this cofactor in vivo. Whereas all hydrazides preferentially inhibit GAD and cause a fall in brain GABA levels, other carbonyl trapping agents, for example hydroxylamine, attack GABA-transaminase in preference, which results in a rise in GABA levels.[134]

A wide range of compounds other than antipyridoxal agents inhibit GABA transaminase. Two alkyl-GABA analogues, γ-vinyl GABA and γ-acetylenic GABA, are among the most potent. Others are aminooxyacetic acid, ethanolamine-O-sulfate, and sodium dipropylacetate (which also inhibits succinic semialdehyde dehydrogenase).

Inhibition of GABA catabolism in this fashion leads to a rise in tissue GABA content (by as much as fivefold to sixfold in the case of γ-alkyl GABA). The rise in GABA causes augmentation of inhibitory neural processes and is associated with quiet and withdrawn behavior, sleep, and protection against epileptic seizures. This implies that raised tissue GABA levels lead to greater extracellular concentrations, allowing GABA to exert its inhibitory actions through specific GABA receptor systems. Raised γ-alkyl GABA levels in the cerebrospinal fluid and in the cortical extracellular fluid have been demonstrated after injection of γ-alkyl GABA into the peritoneal.[136,137] Sodium dipropylacetate (Epilim, Roquentin) is widely used as a clinically effective anticonvulsant, and GABA transaminase inhibitors in general constitute a valuable source for the development of anticonvulsant drugs (see Chapter 9).

To some extent, blocking the uptake of GABA after its release to the extracellular space also leads to higher extracellular concentrations of this inhibitory agent. Not surprisingly, many GABA analogues are effective blockers of GABA transport. Glial cells and neurons apparently have different high-affinity uptake systems; both autoradiography and experiments with purified cell preparations show that uptake to glial cells is inhibited by β-alanine whereas uptake to neuron is prevented by diaminobutyric acid or by cis-3-aminocyclohexane.

Agents that display specific interaction with the active sites of enzymes employing GABA as substrate can be expected to show a propensity to interact with other GABA-recognizing molecules, such as specific receptor systems. Examples are nipocotic acid versus isonipocotic acid, and guavacine versus isoguavacine (Table 4.9).

TABLE 4.9   Selected pharmacological agents affecting glycine and GABA systems

Compound	Structure	Action
	Glycine system	
Glycine	NH_2CH_2COOH	Agonist
Strychnine	Complex alkaloid	Antagonist
	GABA system	
GABA	$NH_2(CH_2)_3COOH$	Agonist
Muscimol	NH_2-CH_2-	Agonist
Isonipocotic acid		Agonist
Isoguvacine		Agonist
p-Chlorophenyl GABA	$NH_2 \cdot CH_2 \cdot CH(C_6H_5) \cdot CH_2 \cdot COOH$	Agonist
Bicuculline	Alkaloid	Antagonist
Picrotoxin	Alkaloid	Antagonist
β-Alanine	$NH_2(CH_2)COOH$	Uptake blocker
Diaminobutyric acid	$NH_2 \cdot CH_2 \cdot CH(NH_2) \cdot CH_2 \cdot COOH$	Uptake blocker
Nipocotic acid		Uptake blocker
Guvacine		Uptake blocker
γ-Vinyl GABA	$CH_2{=}CH-NH-(CH_2)_3 \cdot COOH$	GABA transaminase blocker
γ-Acetylenic GABA	$CH{\equiv}C-NH \cdot (CH_2)_3 \cdot COOH$	GABA transaminase blocker
Sodium valproate		Succinic semialdehyde dehydrogenase blocker GABA transaminase blocker
Ethanolamine-O-sulphate	$SO_4-O-CH_2 \cdot CH_2 \cdot NH_2$	GABA transaminase blocker
GABAculine		GABA transaminase blocker
Thiosemicarbazide	$NH_2-CS \cdot NH \cdot NH_2$	Inhibits glutamate decarboxylase
Semicarbazide (and other pyridoxal antagonists)	$NH_2 \cdot CO \cdot NH \cdot NH_2$	Inhibits glutamate decarboxylase
3-Mercaptopropionic acid	$SH-CH_2-CH_2-COCH$	Inhibits glutamate decarboxylase

Receptor binding leads to either antagonism if GABA is displaced by an agent showing higher binding affinity or to agonism if the successfully competing compound is a GABA mimetic. Muscimol, an isoxazole compound isolated from the mushroom *Amanita muscaria*, is the most potent GABA agonist known (Table 4.9), and its radiolabeled form is commonly used in ligand-binding assays to detect specific GABA receptors. The general conclusion about structure–activity relationships from studies with transport inhibitors, isoxazoles such as muscimol, and other specific ligands used as probes is that GABA is transported in a folded conformation but is extended when interacting with its receptors.[M]

Other well-known compounds that interact with GABA receptors as potent antagonists include picrotoxin and bicuculline, both of which act on postsynaptic GABA$_A$ receptors. Picrotoxin ($C_{30}H_{34}O_{13}$), a nitrogen-free substance extracted from the dried berry of the climbing shrub *Anamirta cocculus*, consists of two dilactones (picrotin and picrotoxinin) linked together. Picrotoxin and its derivatives appear to work by blocking GABA-mediated chloride channels. Bicuculline is a phthalide isoquinoline alkaloid ($C_{20}H_{17},O_6N$) isolated from the fumariaceous Dutchman's-breeches, *Dicentra cucullaria*, and has much the same potency and specificity of action as picrotoxin. Both are powerful convulsants and antagonize GABA-mediated inhibition throughout the vertebrate and the invertebrate nervous system.

Both GABA and glycine receptors appear to control Cl^- currents across the membrane; their activation leads mostly to hyperpolarization and therefore postsynaptic inhibition. Electrophysiological studies indicate that in the spinal cord, GABA often mediates inhibition of longer duration (0.01–0.1 msec) than that attributable to the release of glycine.[138,G]

GABA receptors have been further subdivided into two major subpopulations based on their pharmacological specificity.[200,201,202] Thus, GABA$_A$ receptors interact with GABA, muscimol, and isoguavacine as agonists, and are present on central neurons and on peripheral sympathetic neurons. These actions are prevented by bicuculline and its methohalide salts and also by picrotoxin. In contrast, the agonist baclofen (the β-chlorophenyl derivative of GABA; see Table 4.9), is inactive at these bicuculline-sensitive GABA$_A$ receptors despite causing marked neuronal depression similar to that caused by GABA itself. Baclofen is, however, a potent agonist at so-called GABA$_B$ receptors, which often appear to be presynaptically located as judged by lesioning of nerve tracts, by autoradiography, and by GABA release experiments. Thus, activation of GABA$_B$ (e.g., by baclofen) receptors leads to a reduction of neurotransmitter release[203,279] (dopamine, norepinephrine), an effect that is not produced by isoguavacine and only weakly by muscimol. Since these responses are not blocked by bicuculline or other GABA antagonists, they appear to be due exclusively to GABA$_B$ receptors. Ligand-binding studies using [^{3}H]GABA to label GABA$_B$ receptors have shown that physiological concentrations of Ca^{2+} or Mg^{2+} are required to promote binding to $GABA_B$ receptors whereas the $GABA_A$ subpopulation is not dependent on the presence of a particular ion. And, whereas GTP and GDP decreased saturable binding of [³H]baclofen or [³H]GABA to the $GABA_B$ subtype, it does not influence binding to $GABA_A$ receptors, which indicates a link between $GABA_B$ receptors and adenyl cyclase via the agency of guanine nucleotide-binding proteins.[204,262,263] These membrane proteins mediate the interactions of many neurotransmitter-receptors and hormone-receptors with adenylate cyclase (Figure 4.3). In the case of $GABA_B$ receptors, there is inhibition of adenyl cyclase activity, which would normally lead to a reduction in the level of cyclic AMP inside the nerve terminals bearing these receptors. This, in turn, could lead to a change in the phosphorylation state of specific proteins, some of which could be linked to the mechanisms of neurotransmitter release.[263] This is one possible mechanism by which $GABA_B$ receptors could control (reduce) GABA release.

Another difference between the two subclasses of GABA receptors is that $GABA_A$ receptors appear to be coupled to chloride channels, while the $GABA_B$ type are probably linked to Ca^{2+} channels.[204] Thus, presynaptically located $GABA_B$ receptors might well control neurotransmitter release by directly influencing Ca^{2+} entry during nerve terminal depolarization.

The pre- and postsynaptic localization of $GABA_B$ and $GABA_A$ receptors, respectively, is far from being definitive or universal.[271] Thus, in the dorsal horn of the spinal cord both receptor categories appear on primary afferent terminals, although $GABA_B$ receptors are not usually found on GABAergic terminals and are therefore not "autoreceptors" (see Chapter 7).[203]

GABA and Presynaptic Inhibition   GABA, and not glycine, appears to be the neurotransmitter that operates the mechanism known as presynaptic inhibition (see Chapter 6), principally observed in relay centers such as the spinal cord, the cuneate and gracile nuclei, and the thalamus. The process is characterized by primary afferents depolarization (PAD) after arrival of an afferent volley through the inhibitory input (Figure 4.43, axon 2). The lowered membrane potential of these primary afferent terminals causes a reduction in the amplitude of the action potentials invading the terminals (axon 1) and therefore a reduction in the amount of neurotransmitter they release onto neuron 3. Where the primary transmitter in axon 1 is excitatory in action, the net result is inhibition. The effect has been attributed to axoaxonic or multiple synapses acting one on the next in cascade (see Chapter 1). Since PAD is prevented by picrotoxin and bicuculline, but not by strychnine or other neurotransmitter antagonists, it seems that GABA is the agent mediating this inhibition, which is, in fact, the result of a depolarizing action. The ionic mechanism involved again appears to be Cl^- gating, since raising extracellular GABA

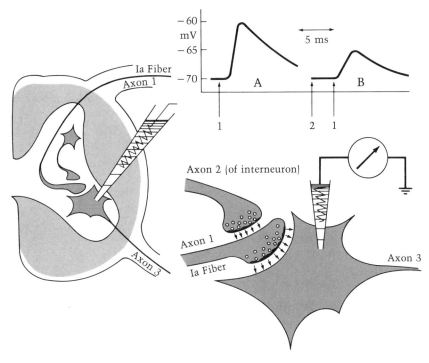

FIGURE 4.43 Presynaptic inhibition of a motoneuron. Axon 1 forms an axosomatic synapse with neuron 3, and is presynaptic to neuron 3 and postsynaptic to axon 2. Activation of axon 1 alone invokes an excitatory postsynaptic potential (EPSP) of approximately 10 mV amplitude in neuron 3 (trace A). Activation of axon 2 before axon 1 causes the amplitude of the EPSP to fall to 5 mV (trace B). Axon 2 has a depolarizing action on axon 1 (primary afferent depolarization, PAD).

concentrations to high levels relative to normal neuronal levels changes the PAD to a hyperpolarization, presumably because of an influx rather an efflux of Cl^- ions. Thus, nerve terminals that respond to a presynaptic input with depolarization may well contain high Cl^- concentrations. Alternatively,[246] these results might be explained by some control of the Ca^{2+} flux into axon 1 owing to activity of the inhibitory axon 2 (Figure 4.43).

GABA and Benzodiazepines   An interesting synergism of action appears to exist between the GABA inhibitory system and the anxiolytic benzodiazepine drugs (e.g., diazepam, clonazepam, nitrazepam; see Table 4.7). In fact, facilitation of GABAergic neurotransmission is likely to be a principal mode of action of the benzodiazepines. Experimentally, this can be detected by a range of interactions between the two systems.[139,230] For instance, benzodiazepines relieve convulsions caused by inhibition of brain GABAergic neurotransmission. Behavioral tests, too, have revealed this interaction—for example, when rats were given electrical shocks for choosing the incorrect route in a maze. In this test, benzodiazepines caused an increase in the number of incorrect (punished) choices, presumably by enhancing GABAergic inhibitory neurotransmission. Picrotoxin (a specific GABA antagonist) prevented this influence of benzodiazepines. These observations demonstrate the requirement for functionally active GABAergic synapses for certain aspects of benzodiazepine action.

Again, benzodiazepines enhance GABA-mediated segmental *presynaptic* inhibition, and they mimic the *presynaptic* inhibitory action of GABA on preganglionic nerve terminals. These drugs also show agonistic action on *postsynaptic* GABAergic mechanisms. For instance, when Golgi, stellate, and basket cells, which are all GABAergic, exert their postsynaptic inhibitory effect in the cerebellum, they cause a decrease in cyclic GMP levels. Blockade of their activity is accompanied by a rise in cyclic GMP, but this rise is substantially attenuated by benzodiazepines. In general, benzodiazepines act similarly to GABA on all inhibitory mechanisms known to be mediated physiologically by GABA, although through some direct enhancing action rather than as a direct GABA agonist.

In vitro, micromolar concentrations of benzodiazepines, increase the affinity of GABA for GABA receptors located on synaptic membranes. GABA and its analogue muscimol increase the affinity of the benzodiazepine binding sites. Moreover, GABA and benzodiazepine receptors are always located close together.[139,230] The precise mode of interaction of these two systems remains obscure, but the hypothesis has been proposed that an unidentified "endogenous benzodiazepinelike substance" influences the postsynaptic actions of GABA. This endogenous benzodiazepine ligand is postulated to be coreleased with GABA (or released from some more distant site) and, by attaching to a contiguous postsynaptic receptor, to work together with the GABA receptor to modulate the flux of Cl^- through the associated Cl^- ionophore (see Figure 7.19). As yet, no convincing candidate for the role of endogenous benzodiazepinelike substance has been isolated.

Isolation of GABA Receptors

Considerable progress has been made in the isolation of a GABA-benzodiazepine receptor complex.[227] Complex taken from bovine brain has been purified to homogeneity from detergent extracts using an immobilized benzodiazepine ligand for affinity chromatography with specific displacement by another benzodiazepine compound, followed by ion-exchange chromatography.[228] The product, 1800-fold enriched in binding sites, carries high-affinity binding sites for GABA$_A$ ligands and benzodiazepines. As judged by gel electrophoresis, it consists of alpha and beta protein subunits (mol wt of alpha, 53,000; beta, 57,000).[228] When prepared with a zwitterionic detergent, the receptor complex retained a high-affinity binding capacity for agents known to block the GABA-controlled Cl⁻ channel (e.g., by picrotoxinin analogues) and displayed barbiturate enhancement of benzodiazepine binding, both properties of the $GABA_A$ receptor in situ.[229]

In both the brain membranes and the purified product, the alpha subunit could be strongly labeled by photoaffinity reaction with the photosensitive benzodiazepine ligand [³H]flunitrazepam, suggesting that the benzodiazepine-binding site resides at least partly on the α subunit.[228] These various properties suggest that the isolated complex carries at least three interacting high-affinity sites for the different key ligand classes (GABA, benzodiazepines, barbiturates) plus at least part of the Cl⁻ ion channel. An active receptor complex with properties of this kind has been shown to be inserted into the membranes of *Xenopus* oocytes following microinjection of mRNA from chick optic lobe or rat brain, showing that the entire receptor complex can be correctly assembled in the oocyte membrane following translation of the mRNA fractions.[231,232]

A high-affinity [³H]strychnine (K_D = 9 nM) binding protein (molecular weight 246,000) has been isolated from detergent extracts of spinal cord by affinity chromatography on aminostrychnine–agarose gels and other purification procedures.[233] It has been tentatively identified as the glycine receptor protein. It appears to consist of three subunits (mol wt 48,000, 58,000 and 93,000). [³H]strychnine was irreversibly incorporated only into the 48,000 mol wt polypeptide by ultraviolet irradiation. Glycine and other receptor agonists (β-alanine, taurine) inhibited the binding of [³H]strychnine to the purified receptor.[233]

Location of Pathways Employing GABA and Glycine

Neurons synthesizing and releasing GABA are found throughout the vertebrate and invertebrate nervous systems. Many arthropod neuromuscular junctions, such as those of the locust and lobster, have excitatory (usually glutamatergic) *and* inhibitory inputs, the latter operated by GABA, and certain well-characterized molluscan neurons show inhibitory responses to GABA.[A] Lobster inhibitory fibers innervating muscle contain tenfold the GABA content of excitatory fibers, and they release this GABA when activated.[140]

Within the vertebrate nervous system, GABA operates pathways at certain sites in the spinal cord and brain stem and in several defined regions of cerebellum, hippocampus, thalamus, hypothalamus, olfactory bulb, and basal ganglia as well as in the retina. The cerebral cortex, too, contains GABAergic pathways, including terminals in Brodman layer IV, as indicated by the presence of high concentrations of GAD and the pattern of sensitivity to iontophoresed GABA.[A,B] The substantia nigra, the center located in the brain stem that projects dopaminergic neurons to the corpus striatum, shows the richest content of GABA and its metabolizing enzymes. They are located in terminals projected to the substantia nigra from other regions of the basal ganglia that control the activity of the dopaminergic cell bodies, such as the globus pallidus (Figure 9.5).

The famous Purkinje cells of the cerebellum, which convey the principal cerebellar output to the medulla, operate by releasing GABA, as do the basket cells whose terminals surround Purkinje cell bodies.

Glycine-operated pathways appear to be restricted to spinal cord and medulla, though synaptic glomeruli in the cerebellum show high-affinity uptake of glycine, indicating a transmitter role at inhibitory Golgi cell terminals. Indeed, other strychnine-sensitive glycine-mediated inhibition has been demonstrated in the cerebellum.[G] Renshaw interneurons in the spinal cord are glycinergic. These cells receive cholinergic collateral fibers from spinal motoneurons and form part of a feedback pathway which inhibits the output of these same motoneurons.[272]

Modern approaches to revealing the presence of a particular neurotransmitter system in particular neural pathways involves a multilevel analysis. Where synthesis enzymes have been isolated and purified, antisera can be raised to them and there allow pathway visualization through immunohistochemistry. For GABA, the antisera to GAD serve this purpose well. The presence of receptor systems specifically binding the appropriate ligand (e.g., muscimol for GABA, strychnine for glycine) provides complementary evidence. Also, as discussed earlier in this chapter, the localization of the appropriate high-affinity transport system in neurons always correlates with the presence of neurons employing a specific transmitter. This test must be combined with a search for depolarization-induced, Ca^{2+}-dependent neurotransmitter release and for raised tissue concentrations of the transmitter that diminish on sectioning of input circuits. Taken together, such tests usually serve well to indicate which neurotransmitters are active in the pathway concerned. Such approaches even allow quantitation of the extent of their participation.

Within the basal ganglia complex, for instance, both GABA release *and* high-affinity transport were found in synaptosomes isolated from caudate putamen, globus pallidus, and substantia nigra. Glutamate and aspartate showed different patterns in this respect, and glycine showed neither depolarization-induced release nor high-affinity uptake.[138]

TAURINE: A DOUBTFUL NEUROTRANSMITTER

A great deal of research effort has gone into collecting evidence to judge whether the sulfonic amino acid taurine serves a role in controlling excitability in the nervous system.[L,157,158,159] It is found very concentrated in neural tissue as in other organs, being present in free solution at 2–10 mM and rising to 60–110 mM in the pineal gland and the posterior pituitary.[145]

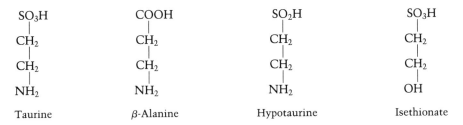

FIGURE 4.44   The structure of taurine and some selected analogues.

This neutral sulfur-containing amino acid (Figure 4.44) is an analogue of β-alanine which does not naturally occur in the nervous system. In view of the powerful excitatory effects of the two sulfonic acids cysteic and homocysteic acids, which are analogues of glutamatic and γ-amino adipic acids, it could be expected that taurine would mimic the inhibitory actions of β-alanine, glycine, and GABA. The general conclusion seems to be that GABA, glycine, and taurine have roughly comparable inhibitory effects on spinal neurons, both in mammals and in amphibians, and probably act through modulating chloride ion conductance.[248,L] However, the inhibition mediated by taurine and β-alanine is glycinelike, rather than GABA-like, as judged by the fact that it is blocked by strychnine but not by bicuculline.

Compared with the strong action of GABA, only a relatively weak depressant action of taurine has been detected in cerebral cortex. Again, this is in keeping with the low efficacy of glycine as an inhibitor in cortex, though the action of taurine here is apparently blocked by both strychnine and bicuculline. Similar properties have been detected in the thalamus, cerebellum (Purkinje cells)[258] and retina. The lack of development of specific antagonists or agonists for taurine has retarded further development of the electrophysiological case for taurine as a neurotransmitter, though a relatively specific taurine antagonist, 6-amino methyl-3-methyl-4H-1,2,4-benzothiadiazine-1,1-dioxide hydrochloride (TAG), has recently been described.[247]

Taurine Synthesis

The key enzyme in taurine synthesis is cysteine sulfinate decarboxylase (EC 4.1.1.29). It produces first hypotaurine and then taurine from cysteine sulfinate by a pyridoxal phosphate-requiring mechanism (Figure 4.45). The step from hypotaurine to taurine is achieved by an NAD^+-mediated action. Cysteine sulfinate is itself formed by enzymatic oxidation of cysteine.[L] It is also likely that a significant part of the taurine found in brain and other organs is, in fact, dietary in origin, which would correlate with its relatively low rates of formation.[145] Another part is undoubtedly formed from the pool of sulfur amino acids including cysteine and methionine, though the precise reaction sequences are not yet clear.

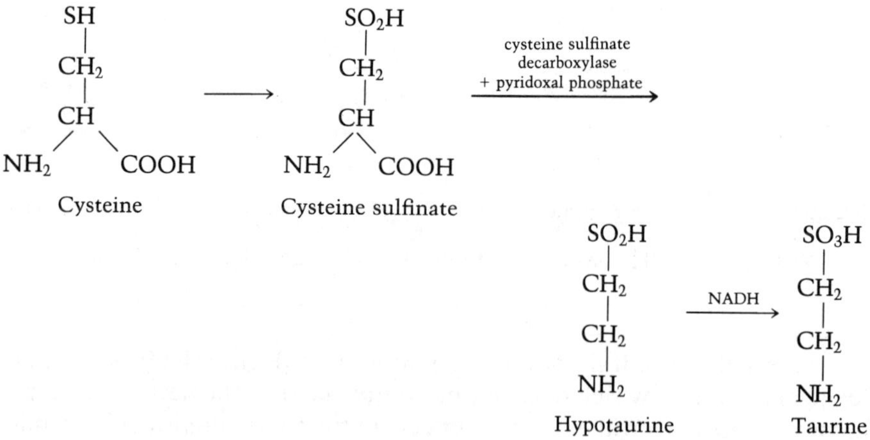

FIGURE 4.45   Enzymatic formation of taurine.

There is a "taurine-cycle" in which taurine is absorbed from the gut, converted to bile acids in the gallbladder, and re-secreted into the gut.[L]

Taurine concentrations are high, ranging from 2 to 8 mM from spinal cord to corpus striatum.[L] They are especially high in synaptic vesicles.[100] Muscle, too, contains high levels of taurine (mouse cardiac muscle has 40 mM) and so do lung, liver, kidney, seminal vesicles, prostate gland, and many other organs, depending on species.[L] Such high levels suggest that this amino acid serves some function distinct from that of neurotransmitters. Taurine levels in the brain are very high at birth or prenatal state and decrease during development;[159] the significance of these changes remains unclear. Until very recently it was thought possible that cysteine sulphinate decarboxylase and glutamate decarboxylase were the same enzyme. However, the two enzymes have now been separated and purified to homogeneity.[277] Antibodies against cysteine sulphinate decarboxylase do not cross react with glutamate decarboxylase and vice versa, and while the former cannot use glutamate as substrate, glutamate decarboxylase can use both cysteine and cysteine sulphinic acid (K_m 5.4, 5.2 mM) as well as glutamate (K_m 1.6).[277] Therefore, it has been concluded that the enzymes are two distinct entities and are responsible for the biosynthesis of taurine and GABA respectively.

Cysteine sulfinate decarboxylase has been shown by immunohistochemical techniques to occur only in certain Purkinje, stellate, basket, and golgi cells of the cerebellar cortex, arranged in saggital microbands.[149,250] This suggests that cerebellar neurons are chemically heterogeneous, with each of the different subclasses containing different neuroactive substances, including neurotransmitters.[250]

Taurine is characterized by its low turnover rates, with both its formation and further metabolism to isethionic acid (Figure 4.44) being at rates that are orders of magnitude lower than turnover rates for most other neurotransmitters, putative or established.[145] This implies that inactivation of any taurine released during synaptic transmission is unlikely to be by enzymatic conversion to the physiologically inactive isethionate. More likely this would be by reuptake transport processes. However, only moderately high-affinity systems have so far been detected[146,150] (K_m = 50–60 μM compared with 5–10 μM for glutamate and GABA), and whether these systems are exclusively neuronal or wholly or partly glial is not certain. Taurine is transported into cultured glial cells by the same process as for β-alanine.[155,160]

Release of Taurine

Depolarization-induced release of taurine has been demonstrated for cortex slices[150,155] and retina[249,157] in vitro and for cerebral cortex in vivo,[156] but the calcium-dependence of these responses are at best partial, though retinal release shows a clearer dependence on calcium.[249] Thus, stimulus-coupled secretion of taurine appears to be an established phenomenon, but its calcium-dependence remains in doubt.

What, then, is the evidence that allows taurine to be put forward as a neurotransmitter candidate? It may be summarized as follows:[157,158,249]

1   Taurine depresses neuronal activity in many CNS regions,[147] and this action is blocked by strychnine or bicuculline. It causes an increase in Cl^- conductance.[248]

2   Taurine is found in nerve terminals and synaptic vesicles.[100,281]

3   Its principal enzyme of synthesis, cysteine sulfinate decarboxylase, is found concentrated in nerve terminals[148] and is present in certain subcategories of cerebellar Purkinje cells and Stellate cells.[149,250] It is also present in the inner nuclear layer and in the ganglion cell layer of the retina.[249]

4   It is released from nerve terminals in vitro and in vivo by partially calcium-dependent processes[150,100] and its release from retina is evoked by light flashes, electrical stimulation,[158,151,L] or by potassium ions,[249] probably from inhibitory synapses of the inner plexiform layer.[158]

5   A moderately high-affinity uptake system exists for taurine that could inactivate the released compound.[150,146]

6   A synaptosomal subpopulation has been reported to specifically accumulate taurine.[152]

There remains a firm reluctance to consider taurine as a neurotransmitter of the established type, which is released from specific neurons and operates through specific receptor systems, partly because no specific antagonists on agonists have been developed that would allow detection of taurine receptors and partly because of the surprising sensitivity of taurine-mediated effects to strychnine and bicuculline, the specific antagonists for inhibitory systems mediated by glycine and GABA respectively. Some of the neurochemical parameters (e.g., relatively low-affinity transport systems) are not as clear-cut as they are for established transmitter systems.

Instead, interest has focused on the influence of taurine on movements of calcium and other ions in cardiac muscle and brain.[L,162] It has been shown to inhibit inward calcium flux in these tissues and to promote the cellular accumulation of potassium ions. It prevents norepinephrine-induced premature ventricular contractions of the dog heart, and it converts the arrhythmias of acute or chronic digoxin toxicity to a normal sinus rhythm.[L] These effects have been correlated with the actions of taurine on the calcium movements of cardiac muscle.[161,162]

Influences of taurine on calcium movements in the nervous system could form a basis for its synaptic inhibitory actions. In this case, the mechanism is of a more generalized and indirect kind and does not involve specific taurine receptors. In this sense taurine is seen to be serving a *neuromodulator* function. Such effects have been reported, with taurine inhibiting ^{45}Ca influx to synaptosomes, increasing spontaneous GABA efflux, and inhibiting its stimulus-induced release.[162] Certainly a generalized inhibitory effect of taurine is implied by its reported anticonvulsant action in both human[152] and experimental epilepsy,[153,154] including seizures induced by a wide range of convulsant chemicals, though not all of these effects have always been readily repeated.[L]

In conclusion, the actions of taurine suggest that it should be categorized as a modulator of neuronal activity rather than as a classical neurotransmitter.

THE PURINES ATP AND AMP:
UNEXPECTED NEUROTRANSMITTERS

The potent extracellular actions of purine nucleosides on excitable membranes have been known since the 1920s and 1930s, when adenosine from yeast was shown to cause slowing of the heart rate and general lowering of the arterial pressure.[165] Most of the early studies on purine sensitivity were confined to smooth muscle and the peripheral nervous system, but in the last decade, purinergic systems and purinergic receptors have been identified in the central nervous system and in various cell types such as mast cells and platelets, with membrane-bound adenylate cyclase systems being reported to occur widely. The central involvement of cyclic AMP in mediating the actions of many neuroactive substances as a second messenger is firmly established (see page 161). Cyclic AMP is formed from ATP by the enzyme adenylate cyclase, which appears to be an integral part of the receptor systems for many neuroactive substances. The idea that ATP, ADP, and AMP might themselves be released from nerve terminals during stimulation and might function directly as neurotrans-

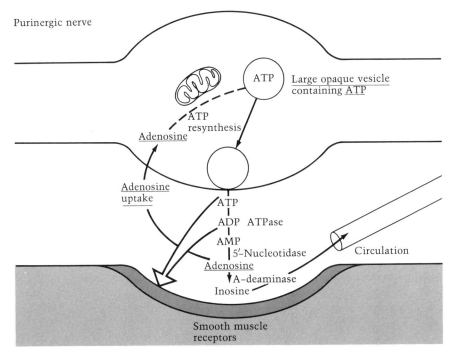

FIGURE 4.46   Schematic representation of synthesis, storage, release, and inactivation of ATP at a purinergic neuromuscular junction. [From ref. 166]

mitters now has considerable experimental backing,[166,167,168] (Figure 4.46) including studies of their release from both central and peripheral neural preparations during stimulation. These compounds are thought to function as neurotransmitters in some of the noncholinergic, nonadrenergic inhibitory nerve pathways of the peripheral autonomic system. These neural pathways are involved in reflex relaxation of the lower esophageal and internal sphincters, "receptive relaxation" of the stomach, and "descending inhibition" of the intestine during peristalsis. They also mediate bronchodilation and vasodilation of a number of blood vessels and supply excitatory nerves to the urinary bladder.

Adenosine breakdown products of adenine nucleotides appear in the venous efflux from the stomach of guinea pigs and toads following stimulation of the vagus nerves. The pattern of release was due to stimulation of the nonadrenergic inhibitory fibers in the vagal roots of the parasympathetic nerves. Stimulation of the cholinergic gastric fibers in the cervical sympathetic branch to the vagus nerves did not cause this increase in nucleoside efflux. The possibility that the purines were released secondarily from the innervated muscle during stimulation has been experimentally ruled out[167] (e.g., purine nucleotides are still released from portions of Auerbach's plexus from turkey gizzard when it is dissected free from muscle). This plexus is heavily innervated by possible "purinergic" inhibitory nerves. ATP is the agent for which the best evidence for a neurotransmitter role exists. Exogenously applied ATP closely mimics the effects of stimulation of nonadrenergic inhibitory nerves supplying many visceral and vascular organs.[166,167] For instance, relaxations of the gut produced by ATP or by nerve stimulation rapidly reach a maximum and decline quickly. Relaxations produced by norepinephrine or sympathetic nerve stimulation reach a maximum more slowly and are maintained for a longer time, which seems to eliminate the adrenergic system as being responsible for these effects. Of course, other neurotransmitters (e.g., peptides) could be operating these and other proposed purinergic pathways. The question is best resolved by the use of very specific receptor antagonists for the purine nucleotides and nucleosides, but they have not yet been developed.

Purinergic Receptors

Two types of purinergic receptor have been proposed,[169] distinguishable according to four criteria: (1) relative potencies of agonists, (2) relative potencies of competitive antagonists, (3) changes in level of cyclic AMP, (4) induction of prostaglandin synthesis. The last point is especially important because ATP is a potent inducer of prostaglandin synthesis and ATP released from purinergic nerves may be linked with prostaglandins in controlling peristalsis. The two classes of purinergic receptor are called P_1 and P_2.

P_1 receptors. Receptors of the first type are more responsive to adenosine and AMP than to ATP or ADP and are antagonized by theophylline. Occupation of a P_1 receptor leads to activation of the adenylate cyclase system with resultant increases in intracellular cyclic AMP. P_1 receptors are more widely distributed than P_2 and they almost exclusively mediate inhibition of smooth muscle systems.

P_2 receptors. Receptors of the second type are more responsive to ATP and ADP than to adenosine or AMP. They are antagonized by 2,2-pyridylisatogen tosylate and by high concentrations of quinidine and the imidazolines. Occupation of a P_2 receptor leads to the synthesis of prostaglandins in certain tissues. P_2 receptors are less widely distributed than the P_1 type. ATP, when activating P_2 purinoreceptors in smooth muscle, can cause either inhibition (e.g., in the gut or coronary artery) or excitation (e.g., in the bladder or saphenous vein).

There is clear evidence for the existence of presynaptic inhibitory autopurinoreceptors of the P_1 type that act to reduce further release of ATP. These same P_1 receptors appear to be located on both adrenergic and cholinergic terminals of peripheral nerves, where they influence the extent of the release of norepinephrine and acetylcholine (See Figure 4.47).

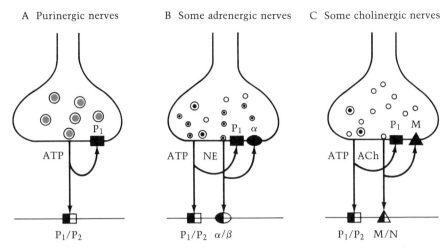

FIGURE 4.47 Schematic representation of presynaptic neuromodulation of transmitter release. (A) Reduction of ATP release by activation of presynaptic P_1 purinoceptors. (B) Reduction norepinephrine (*NE*) release by activation of presynaptic α-adrenoceptors and P_2 purinoceptors. (C) Reduction of ACh release by activation of presynaptic muscarinic cholinergic and P_2 purinoceptors. [From ref. 166]

It is now well established that ATP is released together with catecholamines from adrenergic nerves and from the cholinergic terminals of the *Torpedo* electric organ and other cholinergic terminals.[167]

Synthesis, Storage, and Inactivation of Purines

The mitochondrial enzyme systems that produce ATP for energy-linked reactions are, of course, ubiquitous, as is ATP itself. No special synthetic route for ATP as a purinergic neurotransmitter has been proposed or detected. Other nucleotides and adenosine can be produced both intracellularly and extracellularly in the synaptic cleft by hydrolysis after release of ATP.

It is not certain whether ATP and related nucleotides or nucleosides are stored in special granules in purinergic nerve endings. But quinacrine binds to ATP and gives positive staining of cells known to contain high levels of ATP, and it stains nerve cell bodies and varicose fibers in tissues where pharmacological evidence for purinergic transmission has been presented.[166,167]

Smooth muscle recovers rapidly after application of ATP or stimulation of the noncholinergic, nonadrenergic nerves, and there is no long-lasting action during continued stimulation. These observations indicate the presence of an efficient inactivation mechanism for the purine neurotransmitter. This is likely to be enzymatic breakdown of ATP to compounds with greatly reduced potency, such as adenosine and AMP. High-affinity uptake mechanisms for adenosine have been demonstrated.

In summary, ample experimental evidence supports the original proposal[170] that purinergic systems operate widely throughout the peripheral and central nervous systems employing as neurotransmitters, or neuromodulators, ATP, ADP, AMP, and adenosine either in purely purinergic terminals or as cotransmitters with acetylcholine, norepinephrine, or other categories of neurons. (See Chapter 7, page 386). Thus, in common with amino acid systems, purinergic nerves employ as neurotransmitters substances that are found in every cell, where they are involved in a broad range of metabolic reactions.

HISTAMINE: A NEUROTRANSMITTER ROLE FOR AN OLD COMPOUND?

A huge literature on the occurrence, distribution, and actions of histamine has accumulated since the beginning of the century, but only relatively recently has histamine become seriously considered as a neurotransmitter in the classical sense.[171,172]

Synthesis and Storage of Histamine

Histamine has long been known to occur in a complex with heparin and proteins, contained in granules within mast cells, platelets, and basophils. It is released from these cells in response to allergic reactions and tissue damage. It can be released from its storage sites in the central nervous system, gastric mucosa, lungs, and skin in response to hormonal and neural signals. Within the mammalian brain, histamine content is highest in the hypothalamus, especially the mammillary bodies and supraoptic nucleus, and lowest in the cerebellum and medulla pons.

Histidine is the precursor of histamine. It is converted to histamine within the brain by the enzyme histidine decarboxylase (EC 4.1.1.22), which, like most decarboxylases, requires pyridoxal phosphate as cofactor. The nonspecific enzyme L-aromatic acid decarboxylase can also catalyze this reaction. Histidine decarboxylase also occurs in high concentrations in the hypothalamus (Figure 4.48) and throughout the brain.

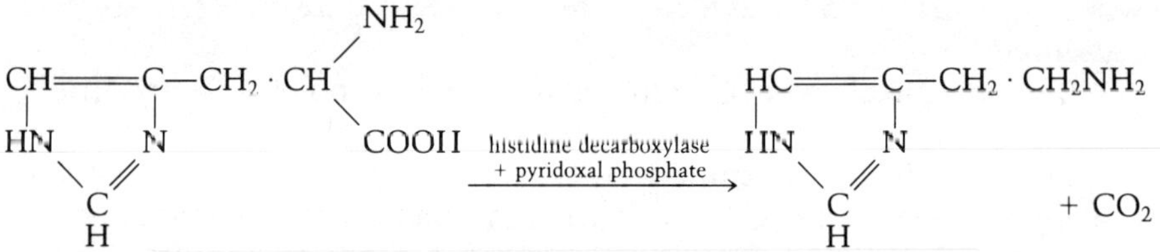

FIGURE 4.48   Enzymatic synthesis of histamine.

The kinetics and properties of the biosynthetic process indicate that this enzyme, and not the nonspecific decarboxylase, is responsible for histamine biosynthesis. This means that histamine is not formed in monoaminergic neurons but in quite separate pathways.[172] An irreversible inhibitor, α-fluromethylhistidine, produces complete and long-lasting inactivation of the enzyme.[173]

Histidine decarboxylase is located in the cytoplasm of nerve endings. In rat cerebral cortex, about 50 percent of total histamine is located in neurons, whereas the decarboxylase appears to be entirely neuronal.[172,174]

Release, Turnover, and Inactivation of Histamine

Calcium-dependent release of endogenous histamine from brain slices is readily evoked by high potassium concentrations. Reserpine, which releases monoamines, will release histamine from brain slices when presented in micromolar concentrations, suggesting similarities between the forms of storage of monoamines and histamine. The in vivo release from cerebral cortex appears to be under the influence of neural activity in afferent neurons that project into the cortex from the medial forebrain bundle.[174] Axotomy of this pathway causes a strong elevation of histamine in the cortex.

Histamine has a rapid turnover. It has a half-life of less than one hour in rodent brain, similar to the half life of acetylcholine. Turnover rates are grossly modified by behavioral manipulations (e.g., stress) and by anesthesia and sedation, which also affect the turnover of monoamines.[171,172]

As yet, no high-affinity transport system for histamine has been detected, and enzymatic transformation seems to be the mode of inactivation. This is most likely to involve its N-methylation by histamine N-methyltransferase followed by its oxidative deamination to 3-methylimidazole–acetic acid (Figure 4.49). The key role of histamine N-methyltransferase in histamine breakdown can be deduced from the rise in histamine levels in the brain following inhibition of the enzyme by antimalarial drugs.[172] Similarly, the role of monoamine oxidases in the second stage can be demonstrated by the rise in the level of 3-methylhistamine in the brains of animals treated with monoamine oxidase inhibitors.[172]

Pharmacology of Histamine

Selectively acting drugs have allowed the characterization of two subpopulations of histamine receptors that mediate the biological actions of histamine.[171] Type H_1, which mediates smooth muscle contraction, is blocked by the classical antihistamine drugs of the mepyramine type that are used in the treatment of hay fever and urticaria and to induce sedation. These H_1 receptors are stimulated by 2-methylhistamine or 2-

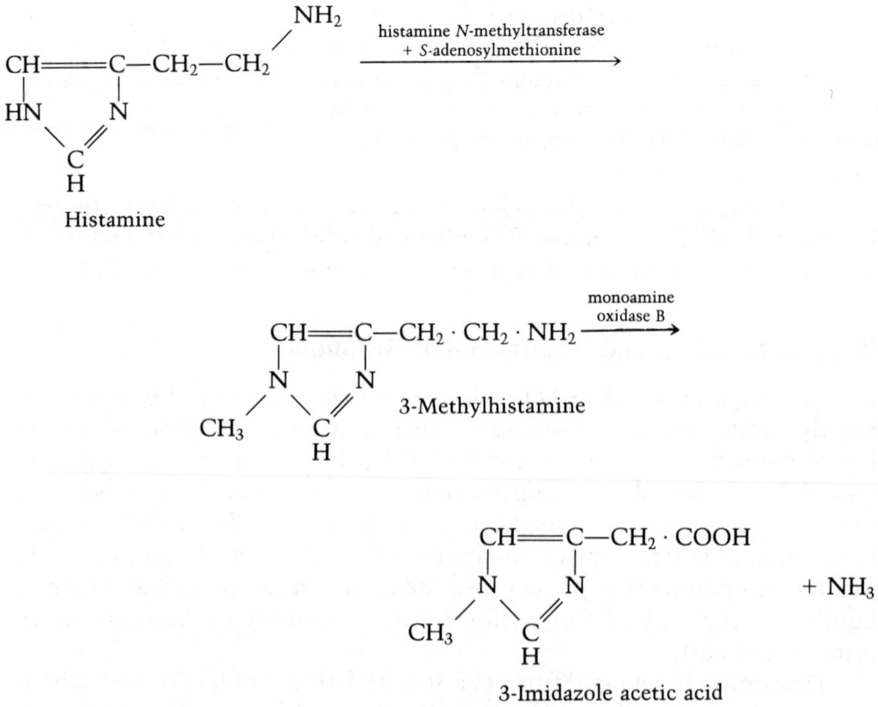

$$\underset{\substack{HN \quad\quad N \\ \diagdown \;\; \diagup \\ C \\ H}}{CH\!=\!\!=\!C}\!-\!CH_2\!-\!CH_2\!-\!\overset{\displaystyle NH_2}{\diagup} \qquad \xrightarrow[\quad\quad\quad\quad\quad]{\substack{\text{histamine } N\text{-methyltransferase} \\ + \; S\text{-adenosylmethionine}}}$$

Histamine

$$\underset{\substack{N \quad\quad N \\ \diagup \;\;\; \diagdown \;\; \diagup \\ CH_3 \quad\;\; C \\ H}}{CH\!=\!\!=\!C}\!-\!CH_2 \cdot CH_2 \cdot NH_2 \xrightarrow[\quad\quad\quad]{\substack{\text{monoamine} \\ \text{oxidase B}}}$$

3-Methylhistamine

$$\underset{\substack{N \quad\quad N \\ \diagup \;\;\; \diagdown \;\; \diagup \\ CH_3 \quad\;\; C \\ H}}{CH\!=\!\!=\!C}\!-\!CH_2 \cdot COOH \qquad\qquad + \; NH_3$$

3-Imidazole acetic acid

FIGURE 4.49   Pathway of inactivation and catabolism of histamine.

thiazolylethylamine. On the other hand, H_2 receptors, which are coupled to histamine-mediated gastric secretion, are selectively blocked by cimetidine or metiamide (cyanoguanidine and thiourea-substituted forms of histamine; see Figure 4.50), drugs used to prevent gastric acid secretion induced by the vagus nerve or the hormone gastrin. These H_2 receptors are selectively stimulated by dimaprit (dimethylaminopropylisothiourea) and impromidine[172] (Figure 4.50).

The H_2 receptors are also selectively coupled to a histamine-sensitive adenylcyclase, which has been characterized in cell-free preparations of hippocampus. It was found to be markedly enhanced by GTP. In addition to typical H_2 receptor agonists, several other neuroactive agents interact with H_2 receptors, including clonidine (α-adrenoreceptor agonist), D-LSD (but not L-LSD), and various tricyclic antidepressants.[171,175]

Cyclic AMP formation can also be induced via interactions with H_1 histamine receptors in intact preparations such as brain slices, though this appears to be indirect and does not involve a coupling with adenylcyclase. It is suggested that this and many other actions mediated by H_1 receptors are linked to translocation of Ca^{2+}, since external calcium is always required for these responses.[172]

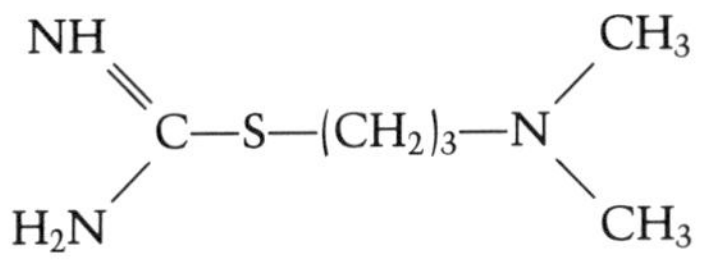

Dimaprit
(dimethylaminopropylisothiourea)

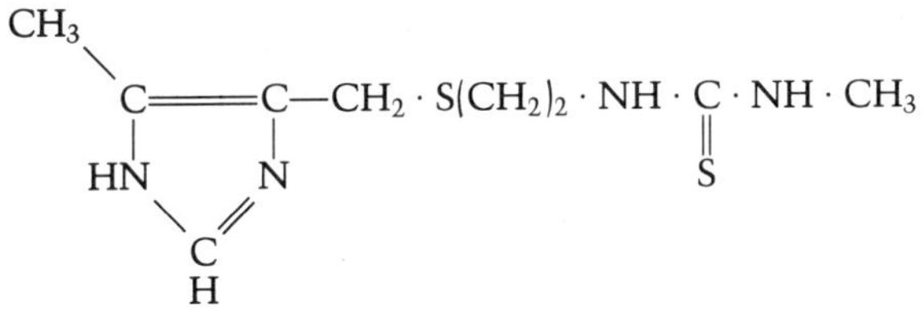

Metiamide

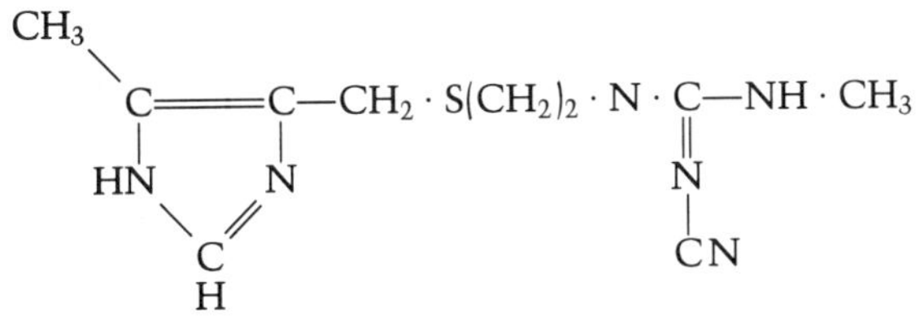

Cimetidine

FIGURE 4.50 Histamine agonists and antagonists.

It has proved possible to use [^{3}H]mepyramine to label and visualize H_1 histamine receptors in the nervous system since this agent binds to these sites with a K_D in the nanomolar range, that is, it has very high affinity binding.

There are now reports in the literature[216] of a third class of histamine receptor, H_3, found in the central nervous system. These receptors are pharmacologically distinct from H_1 and H_2 and appear to be presynaptically located, allowing histamine to participate in the control of its own release at nerve terminals.[216]

Location of Histamine Pathways

There is strong evidence for an ascending histaminergic pathway traveling along the median forebrain bundle to spread widely into the telencephalon. Section of the medial forebrain bundle causes a decrease in the levels of histidine decarboxylase in all telencephalic areas.[174] The enzyme can be reduced to almost undetectable levels by multiple deafferentiation, showing that it is located within nerve terminals from extrinsic neurons. There is also electrophysiological evidence for this pathway, since partial blockade of cortical or hippocampal responses to stimulation of the medial forebrain bundle can be effected by histamine antagonists.[172] The neuronal cell bodies projecting putative histaminergic fibers to the cerebral cortex are likely to be positioned in the mesancephalic reticular formation, whereas those projecting to the corpus striatum are probably in the mamillary bodies.[172]

The existence of reticulocortical pathways employing histamine as neurotransmitter would correlate with well-known central actions of this neuroactive substance in influencing states of sleep, arousal, and self-stimulation behavior.[171,172] Both in location and action, such a pathway would emphasize the similarities between histamine and the other monoamines—serotonin, norepinephrine, and dopamine—with which it shares a number of physiological and pharmacological properties.

REFERENCES

General Texts and Review Articles

A Phillis, J. W. (1970) *The Pharmacology of Synapses*. Pergamon Press, Oxford.

B Burnstock, G., and M. Costa. (1975) *Adrenergic Neurons*. Chapman and Hall, London.

C Tucek, S. (1978) *Acetylcholine Synthesis in Neurons*. Chapman and Hall, London.

D McGeer, P. L., J. C. Eccles, and E. G. McGeer. (1978). *Molecular Neurobiology of the Mammalian Brain*. Plenum Press, New York.

E Davidson, N. (1976) *Amino Acid Neurotransmitters*. Academic Press, London and New York.

F Silver, A. (1974) *The Biology of Cholinesterases*. Elsevier/North Holland, Amsterdam.

G Tebecis, A. K. (1974) *Transmitters and Identified Neurons in the Mammalian Central Nervous System*. Scientechnica, Bristol.

H Kruk, Z. L., and C. J. Pycock. (1979) *Neurostransmitters and Drugs*. Croom Helm, London.

I Calne, D. B. (1975) *Therapeutics in Neurology*. Blackwell Scientific, Oxford.

J Curtis, D. R., and G. A. R. Johnstone. (1974) *Ergeb. Physiol.* (Reviews of Physiology) 69: 97–188.

K Krnjevic, K. (1974) *Physiol. Rev.* 54: 418–540.

L Huxtable, R., and A. Barbeau, eds. (1976) *Taurine*. Raven Press, New York.

M Fonnum, F., ed. (1978) *Amino Acids As Chemical Transmitters, NATO ASI Series*, vol. 48. Plenum Press, New York.

N Dichiara, G., and G. L. Gessa, eds. (1981) *Glutamate As a Neurotransmitter*. Raven Press, New York (*Adv. Biochem. Psychopharmacol.*, vol. 27).

O Watkins, J. C. (1980) *Trends Neurosci.* 3: 61–64.

P Bradford, H. F., ed. (1982) *Interaction and Compartmentation of Neurotransmitters*. NATO ASI Series, vol. 48. Plenum Press, New York.

Q Guldberg, H., and C. A. Marsden. (1975) "Catechol-*O*-Methyl Transferase: Pharmacological Aspects and Physiological Role," *Pharmacol. Rev.* 27: 135–206.

R Horn, A. S., J. Korf, and B. H. C. Westerink, eds. (1979) *The Neurobiology of Dopamine.* Academic Press, London.

S Cooper, J. R., F. E. Bloom, and R. H. Roth. (1982) *The Biochemical Basis of Neuropharmacology,* 4th ed. Oxford University Press, New York.

T Pycock, C. J., and P. V. Taberner, eds. (1981) *Central Neurotransmitter Turnover.* Croom Helm, London.

U Hertz, L., E. Kvamme, E. G. McGeer, and A. Schousboe. (1983) *Glutamine, Glutamate, and GABA in the Central Nervous System.* Alan R. Liss, New York.

V Marsden, C. A., ed. (1984) *Measurement of Neurotransmitter Release* In Vivo. IBRO Handbook, vol. 6. Wiley, Chichester.

Additional Citations

1 Hall, L. W., J. G. Hildebrand, and E. A. Kravitz. (1971) *Chemistry of Synaptic Transmission: Essays and Sources.* Chiron Press, Portland, Oreg.
2 Dale, H. (1935) *Proc. R. Soc. Med.* 28: 319–332.
3 Von Euler, U. S. (1956) *Noradrenaline.* Thomas, Springfield, Ill.
4 Carlsson, A. (1959) *Pharmacol. Rev.* 11: 300–304.
5 Twarog, B. M., and I. H. Page. (1953) *Am. J. Physiol.* 175: 157–161.
6 Emson, P. C. (1979) *Prog. Neurobiol.* 13: 61–116.
7 Orrego, F. (1979) *Neuroscience* 4: 1037–1057.
8 Gibson, G. E., and V. P. Blass. (1976). *J. Neurochem.* 27: 37–42.
9 Kuhar, M. J., and L. C. Murrin. (1978) *J. Neurochem.* 30: 15–22.
10 Rossier, J. (1977) *Int. Rev. Neurobiol.* 20: 284–337.
11 Rossier, J. (1983) *Trends Neurosci.* 6: 201–202.
12 Roskoski, R., C. T. Lim, and L. M. Roskoski. (1975) *Biochemistry* 14: 5105–5110.
13 Chao, L. P., and F. Wolfgram. (1975) *J. Neurochem.* 20: 1075–1081.
14 Polak, R. L., L. C. Sellin, and S. Thesleff. (1981) *J. Physiol.* (Lond.) 319: 253–259.
15 Whittaker, V. P., and H. Stadler. (1980). In *Proteins of the Nervous System,* 2d ed., (eds. R. A. Bradshaw and D. M. Schneider), pp. 231–255. Raven Press, New York.
16 Abdul-Ghani, A. S., J. Coutinho-Netto, and H. F. Bradford. (1980) *Biochem. Pharmacol.* 29: 2179–2182.
17 Rosenberry, T. L. (1975) *Adv. Enzymol.* 43: 103–218.
18 Silman, I. (1976) *Trends Biochem. Sci.* 1: 225–227.
19 Massoulié, J. (1980) *Trends Biochem. Sci.* 5: 160–164.
20 Dudai, Y., M. Herzberg, and I. Silman. (1973) *Proc. Natl. Acad. Sci.* (USA) 70: 2473–2476.
21 Brimijoin, S. (1983) *Prog. Neurobiol.* 21: 291–322.
22 Massoulié, J., and S. Bon. (1982) *Ann. Rev. Neurosci.* 5: 57–106.
23 Reiger, F., S. Bon, J. Massoulié, and J. Cartaud. (1973) *Eur. J. Biochem.* 34: 539–547.

24 Koelle, G. B., and J. S. Friedenwald. (1949) *Proc. Soc. Exp. Biol. Med.* (N.Y.) 70: 617–622.
25 Koelle, G. B. (1963) *Handb. Exp. Pharmakol.* 15: 187–298.
26 Krnjevic, K. (1965) In *Studies in Physiology* (eds. D. R. Curtis and A. K. McIntrye), pp. 144–151. Springer-Verlag, Berlin.
27 Greengard, P. (1976) *Nature* (Lond.) 260: 101–108.
28 Eldefrawi, M. E., and A. T. Eldefrawi. (1977) In *Receptors and Recognition,* Series A, vol. 4 (eds. P. Cuatrecasas and M. F. Greaves), pp. 197–258. Chapman and Hall, London.
29 Conti-Tronconi, B. M., and M. A. Raftery. (1982) *Ann. Rev. Biochem.* 51: 491–530.
30 Dolly, O. J., and E. A. Barnard. (1977) *Biochemistry* 16: 5053–5060.
31 Grünhagen, H. H., and J. P. Changeux. (1976) *J. Mol. Biol.* 106: 517–535.
32 Barnard, E. A., R. Miledi, and K. Sumikawa. (1982) *Proc. R. Soc.* (Lond.), *Ser. B* 215: 241–246.
33 Sumikawa, K., M. Houghton, J. C. Smith, L. Bell, B. M. Richards, and E. A. Barnard. (1982) *Nucleic Acids Res.* 10: 5809–5822.
34 Noda, M., H. Takahashi, T. Tanabe, M. Toyosato, Y. Furutani, T. Hirose, M. Asai, S. Inayama, T. Miyata, and S. Numa. (1982) *Nature* (Lond.) 299: 793–797.
35 Noda, M., H. Takahashi, T. Tanabe, M. Toyosato, S. Kikytoni, T. Horise, M. Asai, H. Takashima, S. Inayama, T. Miyata, and S. Numa. (1983) *Nature* (Lond.) 301: 251–254.
36 Lindstrom, J. (1978) In *Biochemistry of Myasthenia Gravis and Muscular Dystrophy* (eds. G. G. Lunt and R. M. Marchbanks), pp. 135–156. Academic Press, London.
37 Eldefrawi, A. T., M. E. Eldefrawi, N. Mansour, J. Daly, B. Witkop, and E. X. Albequerque. (1978) *Biochemistry* 17: 5474–5484.
38 Blashko, H. (1975) *Biochem. Soc. Trans.* 3: 27–37.
39 Weiner, N. (1979) In *Aromatic Amino Acid Hydroxylases and Mental Disease* (ed. M. B. H. Youdim), pp. 141–190. Wiley, Chichester.
40 Roth, R. H. (1979) In *The Neurobiology of Dopamine* (eds. A. S. Horn, J. Korf, and B. H. C. Westerink), pp. 101–122. Academic Press, London.
41 Mestikawy, S. E., J. Glowinski, and M. Hamon. (1983) *Nature* (Lond.) 302: 820–832.
42 Conti-Tronconi, B. M., C. M. Gotti, M. W. Hunkapiller, and M. A. Raftery. (1982) *Science* 218: 1227–1229.
43 Raese, J. D., A. M. Edelman, M. A. Lazar, and J. D. Barchas. (1977) In *Structure and Function of Monoamine Enzymes* (eds. E. Usdin, N. Weiner, and M. B. H. Youdim), pp. 383–421. Marcel Dekker, New York.
44 Lancaster, G. A., and T. L. Sourkes. (1972) *Can. J. Biochem.* 50: 791–794.
45 Hokfelt, T., K. Fuxe, and M. Goldstein. (1973) *Brain Res.* 53: 175–180.
46 Wallace, E. F., M. J. Krantz, and W. Lovenberg. (1973) *Proc. Natl. Acad. Sci.* (USA) 70: 2253–2255.
47 Tong, J. H., and S. Kaufman. (1975) *J. Biol. Chem.* 250: 4152–4158.
48 Youdim, M. B. H., M. Hamon, and S. Bourgoin. (1975) *J. Neurochem.* 25: 407–414.
49 Youdim, M. B. H. (1979) In *Aromatic Amino Acid Hydroxylases* (ed. M. B. H. Youdim), pp. 233–297. Wiley, Chichester.
50 Carlsson, A., and M. Lindqvist. (1972) *J. Neural Trans.* 33: 23–43.

51 Kettler, R., G. Bartholini, and A. Pletscher. (1974) *Nature* 249: 476–478.

52 Harris, J. E., R. J. Baldessarini, V. H. Morgenroth, and R. H. Roth. (1975) *Proc. Natl. Acad. Sci.* (USA) 72: 789–793.

53 Van Orden, L. S., K. G. Bensch, and N. J. Giarman. (1967) *J. Pharmacol. Exp. Ther.* 155: 428–439.

54 Fillenz, M. (1978) *Prog. Neurobiol.* 8: 251–278.

55 Fillenz, M. (1971) *Philos. Trans R. Soc.* (Lond.), *Ser. B* 261: 319–324.

56 Baldessarini, R. J. (1975) In *Handbook of Psychopharmacology, vol. 3* (eds. L. L. Iversen, S. D. Iversen, and S. H. Synder), pp. 37–137. Plenum Press, New York and London.

57 Descarries, L., K. C. Watkins, and Y. Lapierre. (1977) *Brain Res.* 133: 197–222.

58 Beaudet, A., and L. Descarries. (1978) *Neuroscience* 3: 851–860.

59 Glowinski, J. (1982) In *Neurotransmitter Interaction and Compartmentation* (ed. H. F. Bradford), p. 219–234. Plenum Press, New York.

60 Achee, F. M., S. Gabay, and K. F. Tipton. (1978) *Prog. Neurobiol.* 8: 325–348.

61 Musachio, J. M. (1975) In *Handbook of Psychopharmacology, vol. 3* (eds. L. L. Iversen, S. D. Iversen, and S. H. Synder,), pp. 1–36. Plenum Press, New York, London.

62 Iversen, L. L. (1978) In *Handbook of Psychopharmacology, vol. 3* (eds. L. L. Iversen, S. D. Iversen, and S. H. Synder), pp. 381–442. Plenum Press, New York and London.

63 Levitski, A. (1976) In *Receptors and Recognition*, Series A, vol. 2 (eds. P. Cuatrecasas, and M. F. Greaves). Chapman and Hall, London.

64 Snyder, S. H., and R. R. Goodman. (1980) *J. Neurochem.* 35: 5–15.

65 Kebabian, J. W., and D. B. Calne. (1979) *Nature* (Lond.), 277: 92–96.

66 Iversen, L. L. (1975) *Science* 188: 1084–1089.

67 Langer, S. Z. (1979) In *The Release of Catecholamines from Adrenergic Neurons* (ed. D. M. Paton), pp. 59–86. Pergamon Press, Oxford.

68 Starke, K. (1979) In *The Release of Catecholamines from Adrenergic Neurons* (ed. D. M. Paton), pp. 143–184. Pergamon Press, Oxford.

69 De Belleroche, J. S., and H. F. Bradford. (1981) *Brit. J. Pharmacol.* 72: 427–433.

70 De Belleroche, J. S., and H. F. Bradford. (1978) *Brain Res.* 142: 53–68.

71 De Belleroche, J. S., and H. F. Bradford. (1980) *J. Neurochem.* 35: 1227–1234.

72 Falk, B., N. A. Hillarp, G. Thieme, and A. Torp. (1962) *J. Histochem. Cytochem.* 10: 348–354.

73 Dahlstrom, A., J. Haggendal, and C. Atack. (1973) In *Serotonin and Behavior* (eds. J. Barchas and E. Usdin), pp. 87–96. Academic Press, New York.

74 Swanson, L. W., and B. K. Hartman. (1975) *J. Comp. Neurol.* 163: 467–505.

75 Jones, B. E., and R. Y. Moore. (1977) *Brain Res.* 127: 23–53.

76 Dahlstrom, A., and K. Fuxe. (1964) *Acta Physiol. Scand.* 62 *Suppl.* 232: 1–55.

77 Ungerstedt, U. (1971) *Acta Physiol. Scand. Suppl.* 367: 49–67.

78 Lindvall, O., and A. Bjorklund. (1974) *Acta Physiol. Scand. Suppl.* 412: 1–48.

79 Moore, R. T., and L. W. Kromer. (1977) In *Neuropharmacology and Behavior* (eds. B. Haber and M. Aprison). Plenum Press, New York.

80 Hökfelt, T., K. Fuxe, M. Goldstein, and O. Johansson. (1974) *Brain Res.* 66: 235–251.

81 Fuller, R. W. (1982) *Ann. Rev. Pharmacol. Toxicol.* 22: 31–55.

82 Scatton B., F. Pelayo, M. L. Dubocovich, S. Z. Langer, and G. Batholini. (1979) *Brain Res.* 176: 197–201.

83 Burgess, S. K., and R. E. Tessel. (1980) *Brain Res.* 194: 259–262.

84 Dietl, H., J. N. Sinha, and A. Phillippu. (1981) *Brain Res.* 208: 213–218.

85 Fuxe, K., and G. Jonsson. (1974) *Adv. Biochem. Psychopharmacol.* 10: 1–12.

86 Nobin, A., and A. Bjorklund. (1973) *Acta Physiol. Scand. Suppl.* 388: 1–40.

87 Muscholl, E. (1979) In *The Release of Catecholamines from Adrenergic Neurons* (ed. D. M. Paton), pp. 87–110. Pergamon Press, Oxford.

88 Langer, S. Z., K. Starke, and M. L. Dubocovich. (1979) *Presynaptic Receptors*, Advances in the Biosciences, vol. 18. Pergamon Press, Oxford.

89 Kvamme, E., B. Tveit, and G. Svenneby. (1970) *J. Biol. Chem.* 245: 1871–1877.

90 Svenneby, G., I. A. Torgner, and E. Kvamme. (1973) *J. Neurochem.* 20: 1217–1224.

91 Kvamme, E., and G. Svenneby. (1974) In *Research Methods in Neurochemistry* (eds. N. Marks and R. Rodnight), pp. 277–290. Plenum Press, New York.

92 Kvamme, E., and I. A. Torgner. (1975) *Biochem. J.* 149: 83–91.

93 Salganicoff, L., and E. De Robertis. (1965) *J. Neurochem.* 12: 287–309.

94 Bradford, H. F., and H. K. Ward. (1976) *Brain Res.* 110: 115–125.

95 Kvamme, E., and B. E. Olsen. (1980) *J. Neurochem.* 36: 1916–1923.

96 Hamberger, A. C., G. H. Chiang, E. S. Nylén, S. W. Scheff, and C. W. Cotman. (1979) *Brain Res.* 168: 513–530.

97 Bradford, H. F., H. K. Ward, and A. J. Thomas. (1978) *J. Neurochem.* 30: 1453–1459.

98 Shank, R. P., and G. Le M. Campbell. (1983) In *Handbook of Neurochemistry* 2d ed., vol. 3 (ed. A. Lajtha), pp. 381–404. Plenum Press, New York.

99 Rassin, D. K. (1972) *J. Neurochem.* 19: 139–148.

100 Bradford, H. F. (1975) In *Handbook of Psychopharmacology*, vol. 1, (eds. C. C. Iversen, S. D. Iversen, and S. H. Synder), pp. 191–252. Plenum Press, New York.

101 Fagg, G. E., and J. D. Lane. (1979) *Neuroscience* 4: 1015–1036.

102 Fonnum, F. (1984) *J. Neurochem.* 42: 1–11.

102a Naito, S., and T. Ueda. (1985) *J. Neurochem.* 44: 99–109.

103 Abdul-Ghani, S., J. Coutinho-Netto, and H. F. Bradford. (1980) In *Glutamate: Neurotransmitter in the CNS* (eds. G. A. R. Johnston and P. R. Roberts), pp. 155–203. Plenum Press, New York.

104 Snyder, S. H., A. B. Young, J. P. Bennett, and A. J. Mulder. (1973) *Fed. Proc.* 32: 2039–2047.

105 Ward, H. K., and H. F. Bradford. (1979) *J. Neurochem.* 33: 339–342.

106 Duce, I. R., and P. Keen. (1983) *Neuroscience* 8: 861–866.

107 Steiner, F. A., and K. Ruf. (1966) *Helv. Physiol. Acta* 24: 181–192.

108 Watkins, J. C. (1978) In *Kainic Acid* (eds. E. G. McGeer, J. W. Olney, and P. L. McGeer), pp. 37–70. Raven Press, New York.

109 Davies, J., and J. C. Watkins. (1981) *Brain Res.* 206: 172–177.

110 Fagg, G. E. (1985) *Trends Neurosci.* 8: 207–210.

111 Davies, J., and J. C. Watkins. (1982) *Exp. Brain Res.* 49: 280–290.

112 White, W. F., J. V. Nadler, A. Hamberger, C. W. Cotman, and J. T. Cummins. (1977) *Nature* (Lond.) 270: 356–357.

113 McGeer, E. G., J. W. Olney, and P. L. McGeer. (1978) *Kainic Acid.* Plenum Press, New York.

114 Reif-Lehrer, L. (1977) *Fed. Proc.* 36: 1617–1623.

115 Norenberg, M. D., and A. Martinez-Hernandez. (1979) *Brain Res.* 161: 303–310.

116 Balcar, V. J., G. A. R. Johnston, and B. Twitchin. (1977) *J. Neurochem.* 28: 1145–1146.

117 Johnston, G. A. R., S. M. E. Kennedy, and B. Twitchin. (1979) *J. Neurochem.* 32: 121–127.

118 Michaelis, E. K., E. K. Michaelis, M. L. Michaelis, T. M. Storeman, W. L. Chittenden, R. D. Grubbs. (1983) *J. Neurochem.* 40: 1742–1753.

119 de Plazas, S. F., E. de Robertis, and G. G. Lunt. (1977) *Gen. Pharmacol.* 8: 133–137.

120 Michaelis, E. K., W. L. Chittenden, B. E. Johnson, N. Galton, and C. Decedue. (1984) *J. Neurochem.* 42: 379–406.

121 Taxt, T., and J. Storm-Mathiesen. (1979) *J. Physiol.* (Paris) 75: 677–684.

122 Sharif, N. A., and P. J. Roberts. (1980) *J. Neurochem.* 34: 779–784.

123 Baudry, M., and G. Lynch. (1981) *J. Neurochem.* 36: 811–820.

124 Slevin, J. T., and J. T. Coyle. (1981) *J. Neurochem.* 37: 531–533.

125 McLaughlin, B. J., J. G. Wool, K. Saito, E. Roberts, and J.-Y. Wu. (1974) *Brain Res.* 85: 377–391.

126 Wu, J.-Y., T. Matsuda, and E. Roberts. (1973) *J. Biol. Chem.* 248: 3029–3034.

127 Bridgers, W. F. (1969) *Arch. Biochem. Biophys.* 133: 201–207.

128 Shank, R. P., and M. H. Aprison. (1970) *J. Neurochem.* 17: 1461–1475.

129 Uhr, M. L., and M. K. Sneddon. (1972) *J. Neurochem.* 19: 1495–1500.

130 Shank, R. P., M. H. Aprison, and C. F. Baxter. (1973) *Brain Res.* 52: 301–308.

131 de Belleroche, J. S., and H. F. Bradford. (1973) *J. Neurochem.* 21: 441–445.

132 Abdul-Ghani, A.-S., H. F. Bradford, D. Cox, and P. R. Dodd. (1978) *Brain Res.* 171: 55–66.

133 Gauchy, C., M. L. Kemel, J. Glowinski, and M. J. Besson. (1980) *Brain Res.* 193: 129–141.

134 Roberts, E., J. Wein, and D. G. Simonsen. (1964) *Vitam. Horm.* 22: 503–559.

135 Young, A. B., and S. H. Synder. (1973) *Proc. Natl. Acad. Sci.* (USA) 10: 2832–2836.

136 Abdul-Ghani, A.-S., J. Coutinho-Netto, and H. F. Bradford. (1980) *Brain Res.* 191: 471–481.

137 Löscher, W. (1979) *J. Neurochem.* 32: 1587–1895.

138 Hardy, J. A., J. S. de Belleroche, D. Border, and H. F. Bradford. (1980) *J. Neurochem.* 34: 1130–1139.

139 Costa, E., and A. Guidotti. (1979) *Ann. Rev. Pharmacol. Toxicol.* 19: 531–545.

140 Otsuka, M., L. L. Iversen, Z. W. Hall, and E. A. Kravitz. (1966) *Proc. Natl. Acad. Sci. USA* 56: 1110–1115.

141 Storm-Mathisen, J. (1977) *Brain Res.* 120: 379–386.

142 Fagg, G. E., and A. C. Foster. (1983) *Neuroscience* 9: 701–719.

143 McGeer, P. L., E. G. McGeer, U. Scherer, and K. Singh. (1977) *Brain Res.*
     128: 369–373.
144 Osborne, R. H., H. F. Bradford, and D. G. Jones. (1973) *J. Neurochem.* 21:
     407–419.
145 Peck, E. J., and J. Awapara. (1967) *Biochem. Biophys. Acta* (Amst.) 141: 499–
     506.
146 Lahdesmaki, P., and S. S. Oja. (1975) *J. Neurochem.* 25: 675–680.
147 Haas, H. L., and L. Hosli. (1973) *Brain Res.* 52: 399–402.
148 Agrawal, H. C., A. L. Davison, and L. K. Kaczmerak. (1971) *Biochem. J.* 122:
     759–763.
149 Chan-Palay, V., S. L. Palay, and J-Y. Wu. (1982) *Proc. Natl. Acad. Sci.* (USA)
     79: 4221–4225.
150 Kaczmerak, L. K., and A. L. Davison. (1972) *J. Neurochem.* 19: 2355–2362.
151 Pasantes-Morales, H., J. Klethi, P. F. Urban, and P. Mandel. (1974) *Exp.
     Brain Res.* 19: 131–141.
152 Seighart, W., and M. Karobath. (1974) *J. Neurochem.* 20: 771–782.
153 Bergamini, L., R. Mutani, M. Delsedime, and L. Durelli. (1974) *Eur. J.
     Neurol.* 11: 261–269.
154 Van Gelder, N. M. (1972) *Brain Res.* 47: 157–165.
155 Wheler, G. H. T., H. F. Bradford, A. W. Davison, and E. J. Thompson. (1979)
     J. Neurochem. 33: 331–337.
156 Kaczmerak, L. K., and W. R. Adey. (1974) *Brain Res.* 76: 83–94.
157 Mandel, P., and H. Pasantes-Morales. (1976) *Adv. Biochem. Pharmacol.* 15:
     141–151.
158 Collins, C. G. S. (1977) *Essays Neurochem. Neuropharmacol.* 1: 43–72.
159 Rassin, D. W., and G. E. Gaull. (1978) In *Amino Acids As Chemical Trans-
     mitters* (ed. F. Fonnum), pp. 571–597. Plenum Press, New York.
160 Schousboe, A. (1978) In *Dynamic Properties of Glia Cells* (eds. Schoffeneils,
     E., G. Franck, L. Hertz, and D. B. Tower), pp. 173–182. Pergamon Press,
     Oxford.
161 Dolara, P., A. Agnesti, A. Giotti, and G. Pasquini. (1973) *Eur. J. Pharmacol.*
     24: 352–358.
162 Pasantes-Morales, H., and J. Morán. (1981) In *Regulatory Mechanisms of
     Synaptic Transmission* (ed. R. Tapia and C. W. Cotman), pp. 141–154.
     Plenum Press, New York.
163 De Belleroche, J. S., and H. F. Bradford. (1972) *J. Neurochem.* 19: 1817–1819.
164 Tamir, H., and K.-P. Liu. (1982) *J. Neurochem.* 38: 135–141.
165 Green, H. N., and H. B. Stoner. (1950) *Biological Actions of the Adenine
     Nucleotides.* Lewis, London.
166 Burnstock, G., T. Hokfelt, M. D. Gershon, L. L. Iversen, H. W. Kosperlitz,
     and J. H. Szurszewski. (1979) *Neurosci. Res. Prog. Bull.* 17 (3): 379–519.
167 Burnstock, G. (1981) *J. Physiol.* (Lond.) 313: 1–35.
168 McIlwain, H. (1977) *Neuroscience* 2: 357–372.
169 Burnstock, G., ed. (1981) *Purinergic Receptors.* Receptors and Recognition,
     Series B, vol. 12. Chapman and Hall, London.
170 Burnstock, G. (1971) *Pharmacol. Rev.* 21: 247–324.
171 Hirschowitz, B. I. (1979) *Ann. Rev. Pharmacol. Toxicol.* 19: 203–244.
172 Schwartz, J. C., H. Pollard, and T. T. Quach. (1980) *J. Neurochem.* 35: 26–
     33.
173 Garbarg, M., G. Barbin, E. Rodergas, and J. C. Schwartz. (1980) *J. Neuro-
     chem.* 35: 1045–1052.

174 Garbarg, M., G. Barbin, E. Rodergas, and J. C. Schwartz. (1976) *Brain Res.* 106: 333–348.

175 Kanof, P. D., and P. Greengard. (1979) *J. Pharmacol. Exp. Ther.* 209: 87–96.

176 Stevens, C. F. (1979) *Sci. Am.* 241: 55–65.

177 Fonnum, F., A. Soreide, I. Kvale, J. Walker, and I. Walaas. (1981) In *Glutamate As a Neurotransmitter* (eds. G. Di Chiara and G. L. Gessa), pp. 29–42. Raven Press, New York.

178 Greengard, P. (1979) *Fed. Proc.* 35: 2208–2217.

179 Ariano, M. M. (1983) *Neuroscience* 10: 707–723.

180 Ariano, M. A., J. A. Lewicki, H. J. Brandwein, and F. Murad. (1982) *Proc. Natl. Acad. Sci.* (USA) 79: 1316–1320.

181 Lohmann, S. M., U. Walter, P. E. Miller, P. Greengard, and P. De Camilli. (1981) *Proc. Natl. Acad. Sci.* (USA) 78: 653–657.

182 Nakane, M., M. Ichikawa, and T. Deguchi. (1983) *Brain Res.* 273: 9–15.

183 Cumming, R., Y. Koide, M. R. Krigman, J. A. Beavo, and A. L. Steiner. (1981) *Neuroscience* 6: 953–961.

184 Michell, R. H., C. J. Kirk, L. M. Jones, C. P. Downes, and J. A. Creba. (1981) *Phil. Trans. R. Soc.* (Lond.) *Ser. B.* 296: 123–137.

185 Brown, E., D. A. Kendall, and S. R. Nahorski. (1984) *J. Neurochem.* 42: 1379–1387.

186 Kendall, D. A., and S. R. Nahorski. (1984) *J. Neurochem.* 42: 1388–1394.

187 Berridge, M. J. (1981) *Mol. Cell Endocrinol.* 24: 115–140.

188 Berridge, M. J., C. P. Downes, and M. R. Hanley. (1982) *Biochem. J.* 206: 587–595.

189 Streb, H., R. F. Irvine, M. J. Berridge, and I. Schultz. (1983) *Nature* (Lond.) 306: 67–68.

190 Burgess, G. M., P. P. Godfrey, J. S. McKinney, M. J. Berridge, R. F. Irvine, and J. W. Putney. (1984) *Nature* (Lond.) 309: 63–66.

191 Berridge, M. J. (1984) *Biochem. J.* 220: 345–360.

192 Cuello, A. C., and M. V. Sofroniew. (1984) *Trends Neurosci.* 7: 74–78.

193 Sofroniew, M. V., F. Eckenstein, H. Thoenen, and A. C. Cuello. (1982) *Neurosci. Lett.* 33: 7–12.

194 Mesulam, M.-M., E. J. Mufson, B. H. Wainer, and A. I. Levey. (1983) *Neurosci.* 10: 1185–1201.

195 Foster, A. C., and G. E. Fagg. (1984) *Brain Res. Rev.* 7: 103–164.

196 Susz, J. P., B. Haber, and E. Roberts. (1966) *Biochemistry* 5: 2870–2877.

197 Miller, L. P., and D. L. Martin. (1973) *Life Sci.* 13: 1023–1032.

198 Kanazawa, I., L. L. Iversen, and J. S. Kelly. (1976) *J. Neurochem.* 27: 1267–1269.

199 Wu, J.-Y., O. Chude, J. Wein, E. Roberts, K. Saito, and E. Wong. (1978) *J. Neurochem.* 30: 849–857.

200 Hill, D. R., and N. G. Bowery. (1981) *Nature* (Lond.) 290: 149–152.

201 Bowery, N. G., A. Doble, D. R. Hill, A. C. Hudson, J. Shaw, and M. J. Turnbull. (1981) *Eur. J. Pharmacol.* 71: 53–70.

202 Bowery, N. G., D. R. Hill, and A. L. Hudson. (1983) *Br. J. Pharmacol.* 78: 191–206.

203 Bowery, N. G., A. Doble, D. R. Hill, A. L. Hudson, D. N. Middlemiss, J. S. Shaw, and M. J. Turnbull. (1980) *Nature* (Lond.) 204: 92–94.

204 Hill, D. R., N. G. Bowery, and A. L. Hudson. (1984) *J. Neurochem.* 42: 652–657.

205 Dairman, W., J. C. Christenson, and S. Udenfriend. (1972) *Pharmacol. Rev.* 24: 269–289.

206 Nyback, H. (1972) *Acta Physiol. Scand.* 84: 54–64.

207 Bannon, M. J., and R. H. Roth. (1983) *Pharmacol. Rev.* 35: 53–68.

208 Vulliet, P., T. Langan, and N. Weiner. (1980) *Proc. Natl. Acad. Sci.* (USA) 75: 4744–4748.

209 Builder, S. E., J. A. Beavo, and E. G. Krebs. (1980) *J. Biol. Chem.* 255: 2350–2354.

210 Young III, W. S., and M. J. Kuhar. (1979) *Brain Res.* 179: 255–270.

211 Hamon, M., S. Bourgoin, F. Artand, and J. Glowinski. (1979) *J. Neurochem.* 33: 1031–1042.

212 Monaghan, D. T., V. R. Holets, D. W. Toy, and C. W. Cotman. (1983) *Nature* (Lond.) 306: 176–178.

213 Tapia, R. (1983) In *Handbook of Neurochemistry*, vol. 3 (ed. A. Lajtha), pp. 423–466. Plenum Press, New York.

214 Wu, J.-Y. (1983) In *Handbook of Neurochemistry*, vol. 4 (ed. A. Lajtha), pp. 111–131. Plenum Press, New York.

215 Kvamme, E. (1983) In *Handbook of Neurochemistry*, vol. 3 (ed. A. Lajtha), pp. 405–422. Plenum Press, New York.

216 Arrang, J.-M., M. Garbarg, and J. C. Schwartz. (1983) *Nature* (Lond.) 302: 832–837.

217 Kuhn, D. M., and W. Lovenberg. (1983) In *Handbook of Neurochemistry*, vol. 4 (ed. A. Lajtha), pp. 133–150. Plenum Press, New York.

218 Monaghan, D. T., M. C. McMills, A. R. Chamberlin, and C. W. Cotman. (1983) *Brain Res.* 278: 137–144.

219 Butcher, L. L., and R. Marchand. (1978). *Eur. J. Pharmacol.* 52: 415–417.

220 Shute, C. C. D., and P. R. Lewis. (1967) *Brain Res.* 90: 497–520.

221 Karlin, A., R. Cox, R. R. Kaldany, P. Lobel, and E. Holtzman. (1983) *Cold Spring Harbor Symp. Quant. Biol.* 48: 1–8.

222 Fairclough, R. H., J. Finer-Moore, D. Love, D. Kristofferson, P. J. Desmeules, and R. M. Stroud. (1983) *Cold Spring Harbor Symp. Quant. Biol.* 48: 9–20.

223 Changeux, J.-P. (1981) *Harvey Lect.* 75: 85–97.

224 Changeux, J.-P., F. Bon, J. Cartand, A. Devillers-Thiery, J. Giraudat, T. Heidmann, B. Holton, H.-O. Nghiem, J. L. Popot, R. Van Rapenbusch, and S. Tzartos. (1983) *Cold Spring Harbor Symp. Quant. Biol.* 48: 35–52.

225 Raftery, M. A., S. M. J. Dunn, B. M. Conti-Tronconi, D. S. Middlemas, and R. D. Crawford. (1983) *Cold Spring Harbor Symp. Quant. Biol.* 48: 21–33.

226 Linstrom, J., S. Tzartos, W. Gullick, S. Hochenschwander, L. Swanson, P. Sargent, M. Jacob, and M. Montal. (1983) *Cold Spring Harbor Symp. Quant. Biol.* 48: 89–99.

227 Olsen, R. W. (1983) *Ann. Rev. Pharmacol. Toxicol.* 22: 245–277.

228 Sigel, E., A. Stephenson, C. Mamalaki, and E. A. Barnard. (1983) *J. Biol. Chem.* 258: 6965–6971.

229 Sigel, E., and E. A. Barnard. (1984) *J. Biol. Chem.* 259: 7219–7223.

230 Olsen, R. W. (1981) *J. Neurochem.* 37: 1–87.

231 Miledi, R., I. Parker, and K. Sumikawa. (1982) *Proc. R. Soc.* (Lond.) *Ser. B* 216: 509–515.

232 Smart, T. G., A. Constanti, G. Bilbe, D. A. Brown, and E. A Barnard. (1983) *Neurosci. Lett.* 40: 55–59.

233 Pfeiffer, F., D. Graham, and H. Betz. (1982) *J. Biol. Chem.* 257: 9389–9393.

234 Shorr, R. G. L., M. W. Stohsacker, T. N. Lavin, R. J. Lefkowitz, and M. G. Caron. (1982) *J. Biol. Chem.* 257: 12341–12350.

235 Strader, C. D., V. M. Pickel, T. H. Joh, M. W. Strohsacker, R. G. L. Shorr, R. J. Lefkowitz, and M. G. Caron. (1983) *Proc. Natl. Acad. Sci.* (USA) 80: 1840–1844.

236 Strange, P. G. (1983) In *Cell Surface Receptors* (ed. P. G. Strange), pp. 82–100. Ellis Horwood, Chichester.

237 B. Ilien, H. Gorissen, and P. M. Laduron. (1982) *Mol. Pharmacol.* 22: 243–249.

238 Laduron, P. M., and B. Ilien. (1982) *Biochem. Pharmacol.* 31: 2145–2151.

239 Nagatsu, T., M. Levitt, and S. Udenfriend. (1964) *J. Biol. Chem.* 239: 2910–2917.

240 Petrack, P., F. Sheppy, F. Fetzer, T. Manning, H. Chertock, and D. Ma. (1972) *J. Biol. Chem.* 247: 4872–4878.

241 Hoeldtke, R., and S. Kaufman. (1977) *J. Biol. Chem.* 252: 3160–3169.

242 Kuhn, D. M., and W. Lovenberg. (1983) In *Handbook of Neurochemistry* (ed. A. Lajtha), vol. 4, pp. 133–150. Plenum Press, New York.

243 Sourkes, T. L. (1979) In *The Neurobiology of Dopamine* (eds. A. S. Horn, J. Korf, and B. H. C. Westerink), pp. 123–132. Academic Press, New York.

244 Nestler, E. J., and P. Greengard. (1984) *Protein Phorphorylation in the Nervous System.* John Wiley, Chichester.

245 Rodnight, R. (1983) In *Handbook of Neurochemistry,* 2d ed. vol. 4 (ed. A. Lajtha), pp. 195–217. Plenum Press, New York.

246 Dennis, S. G. C., and J. B. Clark. (1978) *J. Neurochem.* 31: 673–680.

247 Yarbrough, G. G., D. K. Singh, and D. A. Taylor. (1981) *J. Pharmacol. Exp. Ther.* 219: 604–613.

248 Okamoto, K., H. Kimura, and Y. Sakai. (1983) *Brain Res.* 260: 261–269.

249 Lin, C.-T., Y.-Y.-T. Su, G.-X. Song, and J.-Y. Wu. (n.d.) *Brain Res.,* in press.

250 Chan-Palay, V., C.-T. Lin, S. Palay, M. Yamamoto, and J.-Y. Wu. (1982) *Proc. Natl. Acad. Sci.* (USA) 79: 2695–2699.

251 Strange, P. G. (1983) *Trends Pharmacol. Sci.* 4: 188–190.

252 Cerione, R. A., B. Strulovici, J. L. Benovic, R. J. Lefkowitz, and M. G. Caron. (1983) *Nature* (Lond.) 306: 562–566.

253 Ilien, B., A. Schotte, and P. M. Laudron. (1982) *FEBS Lett.* 138: 311–315.

254 Creese, I. (1982) *Trends Neurosci.* 5: 40–43.

255 Sokoloff, P., M. P. Matres, and J.-C. Schwartz. (1980) *Naunyn-Schmiedeberg's Arch. Pharmacol.* 315: 89–102.

256 Seeman, P. (1980) *Pharmacol. Rev.* 32: 229–313.

257 Mena, E. E., G. E. Fagg, and C. W. Cotman. (1982) *Brain Res.* 243: 378–381.

258 Fagg, G. E., A. C. Foster, E. E. Mena, and C. W. Cotman. (1983) *Eur. J. Pharmacol.* 88: 105–110.

259 Baudry, M., and G. S. Lynch. (1983) *Eur. J. Pharmacol.* 90: 161–168.

260 Monaghan, D. T., M. C. McMills, A. C. Chamberlin, and C. W. Cotman. (1983) *Brain Res.* 278: 137–144.

261 Watkins, J. C. (1984) *Trends Pharmacol. Sci.* 5: 373–376.

262 Wojcik, W. J., and N. H. Neff. (1983) *Mol. Pharmacol.* 25: 24–28.

263 Dolphin, A. C. (1984) *Trends Neurosci.* 7: 363–364.

264 Dunlop, K. (1981) *Brit. J. Pharmacol.* 74: 579–585.

265 Berridge, M. J., and R. F. Irvine. (1984) *Nature* (Lond.) 312: 315–321.

266 Dolly, J. O., and E. A. Barnard. (1984) *Biochem. Pharmacol.* 33: 841–858.

267 Whittaker, V. P. (1984) *Biochem. Soc. Trans.* 12: 561–576.

268 Wenthold, R. J., and R. A. Altschuler. (1983) In *Glutamine, Glutamate and GABA in the Central Nervous System.* (eds. L. Hertz, E. Kvamme, E. G. McGeer, and Schousboe, A.), pp. 33–50. Alan Liss, New York.

269 Ottersen, O. P., and J. Storm-Mathisen. (1984) *J. Comp Neurol.* 229: 374–392.

270 Watkins, J. C., and R. H. Evans. (1981) *Annu. Rev. Pharmacol. Toxicol.* 21: 165–204.

271 Bowery, N. G., G. W. Price, A. L. Hudson, D. R. Hill, G. P. Wilkin, and M. J. Turnbull. (1984) *Neuropharmacol.* 23: 219–231.

272 Belcher, G., J. Davies, and R. W. Ryall. (1976) *J. Physiol.* (Lond.) 256: 651–662.

273 Bradford, H. F., and A. J. Thomas. (1969) *J. Neurochem.* 16: 1495–1504.

274 Baldessarini, R. J., and C. Yorke. (1974) *J. Neurochem.* 23: 839–848.

275 Benjamin, A. M., Z. H. Verjee, and J. H. Quastel. (1980) *J. Neurochem.* 35: 67–77.

276 Potashner, S. J. (1978) *J. Neurochem.* 31: 187–195.

277 Wu, J.-Y. (1982) *Proc. Natl. Acad. Sci.* (USA) 79: 4270–4274.

278 Cuello, C., J. V. Priestley, and M. V. Sofroniew. (1983) *Quart. J. Exp. Physiol.* 68: 545–578.

279 Potashner, S. J. (1979) *J. Neurochem.* 32: 103–113.

280 Roberts, P. J., J. Storm-Maltrisen, and H. F. Bradford, eds. (1985) *Excitatory Amino Acids.* Macmillan, London.

281 Oja, S. S., and P. Kontro. (1983) In *Handbook of Neurochemistry,* 2d ed., vol. 3. (ed. A. Lajtha), pp. 501–533. Plenum Press, New York.

282 Steinbusch, H. W. M. (1981) *Neurosci.* 6: 557–618.

283 Geffard, M., O. Kah, B. Onteniente, P. Segula, M. Le Moal, and M. A. DeLaage. (1984) *J. Neurochem.* 42: 1593–1599.

284 Geffard, M., A. M. McRae-Degueurce, and M. L. Souan. (1985) *Science* 229: 77–79.

285 Pin, J.-P., J. Bockaert, and M. Recasens. (1984) *FEBS Lett.* 175: 31–36.

5
Neuropeptides:
The Rising Generation
of Neuroactive Substances

A great deal of information has been gathered on the properties of neuroactive peptides in the past few years as a consequence of the large increase in research effort in this area of neurochemistry. In some cases this new knowledge has represented an extension, in terms of detail, of a basic picture conceived 30 to 50 years ago. Substance P is an example. For other peptides, the new knowledge has led to radical departures from conceptions of their function established decades ago. Examples are oxytocin, vasopressin, and hypothalamic trophic hormone–releasing hormones). In the latter group, peptides discovered through their endocrinological properties and thought to exist largely localized to hypothalamus and pituitary system have become major candidates as neurotransmitters in various regions of the central nervous system, and some of these peptides exist in by far the greater proportion in regions other than the hypothalamus. Again, several peptides originally established as gastrointestinal hormones were subsequently found to be present in the central nervous system, where they display potent physiological activity and other properties consistent with a neurotransmitter or neuromodulator role. Examples of this group are VIP and CCK. Other neuroactive peptides, the enkephalins and endorphins, were isolated from the central nervous system through their potent capacity to mimic some of the key actions of morphine, and they were dubbed "endogenous opiates." They are now known to serve widely throughout the body as neuroregulatory substances.

We would be overconfident to assume that all the major classes of substances functioning in the nervous system as classical neurotransmitters or neuromodulators have now been discovered. The neuropeptides are the most recent group of compounds to be given the title "putative neurotransmitters" and to come under intensive scrutiny as chemical agents mediating neural activity. Indeed, the peptidergic neuron is now an established concept,[B,K] much probed and tested.

PEPTIDERGIC SYSTEMS

Peptides known for many years to serve specialized endocrinological functions both within and without the nervous system (e.g., vasopressin, oxytocin, and hypothalamic releasing hormones), and an entirely new group of peptides (e.g., enkephalins and endorphins) have proved to be strong candidates for a neurotransmitter role throughout the nervous system.[A,B,J] Improved methods for extracting, stabilizing, and measuring peptides, including radioimmunoassay and high-performance liquid chromatography (HPLC), have created a flourishing interest in neural peptides. Immunohistochemistry has provided techniques for mapping peptidergic pathways in the nervous system, although controversy does often occur over whether the immunoreactive material in the tissue is in fact identical with the isolated peptide in question and not another structure containing an immunoreactive amino acid sequence.[40,76]

Special Features

Peptides differ in several respects from other neurotransmitters discussed so far. They are extremely potent and are present in very much smaller quantities.[51,J] (See Table 5.1) Their mode of synthesis appears to be quite different and tends to follow the pattern previously described for protein hormones[R,202] and other secretory proteins. Usually, much larger "mother" peptides or "pre-proproteins" are synthesized in the cell body and routed to the cisternae of the endoplasmic reticulum where the "pre" (signal) sequence is cleaved off to produce proprotein. The "proproteins" are then transported into the Golgi apparatus where they may undergo their first cleavage. The final steps of posttranslational processing occur after packaging into neurosecretory granules and during transport to nerve terminals for storage and release. These final stages typically involve limited proteolytic cleavage to produce shorter peptides. In addition, there may be C-terminal amidation and N-terminal acetylation as well as cyclization of glutamate to form pyroglutamate, disulphide bond formation, glycosylation, phosphorylation, or sulfation.[R] Many neuroactive peptides occur in families of related structures that differ in size but contain certain peptide sequences essential for their activity. Each mem-

TABLE 5.1 Content of peptides and neurotransmitters in different
brain regions

	Caudate nucleus	Cerebral cortex (temporal lobe)	Hypothalamus
Peptides *(immunoreactive, pmol/g wet wt)*			
Enkephaline			
Met-	90	100	110
Leu-	450	100	270
Substance P	138 ± 14	104 ± 29	112 ± 19
Neurotensin	3.7 ± 1.2	11.2 ± 5.6	54.9 ± 5.6
Cholecystokinin	223 ± 26	676 ± 52 (2300)*	255 ± 34
VIP	–	13.8 ± 1.5	23.1 ± 6
TRH	4.7 ± 1.3	3.8 ± 0.5	109 ± 31
Monoamines *(nmol/g wet wt)*			
Dopamine	52.9	0.66	4.96
Norepinephrine	1.32	1.32	12.34
Serotonin	9.09	1.36	11.36
Acetylcholine *(nmol/g wet wt)*	22.2 ± 1.4	5.0 ± 0.2	8.8 ± 11.5
GABA *(μmol/g wet wt)*	3.20	2.09	6.19

Note: Peptide values are for human brain; from many sources quoted in ref. 0. Monoamine values are from brain of various species; from ref. 182. Acetylcholine values are for cat brain; from ref. 183. GABA data are from rhesus monkey brain; from ref. 184. Enkephalin data are from bovine brain; from ref. 22.
* From ref. Q.

ber of the family may be enriched in a different organ or brain region, and this is determined by the processing enzymes present (e.g., CCK-8 in brain, CCK-33 in gut; see page 293).

Since no effective membrane transport systems for these compounds have yet been detected, their mode of inactivation appears to be through the unusual agency of active peptidases. It follows, therefore, that they are not recaptured by nerve endings and recycled, and that the supply of peptides for release as neurotransmitters depends on adequate stores of their precursor proteins. Many of these putative peptide neurotransmitters have long been known to be synthesized and released by central nervous system neurons and to serve a well-defined hormonal role in the pituitary gland (e.g., hypothalamic releasing hormones and posterior pituitary hormones). What is new about them is the extension of their function to that of neurotransmitter in other regions of the nervous system.

Coexistence of Peptides Additional interest attaches to neuroactive peptides through the possibility that some may coexist in neurons with other peptides or neurotransmitters[40,41] (e.g., monoamines), the peptides

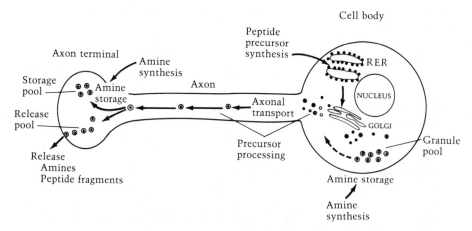

FIGURE 5.1   Diagram illustrating the biosynthesis, packaging, and release of peptides and amines in neurons. The peptides are generated from large precursor molecules (pre-proproteins) produced in the rough endoplasmic reticulum (*RER*) and after removal of the signal sequence are packaged in secretory granules or vesicles in the Golgi stacks. The granules are transported out of the cell body (*Axonal transport*) to the terminals, where upon stimulation, they release their contents by exocytosis. Amines are produced in the cytosol of the cell body, axon, and terminal, and they are packaged by uptake into preformed granules or vesicles. [From ref. 24]

being stored in separate granules or dense-cored vesicles. On release of both substances, the peptides may modulate the action of the "classical" neurotransmitter or even act on a separate group of postsynaptic cells, (e.g., VIP and ACh in salivary-gland nerve terminals; see page 388), thus adding sophistication to the action of the classical neurotransmitter itself[47] (see Figures 4.2 and 5.1). The main cautionary note on the question of coexistence of neuropeptides with other neurotransmitters concerns the possibility that antisera used in these immunohistochemical studies may cross-react with other polypeptide sequences present in the neurons concerned.[40,76] This uncertainty can only be overcome by isolation and chemical characterization of the immunoreactive peptide present in the tissue.

Neurotransmitter, Neuromodulator, or Hormone?   The precise term that should be applied to neuroactive substances secreted by one neuron and acting on another is no longer clear-cut. Essentially, a *neurotransmitter* is released, usually into a synaptic cleft, and acts by binding to presynaptic or postsynaptic receptors in the immediate vicinity of the site of release. A *neuromodulator* is released from a synapse or varicosity and acts at a local or distant postsynaptic site to modify the efficiency with which a neurotransmitter acts on the same postsynaptic site, thereby amplifying or attenuating the neurotransmitter action. A neuromodulator may also exert its influence by modifying the rate of release of a neurotransmitter from its presynaptic site, and it may both coexist and be coreleased with the neurotransmitter (see Chapter 7). A *hormone* (or, neurohormone) acts on a distant receptor population after being released from its site of synthesis (including the presynapse) directly into blood or extracellular fluid, in which it is carried to the target cells (neurons) on which it acts. A given substance may, however, act as a neurotransmitter in one synapse but have modulator, or even hormonelike, actions (paracrine actions) at other, local, cellular sites that bear its receptors.

Many neuropeptides that occur in the brain are also present in the gastrointestinal system, where they are often known to function as hormones or peripheral neurotransmitters. Cholecystokinin (CCK), vasoactive intestinal peptide (VIP), neurotensin, substance P, somatostatin, enkephalins, and bombesin are all established neurogastrointestinal peptides. The significance of the dual existence of these peptides is not yet clear, but specific receptors for them exist in both brain and gastrointestinal systems, allowing the possibility of brain–gut interactions (brain–gut axis).

Peptides and Behavior Many potent behavior-modifying or emotion- and mood-enhancing properties have been reported for the peptides when administered to animals or humans (see Chapter 8). Together with properties such as the morphinelike analgesic effects of the enkephalins and endorphins, these behavior-modifying properties testify to the powerful functional roles of these peptides in the nervous system.[C,185] Whether these functions are mediated by peptides acting as hormone, as neuromodulator, or as neurotransmitter awaits clarification.

Substance P

The pharmacological properties of substance P were first demonstrated in the early 1930s, when it was found to display powerful hypotensive properties that were not blocked by atropine.[2,3] It was not recognized as a peptide until 1936, when it was found to be inactivated by trypsin.[4] This work established substance P as the first neuroactive peptide, and it became the first peptide to be proposed as a neurotransmitter.[E,1] Standard powdered extracts were used in the pharmacological tests, and the letter P in substance P was originally an abbreviation for powder.[2] Some 34 years elapsed between identifying the peptide nature of the compound and determining its structure. It is now known to contain 11 amino acids[9,10] (Figure 5.2). Bioassay and immunohistochemistry have shown the presence of substance P in brain and certain other organs in a wide range of species.[E,A]

Arg-Pro-Lys-Pro-Gln-Gln-Phe-Phe-Gly-Leu-Met-NH$_2$

FIGURE 5.2   The structure of substance P.

Pharmacological Properties   Substance P stimulates contraction of vascular and extravascular smooth muscle (tachykinin activity), causes powerful enhancement of salivation (sialogogue-activity), and has many other central actions. Radioimmunoassay and immunohistochemical techniques developed with the aid of the synthetic compound have shown the presence of substance P in gut, salivary glands, and in most areas of the central nervous system, with high concentrations in the hypothalamus and in the dorsal horn of the spinal cord[11,12] and very high concentrations in the substantia nigra[5,6] (Table 5.2). Its localization in the nerve endings of the striatonigral tract and in the dorsal horn has also been demonstrated by the fall in substance P level following lesion of the appropriate efferent pathway,[7] and it has been proposed as an excitatory neurotransmitter substance in these regions[A,E,12] (Table 5.3).

TABLE 5.2   Regional distribution of substance P in the rat and human brain

	Substance P (nmol/g protein)	
Region	Human	Rat
Frontal cortex	2.07 ± 0.3	0.23 ± 0.02
Caudate	3.7 ± 0.8	2.37 ± 0.26
Putamen	3.3 ± 0.6	–
Substantia nigra	45.0 ± 4.8	16.60 ± 1.01
Hypothalamus	5.2 ± 1.1	5.22 ± 0.50
Amygdala	3.4 ± 0.5	3.6 ± 0.82
Cerebellum	0.2 ± 0.6	0.09 ± 0.03
Dorsal horn	–	10.3 ± 1.5

Note: Data are from sources quoted in ref. A.

TABLE 5.3   Substance P pathways in the rat central nervous system

Origin	Projection	Possible function
Primary sensory neurons	Dorsal horn of spinal cord/trigeminal nucleus	Passage of sensory "pain" information in small-diameter C and A delta fibers
Dorsal and median raphe nuclei	Diffuse ascending projection to telencephalon	Unknown
Raphe pallidus/raphe magnus	Descending substance P to spinal cord	Involved in mechanism of analgesia?
Medial habenular nucleus	Interpenduncular nucleus/lateral habenular nucleus	Unknown
Medial amygdaloid nucleus	Central amygdaloid nucleus	Unknown
Bed nucleus of stria terminalis	Medial preoptic area	Unknown
Caudate nucleus	Globus pallidus/substantia nigra	Extrapyramidal motor function

Note: From ref. A.

Substance P As a Primary Afferent Neurotransmitter   Substance P levels in the dorsal horn of the spinal cord become very much reduced after sectioning of the dorsal spinal root. Immunohistochemistry has shown the presence of densely packed substance P–containing axodendritic synapses on dorsal horn cells.[1] Stimulus-coupled, calcium-dependent release of substance P from various spinal cord preparations in vitro has been established.[8,12,35]

Substance P has a potent depolarizing action when applied to motoneurons in the spinal cord. The characteristics of this response are such that endogenous substance P is now thought to be responsible for the slow ventral root potentials observed to follow stimulation[I,K] of the primary afferent inputs entering through the dorsal roots (Figure 5.3). These observations strongly support the proposal that substance P functions as a primary afferent (sensory) neurotransmitter in the spinal cord.

Sympathetic ganglia also appear to employ substance P as a neurotransmitter at synapses driven by fibers of sensory origin; since sectioning preganglionic nerve fibers causes the loss of most of the considerable content of ganglionic substance P that appears to be localized in nerve

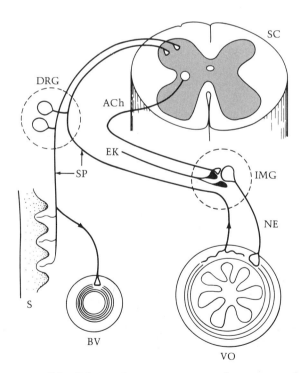

FIGURE 5.3   Schematic representation of somatic and visceral primary afferent neurons and the sympathetic nervous system. *ACh*, acetylcholine; *NE*, norepinephrine; *EK*, enkephalins; *SP*, substance P; *BV*, blood vessel; *DRG*, dorsal root ganglion; *IMG*, inferior mesenteric ganglion; *S*, skin; *SC*, spinal cord; *VO*, visceral organ. [From ref. 13]

terminals (Figure 5.3). The peptide is released by a calcium-dependent process when the ganglion is perfused with medium containing depolarizing levels of potassium. If substance P is added to ganglion cells in low concentrations, it has a depolarizing action of a type that suggests it is responsible for the noncholinergic slow excitatory postsynaptic potentials (EPSPs) produced when the preganglionic fibers are stimulated. The time course of these potentials is similar to the slow ventral root potentials occurring in spinal cord. Both types of synaptic potential, together with the action of added substance P itself, are selectively blocked or reduced by the substance P antagonist [D-Arg^1, D-Pro^2, D-Tryp^{7,9}, Leu^{11}]substance P.[K,I]

Capsaicin, the pungent agent in red pepper, is a specific neurotoxin that destroys a morphologically well-defined population of primary sensory neurons that transmit nociceptive (pain) impulse traffic, but its action is not specifically towards substance P–containing neurons.[14] It causes depletion of substance P from spinal cord and sympathetic ganglia, and at the same time it blocks the slow postsynaptic potentials in these two systems.[O]

Inactivation of substance P released to serve a neurotransmitter function is likely to occur through the action of a membrane-bound neutral metalloendopeptidase (EC 3.4.24.11) that is selective towards substance P and shows a parallel subcellular distribution with substance P and its receptors.[12,I,K] Taken together, there is a persuasive body of evidence for a primary sensory excitatory neurotransmitter function for substance P in spinal cord and sympathetic ganglia.

Substance P as a Neurotransmitter in Other Regions of the Nervous System   There are indications that substance P acts as a neurotransmitter in several central nervous system and peripheral nervous system components, including iris, skin, and the nigrostriatal pathway (Table 5.3). Its concentration in the substantia nigra is by far the greatest detected (Table 5.2), and it is released from slices of substantia nigra by depolarizing stimuli.[8] Substance P release from these slices is suppressed by a picrotoxin-sensitive action of GABA, indicating presynaptic control through GABAergic receptors.[8]

Coexistence of Substance P   It was thought until recently that substance P and dopamine might coexist in dopaminergic terminals of the prefrontal cortex since their distribution is so similar.[40] But careful lesion studies have shown that loss of one of the pair can occur without loss of the other, precluding coexistence in the same nerve terminals.[A] Certainly, there is firm evidence of *interaction* of substance P–containing neurons with dopaminergic systems.

Another well-studied case is the coexistence of 5-HT with substance P in the medullary raphe nuclei.[40,41,42] In fact, there is evidence for coexistence of these two substances with thyrotropin-releasing hormone

(TRH; see below). Depletion of all three substances occurs in parallel following destruction of the raphe neurons with the neurotoxin 5,7-dihydroxytryptamine.[43]

Receptors for Substance P Specific receptors for substance P have been detected both by ligand-binding studies and by bioassay techniques.[A,D] The receptor-mediated actions of substance P are relatively slow and persistent and are often elicited by very low (nanomolar) concentrations of the peptide, indicating the existence of a high-affinity recognition site. Receptor interaction with substance P involves changes in phosphatidyl inositol turnover and mobilization of cellular calcium, but this does not appear to be mediated by cyclic nucleotides.[D,E] Tritium-labeled substance P itself has been employed as the ligand in receptor-binding studies.[I] Substance P is synthesized in the neuronal cell body and transported in granule-packaged form to nerve terminals for release. This transport has been studied in the striatonigral system where the appropriate cell bodies in the striatum are well separated from the nerve terminals in the substantia nigra.[44]

Substance P in Pain Mechanisms In the spinal cord, and elsewhere, substance P also appears to be operating in pathways concerned with pain mechanisms[172,K,I] (Table 5.3). It is present in so-called C fibers, certain small-diameter primary afferents in the spinal cord believed to be important in the neurotransmission of pain. Thus, substance P may be one of the neurotransmitters involved in these pain pathways. Some antagonists of substance P injected into the spinal cord have an analgesic action, and substance P itself reduces the reaction time to pain-inducing stimuli and produces other behavioral responses. These observations are consistent with its being released from nerve terminals mediating pain reception.[I,K]

Since the opiate analgesics morphine and the endorphins (see next section) inhibit the release of substance P from slices of trigeminal nucleus, a brain stem nucleus involved in the transmission of nociceptive information, it seems that there is some interaction between substance P and endogenous opioid neurotransmitters known to be involved in processes generating analgesia.[35] Taken together, there is a range of evidence from different sources suggesting that substance P serves a key function in the neurotransmission of pain in the central nervous system.

Enkephalins and Endorphins

Highly sensitive bioassays for morphine and related opiates specifically antagonized by naloxone, a powerful morphine antagonist, provided the detection system that resulted in the isolation of endogenous opioid compounds from pig brain extracts.[15,16] In these bioassays contractions of guinea pig ileum and mouse vas deferens induced by electrical stimulation are inhibited by the opioid compounds. An assay based on competi-

Leucine enkephalin   Try-Gly-Gly-Phe-Leu-OH

Methionine enkephalin   Try-Gly-Gly-Phe-Met-OH

β-Endorphin   Try-Gly-Gly-Phe-Met-Thr-Ser-Glu-Lys-Ser-Gln-Thr-Pro-
                            Leu-Val-Thr-Leu-Phe-Lys-Asn-Ala-Ile-Val-Lys-Asn-Ala-
                            His-Lys-Gly-Gln-OH

FIGURE 5.4   The structure of enkephalins and β-endorphin.

tion for receptor binding allowed the same compounds to be isolated from
calf brain.[21] These compounds were subsequently characterized as a mixture of two pentapeptides, methionine (met-) enkephalin and leucine
(leu-) enkephalin[17] (Figure 5.4). Two years earlier, specific binding sites
for [³H]naloxone and [³H]etorphine, a close relative of morphine, were
shown to exist on membrane fragments isolated from brain.[D,18,19,20] It
seemed likely that endogenous opioid compounds were interacting with
these receptors and serving some neurophysiological function, possibly
connected with the reception and control of pain.

The discovery of β-endorphin followed shortly afterwards, when it
was recognized that the peptide sequence of a carboxy-terminal fragment
(C-fragment, residues 61–91) of a pituitary peptide called β-lipotropin
contained met-enkephalin as residues 61 to 65. The complete fragment,
residues 61–91, was called β-endorphin and was found to be a potent
agonist of the opiate receptor.[A,D] These findings also indicated a possible
precursor relationship between β-endorphin and met-enkephalin, though
clearly not with leu-enkephalin. Apparently a large common mother precursor protein pro-opiomelanocortin, with 239 amino acid residues and a
molecular weight of 29,259, gives rise to both β-lipotropin (residues 1 to
91) and ACTH (residues 1 to 39) as well as to α-endorphin (residues 61 to
76), which also possesses opioid activity. The messenger RNA for this
mother protein from pituitary cells has been used to prepare complementary DNA (cDNA) by reverse transcription. The cDNA was inserted into
a plasmid vector, which was cloned, the 1091-base pair product was re-
isolated, and its bases were sequenced.[23,24] This feat of genetic engineer-
ing allowed the complete elucidation of the amino acid sequence of the
mother protein as shown in Figure 5.5. This mother protein was shown to
contain other known pituitary peptides including α-melanocyte-stimulating hormone (α-MSH; residues 1 to 13) and β-MSH (residues 41 to 58),
which occur *within* ACTH and β-lipotropin respectively. Another peptide, γ-MSH, contains an amino acid sequence similar to those of α- and
β-MSH.

Two other large precursor peptides for opioid peptides have now been
described: pre-proenkephalin, which was characterized from the adrenal
medulla, and pre-prodynorphin isolated and sequenced from the hypothalamus.[38,39,202,203,K] These two large precursors, together with pro-
opiomelanocortin, can give rise to a total of 18 opioid peptides.[25,202] Pro-

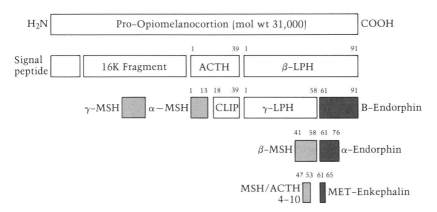

FIGURE 5.5   The opiocortins are a family of peptides sharing a common large-molecular-weight precursor, pro-opiomelanocortin. This common precursor is cleaved to yield equimolar amounts of ACTH and β-LPH in addition to signal peptides and 16K fragment. ACTH is further cleaved to α-melanocyte-stimulating hormone (α-MSH) and corticotropinlike intermediate-lobe peptide (CLIP); β-LPH is cleaved to yield γ-LPH and β-endorphin. Note the recurring MSH sequence (▨) found in 16K fragment, ACTH, and β-LPH. The sequence of met-enkephalin is contained within that of β-LPH and α- and β-endorphin (opioid peptides ■). Present evidence suggests, however, that neither substance is a precursor and that met-enkephalin arises from a separate precursor molecule, proenkephalin [From ref. 24]

opiomelanocortin is likely to be the principal precursor of the endorphins. The other two precursors can each give rise to both enkephalins. In practice, however, it is likely that met-enkephalin derives mainly from proenkephalin and that leu-enkephalin may also be produced from prodynorphin. The proteolytic processing of the precursors and their cleavage products is regulated in a different fashion in each tissue or brain region, and in this way the mixture of active opioid peptides in these various regions is controlled.[24,25,K] Thus, leu-enkephalin may be produced from either prodynorphin or proenkephalin.[56,202]

The opioid peptides were the first of the new families of neuropeptides to be isolated and characterized, and their interrelationships have provided a model or pattern for other neuropeptide families. For instance, a certain key sequence of amino acids (e.g., met-enkephalin) is often the common denominator in neuropeptides of different length that constitute the family. Which member of the family is enriched in different regions of the nervous system is determined by the presence of specific proteolytic enzymes which perform the posttranscriptional processing of the mother protein or peptide.

Distribution of Opioid Peptides   Studies on the distribution of these peptides and their precursors in the central nervous system by direct assay and by immunohistochemistry have shown that enkephalins are widely distributed in discrete pathways and that β-endorphin is limited to a single hypothalamic cell group in the tuberal zone with long ascending projections to the ventral septum, nucleus accumbens, and paraventricular nucleus of the thalamus. Long, descending fibers of this β-endorphin system enter the brain stem, periaqueductal gray matter, locus ceruleus, and reticular formation.[D,26,27] Certain areas such as the corpus striatum, caudate, and globus pallidus contain much more enkephalin than β-endorphin, and a long enkephalin-containing pathway projects from caudate-putamen to globus pallidus. In the spinal cord, laminae I and II contain high levels of enkephalin but very little β-endorphin.

In general, brain regions rich in enkephalins are also rich in monoamine neurotransmitters (norepinephrine, dopamine, and serotonin) and substance P. The globus pallidus, in fact, seems to be the area richest in enkephalin (Figure 5.6), whereas the pituitary gland contains little enkephalin and is rich in β-endorphin. Within the hypothalamus, the peptides are clearly located in different neuronal systems; β-endorphin, β-lipotropin, and ACTH are found in one cell group and enkephalins in another. Lesions in the basal hypothalamus caused losses of β-endorphin and β-lipotropin in anterior and posterior projections without change in enkephalin levels.[27]

β-Endorphin, to a much greater extent than enkephalins, however, is found in large quantities widely distributed in many different tissues and organs outside the nervous system[D,28] (e.g., adrenal medulla, pituitary). Enkephalins have been detected in amacrine cells of the retina,[178,H] and there is a rich enkephalinergic innervation of the gastrointestinal tract, emphasizing that they, too, are found in many different regions of the nervous system and particularly in those regions concerned with sensory transmission, endocrine control, respiration, motor activity, and behavior. Enkephalin-containing neurons are often short, intrinsic interneurons, such as those found in the substantia gelatinosa of the spinal cord.[B]

Functions of Opioid Peptides   The current overall view of β-endorphin function, apart from any neurotransmitter or neuromodulator role it may have in the hypothalamus, is that it may act as a circulating hormone. It is probably released from the pituitary gland, where it is present in abundant quantities. Its principal target organs have not yet been identified. It

FIGURE 5.6   Immunofluorescence micrographs showing the intense enkephalin staining of the rat globus pallidus. Note the relatively weak staining of the caudate-putamen (CP) and enkephalin-positive neuronal cell bodies (arrows) in this area. Lesion studies indicate that the enkephalin terminals in globus palladus originate from the enkephalin-positive neurons in the caudate-putamen. [From ref. A; photomicrograph courtesy of Dr. S. Hunt]

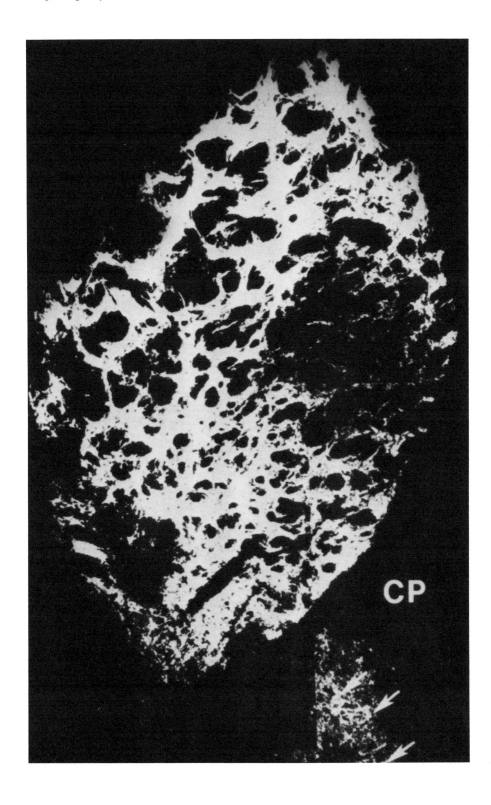

CP

increases the release of prolactin, growth hormone, and γ-MSH, and it decreases the release of thyroid-stimulating hormone and vasopressin from the pituitary gland.[D] Circulating β-endorphin causes analgesic effects and prolonged hypertension, effects that are blocked by serotonin antagonists. β-Endorphin increases the synthesis of corticosterone in isolated adrenal cells in vitro,[29] and therefore the adrenal cortex may be a target organ. The cerebrospinal fluid also contains β-endorphin that probably originates in the central nervous system itself. While circulating in the fluid, β-endorphin may modulate neuronal activity in groups of accessible neurons.[54]

Enkephalins, on the other hand, have a very short half-life in blood because of the action of peptidases, and they have not been thought likely to function as circulating agents. Instead, attention is concentrated on their likely neurotransmitter role. Enkephalins exist in nerve fibers and cell bodies in sympathetic ganglia of rat and guinea pig and in adrenal medulla endocrine cells. They appear to be present throughout the gastrointestinal tract of animals, including humans, with the myenteric plexus containing the highest concentrations.

Enkephalins have been detected in normal human blood (14–140 pg/ ml) and human cerebrospinal fluid (5–29 pg/ml) by employing a special and specific radioimmunoassay for met-enkephalin. The high content of met-enkephalin in adrenal medulla and adrenal venous blood suggests that the adrenal medulla could be the source of the peptide found circulating in the blood.[28,55]

Opioid Receptors   The brain contains at least three separate opioid receptor subcategories, mu, delta, and kappa.[177,K] Current evidence suggests that each of the three large protein precursors, pro-opiomelanocortin, proenkephalin, and prodynorphin, produces opioid peptides that interact differentially with mu, delta, and kappa receptors respectively. However, this is in no way exclusive. For instance, enkephalins that appear to interact predominantly with delta receptors can also interact with mu receptors that mediate analgesia. Also, it seems that mu and delta receptors have an equal affinity for β-endorphin, which has a strong analgesic action.[B,K] In fact, β-endorphin could well exert its actions through its own specific category of receptor (epsilon receptors).

Analgesia   Opioid peptides were first suspected of being endogenous compounds possessing the analgesic properties of morphine when it was shown that, like morphine, they potently inhibited the contraction of various smooth muscle preparations (e.g., ileum, vas deferens) via receptor-mediated mechanisms, which were themselves specifically inhibited by the drug naloxone. Thus the enkephalins and endorphins were predicted to serve at least partly as endogenous analgesics. However, enkephalin itself has been found to be a weak analgesic except when injected intracerebrally,[30] probably owing to its rapid degradation by peptidases. D-Ala²-Met-enkephalin, which is resistant to this enzymatic attack, is effective when administered intravenously and more potent than met-enkephalin when given intracerebrally. Other peptidase-resistant enkephalin analogues are also effective analgesics. These compounds include FK-33-824, which is 30,000 times more potent than met-enkephalin[31] and 1000 times more effective than morphine. In FK-33-824, D-alanine is substituted for glycine², phenylalanine is N-methylated, and in the methionine residue, the sulfur is oxidized to sulfoxide and the carboxyl reduced to a carbinol. Other stable analogues are D-Ala², D-Leu⁵-enkephalinamide, and D-Ala², D-Leu⁵-enkephalin.[32] β-Endorphin itself is 1,000 times more effective than met-enkephalin as an analgesic, though repeated administration of these agents leads to tolerance and physical dependence (addiction).

Other evidence also points to a function for these peptides in pain mechanisms. Enkephalins inhibit the activity of nociceptive neurons, and immunoreactive β-endorphin is found in increased amounts in the ventricular fluid of patients experiencing stimulation-induced analgesia.[33]

β-Endorphin has been implicated in the mechanisms underlying acupuncture-induced analgesia. Low-frequency electroacupuncture in patients with recurrent pain has been reported to increase β-endorphin levels in cerebrospinal fluid, while met-enkephalin levels were unchanged.[57] On the other hand, increased levels of met-enkephalin were seen in the plasma and cerebrospinal fluid of heroin addicts receiving relief from withdrawal symptoms by electroacupuncture.[58] More general leads come from the observation that naloxone reverses acupuncture mediated by low-frequency but not high-frequency stimulation.[59] The low-frequency version causes the rise in cerebrospinal fluid β-endorphins mentioned above.[57] The high-frequency version may involve activation of a serotonergic system, since it is prevented by parachlorophenylalanine, a serotonin synthesis blocker.[59] There are reports from China[60] that analgesia can be produced in recipient animals after transfer of postacupuncture cerebrospinal fluid.[28,57]

Behavioral Effects   When given intracerebrally, β-endorphin causes profound muscular rigidity and immobility (catatonia) lasting several hours. It can also cause locomotor and behavioral stimulation, catalepsy, hyperactivity, seizures, or sedation depending on the site and conditions of administration (Table 5.4).[4,32] Enkephalins are much less potent than β-endorphin in producing these effects, and enkephalin analogues tend to produce hyperactivity more often than β-endorphin. All these effects are prevented by naloxone, indicating the mediation of opioid receptors. Other studies have shown the appearance of a number of withdrawal symptoms (e.g., "wet-dog shakes") when naloxone, which mimics cessation of opioid administration, is injected into animals after chronic intracerebral administration of either enkephalin or β-endorphin.[L,M] The

---

TABLE 5.4   Central actions of endorphins and enkephalins

Muscular rigidity and immobility (catatonia)

Sedation

Analgesic effects

Enhanced locomotor activity

Seizure activity

Behavioral arousal

Stereotyped behavior

Effects on eating and drinking

Effects on blood pressure

Action on core body temperature

Note: Actions depend on the site of injection and dosage employed; they are prevented by naloxone.

---

enkephalins are much less potent in causing the withdrawal symptoms.[33,D] Researchers have reported fairly complex and sophisticated behavioral effects of these peptides and their analogues in rats and other experimental animals.[34]

Thymic and adrenal tumors (pheochromocytomas) and many other tumors contain greatly increased quantities of β-endorphin or met-enkephalin and appear to secrete these peptides. Pheochromocytomas contain 20 to 50 times more met-enkephalin than normal tissue.[28] Such peptides might suppress local pain due to the tumor and, in addition, might mediate some of the psychiatric disturbances seen during malignant disease in the absence of cerebral metastases.[28]

Enkephalin Release and Degradation   Using synaptosomes or slices of regions rich in enkephalins, including corpus striatum and globus pallidus, researchers have demonstrated stimulus-coupled, calcium-dependent release of enkephalins from neural tissue.[D,49,50] Others have followed release from guinea pig ileum by monitoring their naloxone-sensitive inhibition of contractions due to high frequency field-stimulation or have radioimmunoassayed substances released from incubated muscle strips.

Like acetylcholine, enkephalins are inactivated by enzymatic breakdown rather than by reuptake. This is probably true of all peptide neurotransmitters. Enkephalins are rapidly degraded and effectively inactivated by *nonspecific aminopeptidases* (by splitting off *N*-terminal tyrosine). The enzymes are present in abundance in neural tissue. A dipeptidase with a high affinity ($K_m = 1$ μM) for enkephalin has also been detected in striatum;[45] it is not an aminopeptidase, since it hydrolyses the glycine^3-phenylalanine^4 bond of the enkephalins. It has been called enkephalinase, though it attacks a variety of other peptides, including β-endorphin and substance P. This neutral endopeptidase has now been purified. It is found concentrated in synaptic membranes from caudate nucleus, which suggests that it has a location in or close to the synapse.[46]

Interactions with Other Neurotransmitters   Opioid peptides, like other
neurotransmitters, may exert their effect partly through presynaptic
actions on the release of other neurotransmitters. This effect appears to
be largely an inhibitory modulatory action. Substance P might be one of
the pain transmitters in the spinal cord,[172] and this gives a lead to the
possible analgesia-mediating mechanisms involving the endogenous
opioid peptides (Figure 5.7). For instance, opiates, by a naloxone-sensitive
mechanism, selectively block release of substance P from slices of substantia gelatinosa of the trigeminal nucleus,[35] and D-ala-enkephalin inhibits potassium-induced release of substance P from cultured sensory
neurons. Morphine itself blocks the release of acetylcholine and amino
acid neurotransmitter, which may be induced in vivo from the sensorimotor cortex by chemical or sensory stimulation, and naloxone extin-
guishes the morphine block.[36,37] Many investigators have reported inhibition of neurotransmitter release by morphine[35,D] and depression of the
electrical activity of a range of neurons by enkephalins, all these effects
being preventable or extinguishable by naloxone.[D] Considerable evidence
links the enkephalins with the catecholamine system. Locus ceruleus
terminals, containing enkephalin, synapse with catecholaminergic neurons,[48] and when administered, enkephalin prevents the stimulus-evoked
release of norepinephrine from mouse vas deferens and from cerebral
cortex in vitro.

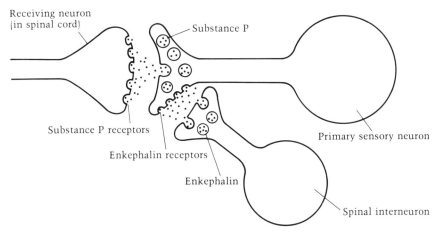

FIGURE 5.7   Proposed gating mechanism at the first synaptic relay in the
spinal cord may regulate the transmission of pain information from the
peripheral pain receptors to the brain. In the dorsal horn of the spinal
cord, interneurons containing the peptide transmitter enkephalin make
synapses with the axon terminals of the pain neurons, which utilize
substance P as their neurotransmitter. Enkephalin released from the
interneurons inhibits the release of substance P, so that the receiving
neuron in the spinal cord receives less excitatory stimulation and hence
sends fewer pain-related impulses to the brain. Opiate drugs such as
morphine appear to bind to unoccupied enkephalin receptors, mimicking
the pain-suppressing effects of enkephalin system. [From ref. 173]

Somatostatin

The compound somatostatin is one of a group of peptides, originally detected in the hypothalamus, having profound actions on the endocrine function of the pituitary gland by controlling the release of its trophic and other hormones. Many of these hypothalamic peptides have subsequently been found outside the hypothalamus in large amounts, where they may well serve a neurotransmitter or neuromodulator function (see Table 7.1).

Somatostatin consists of 14 amino acid residues forming a ring through the disulfide bridge formed between two cysteine residues (Figure 5.8). It received its name on the basis of its ability to inhibit the release of growth hormone from the anterior pituitary gland; this activity was used to bioassay the substance during its isolation.[53] Somatostatin has since been shown to exert a similar inhibitory action on the release of many other functionally active peptides from peripheral tissues (e.g., insulin, thyrotropin, parathyroid hormone, gastrointestinal hormones)[D] and the name *panhibin* has also been proposed for this peptide.[N]

Somatostatin should be distinguished from Growth Hormone Releasing Factor (GRF, somatocrinin), which *stimulates* growth hormone release from the anterior pituitary gland.[R]

Somatostatin - 28   Ser-Ala-Asn-Ser-Asn-Pro-Ala-Met-Ala-Pro-Arg-
Glu-Arg-Lys-Ala-Gly-Cys-Lys-Asn-Phe-
Phe-Tryp-Lys-Thr-Phe-Thr-Ser-Cys-OH

Somatostatin - 14   H-Ala-Gly-Cys-Lys-Asn-Phe-Phe-Trp-Lys-Thr-Phe-Thr-Ser-Cys-OH

FIGURE 5.8   The structure of somatostatin peptides. Somatostatin-25, not shown, has the same structure as somatostatin-28 minus the first three amino terminal peptides. The two cysteine residues in both molecules are linked by a disulfide bridge (not shown for somatostatin-28).

Distribution of Somatostatin   It is now established that about 90 percent of the somatostatin in the brain is widely distributed outside the hypothalamus.[B,J,N] Immunohistochemistry, radioimmunoassay, and bioassay all show it to be in cell bodies of the amygdaloid complex, anterior periventricular area, zona incerta interpeduncular area, limbic system, neocortex, hippocampus cerebral cortex, and in dorsal root ganglia,[1,40,61,62] though the highest concentrations are in the hypothalamus itself. Nerve terminals containing positive somatostatin immunoreactivity have been demonstrated in many other areas. In addition, somatostatin is widely distributed outside the nervous system (in endocrine cells of the pancreas, in A or D cells of the islets of Langerhans, in the gastrointestinal tract, the thyroid gland, and the retina.[62,178,H] In fact, it appears to coexist with β-endorphin in the D cells of the pancreatic islets.

Synthesis of Somatostatin The source of somatostatin appears to be a large precursor protein, prosomatostatin. Judged by recombinant DNA studies, pre-prosomatostatin of molecular weight 18,000 is likely to be an earlier precursor. The precursor protein is proteolytically cleaved and otherwise processed to produce somatostatin—and possibly other biologically active peptides.[63,64,65] Indeed, two forms of somatostatin, extended at the amino terminal end to 28 or 25 amino acid residues, have been isolated and synthesized[65] (see Figure 5.8), and these have greater biological activity than the tetradecapeptide form of somatostatin in inhibiting growth-hormone secretion from pituitary gland. Thus, a whole family of somatostatinlike peptides is likely to exist, each member showing a differing potency of biological activity.

Somatostatin As a Neurotransmitter The presence of somatostatin in nerve terminals within the substantia gelatinosa of the dorsal horn of the spinal cord and in small-diameter neurons of dorsal root ganglia suggests that it could be a primary afferent neurotransmitter in the spinal cord. Its apparent coexistence with catecholamines in certain sympathetic neuronal cell bodies is another example of evidence contrary to Dale's principle of "one neurotransmitter, one neuron" (see Chapter 7).

A large fraction of central nervous system somatostatin is recoverable in synaptosomes and synaptic vesicles. Calcium-dependent release of the peptide from hypothalamic synaptosomes and slices, and from slices of amygdala, have been reported.[66,67] It is rapidly degraded by peptidases present in brain homogenates.[A,97] These observations show that several of the criteria for the neurotransmitter function of somatostatin are satisfied. In addition to its potent action inhibiting growth-hormone release from the pituitary, somatostatin is known to have a powerful central behavioral effect suggestive of a depressant action (Table 5.5). It prolongs the sedative and hypothermic actions of barbiturates, it causes reduced motor activity, and it inhibits the firing of many neurons in various brain regions when applied iontophoretically.[D] Many of its

---

TABLE 5.5   Central actions of somatostatin

Increased motor activity in dopa potentiation test
Enhancement of pentobarbital-induced sedation, hypothermia, and mortality
Blockade of bombesin-induced hyperglycemia
Analgesic effects
Inhibition of salivary secretion
Disturbances of sleep
Seizure activity
Motor uncoordination and akinesia
"Wet-dog shakes" and other symptoms resembling opiate withdrawal

Note: From ref. D.

actions are opposite to those of thyrotropin-releasing hormone (see below), another hypothalamic peptide with behavioral and central actions.

Somatostatin given peripherally suppresses gastric acid and pepsin secretion as well as gastric emptying. At the same time, release of all known gastrointestinal hormones is inhibited.[68]

A few convincing full studies of somatostatin binding to neural tissue have now appeared,[70,O] and the importance of its amino acid sequence for its receptor-mediated biological actions is becoming well established. Deletion of amino acids within the molecule's ring sequence (3 to 14) reduces the activity of somatostatin analogues, indicating the importance of specific intramolecular associations for receptor recognition.[71,1]

Interactions with Other Neurotransmitters   Catecholamines and acetylcholine appear to influence the release of somatostatin and may exert a control over it in vivo. Low doses of dopamine and higher levels of noradrenaline were reported to release somatostatin from the median eminence; these responses were blocked by pimozide, a dopamine receptor-blocker. and by phentolamine, an alpha-adrenoreceptor blocker.[72]

Somatostatin levels in the hypophyseal blood are increased in vivo by intracerebral administration of dopamine, norepinephrine, or acetylcholine but not by serotonin.[73]

Neurotensin and GABA may also exert control over somatostatin release through receptor-mediated processes.[N] Presynaptic receptors on neurons releasing somatostatin could be involved in these control processes (see Chapter 7).

Thyrotropin-Releasing Hormone

Thyrotropin is the trophic hormone that activates the thyroid gland after release from the anterior pituitary. Thyrotropin-releasing hormone (TRH), originally isolated from the hypothalamus, releases thyrotropin from the pituitary, hence its more recent name, thyroliberin. It has a tripeptide structure (Figure 5.9), and it is one of a group known as the hypothalamic-releasing hormones that have been isolated and characterized since 1970 on the basis of their ability to release various trophic hormones from anterior pituitary tissue. These releasing hormones are synthesized in hypothalamic neurons (medial basal region) and released into the local blood supply, the hypothalamo-hypophyseal portal system. This system is a bed of fine capillaries which anastomoses into both the hypothalamus and the anterior pituitary (Figure 5.10). The releasing hormones are carried through this portal system into the anterior pituitary gland, where they act on local pituitary cells and cause them to release

pyro-Glu-His-Pro-NH$_2$

FIGURE 5.9   The structure of thyrotropin-releasing hormone (TRH).

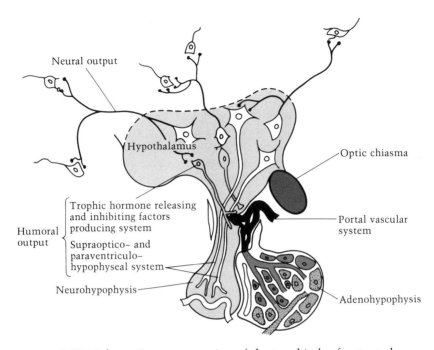

FIGURE 5.10   Schematic representation of the two kinds of output channels of the hypothalamus. The supraoptic paraventricular neurons project directly into the posterior pituitary (neurohypophysis). The parvicellular neurons secrete trophic hormone–releasing hormones into the portal vascular system, which delivers them into the anterior pituitary gland (adenohypophysis). [From ref. 74]

trophic hormones into the bloodstream. Thyrotropin-releasing hormone (TRH) releases both thyrotropin and prolactin from the anterior pituitary gland.

Distribution of TRH   About 70 percent of thyrotropin-releasing hormone occurs outside the hypothalamus. Immunohistochemistry has allowed visual detection of TRH-containing nerve terminals in the cerebral cortex and in brain stem nuclei of several major cranial nerves, including the facial, trigeminal, and hypoglossal nerves. Within the spinal cord, immunohistochemically positive nerve terminals are seen in the ventral horn surrounding large motor neurons, and they are seen in the nucleus accumbens, lateral septal nuclei, and bed nucleus of the stria terminalis.[77,83] The retina, too, appears to contain TRH.[178,H] This extrahypothalamic TRH originates in these regions and is not transported there via axons of hypothalamic neurons—surgical isolation of the medial basal hypothalamus does not result in changes in levels of TRH in these nuclei. Whether all of this immunoreactive material is identical to the tripeptide TRH has been challenged.[76] In addition to its wide occurrence in brain and spinal cord of mammals, TRH occurs in high concentrations in amphibian blood, brain, and skin,[78] and in mollusk brain; it serves no known function in the endocrine system of these animals.

TRH As a Neurotransmitter   Evidence suggests that the TRH peptide, too, may serve as a neurotransmitter or neuromodulator in the nervous system, with its hormonal role representing a late evolutionary development in vertebrates. The evidence for such a transmitter function for TRH continues to accumulate. Its high content and Ca^{2+}-dependent release from hypothalamic synaptosomes, which can be modulated by low levels of dopamine and serotonin,[80] suggests that intrahypothalamic terminals secrete TRH. But whether the TRH is acting as an endocrine or paracrine agent, or as a neuron-to-neuron transmitter agent, remains uncertain.

Following the pattern for many other neuropeptides, thyroliberin appears to be derived from a larger prohormone synthesized conventionally by ribosomal protein synthesis[j] and transported to terminals. Again, peptidases appear to constitute the principal method of inactivation of the released peptide. A pyroglutamyl peptidase that attacks peptidylamides is implicated.[97,81]

The neuroactive potency of TRH is reflected in its numerous central actions, mostly involving increase in arousal. Some are listed in Table 5.6. Taken together, the evidence strongly suggests the likelihood of a role for TRH beyond its established endocrine function. This assumption is further supported by the finding of high affinity [³H]TRH binding sites in mammalian forebrain, amygdala, septum, hypothalamus, and spinal cord in addition to the binding sites detected in pituitary gland and cul-

---

TABLE 5.6   Central actions of thyroliberin

Increases spontaneous motor activity
  Alters sleep patterns
  Produces anorexia
  Inhibits conditioned avoidance behavior
  Causes head-to-tail rotation

Opposes actions of barbiturates on sleeping time, hypothermia, and lethality
  Opposes actions of ethanol, chloral hydrate, chlorpromazine, and diazepam on sleeping time and hypothermia
  Enhances convulsion time and lethality of strychnine
  Increases motor activity in morphine-treated animals

Potentiates dopa-pargyline hyperactivity syndrome

Ameliorates human behavioral disorders?

Alters brain cell membrane electrical activity

Increases norepinephrine turnover
  Releases norepinephrine and dopamine from synaptosomal preparations
  Enhances disappearance of norepinephrine from nerve terminals

Potentiates excitatory actions of acetylcholine on cerebral cortical neurones

---

Note: From ref. D.

tured pituitary cells.[D,82,83] The close similarity of the kinetic and equilibrium properties of [^{3}H]TRH binding to sheep anterior pituitary, nucleus accumbens–septal area, and retina enhances the probability that biologically active extrapituitary receptors exist.[82,D] Studies with several TRH analogues more potent than TRH itself (e.g., [3-Me-His^2]TRH) lead to a similar conclusion.

Luteinizing Hormone–Releasing Hormone

Another compound in the family of hypothalamic peptides that act as trophic hormone releasing agents is luteinizing hormone–releasing hormone (LHRH). It has previously been studied principally in the light of its endocrine function,[A,B,D,F] and its peptide sequence is fully established (Figure 5.11).

pyro-Glu-His-Trp-Ser-Tyr-Gly-Leu-Arg-Pro-Gly-NH$_2$

FIGURE 5.11   The structure of luteinizing hormone-releasing hormone (LHRH).

Distribution of LHRH   LHRH releases luteinizing hormone (LH) and follicle-stimulating hormone (FSH) form the anterior pituitary gland. These hormones activate the gonads to produce sex steroid target-hormones. Rabbits immunized with synthetic LHRH develop infertility and gonadal atrophy. Immunohistochemistry with antisera raised against LHRH shows immunoreactive neurons in the arcuate nucleus and in the preoptic area, which suggests that extrahypothalamic systems for this neuroactive peptide exist.[89] Some researchers have claimed that 90 percent of hypothalamic LHRH is in the fibers and nerve endings projecting into the hypothalamus from external nuclei, since deafferentiation results in a large fall in hypothalamic LHRH content.[86,N] However, this claim remains strongly contested.[A] As with TRH and other peptides detected in various neural pathways by immunohistochemistry, the possibility of a cross-reactivity artifact requires that isolation and biochemical characterization of the peptide precede any firm conclusions concerning its true presence in these systems.

LHRH As a Neurotransmitter   In the sympathetic ganglia of the bullfrog, LHRH appears to coexist with acetylcholine in the same preganglionic nerve terminals.[87] Stimulation of preganglionic fibers or depolarizing levels of potassium evoke calcium-dependent release of LHRH immunoreactivity and probably also of acetylcholine, as judged by the postsynaptic responses on the sympathetic ganglion cells. When applied directly, synthetic LHRH produces "late slow excitatory–postsynaptic potentials" (EPSPs) that are indistinguishable from those produced by nerve stimulation, showing that it mimics the action of the natural neurotransmitter. The actions of both substances are blocked by LHRH antagonists (e.g., the substituted analogues D-pGlu^1, D-Phe^2, D-Trp^{3,6}) without affecting the synaptic potential due to released acetylcholine.[88,K] The slow time-course of the EPSP may reflect the long lifetime, perhaps 30 seconds, of released LHRH. There is good evidence that the LHRH released by conventional synapses at one group of sympathetic neurons (C cells) in the bullfrog ganglia can diffuse about 100 μm from its site of release to induce "slow EPSP's" in adjacent B cells, which do *not* receive direct LHRH-containing fibers or possess conventional synapses. In these actions the LHRH can be seen as exerting a localized hormonal (paracrine) action on the B cells[88,K] (Figure 5.1).

This peptide, as other hyperphysiotropic peptides, shows Ca^{2+}-dependent, stimulus-evoked release from hypothalamic synaptosomes,[80,67,90,91] but this does not distinguish a neurotransmitter role for LHRH as opposed to a neuroendocrine role. LHRH stimulates neuronal firing when iontophoresed into the arcuate nucleus and a variety of other central nervous system regions,[92] indicating the presence of receptors for LHRH. It is rapidly inactivated by peptidases, and inactivation can be prevented by insertion of D-amino acids in key positions in the molecule[71] (see above). These same analogues are also *agonists* of LHRH and show prolonged biological action, in contrast to the D-substitution of terminal amino acids, which tends to produce LHRH *antagonists*.[55,B]

Angiotensin

Angiotensin I is a decapeptide product of the action of the proteolytic enzyme renin on angiotensinogen, a prohormone circulating in the blood (mol wt 50,000–60,000) and derived from the liver (Figure 5.12). Removal of the C-terminal dipeptide (His-Leu) of angiotensin I by angiotensin-converting enzyme (found mainly in lung tissue) yields angiotensin II, the major active species. A third peptide containing seven amino acid residues (angiotensin III) can be formed through the action of aminopeptidases.

The octapeptide angiotensin II has both central and peripheral actions. It potently stimulates thirst and drinking, and it releases vasopressin ACTH and aldosterone in addition to its well-known effect of raising blood pressure, from which its name derives.[84,85] Many of these effects are prevented by saralasin, a peptide which is a potent antagonist of angiotensin II ([Sar^1-Ala^8]angiotensin II; [Sar^1-OMe-Thr^8]angiotensin II; [Sar^1 Ileu^8]angiotensin II).

Distribution of Angiotensin   It now seems well accepted that the brain contains all the enzymes for the biosynthesis of the angiotensins. Studies with labeled antisera to angiotensin II have indicated its presence in

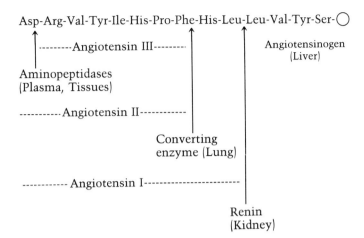

FIGURE 5.12 The renin-angiotensin interconversion system.

many regions of the brain and spinal cord.[D,93,98] It seems likely that locally synthesized and released angiotensin is interacting with receptors that would be inaccessible to circulating angiotensin. Some of the angiotensin-containing brain regions such as the subfornical organ, the organ vasculosum of the lamina terminalis, and the area postrema *are* accessible to circulating angiotensin since they are present in the circumventricular organs possessing fenestrated capillaries that lack the blood-brain barrier. The subfornical organ is implicated in mediating the thirst-stimulating action of the peptide, and the area postrema mediates its central vasopressor effect. (See Chapter 8.)

Angiotensin As a Neurotransmitter   Binding studies with [[125I]]labeled angiotensin II and its analogues have mainly investigated the neurotransmitter-related peripherally "inaccessible" receptors.[94,D] Studies of localized "accessible" receptors (e.g., organum vasculosum of the lamina terminalis) have revealed twofold greater concentrations of angiotensin II in spontaneously hypertensive rats, suggesting a functional response by the receptor population.[95] Radioreceptor assays have detected specific angiotensin II receptors localized in nerve terminals (synaptosomes) and of higher affinity than the receptors of the adrenal cortex.[93,96] In one study, iontophoresed angiotensin II activated the rate of discharge of 75 percent of the neurons in the subfornical organ, the response being dose-related and antagonized by saralasin.[99,100] Other studies report angiotensin activating the neurons of the supraoptic nucleus that synthesize and secrete vasopressin; these observations correlate with the known enhancement of vasopressin release caused by angiotensin. A central nervous system transmitter function for the angiotensin system has yet to be convincingly established, however, and more definitive evidence is awaited.

Neurotensin

Another peptide first isolated from the hypothalamus on the basis of its effects on blood pressure is neurotensin.[101] This peptide contains 13 amino acids (Figure 5.13) and was isolated on the basis of its effects on local vasodilation and hypotension.[101] It has many other peripheral effects such as decreasing motor activity, stimulating uterine contraction, relaxing the duodenum, increasing vascular permeability, and reducing gastric secretion.[170,D] Its central actions are manifold and include analgesia, the potent induction of hypothermia, and enhanced release of the hormones ACTH, FSH, and LH—actions mediated via the hypothalamus rather than the pituitary gland. This suggests that neurotensin either acts as a trophic hormone–releasing hormone[105] or activates the release of other such hormones.

p-Glu-Leu-Tyr-Glu-Asn-Lys-Pro-Arg-Arg-Pro-Tyr-Ile-Leu-COOH

FIGURE 5.13   The structure of neurotensin.

Neurotensin (along with bombesin; see below) is one of the most potent endogenous hypothermic substances known. Rats maintained at low ambient temperatures (4°C) show a 10°C fall in body temperature after intracisternal administration of 1 μg of neurotensin. The essential component of the peptide for these hypothermic properties is the C-terminal end of the molecule.[170] It is interesting that neurotensin's potent analgesic effects are not blocked or reversed by opiate antagonists such as naloxone, which means that the analgesia is not mediated via opioid pathways.

Distribution of Neurotensin   The high concentration of neurotensin in regions known to be involved in pain transmission—such as the substantia gelatinosa of the spinal cord, the trigeminal nerve nucleus, and the periaqueductal gray matter—provides circumstantial evidence for participation of this peptide in nociception. Although most of neurotensin's central effects are produced by intracerebral rather than systemic administration, 95 percent of its total content in the body appears to be localized to the gastrointestinal tract,[106] largely in the mucosa of the ileum.[107] Within the central nervous system it follows the pattern of substance P and somatostatin.[108] It is found in highest concentrations in the substantia gelatinosa, trigeminal nerve nucleus, nucleus accumbens, substantia nigra, hypothalamus, preoptic area, amygdala, and the central gray matter of the mesencephalon. Very little is found in the cerebral cortex or hippocampus. Smaller, biologically active fragments of neurotensin, hexa- or heptapeptides may exist in the brain in addition to the tridecapeptide, and the neurotensin of the stomach appears to be a pentapeptide.[102,A]

There is good evidence that neurotensin interacts with dopaminergic systems[103] and that it produces opposite effects to those of TRH. It is not yet clear whether this is due to a specific antagonist effect of TRH on neurotensin systems.[170] Neurotensin interaction with catecholamine systems is also suggested by the rich content of neurotensin and its receptors in catecholamine nuclei, such as substantia nigra, ventral tegmentum, and locus ceruleus. Indeed, 6-OH-dopamine lesioning (to destroy catecholamine pathways) suggests that some of the neurotensin receptors are located on dopamine-containing neurons.[170,171]

Neurotensin As a Neurotransmitter   Within the nervous system, neurotensin is localized to discrete neural pathways. It is concentrated in synaptosomal subfractions of neural tissue from these regions. The release of neurotensin from hypothalamic tissue slices is evoked through a Ca^{2+}-dependent mechanism by depolarizing levels of potassium.[66] When applied locally, neurotensin selectively inhibits neurons of the locus ceruleus but, in contrast, it excites neurons of the spinal cord.[D,109] It is also rapidly inactivated by endopeptidases,[104] and specific receptor sites have been detected for it by ligand binding assays and by autoradiography using [³H]neurotensin as the radioligand.[110,111] These receptors occur in high density in brain regions enriched in neurotensin itself. It is clear from this evidence that neurotensin is another contender for the role of neurotransmitter or neuromodulator in the central nervous system.

Oxytocin and Vasopressin

Long established in their endocrine function as the major hormones secreted by the posterior pituitary gland, oxytocin and vasopressin (both nonapeptides; see Figure 5.14) have recently been appraised for a wider functional involvement[112,H] and the concept of vasopressinergic and oxytocinergic neurons developed. This concept is derived from the observation that the two peptides have excitatory or inhibitory effects on neurons in the paraventricular and supraoptic nuclei, which synthesize and secrete these hormones.[113] The existence of recurrent collateral fibers that could be feeding back and controlling the activity of these neurons has long been known. These fibers may not be collaterals of the neurose-

Vasopressin H_2N—Cys-Tyr-Phe-Gln-Asn-Cys-Pro-Arg-Gly-NH_2
                └──── s-s ────────────┘

Oxytocin H_2N—Cys-Tyr-Ile-Gln-Asn-Cys-Pro-Leu-Gly-NH_2
                └──── s-s ────────────┘

FIGURE 5.14   The structure of vasopressin and oxytocin.

cretory cells themselves, as none of the endings contain classical neuro-secretory granules. Instead, they may be from local interneurons. What-ever their source, such recurrent collaterals could well serve to control release of vasopressin or oxytocin through inhibition of neuronal activity by negative feedback. Such negative feedback would be an explanation of the observed episodic activity of supraoptic or paraventricular neurons. It would also explain the long-lasting reduction of spontaneous discharges after the osmotically induced stimulation of activity in the supraoptic nucleus, as well as the inhibition of paraventricular neurons after the milk ejection reflex. If these collaterals *were* from the neurosecretory neurons, Dale's hypothesis would predict that they use either vasopressin or oxytocin as the neurotransmitter at their terminals.[113] Certainly, vaso-pressin has been shown to inhibit the firing of supraoptic neurons when applied by iontophoresis, which would support vasopressin as the neuro-transmitter candidate, but against this proposition is the finding that recurrent inhibition and phasic activity of these nuclei is still present in Brattleboro rats, which cannot synthesize vasopressin.[113] Oxytocin seems to be eliminated by the consistently excitatory rather than inhibitory actions it exerts on both supraoptic and paraventricular cells.

Intracerebrally administered vasopressin alters blood pressure and acts as an antipyretic and analgesic agent, suggesting that it is involved in autonomic regulation and nociception. Both oxytocin and vasopressin appear to influence a variety of behavioral performances in experimental animals and in humans.[114] Attention has been focused on their effects on memory and learning when administered either intracerebrally or sys-temically (see Chapter 8). This would correlate with the presence of oxy-tocin- and vasopressin-containing fibers in brain regions known to be concerned with the neural mechanisms of learning, memory, and other higher forms of behavior.

It is now clear that vasopressin and oxytocin are present in many different brain regions,[117,118] but mostly in fibers that originate in the hypothalamus from the magnocellular neurons and the parvocellular neurons located in a number of nuclei. These fibers project to regions such as hippocampus, septum, amygdala, neocortex, and autonomic cen-ters of brain stem and spinal cord.[114,115,116] There is also a high density of vasopressin-containing nerve terminals in the locus ceruleus and sub-stantia nigra,[116,117] two catecholamine nuclei.

Other evidence for separate oxytocinergic and vasopressinergic neu-rons comes from localization studies based on immunohistochemistry employing antisera both to the peptides themselves and to the neurophy-sins. The neurophysins are proteins found in association with oxytocin (neurophysin I) or vasopressin (neurophysin II) in neurosecretory gran-ules. They are products of large precursor proteins which give rise to the two hormones.[113,119,120] Immunoreactivity to neurophysins was detected in autonomic centers of the brain stem and in spinal cord, in pathways apparently projecting from the paraventricular nucleus, and may serve to mediate hypothalamic control of autonomic function.[A,121,122,123] Vasopressin and oxytocin immunoreactivity have been demonstrated in neuronal cell bodies of the paraventricular and supraoptic nuclei and can be seen in fibers running from these nuclei towards the posterior pituitary gland and then coursing throughout the hypothalamus. Oxytocin fibers project to the caudal brain regions (brain stem, spinal cord), whereas vasopressin fibers innervate the more rostral brain regions[179] (e.g., the limbic system). When studied by immunoelectronmicroscopy, nerve terminals containing vasopressin and oxytocin were seen to synapse with dendrites and cell bodies of neurons in these regions.[124,179]

In conclusion, it is clear that vasopressin and oxytocin are present in the fibers and nerve terminals of neuronal systems unrelated to their endocrine functions,[198] and both peptides have profound electrophysiological, pharmacological, and behavioral effects. Whether they are released from these same fibers during neural activity, and whether they serve any significant neurotransmitter, neuromodulator, or paracrine function, remains to be established. There are now reports of their stimulus-coupled calcium-dependent release in vitro from various brain preparations and in vivo from sheep septal region.[125,179]

Cholecystokinin

Cholecystokinin (CCK) is a gut hormone that causes contraction of the gallbladder. This action was known to be a property of an extract of duodenal mucosa, and after the extract was given the name cholecystokinin it was found to be identical with pancreozymin, a gut hormone causing pancreatic secretion. The peptide has been sequenced and synthesized and found to contain 33 amino acid residues (Figure 5.15).[128] This compound, CCK-33, has a sulfated C-terminal octapeptide, CCK-8, to which it gives rise by proteolysis. This is the predominant form of CCK found in both gut (60 percent) and brain (80 percent).[129,135,Q] It is present in substantial quantities in the gut and also in the endocrine pancreas. In the brain, it is the most enriched of all the known neuropeptides (e.g., 1–2 mg in human brain). It happens to contain a C-terminal pentapeptide sequence shared with another gut hormone, gastrin,[175] and with the decapeptide cerulein, a neuroactive peptide found in amphibian skin (Figure 5.15). Antigastrin antibodies interact with the CCK family of peptides, which explains the finding of gastrinlike immunoreactivity in the brain even though gastrin itself appears to be virtually absent[130,A] except in the posterior pituitary and vagal sensory neurons.

Among the brain regions, the cerebral cortex contains the highest content of CCK-8 (676–2,300 pmol/g tissue; see Table 5.1),[O,Q] several thousand-fold greater than the content of other peptide hormones in neural tissue. CCK is also present in retina,[178,H] hippocampus, hypothalamus, amygdaloid nucleus, and substantia gelatinosa of the spinal cord.[126,127,135] It is prominent in cells of the pyriform cortex.[132] The greatest density of

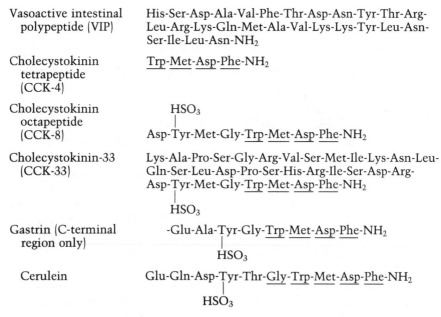

Vasoactive intestinal polypeptide (VIP)	His-Ser-Asp-Ala-Val-Phe-Thr-Asp-Asn-Tyr-Thr-Arg-Leu-Arg-Lys-Gln-Met-Ala-Val-Lys-Lys-Tyr-Leu-Asn-Ser-Ile-Leu-Asn-NH₂
Cholecystokinin tetrapeptide (CCK-4)	Trp-Met-Asp-Phe-NH₂
Cholecystokinin octapeptide (CCK-8)	HSO₃ | Asp-Tyr-Met-Gly-Trp-Met-Asp-Phe-NH₂
Cholecystokinin-33 (CCK-33)	Lys-Ala-Pro-Ser-Gly-Arg-Val-Ser-Met-Ile-Lys-Asn-Leu-Gln-Ser-Leu-Asp-Pro-Ser-His-Arg-Ile-Ser-Asp-Arg-Asp-Tyr-Met-Gly-Trp-Met-Asp-Phe-NH₂ | HSO₃
Gastrin (C-terminal region only)	-Glu-Ala-Tyr-Gly-Trp-Met-Asp-Phe-NH₂ | HSO₃
Cerulein	Glu-Gln-Asp-Tyr-Thr-Gly-Trp-Met-Asp-Phe-NH₂ | HSO₃

FIGURE 5.15   The structure of vasoactive intestinal polypeptide (*VIP*) and the cholecystokinin family of peptides. The chemically identified forms of gastrin are extended at the N-terminus to molecules containing 14, 17, and 34 residues. CCK containing 39 and 58 residues have also been isolated. Residues underlined indicate homologous regions. [See ref. Q]

cells containing CCK-8 was detected in the periaqueductal gray matter at the level of the exit of cranial nerve III.[132] Throughout the brain and in the retina, CCK-8 immunoreactivity appears to be localized to neuronal cell bodies, fibers, and terminals rather than to glial cells,[1,132] and synaptosomal fractions are greatly enriched in this peptide. Slices of cerebral cortex release CCK-8 on stimulation with high K⁺ levels.[1,137]

Iontophoresis of members of the CCK family of peptides into various regions of the nervous system have shown it to depolarize dorsal and ventral roots of the spinal cord and to have an excitatory action on neurons of the myenteric plexus, hippocampus, cerebral cortex, substantia nigra, the ventral tegmental area, and the dorsal horn of the spinal cord.[79,136,Q] Non-sulfated CCK was found to be inactive. The sulfate group appears to be important in receptor interaction (Figure 5.15). In brain tissue, [³⁵S]methionine rapidly incorporates into a large high-molecular-weight precursor of cholecystokinin. It appears to be a mother protein, giving rise by controlled cleavage to five main groups of smaller CCK-like peptides, including CCK-4, CCK-8, and CCK-33.[129,135,138,175] Thus, a whole family of active CCK peptides exists.

As in other families of peptides showing molecular heterogeneity and amino acid sequences of 30 or more, it seems that only a relatively short region of the sequence (e.g., 4–8 consecutive residues) is critical for biological activity when applied directly to the site of action. The contribution of the rest of the sequence to molecular change, solubility, transport, storage, or degradation may be of equal importance.

It remains uncertain whether CCK-8 serves in any neurotransmitter or modulator capacity, and specific antagonists must be developed to assist in answering this question. If one looks at the brain regions where it is localized, a hint may be gleaned of the functions with which it may be associated.[133,Q] CCK has an effect on the neurons of the periaqueductal gray matter involved in pain perception and on the neurons of the medial hypothalamus controlling feeding. In fact, the level of CCK rises in the blood towards the end of a meal and is thought to trigger satiety mechanisms that cause the animal to cease feeding.[131] (See Chapter 8.)

It is possible that in the amygdala and in parts of the cerebral cortex CCK is involved in neurotransmitter systems concerned with the generation and regulation of emotion.[132] It has been reported to coexist with dopamine in certain of the nigrostriatal fibers innervating the corpus striatum and the nucleus accumbens, which indicates some form of synaptic interaction between these two neuroactive substances.[134,135,139]

Vasoactive Intestinal Peptide

The name vasoactive intestinal peptide (VIP) was conferred on a 28 amino acid–containing peptide with a potent vasodilating property (Figure 5.15); subsequently, it was shown to display a wide variety of other effects in the peripheral system. VIP relaxes the trachea and lungs and also relaxes gastric muscle. It inhibits the secretion of gastric enzymes while stimulating the secretion of insulin, glucagon, and somatostatin, and it augments adenylate cyclase as well as bile secretion in the liver.[140] It is related in molecular structure to secretin, glucagon, and gastric inhibitory peptide.[126,174,R]

VIP-like immunoreactivity is found throughout the gastrointestinal tract from esophagus to rectum, as well as in lung, placenta, adrenal gland, and pancreas.[140] Within the nervous system, it is found, like CCK, in relatively large quantities in the cerebral cortex, hippocampus, and amygdaloid nucleus, and in smaller amounts in the hypothalamus,[A,126,135,174] and the amacrine cells of the retina.[178,H]

VIP As a Neurotransmitter or Neuromodulator   VIP appears to occur predominantly in nerve terminals, as judged by its 70 percent recovery in brain synaptosomal fractions.[A] Indeed, it may be released from these fractions by depolarizing stimuli.[A,145,146,174] Within synaptosomal fractions, the synaptic vesicle subfraction is the most enriched in the peptide,[145] suggesting a vesicular localization inside the nerve terminal. Since undercutting of the cerebral cortex does not lead to a diminished VIP content, the bulk of the peptide must be in cortical interneurons.[A]

Central effects of VIP are numerous. When applied iontophoretically it excites hippocampal and cortical neurons,[135,147] and when applied in vitro it stimulates membrane-bound adenylate cyclase activity.[140] Intracerebrally administered VIP causes shivering, hypothermia, and transient hypotension. It increases the release of prolactin, growth hormone, and luteinizing hormone. These responses are all presumably mediated at the hypothalamic level via release of hypothalamic trophic hormone–releasing hormones, though VIP itself appears to be released into the hypothalamo-hypophyseal portal blood system[148] and may act directly (see Figure 5.9). The N-terminal sequence His-Ser-Asp-Gly seems to be obligatory for these central actions of VIP, with potency increasing as chain length increases.[O] Many nerve terminals of VIP-containing intrinsic neurons in the cerebral cortex are found in close association with cerebral blood vessels, suggesting a physiological role for VIP in controlling cerebral blood flow by vasodilation.

Coexistence of VIP and Acetylcholine   Co-existence and corelease of VIP and acetylcholine from the terminals of parasympathetic nerve fibers innervating salivary glands has been demonstrated by immunohistochemistry.[141,180] In addition, there is simultaneous loss of both substances on denervation of the submandibular gland, and simultaneous release on stimulation of the nerve.[141,142] Acetylcholine and VIP are synaptically released from the same terminals but at different impulse frequencies.[142] At low frequencies, (e.g., 2 Hz) only acetylcholine is released, and it initiates salivation and vasodilation. At higher frequencies (e.g., 10 Hz), VIP is also released and causes further vasodilation. The two substances could be acting at postsynaptic sites on different cell types, for example, on secretory (acinus) cell, blood vessel endothelium, or smooth muscle (see Figure 7.20).

Colocalization of VIP and acetylcholine in intrinsic cholinergic neurons in the cerebral cortex has also been reported.[143] VIP is synthesized,[174,A] stored, and released from nerve terminals of the central and peripheral nervous systems and shows a considerable range of neurally mediated physiological actions. Studies with [125I]VIP have shown the presence of specific binding receptors in brain membranes, which in some regions correlates with the stimulatory effect of VIP on adenylate cyclase.[135,140,144] Similar binding properties and linkage to adenylate cyclase activation has been demonstrated for many nonneural tissues.[149] VIP, therefore, satisfies many of the criteria necessary for inclusion in the list of promising brain peptides that may be serving neurotransmitter or neuromodulator functions.

Neuropeptide Y

Another neuropeptide occurring in both the brain and the gastrointestinal system is neuropeptide Y (NPY), so called because it has an amino-terminal tyrosine residue and a carboxy-terminal tyrosine amide residue, and Y is the abbreviation for tyrosine in the single-letter amino acid code. The presence of NPY in mammalian brain was first detected by immuno-histochemistry, using antisera raised against a closely related peptide found in chicken pancreas and called avian pancreatic polypeptide (APP).[186] Subsequently neuropeptide Y was isolated from porcine brain and sequenced.[187] It was found to contain considerable amino acid sequence homology with APP and also with a range of pancreatic polypeptides found in all mammalian species studied, including humans. It therefore seems likely that these groups are members of a family of closely related peptides that occur in pancreas, in brain, or in both organs[187,188] (Fig 5.16). One group, peptide YY (PYY), isolated from porcine intestinal tract, is present in endocrine cells of the gut but is absent from brain;[189] NPY is abundantly present in brain but has not been detected in the gastrointestinal tract.[190]

Both immunohistochemistry and radioimmunoassay have revealed the widespread occurrence of NPY throughout central and peripheral nervous systems. It is present in relatively large quantities equal to or greater than the levels of CCK, the other abundantly occurring neuropeptide (i.e., 600–2000 pmols/g wet wt of brain).[191] Hypothalamus, olfactory tubercle, nucleus accumbens, and amygdala are brain regions especially rich in the peptide, and the spinal cord has two major groups of NPY-containing neurons in the substantia gelatinosa and in the ventral region of the sacral spinal cord. In the cerebral cortex and basal ganglia, NPY-containing neurons correspond closely with somatostatin-containing neurons in both morphology and distribution, suggesting the possible coexistence of these peptides in the same neurons.[188] In some regions, the

	1	2	3	4	5	6	7	8	9	10	11	12	13	14	15	16	17	18
NPY	Tyr	Pro	Ser	Lys	Pro	Asp	Asn	Pro	Gly	Glu	Asp	Ala	Pro	Ala	Glu	Asp	Leu	Ala
PYY	Tyr	Pro	Ala	Lys	Pro	Glu	Ala	Pro	Gly	Glx	Asx	Ala	Ser	Pro	Glx	Glx	Leu	Ser
APP	Gly	Pro	Ser	Gln	Pro	Thr	Tyr	Pro	Gly	Asp	Asp	Ala	Pro	Val	Glu	Asp	Leu	Ile
HPP	Ala	Pro	Leu	Glu	Pro	Val	Tyr	Pro	Gly	Asp	Asn	Ala	Thr	Pro	Glu	Gln	Met	Ala
PPP	Ala	Pro	Leu	Glu	Pro	Val	Tyr	Pro	Gly	Asp	Asn	Ala	Thr	Pro	Glu	Gln	Met	Ala

	19	20	21	22	23	24	25	26	27	28	29	30	31	32	33	34	35	36
NPY	Arg	Tyr	Tyr	Ser	Ala	Leu	Arg	His	Tyr	Ile	Asn	Leu	Ile	Thr	Arg	Gln	Arg	Tyr-NH₂
PYY	Arg	Tyr	Tyr	Ala	Ser	Leu	Arg	His	Tyr	Leu	Asn	Leu	Val	Thr	Arg	Gln	Arg	Tyr-NH₂
APP	Arg	Phe	Tyr	Asp	Asn	Leu	Gln	Gln	Tyr	Leu	Asn	Val	Val	Thr	Arg	His	Arg	Tyr-NH₂
HPP	Gln	Tyr	Ala	Ala	Asp	Leu	Arg	Arg	Tyr	Ile	Asn	Met	Leu	Thr	Arg	Pro	Arg	Tyr-NH₂
PPP	Gln	Tyr	Ala	Ala	Glu	Leu	Arg	Arg	Tyr	Ile	Asn	Met	Leu	Thr	Arg	Pro	Arg	Tyr-NH₂

FIGURE 5.16 The structure of neuropeptide Y and related pancreatic polypeptides. Abbreviations: *NPY*, neuropeptide Y; *PYY*, peptide YY; *APP*, avian pancreatic polypeptide; *HPP*, human pancreatic polypeptide; *PPP*, porcine pancreatic polypeptide. Amino acid residues common to APP and NPY are underlined. *Glx, Asx*, acid or amide undefined.

distribution of NPY also closely follows that of catecholamines, and there is a wide range of other evidence for the coexistence of NPY with these neurotransmitters in certain parts of both central and peripheral nervous systems.[188] For instance, the catecholamine-specific neurotoxin 6-hydroxydopamine depletes, in parallel, both catecholamine stores and NPY-immunoreactivity from a population of noradrenergic fibers present in guinea pig gut.[192] Reserpine has the same action. Within the central nervous system, there is immunohistochemical evidence for the coexistence of NPY in epinephrine-containing neurons of the locus ceruleus and the dorsal medullary and lateral tegmental cell groups.[194]

The richness of NPY-containing cells in the peripheral nervous system may be judged from the finding in the guinea pig that approximately 5 percent of all neurons within the myenteric plexus and 20 percent of neurons in the submucous ganglion are in this category.[192,193] Heart, gut, respiratory tract, urogenital tract, and vascular smooth muscle are all innervated by NPY-containing sympathetic nerves.

NPY displays potent vasoconstrictor activity. Cat cerebral vessels respond to NPY with a characteristic slow, long-lasting vasoconstriction, which is also seen as the typical response of many vascular beds following systemic administration of NPY.[195] In contrast, intracisternal administration of NPY results in a marked hypotensive action that mimics that of epinephrine. These biological effects of NPY could be of physiological significance in view of the presence of NPY in many of the nerve fibers associated with vascular smooth muscle and in epinephrine-containing cells in the brain. Whether NPY functions as a neurotransmitter or neuromodulator remains to be established, but it has been shown to be released from splanchnic nerve during stimulation, and in in vitro experiments it modifies the release of norepinephrine and acetycholine. A strong case is therefore accumulating to suggest that this peptide does, indeed, serve an important neuroregulatory role.[188,196]

Carnosine

Carnosine (β-alanyl histidine (Figure 5.17), a dipeptide is a strong candidate for being the neurotransmitter that functions at the first-stage sensory synapse of the olfactory nerve fibers coming from the nasal epithe-

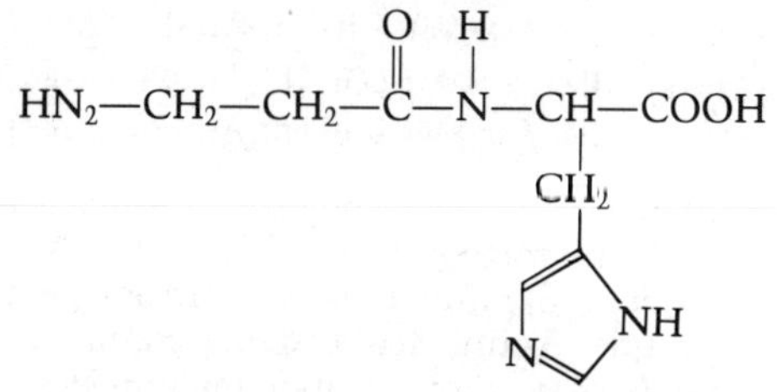

FIGURE 5.17   The structure of carnosine (β-alanylhistidine).

Olfactory bulb

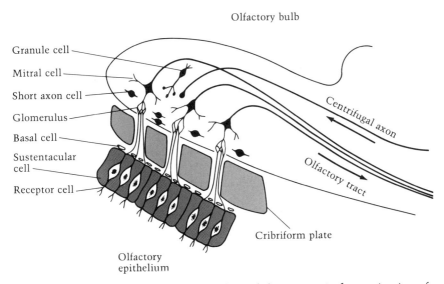

Granule cell

Mitral cell

Short axon cell

Glomerulus

Basal cell

Sustentacular
cell

Receptor cell

Centrifugal axon

Olfactory tract

Cribriform plate

Olfactory
epithelium

FIGURE 5.18 Diagrammatic outline of the anatomical organization of the olfactory bulb and mucosa.

lium[150] to the olfactory bulb. It could be the primary neurotransmitter involved in conveying the sensory information of olfaction from the chemoreceptor cells (Figure 5.18) into the olfactory bulb.[185]

Evidence for this neurotransmitter role is manifold. For instance, denervation of the bulb by destruction of the nasal epithelium with $ZnSO_4$ solutions, or retrograde degeneration caused by cutting the olfactory nerve, caused levels of carnosine in olfactory bulb to greatly diminish, up to 98 percent, while none of the enzymes associated with other neurotransmitter systems changed in activity. Carnosine synthetase, the enzyme that catalyzes the production of carnosine through condensation of β-alanine and histidine in the presence of ATP and Mg^{2+}, also decreases in activity in the bulb following deafferentiation resulting from destruction of the epithelial receptor cells.[168] This enzyme is greatly enriched in the olfactory bulb and its mucosa relative to other tissues, and carnosine itself is present almost exclusively in the bulb.[A,150]

Carnosinase is a peptidase that hydrolyses carnosine into its constituent amino acids in the presence of Mn^{2+} or certain other divalent cations. The enzyme is present in both the olfactory mucosa and the bulb, with the former containing much more of the total activity. It appears to have a nonneuronal location since it does not decline in the mucosa after sectioning of the olfactory nerve (central denervation causing retrograde degeneration). No uptake of carnosine into olfactory tissue has been detected, indicating that, should it serve as a neurotransmitter, cleavage by the peptidase would probably be the mode of inactivation. Uptake and rapid incorporation of histidine and β-alanine into carnosine occur readily, in vitro and in vivo.

Further evidence of a neurotransmitter function for carnosine comes from the finding of saturable, stereospecific high-affinity binding of [³H]carnosine to olfactory bulb membranes (K_D = 770 nM; B_{max} = 300–500 fmol/mg). This binding was highest in olfactory tissue and declined by 90 percent following peripheral deafferentiation[150] or sectioning of the olfactory nerve. There are also reports of a calcium-dependent, stimulus-coupled release of carnosine from crude synaptosome fractions from olfactory bulb.[169]

The neurochemical evidence to date suggests the existence of synapses operated by carnosine and located in the synaptic glomerulus, the primary input point from the chemoreceptor cells of the epithelium (Figure 5.18). More convincing electrophysiological evidence is needed to support this notion,[199] as well as better evidence of a stimulus-coupled release of carnosine from bulb tissue, preferably in vivo employing sensory stimulation via the chemoreceptor cells.

Bombesin

Originally isolated from the skin of the European frog *Bombina bombina*, the bombesin peptide contains 14 amino acids (Figure 5.19) and is associated with six or more related peptides.[151,152] It was later found to be present in several mammalian tissues including lung, brain tissue, and the gastrointestinal tract (stomach, jejunum, and duodenum).[153,154] Bombesin-27 (also called gastrin-releasing peptide), the peptide isolated from porcine gut, contains 27 amino acid residues; its C-terminal fragment is almost identical with the C-terminal decapeptide of bombesin-14 found in frog skin (see Figure 5.19). Studies employing antisera indicate that the hypothalamus is the most richly endowed brain region and that bombesinlike immunoreactivity is low in concentration or absent in cerebellum and spinal cord. However, a bombesinlike peptide showing considerable homology with N-terminal sequences of bombesins has been isolated from spinal cord and termed neuromedin B. It contains 10 amino acid residues[155] and has many of the properties of bombesins.

The peripheral actions of bombesin-14 or bombesin-27 resemble those of angiotensin in their vasocontrictor properties and those of other neuroactive peptides (e.g., CCK, VIP) in stimulating the secretion of many enzymes and hormones such as glucagon, gastrin, and pancreatic amylase, as well as thyroid-stimulating hormone, pituitary prolactin, and growth hormone.[D,156,157,158] In fact, bombesin-27 appears to be identical

Bombesin-27   Ala-Pro-Val-Ser-Val-Gly-Gly-Gly-Thr-Val-Leu-Ala-Lys-
                    Met-Tyr-Pro-Arg-Gly-Asn-His-Trp-Ala-Val-Gly-His-Leu-Met-NH₂

Bombesin-14   pGlu-Gln-Arg-Leu-Gly-Asn-Gln-Trp-Ala-Val-Gly-His-Leu-Met-NH₂

FIGURE 5.19 The structure of bombesin peptides. Residues underlined indicate homologous regions.

TABLE 5.7   Central actions of bombesins

Hypothermia
Analgesia
Hyperglycemia
   Elevated glucagon
   Decreased insulin
   Elevated catecholamines
Stereotyped scratching
Inhibition of immobilization
Inhibition of stress-induced gastric ulcerations
Reduced secretion of gastric acid and pepsia
Inhibition of gastric emptying
Stimulation of mucous and bicarbonate secretion into the stomach
Apparent inhibition of thyroliberin secretion by hypothalamus (in response to cold)

with gastrin-releasing peptide[159] (also termed neuromedin-C or mammalian bombesin). They also stimulate contraction of the gallbladder (cf CCK) and inhibit gastric emptying (Table 5.7) as well as stimulating contraction of smooth muscle in the wall of the gastrointestinal tract.[R]

When bombesins are given intracerebrally, the central effects include a potent capacity to lower body temperature in rats exposed to the cold. This hypothermic effect seems to be due to its action in the anterior hypothalamic preoptic area, since injections into other brain regions are relatively ineffective. Catecholamines do not seem to be intermediates in these actions. Intracerebrally administered bombesins also raise plasma glucose (i.e., induce hyperglycemia), probably by raising adrenal catecholamine secretion from adrenal medulla since adrenalectomy abolishes the effect.[156,157,158] The released epinephrine exerts its influence by decreasing insulin and raising glucagon secretion levels. This hyperglycemic action of bombesin is blocked by somatostatin-14 or -28 but not by other peptides, which indicates that some interaction occurs between these two peptide systems.

Intracerebrally administered bombesins prevent stress-induced gastric ulcer formation by their inhibitory action on gastric acid and pepsin secretion and their stimulatory actions on mucous and bicarbonate secretion. The net result is a rise in stomach pH to neutral levels and above.[158] Peripherally administered bombesins do not produce these responses. The central mechanisms involved remain unclear, though neural rather than hormonal influences appear to be involved.[158]

It has been suggested that bombesin may be a behavioral and biochemical regulator of "nutrient homeostasis" in the body by coordinating the physiological process involved in controlling glucoregulation, gastrointestinal activity, appetite behavior, and thermoregulation (Table 5.7). Such a factor active in the central nervous system indicates the existence of new and unexplored dimensions to the control of carbohydrate metabolism.[157]

When iontophoresed onto cerebral cortical neurons, bombesin causes augmentation of firing rates.[160,161] This action is blocked by tetrodotoxin, which indicates a depolarizing action involving Na^+-channel activation.[B]

Subcellular fractionation studies have shown that bombesinlike immunoreactivity (and bombesin receptors) is concentrated in synaptosome fractions from which it can be released by depolarizing agents in a $Ca^{2\pm}$ dependent fashion.[201] Studies with [^{125}I-Tyr^4] bombesin have shown the existence of saturable high affinity receptor sites in whole-brain membranes (K_D = 6nM; B_{max} = 80 fmol/mg protein[162]). The pattern of distribution of these receptor sites follows that for bombesinlike immunoreactivity, with high concentrations in hypothalamus, thalamus, hippocampus and cerebral cortex and low levels in cerebellum and medulla. Thus, bombesin has some of the cardinal properties required of a neurotransmitter or neuromodulator.

Other Peptides

Many other peptides are currently undergoing rigorous scrutiny for their content, distribution, and pharmacological potency in the nervous system with a view to their possible participation as neurotransmitters, neuromodulators, or neurohormones. They include bradykinin,[132,163] luliberin,[164] corticotropin,[165] prolactin,[166] and the sleep-inducing peptides (or factors).[167,176,200] In addition, motilin, a neuroactive peptide first isolated from the gut and containing 22 amino acid residues, appears to be present in certain cerebellar Purkinje cells as judged by immunohistochemistry,[181] though its function remains unclear.[197,R]

REFERENCES

General Texts and Review Articles

A Emson, P. C. (1979) "Peptides As Neurotransmitter Candidates in the CNS," *Prog. Neurobiol.* 13: 61–116.

B Iversen, L. L., R. A. Nicoll, and W. W. Vale, eds. (1978) *Neurobiology of Peptides. Neurosci. Res. Program Bull.* 16 (2): 211–370.

C Liebeskind, J. C., R. K. Dismukes, J. L. Barker, P. A. Berger, I. Criese, A. J. Dunn, D. S. Segal, L. Stein, and W. W. Vale, eds. (1978) *Peptides and Behavior: A Critical Analysis of Research Strategies. Neurosci. Res. Program Bull.* 16 (4): 490–635.

D Enna, S. J., and H. I. Yamamura, eds. (1980) *Amino Acids, Peptides, and Benzodiazepines Receptors and Recognition, Series B,* vol. 9, part 1. Chapman and Hall, London and New York.

E von Euler, U. S., and B. Pernow. (1977) *Substance P.* Raven Press, New York.

F Cox, B., I. D. Morris, and A. H. Weston, eds. (1978) *Pharmacology of the Hypothalamus.* Macmillan, London.

G Bloom, S. R., ed. (1978) *Gut Hormones.* Churchill Livingstone, Edinburgh.

H Bradford, H. F., ed. (1978) *Interaction and Compartmentation of Neurotransmitters,* NATO ASI Series A, vol. 48. Plenum Press, New York.

I Porter, R., and M. O'Connor, eds. (1982) *Substance P in the Nervous System.* Ciba Foundation Symposium 91, Pitman, London.

J Emson, P. C., and O. Lindvall. (1979) "Distribution of Putative Neurotransmitters in the Neocortex," *Neuroscience* 4: 1–30.

K *Neurotransmitters and Their Actions.* (1983) *Trends Neurosci.* 6 (8). A special issue.

L Beaumont, A., and J. Hughes. (1979) "The Biology of Opioid Peptides," *Ann. Rev. Pharmacol. Toxicol.* 19: 245–268.

M Snyder, S. H., and S. R. Childers. (1979) "Opiate Receptors and Opioid Peptides," *Ann. Rev. Neurosci.* 2: 35–112.

N McCann, S. M. (1982) "Physiology and Pharmacology of LHRH and Somatostatin," *Ann. Rev. Pharmacol. Toxicol.* 22: 491–554.

O Iversen, L. L. (1983) *Ann. Rev. Pharmacol. Toxicol.* 23: 1–27.

P Cuello, A. C. (1982) *Cotransmission.* Macmillian, London.

Q De Belleroche, J. S., and G. Dockray, eds. (1984) *Cholecystokinins (CCK) in the Nervous System.* Ellis Horwood, Chichester.

R Krieger, D. T., M. J. Brownstein, and J. B. Martin, eds. (1983) *Brain Peptides.* Wiley, New York.

Additional Citations

1 Barber, R. P., J. E. Vaughn, J. R. Slemmon, P. M. Salvaterra, E. Roberts, and S. E. Leeman. (1979) *J. Comp. Neurol.* 184: 331–352.

2 von Euler, U. S., and J. H. Gaddum. (1931) *J. Physiol.* (Lond.) 72: 74–87.

3 Gaddum, J. H., and H. Schild. (1934) *J. Physiol.* (Lond.) 83: 1–14.

4 von Euler, U. S. (1936) *Naunyn-Schmeideberg's Arch. Pharmacol.* 181: 181–187.

5 Kanazawa, I., and T. M. Jessell. (1976) *Brain Res.* 117: 362–367.

6 Gale, J. S., E. D. Bird, E. G. Spokes, L. L. Iversen, and T. M. Jessell. (1978). *J. Neurochem.* 30: 633–634.

7 Takahashi, T., S. Konishi, D. Powell, S. E. Leeman, and M. Otsuka. (1974) *Brain Res.* 23: 59–69.

8 Jessell, T. M. (1978). *Brain Res.* 151: 469–478.

9 Chang, M. M., and S. E. Leeman. (1970) *J. Biol. Chem.* 245: 4784–4790.

10 Chang, M. M., S. E. Leeman, and H. D. Niall. (1971) *Nature (New Biol.)* 232: 86–87.

11 Lungdahl, A., T. Hökfelt, and G. Nilsson. (1978) *Neuroscience* 3: 861–976.

12 Otsuka, M., and S. Konishi. (1976) *Cold Spring Harbor Symp. Quant. Biol.* 40: 135–143.

13 Konishi, S., A. Tsunoo, N. Yanihara, and M. Otsuka. (1980) *Biomed. Res.* 1: 528–536.

14 Nagy, J. I. (1982) In *Handbook of Psychopharmacology,* vol. 15 (eds. L. L. Iversen, S. D. Iversen, and S. H. Snyder), pp. 185–235. Plenum Press, New York.

15 Hughes, J. (1975) *Brain Res.* 88: 295–308.

16 Kosterlitz, H. (1979) *Ann. Rev. Pharmacol. Toxicol.* 19: 1–12.
17 Hughes, J., T. W. Smith, H. W. Kosterlitz, L. Fothergill, B. A. Morgan, and H. R. Morris. (1975) *Nature* (Lond.) 258: 577–579.
18 Pert, C. B., and S. H. Snyder. (1973) *Science* 179: 1011–1014.
19 Terenius, L. (1973) *Acta Pharmacol. Toxicol.* 33: 377–384.
20 Simon, E. J., J. M. Hiller, and I. Edelman. (1973) *Proc. Natl. Acad. Sci.* (USA) 70: 1947–1949.
21 Simantov, R., and S. H. Snyder. (1976) *Proc. Natl. Acad. Sci.* (USA) 73: 2515–2519.
22 Simantov, R., and S. H. Snyder. (1976) *Brain Res.* 135: 358–367.
23 Nakanishi, S., A. Inoue, T. Kita, M. Nakamura, A. C. T. Chang, S. N. Cohen, and S. Numa. (1979) *Nature* (Lond.) 278: 423–427.
24 Cooper, P. E., and J. B. Martin. (1980) *Ann. Neurol.* 8: 551–557.
25 Bloom, F. E. (1983) *Ann. Rev. Pharmacol. Toxicol.* 23: 151–170.
26 Bloom, F. E., D. Segal, N. Ling, and R. Guillemin. (1978) *Science* 194: 630–632.
27 Watson, S. J., A. Akil, C. W. Richard, and J. D. Barchas. (1978) *Nature* (Lond.) 275: 226–228.
28 Rees, L. H. (1981) *J. R. Coll. Physicians* (Lond.) 15: 130–134.
29 Shanker, G., and R. K. Sharma. (1979) *Biochem. Biophys. Res. Commun.* 86: 1–5.
30 Chang, J. K., B. T. W. Fong, A. Pert, and C. B. Pert. (1976) *Life Sci.* 18: 1473–1482.
31 Roemer, D., H. H. Buescher, R. C. Hill, J. Pless, W. Bauer, F. Cardinaux, A. Closse, D. Hauser, and R. Huguenin. (1977) *Nature* (Lond.) 268: 547–549.
32 Joyce, E. M., G. F. Koob, R. Strecker, S. D. Iversen, and F. E. Bloom. (1981) *Brain Res.* 221: 359–370.
33 Terenius, L. (1978) *Ann. Rev. Pharmacol. Toxicol.* 18: 189–204.
34 De Weid, D., G. L. Kovacs, B. Bohus, J. M. Van Ree, and H. M. Greven. (1977) *Eur. J. Pharmacol.* 49: 427–436.
35 Jessell, T. M., and L. L. Iversen. (1977) *Nature* (Lond.) 268: 549–561.
36 Coutinho-Netto, J., A.-S. Abdul-Ghani, and H. F. Bradford. (1980) *Biochem. Pharmacol.* 29: 2777–2780.
37 Coutinho-Netto, J., A.-S. Abdul-Ghani, and H. F. Bradford. (1981) *Biochem. Pharmacol.* 31: 1019–1023.
38 Comb, M., P. H. Seeburg, J. Adelman, L. Eiden, and E. Herbert. (1982) *Nature* (Lond.) 295: 663–666.
39 Kakidani, H., Y. Furutani, H. Takahashi, M. Noda, Y. Morimoto, T. Hirose, M. Asai, S. Inayama, and S. Numa. (1982) *Nature* (Lond.) 298: 245–249.
40 Hökfelt, T., O. Johansson, A. Ljungdahl, J. M. Lundberg, and M. Schultzberg. (1980) *Nature* (Lond.) 284: 515–521.
41 Gilbert, R. F. T., and P. C. Emson. (1982) In *Handbook of Psychopharmacology*, vol. 16 (eds. L. L. Iversen, S. D. Iversen, and S. H. Snyder), pp. 519–556. Plenum Press, New York.
42 Johansson, O, T. Hökfelt, B. Pernow, S. L. Jeffcoate, N. White, W. M. Steinbusch, A. A. J. Verhofstad, P. C. Emson, and E. Spindel. (1981) *Neuroscience* 6: 1857–1881.
43 Gilbert, R. F. T., P. C. Emson, S. P. Hunt, G. W. Bennett, C. A. Marsden, B. E. B. Sandberg, H. W. M. Steinbusch, and A. A. J. Verhofstad. (1982) *Neuroscience* 7: 69–87.
44 Pettibone, D. J., and R. J. Wurtman. (1980) *Brain Res.* 186: 409–419.

45　Schwartz, J. C., B. Malfroy, and S. D. L. Baume. (1981) *Life Sci.* 29: 1715–1740.

46　Matsas, R., I. S. Fulcher, A. J. Kenny, and A. J. Turner. (1983) *Proc. Natl. Acad. Sci.* (USA) 80: 3111–3115.

47　Burnstock, G. (1976) *Neuroscience* 1: 239–248.

48　Pickel, V. M., T. H. Joh, D. J. Reis, S. H. Leeman, and R. J. Miller. (1979) *Brain Res.* 160: 387–400.

49　Henderson, G., J. Hughes, and H. W. Kosterlitz. (1978) *Nature* (Lond.) 271: 677–679.

50　Iversen, L. L., S. D. Iversen, F. E. Bloom, T. Vargo, and R. Guillemin. (1978) *Nature* (Lond.) 271: 679–681.

51　Glowinski, J., and L. L. Iversen. (1966). *J. Neurochem.* 13: 655–669.

52　Coté, L. J., and S. Fahn. (1969) In *Progress in Neurogenetics* (eds. A. Barbeau and J. R. Brunette), pp. 311–317. Excerpta Medica, Amsterdam.

53　Efendic, G., T. Hökfelt, and R. Luft. (1978) *Adv. Metab. Disord.* 9: 367–424.

54　McLoughlin, L., P. J. Lowry, S. J. Ratter, J. Hope, G. M. Besser, and L. H. Rees. (1981) *Neuroendocrinology* 32: 209–212.

55　Clement-Jones, V., P. J. Lowry, L. H. Rees, and G. M. Besser. (1980) *J. Endocr.* 86: 231–243.

56　Ikedo, Y., K. Nakao, T. Yoshimasa, N. Yanaihara, S. Numa, and H. Imura. (1982) *Biochem. Biophys. Res. Commun.* 107: 656–662.

57　Clement-Jones, V., L. McLoughlin, S. Tomlin, G. M. Besser, L. H. Rees, and H. L. Wen. (1980) *Lancet* II: 946–948.

58　Clement-Jones, V., P. J. Lowry, L. McLoughlin, G. M. Besser, L. H. Rees, and H. L. Wen. (1980) *Lancet* II: 380–383.

59　Cheng, R. S. S., and B. Pomerantz. (1979) *Life Sci.* 25: 1957–1962.

60　Research Group of Acupuncture and Anesthesia, Peking. (1974) *Scientia Sinica* 17: 112–130.

61　Hökfelt, T., A. Ljungdahl, H. Steinbusch, A. Verhofstad, G. Nilsson, E. Brodin, B. Pernour, and M. Goldstein. (1978) *Neuroscience* 3: 517–538.

62　Elde, R., T. Hökfelt, O. Johansson, M. Schultzberg, S. Efendic, and R. Luft. (1978) *Metabolism* 27: (Suppl. 1), 115–119.

63　Schally, A. V., W. Y. Huang, R. C. Chang, A. Arimura, and T. W. Redding. (1980) *Proc. Natl. Acad. Sci.* (USA) 77: 4489–4493.

64　Benoit, R., P. Bohlen, N. Ling, A. Briskin, F. Esch, P. Brazeau, S.-Y. Ying, and R. Guillemin. (1982) *Proc. Natl. Acad. Sci.* (USA) 79: 917–921.

65　Goodman, R. H., J. W. Jacobs, W. W. Chin, P. K. Lund, P. C. Dee, and J. F. Habener. (1980) *Proc. Natl. Acad. Sci.* (USA) 77: 5869–5873.

66　Iversen, L. L., S. D. Iversen, F. E. Bloom, C. Douglas, M. Brown, and W. Vale. (1978) *Nature* (Lond.) 273: 161–163.

67　Bennett, G. W., J. A. Edwardson, D. Marcano de Cotte, M. Berelowitz, B. L. Pimstone, and S. Kronheim. (1979) *J. Neurochem.* 32: 1127–1130.

68　Schusdziarra, V., E. Ipp, V. Harris, R. E. Dobbs, P. Raskin, L. Orci, and R. H. Unger. (1978) *Metabolism* 27: 1227–1232.

69　Kastin, A. J., D. H. Coy, Y. Jacquet, A. V. Schally, and N. P. Plotnikoff. (1978) *Metabolism* 27: (Suppl. 1): 1247–1252.

70　Reubi, J.-C., M. H. Perrin, J. E. Rivier, and W. Vale. (1981) *Life Sci.* 28: 2191–2198.

71　Vale, W., C. Rivier, and M. Brown. (1977) *Ann. Rev. Physiol.* 39: 473–527.

72　Negro-Vilar, A., S. R. Ojeda, A. Arimura, and S. M. McCann. (1978) *Life Sci.* 23: 1493–1498.

73 Chihara, K., A. Arimura, and A. V. Schally. (1979) *Endocrinology* 104: 1656–1662.

74 Halász, N. (1978) In *Pharmacology of the Hypothalamus* (eds. B. Cox, I. D. Morris, and A. H. Weston), pp. 5–28. Macmillan, London.

75 Hökfelt, T., K. Fuxe, D. Johansson, S. Jeffcoate, and N. White. (1975) *Eur. J. Pharmacol.* 34: 389–392.

76 Youngblood, W. W., J. Humm, and J. S. Kizer. (1979) *Brain Res.* 163: 101–110.

77 Morley, J. E. (1979) *Life Sci.* 25: 1539–1550.

78 Jackson, I. M. D. (1978) *Am. Zool.* 18: 385–399.

79 Phillis, J. W., and J. R. Kirkpatrick. (1980) *Canad. J. Physiol. Pharmacol.* 58: 612–623.

80 Bennett, G. W., J. A. Edwardson, D. Holland, S. L. Jeffcoate, and N. White. (1975) *Nature* (Lond.) 257: 323–325.

81 Jackson, I. M. D., and S. Reichlin. (1979) In *Central Nervous System Effects of Hypothalamic Hormones and Other Peptides* (eds. R. Collu, A. Barbeau, J. R. Ducharme, and J. G. Rochefort), pp. 3–33. Raven Press, New York.

82 Burt, D. R., and R. L. Taylor. (1980) *Endocrinology* 106: 1416–1423.

83 Ogawa, N., Y. Yamakawi, H. Kuroda, T. Ofuji, E. Itoga, and S. Kito. (1981) *Brain Res.* 205: 169–174.

84 Severs, W. B., and A. E. Daniel-Severs. (1973) *Pharmacol. Rev.* 25: 415–449.

85 Fitzsimmons, J. T. (1980) *Proc. R. Soc.* (Lond.), *Ser. B* 210: 165–182.

86 Brownstein, M. J., A. Arimura, A. V. Schally, M. Palkovits, and J. S. Kizer. (1976) *Endocrinology* 98: 662–665.

87 Jan, Y. N., L. Y. Jan, and S. W. Kuffler. (1979) *Proc. Natl. Acad. Sci.* (USA) 76: 1501–1505.

88 Jan, L. Y., and Y. N. Jan. (1982) *J. Physiol.* (Lond.) 327: 219–246.

89 Selmanoff, M. K., P. M. Wise, and C. A. Barraclough. (1980) *Brain Res.* 192: 421–432.

90 Edwardson, J. A., G. W. Bennett, and H. F. Bradford. (1972) *Nature* (Lond.) 240: 554–556.

91 Bradford, H. F., P. N. Cheifitz, and J. A. Edwardson. (1972) *J. Physiol.* (Lond.) 222: 52–53.

92 Moss, R. L. (1977) *Fed. Proc.* 36: 1978–1983.

93 Fuxe, K., D. Ganten, T. Hökfelt, and P. Bolme. (1976) *Neurosci. Lett.* 5: 241–246.

94 Bennett, J. P., and S. H. Snyder. (1980) *Eur. J. Pharmacol.* 67: 11–25.

95 Stamler, J. F., M. K. Raizada, M. I. Phillips, and R. E. Fellows. (1978) *Physiologist* 21 (4): 115.

96 Bennett, J. P., and S. H. Snyder. (1976) *J. Biol. Chem.* 251: 7423–7430.

97 Marks, N. (1977) In *Peptides in Neurobiology* (ed. H. Gainer), pp. 221–250. Plenum Press, New York.

98 Ramsey, D. J. (1979) *Neuroscience* 4: 313–321.

99 Felix, D., and M. I. Phillips. (1978) In *Iontophoresis and Transmitter Mechanisms in the Mammalian CNS* (eds. R. Ryall and J. S. Kelly), pp. 104–106. North Holland/Elsevier, Amsterdam.

100 Phillips, M. I., and D. Felix. (1976) *Brain Res.* 109: 531–540.

101 Carraway, R., and S. E. Leeman. (1975) *J. Biol. Chem.* 250: 1907–1911.

102 Carraway, R., and S. E. Leeman. (1975) *J. Biol. Chem.* 250: 1912–1918.

103 de Quidt, M. E., and P. Emson. (1982) *Brain Res.* 274: 376–381.
104 Goedert, M., K. Pittaway, and P. C. Emson. (1984) *Brain Res.* 299: 164–168.
105 Bissette, G., P. Manberg, C. B. Nemeroff, and A. J. Prange. (1978) *Life Sci.* 23: 2173–2182.
106 Carraway, R., and S. E. Leeman. (1976) *J. Biol. Chem.* 251: 7045–7052.
107 Polak, J. M., S. M. Sullivan, S. R. Bloom, A. M. J. Buchan, P. Facer, M. R. Brown, and A. G. E. Pearse. (1977) *Nature* (Lond.) 270: 183–184.
108 Uhl, G. R., and S. H. Snyder. (1977) *Eur. J. Pharmacol.* 41: 89–91.
109 Young, W. S., G. Uhl, and M. J. Kuhar. (1978) *Brain Res.* 150: 431–435.
110 Young, W. S., and M. J. Kuhar. (1981) *Brain Res.* 206: 273–785.
111 Uhl, G. R., G. R. Bennett, and S. H. Snyder. (1977) *Brain Res.* 130: 299–313.
112 Hayward, J. N. (1977) *Physiol. Rev.* 57: 574–658.
113 Kelly, J. S., and L. P. Renaud. (1978) In *Pharmacology of the Hypothalamus* (eds. B. Cox, I. D. Morris, and A. H. Weston), pp. 63–102. Macmillan, London and Basingstoke.
114 Koob, G. F., and F. E. Bloom. (1982) *Ann. Rev. Physiol.* 44: 571–582.
115 Sofroniew, M. V. (1980) *J. Histochem. Cytochem.* 28: 475–478.
116 Hawthorn, J., V. T. Y. Ang, and J. S. Jenkins. (1980) *Brain Res.* 197: 75–81.
117 Sofroniew, M. V. (1983) *Trends Neurosci.* 6: 467–472.
118 Swanson, W. L., and P. E. Sawchenko. (1983) *Ann. Rev. Neurosci.* 6: 269–324.
119 Pickering, B. T. (1978) *Essays Biochem.* 14: 45–81.
120 Russell, J. T., M. J. Brownstein, and H. Gainer. (1981) *Brain Res.* 205: 299–311.
121 de Vries, G. J., and R. M. Buijs. (1983) *Brain Res.* 273: 307–317.
122 Vandesande, F., K. Dierickx, and J. de Mey. (1975) *Cell Tissue Res.* 156: 189–200.
123 Buijs, R. M. (1980) *J. Histochem. Cytochem.* 28: 357–360.
124 Buijs, R. M., and D. F. Swaab. (1980) *Cell Tissue Res.* 204: 355–365.
125 Cooper, K. E., N. W. Kasting, K. Lederis, and W. L. Veale. (1979) *J. Physiol.* (Lond.) 295: 33–45.
126 Fuxe, K., T. Hökfelt, K. Andersson, L. Ferland, O. Johansson, D. Ganten, P. Eneroth, J.-A. Gustafsson, P. Skett, I. Said, and V. Mutt. (1978) In *Pharmacology of the Hypothalamus* (eds. B. Cox, I. D. Morris, and A. H. Weston), pp. 31–59. Macmillan, London and Basingstoke.
127 Larsson, L. I., and J. F. Rehfeld. (1979) *Brain Res.* 165: 201–218.
128 Ondetti, M. A., J. Pluscec, E. F. Sabo, J. T. Sheehan, and M. Williams. (1970) *J. Am. Chem. Soc.* 92: 195–200.
129 Goltermann, N. R., J. F. Rehfeld, and H. Røigaard-Petersen. (1980) *J. Biol. Chem.* 255: 6181–6185.
130 Vanderhaegen, J. J., J. C. Signeau, and W. Gepts. (1975) *Nature* (Lond.) 257: 604–605.
131 Smith, G. P., C. Jerome, B. J. Cushin, R. Etemo, and K. J. Simansky. (1981) *Science* 213: 1036–1037.
132 Innis, R. B., and S. H. Snyder. (1980) In *Hormones and the Brain* (eds. D. de Wild and P. A. Van Keep), pp. 53–62. MTP Press, Lancaster, U.K.
133 Dockray, G. J. (1982) *Brit. Med. Bull.* 38: 253–258.
134 Marley, P. D., J. P. Nagy, P. C. Emson, and J. F. Rehfeld. (1982) *Brain Res.* 238: 494–498.

135 Emson, P. C., and P. D. Marley. (1982) In *Handbook of Psychopharmacology*, vol. 16 (eds. L. L. Iversen, S. D. Iversen, and S. H. Snyder), pp. 255–306. Plenum Press, New York.

136 Dodd, J., and J. S. Kelly. (1981) *Brain Res.* 205: 337–350.

137 Emson, P. C., C. M. Lee, and J. Rehfeld. (1980) *Life Sci.* 26: 2157–2163.

138 Rehfeld, J. F., L. I. Larsson, N. R. Goltermann, T. W. Schwartz, J. J. Jolst, S. L. Jensen, and J. S. Morley. (1980) *Nature* (Lond.) 284: 33–38.

139 Hökfelt, T., L. Iskirboll, J. F. Rehfeld, M. Goldstein, K. Markey, and O. Dann. (1980) *Neuroscience* 5: 2093–2124.

140 Burt, D. R. (1980) In *Amino Acids, Peptides, and Benzodiazepines* (eds. S. J. Enna and H. I. Yamamura), pp. 149–206. Receptors and Recognition, Series B, vol. 9, part 1. Chapman and Hall, London and New York.

141 Lundberg, J. M., A. Anggard, P. Emson, J. Fahrenkrug, and T. Hökfelt. (1981) *Proc. Natl. Acad. Sci.* (USA) 78: 5255–5259.

142 Lundberg, J. M., A. Anggard, J. Fahrenkrug, G. Lundgren, and B. Holmstedt. (1982) *Acta Physiol. Scand.* 115: 525–528.

143 Eckenstein, F., and R. W. Baughman. (1984) *Nature* (Lond.) 309: 153–155.

144 Stuan-Olsen, P., S. Gammeltoft, J. Fahrenkrug, B. Ottesen, P. D. Bartels, and M. Nielsen. (1982) *J. Neurochem.* 39: 1242–1251.

145 Emson, P. C., J. Fahrenkrug, O. B. Schaffalitzky de Muckadel, T. M. Jessell, and L. L. Iversen. (1978) *Brain Res.* 140: 174–178.

146 Giachetti, A., S. I. Said, C. R. Rolland, and F. C. Koniges. (1977) *Proc. Natl. Acad. Sci.* (USA) 74: 3424–3427.

147 Phillis, J. W., J. R. Kirkpatrick, and S. I. Said. (1978) *Can. J. Physiol. Pharmacol.* 56: 337–340.

148 Said, S. I., and J. C. Porter. (1979) *Life Sci.* 24: 227–230.

149 Gardner, J. D. (1979) *Gastroenterology* 76: 202–214.

150 Hirsch, J. D., M. Grille, and F. L. Margolis. (1978) *Brain Res.* 158: 407–422.

151 Melchioni, P. (1978) In *Gut Hormones* (ed. S. R. Bloom), pp. 534–540. Churchill Livingstone, Edinburgh.

152 Said, S. I. (1978) In *Gut Hormones* (ed. S. R. Bloom), pp. 465–469. Churchill Livingstone, Edinburgh.

153 Brown, M. R., R. Allen, J. A. Villarreal, J. Rivier, and W. Vale. (1978) *Life Sci.* 23: 2721–2728.

154 Villarreal, J. A., and M. R. Brown. (1978) *Life Sci.* 23: 2729–2734.

155 Minamino, N., K. Kangawa, H. Matuso. (1983) *Biochem. Biophys. Res. Comm.* 114: 541–548.

156 Brown, M., Y. Taché, and D. Fisher. (1979) *Endocrinology* 105: 660–665.

157 Brown, M. R., and W. Vale. (1979) *Trends Neurosci.* 2: 95–97.

158 Taché, Y., and M. Brown. (1982) *Trends Neurosci.* 5: 431–433.

159 McDonald, T. J., H. Jörnvall, G. Nilsson, M. Vagne, M. Ghatei, S. R. Bloom, and V. Mutt. (1979) *Biochem. Biophys. Res. Comm.* 90: 227–233.

160 Phillis, J. W., and J. J. Limacher. (1974) *Brain Res.* 69: 158–163.

161 Phillis, J. W., and J. J. Limacher. (1974) *Exp. Neurol.* 43: 414–423.

162 Moody, T. W., C. B. Pert, J. Rivier, and M. R. Brown. (1970) *Proc. Natl. Acad. Sci.* (USA) 75: 5372–5376.

163 Correa, F. M. A., R. B. Innis, and S. H. Synder. (1978) *J. Pharmacol. Exp. Ther.* 192: 670–676.

164 Jan, Y. N., L. Y. Jan, and S. W. Kuffler. (1979) *Proc. Natl. Acad. Sci.* (USA) 75: 6139–6143.

165 Krieger, D. T., A. Liotta, and M. J. Brownstein. (1979) *Proc. Natl. Acad. Sci.* (USA) 74: 648–652.

166 Fuxe, K., T. Hökefelt, P. Eneroth, J. A. Gustafsson, and P. Skett. (1977) *Science* 196: 899–900.

167 Schoenberger, G. A., and M. Monnier. (1977) *Proc. Natl. Acad. Sci.* (USA) 74: 1282–1286.

168 Harding, J., and F. L. Margolis. (1976) *Brain Res.* 110: 351–360.

169 Rochel, S., and F. L. Margolis. (1982) *J. Neurochem.* 38: 1505–1514.

170 Nemeroff, C. B., D. Luttinger, and A. J. Prange. (1980) *Trends Neurosci.* 3: 212–215.

171 Palacios, J. M., and M. J. Kuhar. (1981) *Nature* (Lond.) 294: 587–589.

172 Henry, J. L. (1980) *Trends Neurosci.* 3: 95–97.

173 Iversen, L. L. (1979) *Sci. Am.* 241(3): 134–149.

174 Fahrenkrug, J. (1980) *Trends Neurosci.* 3: 1–2.

175 Rehfeld, J. S. (1980) *Trends Neurosci.* 3: 65–67.

176 Kastin, A. J., G. A. Olson, A. V. Schally, and D. H. Coy. (1980) *Trends Neurosci.* 3: 163–165.

177 Chang, K.-J., E. Hazum, and P. Cuatrecasas. (1980) *Trends Neurosci.* 3: 160–162.

178 Stell, W., D. Marshak, T. Yamada, N. Brecha, and H. Karten. (1980) *Trends Neurosci.* 3: 292–295.

179 Buijs, R. M. (1982) In *Transmitter Interaction and Compartmentation* (ed. H. F. Bradford), pp. 654–664. NATO ASI Series, vol. 48, Plenum Press, New York.

180 Lundberg, J. M., T. Hökefelt, A. Anggard, L. Terenius, R. Elde, J. Kimmel, M. Goldstein, and K. Markey. (1982) *Proc. Natl. Acad. Sci.* (USA) 79: 1303–1307.

181 Chan-Palay, V., G. Nilaver, S. L. Palay, M. C. Beinfeld, E. A. Zimmerman, J.-Y. Wu, and T. L. O'Donohue. (1981) *Proc. Natl. Acad. Sci.* (USA) 78: 7787–7791.

182 Kety, S., and L. Samson. (1967) *Neurosci. Res. Prog. Bull.* 5: 1–119.

183 De Feudis, F. V. (1974) *Central Cholinergic Systems and Behavior*, Academic Press, London.

184 Fahn, S., and L. J. Coté, (1968) *J. Neurochem.* 15: 209–213.

185 Halász, N., and G. M. Shepherd. (1983) *Neuroscience* 10: 579–619.

186 Loren, I., J. Alumets, R. Håkanson, and F. Sundler. (1979) *Cell Tissue Res.* 200: 179–186.

187 Tatemoto, K. (1982) *Proc. Natl. Acad. Sci.* (USA) 79: 5485–5489.

188 Emson, P. C., and M. E. De Quidt. (1984) *Trends Neurosci.* 7: 31–35.

189 Lundberg, J. M., K. Tatemoto, L. Terenius, P. M. Hellström, V. Mutt, T. Hökfelt, and B. Hamberger. (1982) *Proc. Natl. Acad. Sci.* (USA) 79: 4471–4475.

190 Tatemoto, K., M. Carlquist, and V. Mutt. (1982) *Nature* (Lond.) 296: 659–660.

191 Allen, Y. S., T. E. Adrian, J. M. Allen, K. Tatemoto, T. J. Crow, S. R. Bloom, and J. M. Polak. (1983) *Science* 221: 877–879.

192 Furness, J. B., M. Costa, P. C. Emson, R. Håkanson, E. Moghimzadeh, F. Sundler, I. L. Taylor, and R. E. Chance. (1983) *Cell Tissue Res.* 234: 71–92.

193 Sundler, F., E. Moghimzadeh, R. Håkanson, M. Ekelund, and P. C. Emson. (1983) *Cell Tissue Res.* 230: 487–493.

194 Hökfelt, T., J. M. Lundberg, H. Langercrantz, K. Tatemoto, V. Mutt, L. Terenius, J. Polak, S. Bloom, C. Sasek, R. Elde, and M. Goldstein. (1983) *Acta Physiol. Scand.* 117: 315–318.

195 Lundberg, J. M., and K. Tatemoto. (1983) *Acta Physiol. Scand.* 110: 107–109.

196 Allen, N. M., J. Polak, and S. R. Bloom. (1975) In *Monitoring Neurotransmitter Release during Behavior* (eds. M. H. Joseph, M. Fillenz, I. A. Mcdonald, and C. A. Marsden). Ellis Horwood, Chichester.

197 Fox, J. E. T. (1984) *Life Sci.* 35: 695–706.

198 Hawthorn, J., J. M. Graham, and J. S. Jenkins. (1984) *Life Sci.* 35: 277–284.

199 Nicoll, R. A., B. E. Alger, and C. E. Jahr. (1980) *Proc. Roy. Soc. B* 210: 133–149.

200 Koella, W. P. (1983) *Trends Pharmacol. Sci.* 4: 210–211.

201 Moody, T. W., N. B. Thoa, T. L. O'Donahue, and C. B. Pert. (1980) *Life Sci.* 26: 1707–1712.

202 Kitchen, I. (1984) *Prog. Neurobiol.* 22: 345–358.

203 Noda, M., Y. Furutani, H. Takahashi, M. Toyosato, T. Hirose, S. Inayama, S. Nakanishi, and S. Numa. (1982) *Nature* (Lond.) 295: 202–206.

6

The Synaptosome: An In Vitro Model of the Synapse

The value of intact neural tissue for the study of neurotransmitter synthesis, storage, and release is limited by the relatively small portion of neural tissue volume occupied by synaptic terminals in most brain regions (5 percent of the cerebral cortex), although this can rise to higher levels in a few regions such as hippocampus (30 percent).[1,2] This creates a problem of ambiguity in identifying the tissue compartments involved in the storage and release of neurotransmitters. Eliminating glial and nonsynaptic neuronal contributions is difficult. In theory, one should be able to circumvent this problem by applying agents known to modulate synaptic release without influencing other tissue compartments, but unequivocal results are rare in practice.

Progress in our knowledge of presynaptic neurochemistry would have been much slower had researchers not come upon the synaptosome, a remarkable investigative tool. Synaptosome preparations are subcellular fractions of neural tissue comprising largely intact, surviving presynaptic nerve terminals, their membranes having "pinched off" at the point of connection with the axon (Figure 6.1). They are metabolically active.

Synaptosome-enriched neural tissue fractions may be isolated from brain homogenates by using simple density-gradient centrifugation techniques.[3,4,121] Such tissue preparations are, of course, rich in the biochem-

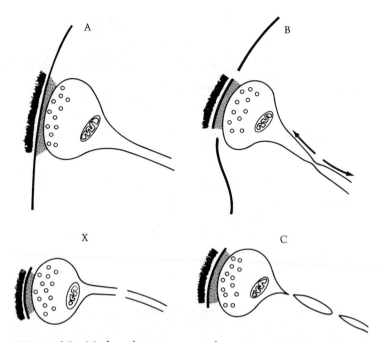

FIGURE 6.1   Modes of synaptosome formation. From A to B to C results in "pinched-off" structures. From A to X results in fenestrated structures whose membranes will subsequently reseal.

ical agents and morphological structures directly mediating synaptic transmission, and they have provided a rich harvest of information on the neurochemical events of the presynaptic nerve terminal.[A–D]

Synaptosomes were first recognized morphologically in the late 1950s and early 1960s as being derived from nerve terminals that had pinched off and survived intact.[3] We have since learned that they mostly comprise completely sealed cytoplasmic sacs of 1–2 μm diameter that retain all the microstructures and cell organelles visible in the intact synapse in situ. Remarkably, these structures seem to sustain little damage during preparation. When incubated, they swell to form beautifully rounded structures able to perform, in a well-organized and integrated fashion, several of the principal activities of the in vivo nerve terminals from which they derive.[B,E]

The mitochondria contained in synaptosomes are often long, thread-like structures. This is readily seen in electron micrographs of negatively stained synaptosomes (Figure 6.2). When positive staining is used, several cross-sectional profiles of what is often one long mitochondrion are commonly seen (Figures 6.3 and 6.4). Negative staining reveals the membra-

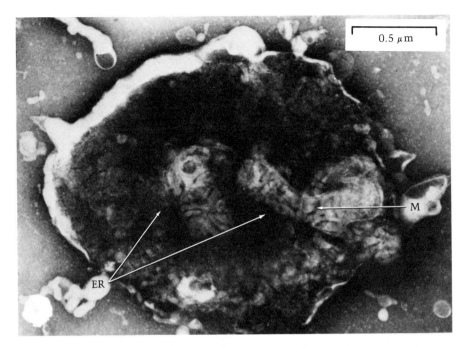

0.5 μm

FIGURE 6.2   Negatively stained synaptosome from guinea pig cerebral cortex (phosphotungstic acid stain). Visible are long coiled mitochondrion (*M*) and tubular, saclike, endoplasmic reticulum (*ER*) [From ref. 96]

nous, saclike, and often tubular endoplasmic reticulum present in the synaptosomal cytoplasm (Figure 6.2), which appears as a granular matrix or series of vacuoles in positively stained sections (Figures 6.3 and 6.4). A small proportion of synaptosomes carry an attached postsynaptic bag, which is formed from the postsynaptic structure (e.g., dendrite, dendrite spine, cell body) by a "pinching-off" process (Figure 6.4). The extent to which these postsynaptic structures are intact and contain cytoplasm remains uncertain. In positively stained synaptosomes, synaptic vesicles, coated vesicles, and dense-cored vesicles may be seen. The synaptic vesicles can be seen to be hollow structures with a clear delimiting membrane (Figure 6.4). Pre- and postsynaptic densities are visible where the plane of section passes through the synaptic cleft (Figures 6.3 and 6.4). Since the typical thickness of a section used for electron microscopy is 50 nm, some 20 sections can be made through each diameter of a synaptosome of 1 μm diameter, in a large number of different planes. Thus, the synaptic cleft region would often be missed (see Figure 6.3). Sometimes incubated synaptosomes show large vacuoles (Figure 6.3), which presumably reflect considerable pinocytic activity by the synaptosomes.

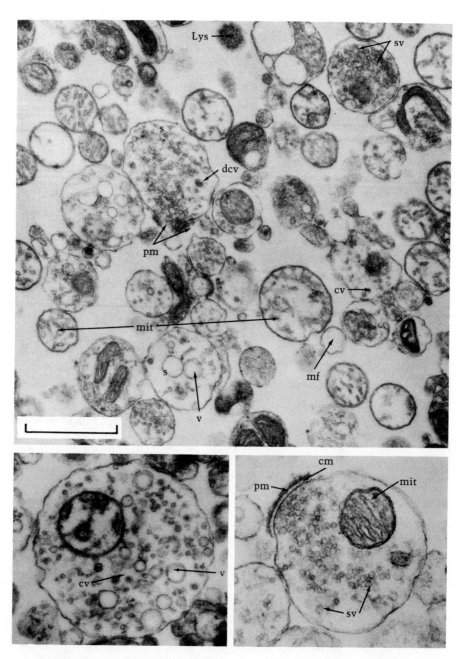

FIGURE 6.3 Typical field of cortical synaptosomes prepared from rat cerebral cortex. Rounded synaptosomes (s) are seen after 60 minutes of incubation at 37°C in Krebs-phosphate-saline medium. Mitochondria (mit), synaptic vesicles (sv), coated (complex) vesicles (cv), dense-cored vesicles (dcv) can be seen together with larger vacuoles (v) inside the synaptosomes. Postsynaptic membranes (pm) and cleft material (cm) are seen adjacent to presynaptic components of a few synaptosomes, and lysosomes (Lys) can be seen together with membrane fragments (mf) as contaminants of the preparation. Osmium tetroxide fixation; uranyl acetate and lead citrate staining. Bar, 1 μm. [From ref. 8]

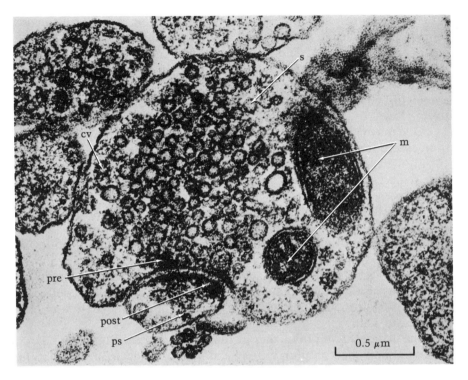

FIGURE 6.4 Synaptosome from rat cerebral cortex after incubation in physiological saline at room temperature for 10 minutes. Key: *s*, synaptic vesicles; *cv*, coated vesicles; *m*, sections through mitochondrion; *pre*, presynaptic thickening; *post*, postsynaptic thickening; *ps*, postsynaptic bag. [From ref. 120]

METABOLIC PROPERTIES
OF THE SYNAPTOSOME

Synaptosomes respire actively when incubated in medium of ionic composition similar to that of plasma or cerebrospinal fluid, that is, much richer in Na^+ ions than in K^+ ions and supplemented with glucose (5–10 mM). Respiration is not only at a high rate but is coupled to the formation of ATP and phosphocreatine at high P : O ratios[5,10,B,C,D] and is maintained at a constant rate over several hours.[11] Such efficient energy metabolism means that all components of the glycolytic sequence are present, well-controlled, and fully linked to both mitochondrial Krebs-cycle activity and oxidative phosphorylation. It also demonstrates that synaptosomes are cytoplasmic units able to maintain viability over long periods; they do not rapidly deteriorate at 37°C because of lysomal activation and autolysis, as was once thought.[6]

Synaptosomes, like cortex slices, rapidly convert radioactive glucose to pyruvate, as evidenced by the formation of quantities of labeled alanine and lactate.[12,13] The pyruvate readily enters the Krebs cycle by way of acetyl CoA, since aspartate, glutamate, and GABA rapidly become labeled together with the carbon dioxide evolved.[13] Thus, the systems for employing glucose as an energy-yielding metabolic substrate are all active and well organized in incubated synaptosomes (see Chapter 3).

One may calculate the rate of utilization of glucose by synaptosomes from estimates of respiratory rates for these structures (50–70 μmol O_2/ 100 mg synaptosome protein/hour)[5,14,15,B] and from rates of lactate formation. Such calculations indicate that each 100 mg of synaptosomal protein uses at least 25 μmol glucose per hour, which is similar to the glucose utilization of cortex slices.[12] Synaptosomal respiration in the absence of added glucose falls to very low levels over a period of 20–30 minutes, and during this period endogenous stores of amino acids are used up as respiratory fuels, their pools becoming greatly diminished.[11,13,B] Radiolabeled glutamate and other amino acids added to incubated synaptosomes are rapidly oxidized to labeled CO_2 and converted to other metabolites, showing the ready capacity of the preparation to oxidize amino acids.[13]

The accumulation of lactate in the incubation medium of synaptosomes or brain slices is an in vitro phenomenon, as lactate accumulation does not occur to any extent in cerebrospinal fluid in vivo, nor does lactate increase in blood flowing out of the brain compared with the inflowing arterial blood[16] (see Chapter 3). The respiratory rates of cortical tissue, however, are two fold higher in vivo, apparently because of the continuous electrical activity and associated Na^+ pump activity in the intact brain.

Since synaptosome preparations are able to extrude Na^+ and accumulate K^+ against concentration gradients,[5,7,17,B] a large proportion of the ATP and phosphocreatine (PhCr) generated from the metabolism of glucose and other endogenous substrates is likely to be used to maintain ionic gradients across the semipermeable and sealed synaptosome membrane. There is a very active and concentrated Na^+K^+-ATPase system in synaptosome preparations.[18,19,20,21,22] Through its continuous generation of ADP, the rate-limiting factor of oxidative phosphorylation, this system must act as a respiratory pacemaker for synaptosomes, as it does for the central nervous system as a whole (Chapter 3).

In nerve terminals and their derived synaptosomes, the ratio of surface area to volume takes on a special importance with respect to respiratory rates because their small size makes this ratio relatively high (Table 6.1). Thus sodium ions, flowing into the small volume of cytoplasm relative to surface membrane area, produce a larger rise in intracellular Na^+ in nerve terminals than in larger structures. This leads to a constantly higher stimulation of the sodium pump, and it causes the respiratory rates to be higher per unit weight in nerve terminals than in larger structures such as neuronal cell bodies.

TABLE 6.1 Comparison of metabolic properties of cortex slices and cortex synaptosomes

	Slice	Synaptosome	Ratio of slice to synaptosome
Respiratory rate (μmol O_2/hour)	60	60 ± 3(20)	1
Glycolytic rate (μmol lactate/hour)	42	16 ± 2(15)	3
Potassium content (equiv.)	73 ± 5(11)	25 ± 2(5)	3
ATP content (nmol)	2000	267 ± 23(8)	5
Phosphocreatine content (nmol)	1800	394 ± 58(8)	

Note: Values are per 100 mg of synaptosomal protein or per g wet weight of cortex slice. Data for synaptosomes are from ref. 14. Data for slices are from refs. 9 and 16.

Similar considerations apply to axons, or to dendrites of similar small diameter, particularly where they are receiving an enhanced Na^+ influx due to depolarization. Such differences are indeed reflected in the ratios of metabolic activity and respiration in cortex slices and synaptosomes. The ratio of respiratory rate to glycolytic rate, and to potassium content, in synaptosomes is threefold greater in synaptosomes, and the ratio to total energy phosphate (ATP + PhCr) is even higher (see Table 6.1).

If we assume that absolute levels of potassium content, glycolytic rates, and energy phosphate content are actually equivalent in synaptosomes and cortex slices, we can estimate that the calculated respiratory rates are probably threefold higher in synaptosomes on a wet-weight basis.[B]

Synaptosomal Membrane Potentials

The current view of synaptosomes as sealed structures carrying a membrane potential is now well established. It comes from several observations. About 80 percent of synaptosomes are impermeable to large extracellular marker substances (e.g., thorium dioxide), indicating the presence of intact plasma membranes.[8,40] The ability of synaptosomes to swell and round off during incubation, and their ready propensity to shrink and swell in media of differing isotonicities, is also direct evidence of the continuity of their outer membrane[8,40,24] (Figures 6.3, 6.4, and 6.9). The active extrusion of Na^+ and accumulation of K^+, reflected in their ionic composition following incubation with glucose, also supports this view. So does their ability to retain, when sedimented, the freely soluble amino acids they generate together with many soluble cofactors and soluble enzymes.[5,7,13,B]

Thus, two of the primary requirements for generating a membrane potential, an intact membrane and an active sodium pump, appear to be present in incubated synaptosomes. Direct evidence for the existence of the membrane potential is more difficult to obtain because of the small size of synaptosomes. Penetration of individual synaptosomes with microelectrodes (tip diameter 0.5 μm) is not possible, and the only direct demonstration of the existence of a synaptosome membrane potential has come about through the use of membrane-soluble fluorescent dyes that change their fluorescence output in the presence of a voltage gradient across the membrane. Fluorescence output changes in the expected direction following application of depolarizing agents.[25,26] These voltage-sensitive fluorescent dyes have previously been used to show the propagation of the action potential along the squid axon.[27]

Since synaptosomes accumulate K^+ and extrude Na^+, it is possible to estimate a likely size for their membrane potential by using the Hodgkin-Huxley equation and assuming values for the permeability constants for these ions. Using this approach, minimal values of around -30 mV have been obtained.[28,29] Values of this size are low compared with the -60 to -70 mV potentials commonly recorded in resting neurons in various parts of the nervous system. One problem associated with this indirect approach concerns the estimation of the intrasynaptosomal space available to the K^+, Na^+, and Cl^- found to be present. Synaptic vesicles and mitochondria are not likely to be equilibrated with the cytoplasmic space and yet they occupy a significant volume (Figures 6.2 to 6.4). Partial or complete exclusion of ions from these structures could, for instance, double the ionic concentration and correspondingly change their contribution to a synaptosomal membrane potential. Attempts have been made to standardize the voltage-detecting capacity of fluorescent dyes, and the fluorescence output of synaptosomes incubated under optimal conditions suggests the presence of a membrane potential in the -60 to -65 mV range.[25]

Other evidence for synaptosomal membrane potentials comes from investigations of ionic equilibration across the membrane. For instance, by measuring the relative accumulation of ions equilibrated across either the synaptosomal plasma membrane[100] (e.g., Rb^{2+}), or of other substances whose properties allow them to report on the presence of a voltage gradient across the intrasynaptosomal mitochondrial membrane (e.g., methyltriphenyl phosphonium or safranine), it has proved possible to monitor both the synaptosomal and the mitochondrial membrane potentials in incubated synaptosomes.[30,31,128]

Specific depolarization of intrasynaptosomal mitochondria by addition to incubated synaptosomes of a proton translocator (e.g., carbonylcyanide p-trifluoromethoxyphenylhydrazone) results in a rapid efflux of Ca^{2+} from the synaptosomes. This has been interpreted as being due to a Ca^{2+} release by mitochondria and a consequent rapid rise in cytosolic free Ca^{2+} in the synaptosomes.[31] Using probes of membrane voltage gradients of this kind, it becomes possible to study the transport and release properties of incubated synaptosomes and monitor in parallel both the plasma membrane potential and the mitochondrial membrane potential.

Indirect approaches also provide compelling evidence for the existence of membrane potentials. When synaptosomes are treated with agents known to cause displacement of neuronal membrane potentials, the agents induce a range of biochemical responses indicating the onset of membrane depolarization. Three such agents—electrical stimulation, high $-K^+$ concentrations, and veratrum alkaloids (e.g., veratridine)—produce increases in respiratory and glycolytic rates (Figure 6.5) that are accompanied by K^+ efflux and Ca^{2+}-dependent transmitter release.[32] All compounds that activate voltage-dependent Na^+ channels are effective depolarizing agents (e.g., scorpion venom toxin, batrachotoxin) and produce similar responses. All the responses are prevented by tetrodotoxin or saxitoxin, agents that potently and specifically block Na^+ channel activation. These observations lead to the conclusion that the evoked biochemical responses are: (1) due to Na^+-channel activation and subsequent

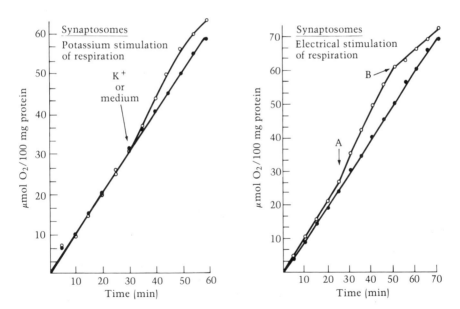

FIGURE 6.5 Graphs showing respiratory response of cortical synaptosomes to potassium or electrical stimulation. Potassium was added at the point indicated to give a final concentration of 56 mM. A, current on; B, current off. O, stimulated sample; ●, control sample. Suspensions of synaptosomes were incubated at 37° C in Krebs-phosphate medium containing glucose (10 mM). Mean stimulating current 15 mA. Note the linearity of synaptosomal respiration over at least 70 minutes. [Data from refs. 7, 32]

membrane depolarization, (2) due exclusively to neuronal elements in the synaptosome preparations since voltage-sensitive Na$^+$ channels are present only in neurons. The metabolic and respiratory response is due to enhanced utilization of ATP by the sodium pump, which expels the Na$^+$ influxing through Na$^+$ channels during depolarization (see Chapter 3).

Purity of Synaptosome Fractions

The purity of standard synaptosome fractions has often been questioned. For instance, studies in which homogenates of radiolabeled cultured glial cells (C6 glioma) were added to brain homogenates before subcellular fractionation commenced have suggested that as much as half of the material in synaptosome fractions may be derived from glial cells rather than from nerve terminals.[79] Such claims clearly put in doubt the relevance of studies employing synaptosome preparations to investigate the properties of the nerve terminals from which they are assumed to be derived. But there also exists a considerable body of evidence suggesting that synaptosomes *do* comprise the majority of cytoplasmic bodies present in these fractions.[80]

Since the commonly used procedures for preparing synaptosome fractions are simple—(differential ultracentrifugation followed by discontinuous density-gradient centrifugation)—a priori it would seem unlikely that synaptosome fractions are very homogenous and free of contaminating material. A wide size-range of membrane fragments certainly does usually contaminate synaptosome preparations, as can be seen in most published electron micrographs of these fractions (e.g., Figure 6.3). Many saclike membrane profiles do not contain structures characteristic of the synapse, such as synaptic thickenings or synaptic vesicles. The absence of the specialized synaptic vesicles, however, can often be due to the plane of section missing these structures; electron microscopy sections are commonly about 50 nm thick and nerve terminals are much larger (1–2 μm) in diameter and appear to contain regions not occupied by vesicles or mitochondria (see Figure 6.3). Empty saclike structures of this kind often contain granular material that could be cytoplasm, and such structures might conceivably be formed from pinched-off glial-cell processes (gliosomes)[75,76] or from dendrites (dendrosomes). Where they contain polysomes, such an origin *is* very likely, since polysomes have not been observed in axons or nerve terminals but are present in dendrites and glial processes. But the occurrence of polysomes is a relatively rare event. Other structures frequently seen in preparations from brain are synaptosomes with an attached bag of postsynaptic dendrite[74] (Figure 6.4). These mixed pre- and postsynaptic structures are very much in the minority, and it is unclear whether the postsynaptic region consists of a *sealed* cytoplasmic body.

The equivalent subcellular fraction prepared from cultured glial cells corresponding to the synaptosome fraction appears quite different morphologically, consisting essentially of empty membrane profiles, mito-chondria, or multilaminar vesicles.[79,81]

Chemical or enzymic markers for glial cells should give the best index of glial contamination of synaptosome fractions, but markers with the required degree of glial specificity are few and are contested. Examples are glutamine synthetase, carbonic anhydrase, glial fibrillary protein, and S-100 protein (see Chapter 2). Comparison between the content of such markers in whole-tissue homogenates and their content in the synaptosome fractions gives an index of contribution of glial-derived material to these fractions. Such studies[81-85] suggest a relatively minor contribution from glial cells.

In an immunological test of the contamination due to glial or other membranes, antisera were raised to membrane fragments isolated from lysed synaptosomes. The antisera were used to stain sections of brain by immunofluorescence, and they were found to stain only synaptic regions.[86] Glial cells, myelin, and nonsynaptic regions of neurons showed no binding of the antibodies, suggesting that they make only a minimal contribution to synaptosome fractions.

But since simple membrane fragments can carry enzyme activity or structural markers without contributing to the metabolic or other cell-like properties that are restricted to organized cytoplasmic particles, such markers do not give an unequivocal index of the *cytoplasmic* contribution from glial cells. A more appropriate test system for this would rely on the presence of intact cytoplasmic bodies. For example, a useful test is to examine specific membrane transport systems that require the presence of an ATP-generating system linked to glucose utilization. An example is GABA uptake. Transport of U-^{14}C–GABA into neurons is blocked by diaminobutyrate (DABA) and by *cis*-3-aminocyclohexane carboxylic acid, whereas its uptake by glial cells is inhibited by β-alanine, the two transport systems being organized in different ways at the molecular level. Measurement of the relative effectiveness of DABA and β-alanine in blocking radiolabeled-GABA uptake to synaptosome preparations should give a rough index of the ratio of "neuron-derived" to "glial-derived" cytoplasmic structures. Such measurements suggest at least a 5 : 1 ratio of synaptosomes (or dendrosomes) to gliosomes in cortical preparations. Since a very high proportion of the surface of the dendritic tree appears to provide sites for synaptic boutons making contact, it seems unlikely that the fragmentation during tissue dispersion (homogenization) would allow dendrosome formation. The dendritic plasma membranes and the dendrite spines probably fragment to form free membranes or part of the synaptosome during tissue disruption (see Figures 6.4 and 1.35).

These various lines of argument, and the data they have generated, suggest that standard synaptosome-enriched preparations contain relatively little (20 percent or less) contamination in the form of cytoplasmic glial elements, even though the simplicity of the separation method might make this seem surprising.

NEUROTRANSMITTER-RELEASING
PROPERTIES

Synaptosome preparations treated with depolarizing agents show a metabolic response accompanied by Ca^{2+}-dependent neurotransmitter release and enhanced neurotransmitter turnover. Since synaptosome preparations from any brain region contain nerve terminals from various neurons employing different neurotransmitters, a whole range of transmitters are released. For instance, cortical,[32] spinal,[33] striatal[34] and hypothalamic synaptosomes[35] release glutamate and aspartate, the putative excitatory amino acid neurotransmitters, and they also release gly-

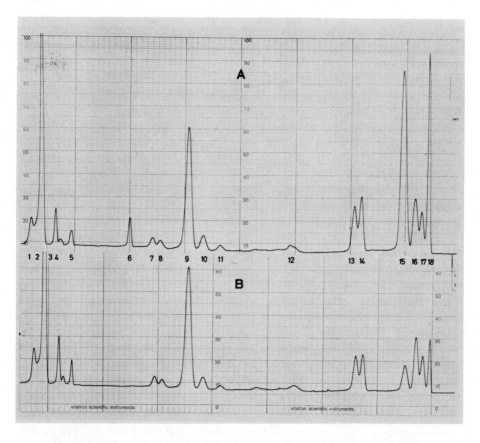

FIGURE 6.6   Chromatograms of amino acids released into the medium by incubated cerebral cortex synaptosomes. Electrical stimulation of synaptosome suspensions. (A) Stimulated sample, (B) control sample. Key: 1, arginine; 2, ammonia; 3, lysine; 4, tryptophan; 5, histidine; 6, GABA; 7, phenylalanine; 8, tyrosine; 9, norleucine (added standard); 10, leucine; 11, isoleucine; 12, valine; 13, alanine; 14, glycine; 15, glutamate; 16, serine; 17, glutamine; 18, aspartate. [From ref. 32]

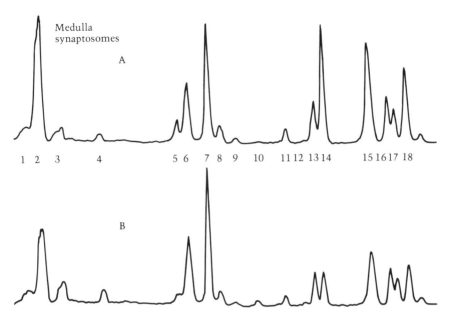

FIGURE 6.7   Chromatogram of amino acids released into the medium by incubated spinal cord and medulla synaptosomes. (A) Electrically stimulated sample; (B) control sample. Key: *1*, arginine; *2*, ammonia; *3*, lysine; *4*, histidine; *5*, GABA; *6*, added tyrosine standard; *7*, added norleucine standard; *8*, leucine; *9*, isoleucine; *10*, methionine; *11*, valine; *12*, cysteine; *13*, alanine; *14*, glycine; *15*, glutamate; *16*, serine; *17*, glutamine; *18*, aspartate. [From ref. 33]

cine and GABA, the inhibitory compounds. Glycine is released in far greater amounts by spinal cord preparations than by cortical preparations[B] (Figures 6.6 and 6.7; see also Figures 4.29 and 4.30). Endogenous pools of these neuroactive amino acids are the source of the material released. Endogenous acetylcholine or monoamines are released by depolarizing agents from cortical,[36] striatal, or whole-brain synaptosomes.

Botulinum toxin, which appears to act presynaptically by preventing neurotransmitter release from cholinergic nerve terminals, grossly reduces the release of acetylcholine by synaptosomes,[37] indicating that the mechanism of neurotransmitter release by synaptosomes has many of the properties of neurotransmitter release by intact nerve terminals.[38]

Endogenous dopamine is released by stimulation of synaptosomes prepared from the corpus striatum, where dopamine is highly concentrated.[34] In many studies, the release of radiolabeled neurotransmitters has been monitored following their preloading into the synaptosome preparation by high-affinity transport during a previous incubation (i.e., from an exogenous source). Although there are many reported discrepancies over detail, overall these two kinds of experiments, that is, studies of endogenous and exogenous neurotransmitters, give much the same results.[49]

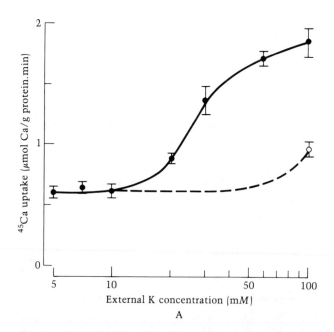

FIGURE 6.8   (A) Effect of external K^+ concentration on rate of ^{45}Ca accumulation by synaptosomes. Incubation with ^{45}Ca was for 2 minutes at 30°C in the presence (●) or absence (○) of K^+ added to the concentrations indicated. Synaptosomes were from whole rat brain. (B) Effect of external K^+ concentration on net Ca content of synaptosomes. Experimental protocol was similar to that used for (A) except that no radioactive tracers were used and incubation was for 4 minutes at 30°C. Calcium content was determined by atomic absorption spectrophotometry. [From ref. 26]

The events accompanying neurotransmitter release from synaptosomes are similar to those known to occur in nerve terminals in situ when they respond to the depolarizing invasion of an action potential. Apart from the metabolic responses already described, there are key events such as Ca^2 influx (Figure 6.8) and the resynthesis of neurotransmitters to restore their pool sizes following their release to the incubation medium.[36,39] The enhanced fluxes of ions across the synaptosomal membrane are accompanied by increased membrane turnover. For instance, there is accelerated endocytosis of membrane vesicles (Figure 6.9) and a decrease in the number of synaptic vesicles in the synaptosomes[40,41] (see Chapter 7).

This use of the synaptosome as a model nerve terminal has been applied to the study of the factors controlling the release and turnover of many neurotransmitters, including a wide range of neuropeptides (see Chapter 5) and neurosecretory agents such as hypothalamic trophic hor-

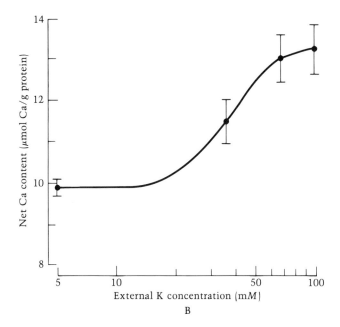

B

mone–releasing hormones.[42,43,B] Neurosecretion is another field in which studies of the dynamic properties of synaptosomes are providing a detailed picture of the factors underlying controlled secretion of biologically active agents.

TRANSPORT PROPERTIES OF SYNAPTOSOMES

High-Affinity and Low-Affinity Transport

The systems operating to actively transport amino acid and other transmitters into synaptosomes have been analyzed kinetically and found to contain two components.[44,56] One of these subsystems works at low-substrate concentration with a half-maximal velocity of 1–20 μM (i.e., low K_m value). This system necessarily has a low capacity (V_{max}) and can transport only small quantities of neurotransmitter per unit time. However, most potent neurotransmitter compounds are likely to be released in small enough quantities to be rapidly taken up by these low V_{max} transport systems.

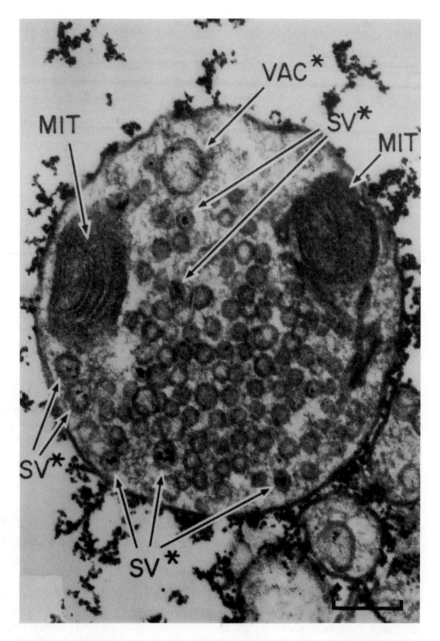

FIGURE 6.9   Effect of depolarization on endocytosis by synaptosomes. Thorium dioxide particles were added to synaptosomes incubated at 30°C in Krebs medium. Veratridine (75 μM) was added for 1 minute before its action was blocked by tetrodotoxin (200 nM). The number of synaptosomes with ThO$_2$-containing particles was twofold to fourfold higher than in controls pretreated with tetrodotoxin to prevent depolarization. SV^*, synaptic vesicle containing ThO$_2$; VAC, ThO$_2$-containing vacuole; MIT, mitochondrion. Bar, 0.2 μM. [From ref. 40]

High-affinity transport apparently evolved to "inactivate" the neurotransmitters by rapid removal from the extracellular space to the interior of nerve terminals—or of glial cells, which also possess these transport mechanisms.[45] Present evidence based on autoradiography and density-gradient separations (see below) leads to the conclusion that each neuron carries only the high-affinity transport system for the neurotransmitter it releases.[46,47] Thus, it recaptures some of its own transmitter after release, and no doubt a reasonable portion is reused. Such an arrangement also allows specific neurons to be visualized by autoradiography after exposure to levels of radiolabeled neurotransmitter in the concentration range of its high-affinity transport mechanism (1–20 μM).

Low-affinity transport systems operate for both transmitter and nontransmitter amino acids (Table 6.2). These are less effective in the low-substrate range but much more effective at higher concentrations, their half-maximal velocities (K_m values) being in the 1–5 mM range. These latter processes, therefore, have a high capacity (V_{max}) and can provide bulk transport of amino acids for protein synthesis or other biosynthetic or metabolic activities.

The advantage of having high-affinity uptake systems "scavenging" for physiologically active compounds is clear, and their efficiency is reflected in the low levels of these substances in cerebrospinal fluid. Gly-

TABLE 6.2   Kinetics of amino-acids uptake into homogenates of rat central nervous system (K_m values $\times$ 10^{-4} M)

Amino acid	Low-affinity uptake	High-affinity uptake
Cerebral cortex		
L-Aspartic acid	3.68 ± 1.94	0.169 ± 0.128
L-Glutamic acid	14.97 ± 9.59	0.362 ± 0.129
Glycine	7.63 ± 0.53	–
L-Alanine	4.53 ± 0.24	–
L-Arginine	6.24 ± 1.11	–
L-Leucine	3.51 ± 0.47	–
L-Serine	2.16 ± 0.35	–
Spinal cord		
L-Aspartic acid	37.62 ± 14.17	0.215 ± 0.077
L-Glutamic acid	16.73 ± 6.48	0.189 ± 0.058
Glycine	9.23 ± 6.38	0.265 ± 0.141
L-Alanine	10.58 ± 0.71	–
L-Arginine	26.79 ± 7.58	–
L-Leucine	91.36 ± 30.18	–
L-Serine	10.81 ± 2.06	–

Note: A minimum of six concentrations of every amino acid were incubated in quadruplicate with homogenates. Data are from ref. 56.

cine, which is a potent inhibitory amino acid in the spinal cord but not in the cerebral cortex, provides striking evidence for such a role. Thus, high-affinity transport systems for glycine are found in spinal cord synaptosome preparations but not in their cortical equivalent.[44,56,107] (See Table 6.2.)

Homoexchange and the Relationship Between Uptake and Release

Using a method that superfuses synaptosomes immobilized on filters, it has been possible to demonstrate the presence of an exchange transport process called *homoexchange*, which operates only for neurotransmitter compounds.[48] It is conceived as involving the inward transit of one molecule obligatorily linked to the exit of another molecule of the same species. Outward flux cannot occur in the absence of the external trigger, and the exchange is one-for-one, allowing no net accumulation of the substance transported. The superfusion system limits reuptake of any released compound. Homoexchange can be demonstrated by applying the nonisotopic compound to superfused synaptosomes previously loaded with the radiolabeled equivalent; such addition causes the appearance of the labeled compound in the superfusion stream as homoexchange is activated.

These methods have so far been most effectively applied to glutamate, an excitatory transmitter, and to GABA,[49] an inhibitory transmitter. The role of homoexchange remains obscure, but it has been suggested that the high-affinity uptake described above is really homoexchange and does not result in net uptake—that is, it would *not* inactivate neurotransmitter by reducing its external concentration. However, many published studies show that net uptake by high-affinity systems does occur, and the consensus is in favor of its serving the inactivation function described above. It has been argued[50,51] that the inward and outward fluxes of homoexchange are not obligatorily coupled and that a continuous shuttling or cycling of neurotransmitter compounds occurs irrespective of the presence of exogenous neurotransmitter. Such a membrane shuttle may underlie nonvesicular release mechanisms (see Chapter 7).

Uptake is usually measured by employing short time-periods of incubation (1–5 minutes) with the radioactively labeled compound, (1) to limit losses of counts due to metabolism (as CO_2 or lactate), which leads to underestimates of the inward transport, and (2) to keep the specific radioactivity of the internal pool as low as possible. The latter stratagem means that any concomitant spontaneous efflux will return minimal counts to the exterior. But when the actions of various agents on uptake processes are tested, it is clear that blocked uptake, and also enhanced release, will lead to a depression in the amounts of radioactivity accumulated in the tissue (synaptosomes) compared with the control. Enhanced release can result if the agent being tested has a *depolarizing* action (e.g.,

Na^+ ionophore) or allows calcium entry (Ca^{2+} ionophore). Uptake and release cannot be completely separated experimentally, but by using short incubation periods, relatively few counts would be released by an agent acting primarily to enhance efflux. And making measurements in the presence and absence of calcium should expose any releasing action due to depolarization. Moreover, preloading synaptosomes with the substance whose transport is under study and then exposing it to the agent for the usual short periods used to study uptake (1 or 5 minutes) would reveal any releasing action as radiolabel appearing in the incubation medium. Thus, uptake and release may well be closely associated processes with a considerable degree of interdependence.

Calcium Transport

Synaptosomes, like neurons, maintain a low (0.4 μM) intracellular Ca^{2+} concentration in spite of the continual entry of calcium ions during neural activity. This low intracellular level is achieved by transport of Ca^{2+} into the intrasynaptosomal mitochondria and endoplasmic reticulum, as well as by efflux across the synaptosomal plasma membrane. This efflux process uses the energy of the Na^+ gradient across the membrane; in addition, there is a Ca^{2+}-stimulated ATPase which pumps Ca^{2+} to the exterior employing the energy of ATP.[123] In general, the data for Ca^{2+} efflux from synaptosomes is consistent with a Ca^{2+} carrier mechanism that extrudes Ca^{2+} in exchange for external Na^+ or Ca^{2+}. The relationship between Ca^{2+} efflux and external sodium concentration is sigmoidal and gives a ratio of $3Na^+$ transported inwardly for each Ca^{2+} effluxed. These transport properties are maintained in synaptic plasma membrane vesicles derived from synaptosomes.[124]

Within synaptosomes, accumulation of Ca^{2+} into the mitochondria and endoplasmic reticulum largely accounts for its removal from the cytosol, other organelles (e.g., synaptic vesicles, coated vesicles) being relatively inactive in this respect.[123] Although nonmitochondrial organelles show a high affinity for Ca^{2+}, with uptake half-saturated at about 0.35 μM Ca^{2+}, they can store only relatively small amounts of Ca^{2+} (2–3 μmol/mg protein). In contrast, mitochondria, which have a lower affinity for Ca^{2+} (uptake is half-saturated at 10–100 μM Ca^{2+}), have a larger storage capacity than nonmitochondrial organelles, about 30 μmol/mg protein).

The distribution of Ca^{2+} entering synaptosomes after depolarization has been studied. In spite of the higher Ca^{2+} storage capacity of mitochondria, about 50 percent of the Ca^{2+} entering the nerve terminals within 10 seconds is recovered in the nonmitochondrial stores, with only 20 percent appearing in mitochondria. This suggests that in the intact brain, transport into the nonmitochondrial organelles (identified as mainly endoplasmic reticulum) is of critical importance in buffering the internal concentration of Ca^{2+} in the nerve-terminal cytoplasm in the shorter term.[125]

EXPERIMENTAL USES OF
SYNAPTOSOME PREPARATIONS

Synaptosome Subpopulations

Synaptosomal preparations are greatly enriched in nerve terminals of mixed origin. Cortical synaptosomes, for example, may contain several separate subpopulations, each releasing one of the various neurotransmitters detectable in the preparation: glutamate, aspartate, GABA, acetylcholine, norepinephrine, dopamine, serotonin, and various peptides.[44,56,107] Some of these neurotransmitters may coexist in some of the terminals and may even be coreleased in response to stimulation (see Chapter 7). It would clearly constitute a substantial advance if subpopulations of synaptosome synthesizing, storing, and releasing individual neurotransmitters (or coexisting groups of neurotransmitters) could be prepared from these mixed populations. Affinity chromatography employing antibodies to specific surface antigens as affinity ligands might well provide us with techniques for achieving this end.[52,53,54,55]

Presynaptic Receptors

Synaptosomes represent the isolated, working presynaptic terminal, but often (possibly always) carry the postsynaptic thickening. Such a structure should allow the study of many, but not all, of the neurotransmitter-linked properties: synthesis, storage, release, and, to some extent, inactivation. Any binding of released neurotransmitter to postsynaptic receptors of synaptosomes cannot be detected as a biological action. However, evidence has accumulated during the past decade for the existence of *presynaptic* receptors. They are situated on the presynaptic terminals and are involved in the control of neurotransmitter release.[58] Synaptosomes necessarily carry these presynaptic receptors and therefore provide an excellent preparation for studying their influence on neurotransmitter release. Three kinds of presynaptic receptor probably exist on synaptosomes from different regions of the nervous system: (1) those due to serial (e.g., axoaxonic) synapses, particularly in relay nuclei such as the cuneate nucleus; (2) those operated by a neurotransmitter or neuromodulating substance different from that being released from the terminal in question; (3) those operated by the neurotransmitter being released from the terminal (Chapter 7).

Although much of the evidence for presynaptic receptors comes from the noradrenergic peripheral nervous system, some examples of their counterpart in the central nervous system have been described. For instance, the release of dopamine from striatal synaptosomes can be enhanced or decreased by acetylcholine acting through nicotinic or muscarinic presynaptic receptors respectively.[59,60]

Phospholipid and Protein Biosynthesis

Both electrical stimulation and addition of acetylcholine at low concentrations have been shown to cause a twofold stimulation in the incorporation of ^{32}P into phosphatidic acid and phosphatidylinositol of incubated synaptosomal preparations.[61–67] There is a stimulated loss of ^{32}P when the phospholipids are prelabeled together with an associated fall in the pool sizes of these phospholipids. These observations have led to the conclusion that the primary action of added acetylcholine is to cause an increase in phosphatidylinositol hydrolysis by phospholipase C, and a consequent increase in its turnover, i.e., the reaction

phosphatidylinositol 4,5-bisphosphate $\rightarrow$

diacylglycerol + D-inositol 1,4,5-trisphosphate

is accelerated as a primary event. The response appears to be mediated by a muscarinic cholinergic receptor and is localized to synaptic vesicles, suggesting that the increased lipid labeling is associated with re-formation of vesicles or some other aspect of a vesicle-linked transmitter release mechanism. For instance, increased availability of diacylglycerol would be able to promote fusion between synaptic vesicles and plasma membranes during an exocytotic transmitter secretion mechanism (see Chapter 7 for further discussion). Phosphatidylinositol hydrolysis is becoming implicated in the mechanisms underlying several membrane-linked processes, including calcium gating,[68,69,119] Ca^{2+} storage, and various neurotransmitter and hormone receptor mechanisms involving the generation of inositol trisphosphate[126] as a second messenger (see Chapter 4). Activation of Ca^{2+} channels (e.g., by Ca^{2+} ionophore A23187) stimulated hydrolysis of prelabeled *triphosphoinositide* and *diphosphoinositide,* which creates a special interest in these highly charged phosphoinositides in relation to synaptic mechanisms. Previous work has implicated the monophosphoinositide as the principal molecular species showing response.[65,66,70,71]

These interesting changes in phospholipid turnover in synaptosomes and their linkage, directly or indirectly, with synaptic transmission mechanisms has exact parallels in a number of other neural preparations,[67] suggesting once again that these isolated nerve terminal preparations are providing a valid in vitro model of their in vivo synaptic counterpart.

Radioactive amino acids are readily incorporated into the proteins of incubated synaptosome fractions.[73,74] There is a long-standing controversy, however, whether synaptosomes themselves or contaminating cytoplasmic structures derived from glia (gliosomes) or from dendrites (dendrosomes, or postsynaptic bags attached to synaptosomes; see Figure 6.4) are responsible for the observed amino acid incorporation, since the process appears to be at least in part mediated by cycloheximide-sensitive ribosomal mechanisms. Electron microscopy studies have indicated that up to 50 percent of radiolabeled proteins are associated with rare membrane sacs containing ribosomes,[72] and 20 percent are located in mitochondria contained with synaptosomes.[74,75,76]

A considerable increase in the rate of amino acid incorporation into proteins can be evoked by treatment of incubated synaptosome preparations with depolarizing agents[73] (e.g., electrical pulses, veratridine). This response is entirely tetrodotoxin-sensitive and concentrated in a subfraction showing enrichment in synaptic complexes—pre- and postsynaptic densities held together by cleft material. Since tetrodotoxin prevents the opening of active Na^+ channels, its action is specifically on neuronal structure, which implies that the responsive elements are either synaptosomes, dendrosomes, or postsynaptic bags.

SPECIALIZED SYNAPTOSOME PREPARATIONS

The techniques for separating synaptosomes from the various major regions of the mammalian central nervous system have been applied in modified forms for separating similar synaptic structures from smaller or more specialized regions of both central and peripheral systems. This includes preparations from the superior cervical ganglion,[87] from mammalian gut wall,[88] and from glomerular synapses of the cerebellum[89,108] (Figure 6.10). Although the yield from these tissues is very small (1–2 mg protein per gram of tissue), the synaptosomes appear to be in good condition and capable of neurotransmitter uptake and organized metabolic activity.[108,B]

Synaptosome Preparations from Nonmammalian Sources

Mollusk Brain Synaptosomes   Synaptosome-enriched preparations from the head ganglion or optic lobes of the squid *Loligo pealei* show high levels of oxygen uptake, which remain constant over long periods.[90,91] Morphologically, the preparations are similar in appearance (Figure 6.11) and size to their mammalian equivalent and have been shown to take up choline by both high-affinity and low-affinity transport systems, each of which requires Na^+. It is assumed that a cholinergic subpopulation is responsible for the choline accumulation.

The supraesophageal ganglia of *Octopus vulgaris* has also been used to produce what appears to be a fraction greatly enriched in intact synaptosomes[92] (Figure 6.12).

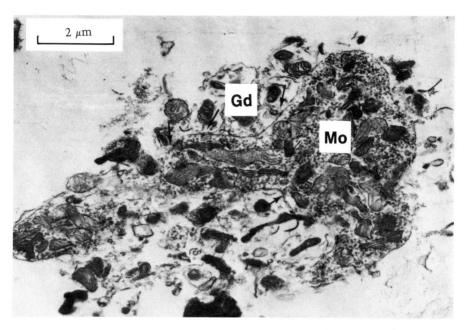

FIGURE 6.10   Synaptic glomerulus particles from cerebellum. The mossy fiber terminal (*Mo*) contains numerous mitochondria and synaptic vesicles and is surrounded by torn-off dendritic digits of granule cells (*Gd*). The arrows point to synaptic contacts. (Cf. Figure 1.31.) Field-view electron micrographs of the preparation shows considerable enrichment in glomerulus particles. [From ref. 108]

Torpedo Synaptosomes   The electric ray *Torpedo mamorata* possesses an electric organ of considerable size and weight (1–2 kg). It has only cholinergic innervation. In essence, the structure is a giant stack of neuromuscular junctions modified to produce high voltages (1–2 kV) for defense. Producing intact and metabolically active synaptosomes from this organ has been difficult because of the nature of the tissue elements being disrupted, their relatively large size, and the hypertonicity required in the isolation media. Nevertheless, intact and relatively pure preparations of *Torpedo* synaptosomes have now been achieved[93,94] (Figure 6.13). They are metabolically active and they synthesize and store acetylcholine, which they release by a calcium-dependent process when exposed to depolarizing stimuli.[95] These are the first "pure" synaptosome preparations in terms of neurotransmitter content and should help considerably in the study of the organization of transmitter pools and acetylcholine release mechanisms. The open question is whether the findings from so specialized a structure will be relevant for other kinds of synapse such as those commonly found in the central nervous system.

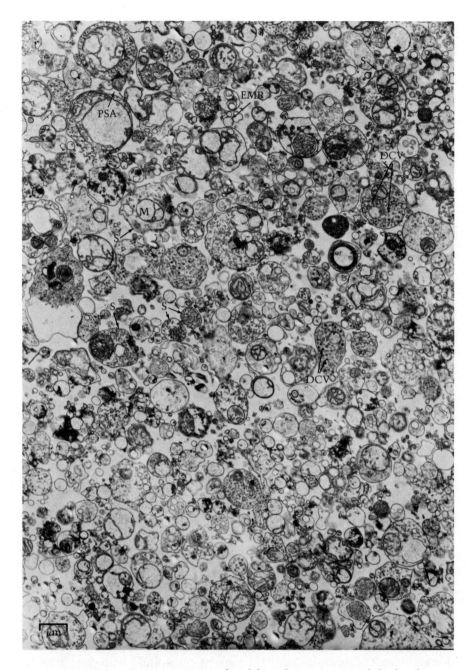

FIGURE 6.11   Synaptosomes isolated from homogenates of the head ganglion of the squid *Loligo pealei* by differential centrifugation. The synaptosomes (S) may bear a postsynaptic adhesion (PSA). They contain dense-
cored vesicles (DCV) and mitochondria (M). There are few free mitochondria but many empty membranous particles (EMP). Bar, 1 μm. [From
ref. 109]

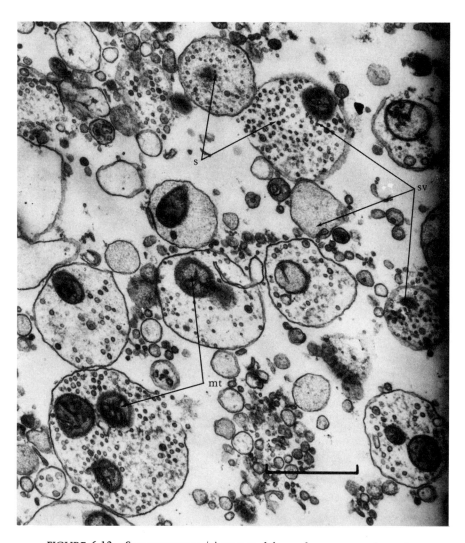

FIGURE 6.12 Synaptosomes (s) prepared from the supraesophageal ganglia of the brain of *Octopus vulgaris* fixed with KMnO₄ and stained with
lead citrate. Note synaptic vesicles (sv) and mitochondria (mt). Bar, 1
μm. [From ref. 92]

Giant Synaptosomes

When rat brain synaptosome preparations are incubated at 23°C in isotonic sucrose, containing a neutral protease, some of the synaptosomes
fuse together to form large structures (giant synaptosomes) with diameters up to 250 μm, which is tenfold to twentyfold larger than individual
synaptosomes. The interior of these particles is devoid of electron-dense
material, though a few mitochondria and occasional vesicle clusters are

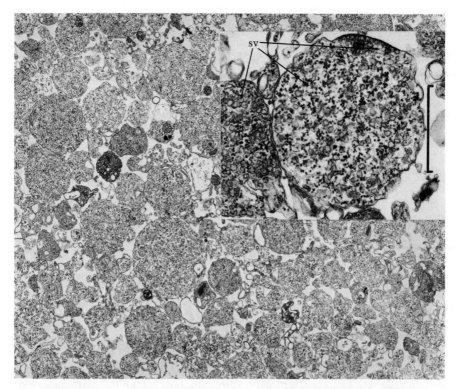

FIGURE 6.13   Purified synaptosomes prepared from the electric organ of
Torpedo. Inset gives detail of a single synaptosome; synaptic vesicles (*sv*)
are surrounded by small densely staining glycogen granules. Attached
postsynaptic membranes are not observed. Bar, 1.5 μm in inset. [From
refs. 93, 94]

seen. Although occurring only rarely, these giant synaptosomes provide
the opportunity to study the electrophysiological characteristics of the
synaptosome membrane, since glass intracellular recording electrodes
can be readily inserted into them.[122] When incubated in physiological
salines resting membrane potential of mean value −55 mV were obtained, indicating that the outer membrane of these giant synaptosomes
was intact. They were readily depolarized by increasing concentrations of
KCl in a fashion indicating that K^+ permeability was the major determi-
nant of the resting membrane potential. Voltage-dependent channels for
both Na^+ and Ca^{2+} could be detected, as well as calcium-activated K^+
conductance. Many neurotransmitter substances (e.g., glutamate, acetylcholine, serotonin) influenced the resting membrane potentials of these
synaptosomes, and the actions of acetylcholine were prevented by the

muscarinic receptor antagonist atropine, which indicates that the effects are receptor-mediated. These responses of individual giant synaptosomes to a wide range of neurotransmitters suggest that they are hybrid structures, formed from the fusion of synaptosomes derived from nerve terminals employing different neurotransmitters (cholinergic, serotonergic, etc.). Nonetheless, they provide a unique opportunity to investigate the electrophysiology of the nerve terminal membrane.[122]

PREPARATION OF NERVE TERMINAL COMPONENTS FROM SYNAPTOSOMES

A few years after their first isolation and identification, it proved possible to dissect synaptosomes into their component parts. Synaptic vesicles, synaptic plasma membranes, and nerve terminal mitochondria can be separated from each other and from the soluble cytoplasm by density-gradient centrifugation after rupture of synaptosomes by exposure to hypotonic media.[96,97] These fractions are proving particularly useful for localizing sites of transmitter synthesis and storage and for investigating whether terminal mitochondria possess special properties. The isolated synaptic plasma membranes can be used to prepare the various synaptic thickenings and other paramembranous specializations that characterize the ultrastructure of the contact zone of the synapse. Preparation is by detergent treatment and further density-gradient centrifugation procedures.

Synaptic vesicle preparations, though not entirely free of membranous contaminants, are highly enriched, as assessed by electron microscopy (Figure 6.14). They are not easily sedimented from sucrose suspensions but may be freed from low-molecular-weight contaminants by gel filtration. It has not proved possible to "load" isolated synaptic vesicles with neurotransmitters, nor has it been possible to rupture them and empty their neurotransmitter content by osmotic lysis. They appear as rather resilient structures. Some vesicle profiles are permeable to the negative staining and others are not (Figure 6.15). It seems, therefore, that they are hollow. This is the clear conclusion from the presence of a well-defined boundary membrane. (See Figures 6.4 and 6.14 to 6.16.)

Detailed studies have been made of the components of the synaptic vesicles isolated from *Torpedo* electric organ and purified to constant composition. The major constituents of this preparation have been reported.[23,127] The 90-nm-diameter vesicle is bounded by a 4-nm-thick lipoprotein membrane consisting of lipid (54 percent), protein (16 percent), and water (30 percent). The highly hydrated core occupies 75 percent of the vesicle and contains a membrane-bound proteoglycan together with acetylcholine (0.9 M) and ATP (0.17 M) in solution. There is also appre-

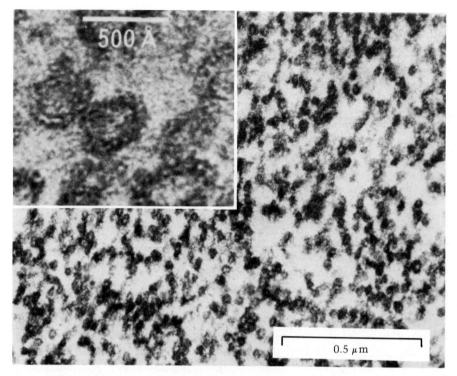

FIGURE 6.14   Section of pellet of a synaptic vesicle preparation isolated from guinea pig cortex. Note the homogeneity of the preparation. Inset shows enlarged synaptic vesicle from cortex of coypu (*Myocastor coypus*), thin-sectioned to show unit membrane structure and hollow interior. [From ref. 110]

ciable calcium (0.25 M) and magnesium (0.06 M). About 90 percent of the water is osmotically active and the rest is bound to small solutes. The membrane proteins include actin, a $Ca^{2+}Mg^{2+}$-activated ATPase, a protein-translocating ATPase, an ATP translocase, and an acetylcholine translocase.

The membrane subfractions can be prepared from osmotically lysed synaptosome preparations. They are not so useful, however, for determining the chemical composition or other properties of nerve terminal membranes because of the likely diverse origins of their membranes. Many membrane fragments of unknown origin are clearly visible in standard synaptosome preparations (see Figures 6.3 and 6.4); they could be derived, for example, from endoplasmic reticulum or fragmented glial plasma membrane. But these synaptic plasma membrane (SPM) preparations are

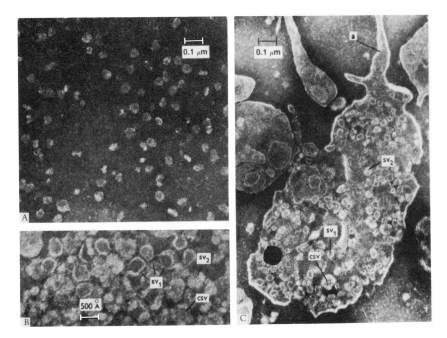

FIGURE 6.15 Synaptosomes and isolated synaptic vesicles negatively stained with phosphotungstic acid. (A) Large numbers of synaptic vesicles, about 50 nm in diameter. (B) Hollow (sv_1), dense-cored (sv_2), and complex (csv) synaptic vesicles. (C) Synaptosome negatively stained under slightly hypo-osmotic conditions, showing synaptic vesicles of the three main types of in situ; a portion of axon (a) remains attached, but this synaptosome does not contain a mitochondrion. [From ref. 96]

enriched in the neurotransmitter receptors of the synaptic junction, and they are often used in studies of specific binding of neurotransmitters.

Synaptic Junctions or Complexes

Whether or not every synaptosome carries an attached postsynaptic thickening remains a controversial question. It has been claimed that many of these structures become detached from the synaptosome during preparation unless Mg^{2+} is included in the isolation media.[98] In section, the postsynaptic thickenings, or densities, appear to be in intimate contact, or even in continuity, with the other components of the synaptic junction: the *presynaptic dense projections*, the *cleft material*, and the *junctional region* of the postsynaptic membrane where the postsynaptic receptors are located (Figure 6.16).

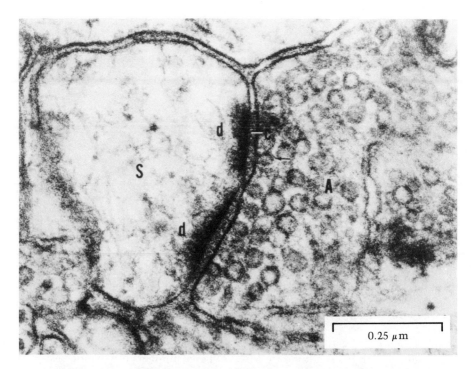

FIGURE 6.16   Synaptic junctions between axon terminals (A) and dendritic spines (S). The pre- and postsynaptic membranes are separated by a cleft (c) containing dense cleft material. On the cytoplasmic face of the postsynaptic membrane are postsynaptic densities (d). [From ref. 111]

FIGURE 6.17   (A) Electron micrograph of intact rat cerebellar cortex fixed with glutaraldehyde and block-stained with ethanolic phosphotungstic acid to show synaptic complexes in situ. Numerous heavily stained synaptic complexes are seen. Bar, 0.1 μm. (B) High magnification of a typical synaptic complex in rat hypothalamus prepared as for the cerebellar cortex. The stained elements of the synaptic complex are the presynaptic projections (P), the cleft material shown as a band structure (arrow), and the postsynaptic material with tufts or whorls (w) stretching into the postsynaptic cytoplasm. Bar, 0.3 μm. [From ref. 112]

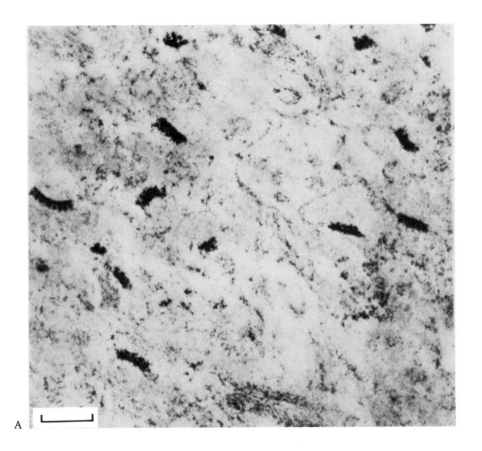

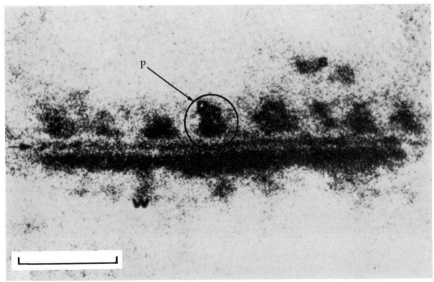

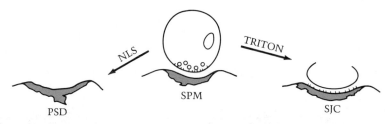

FIGURE 6.18   Preparation of synaptic junctional complexes (*SJC*) and postsynaptic densities (*PSD*). Treatment of synaptic plasma membranes (*SPM*) with Triton X-100 (*TRITON*) solubilizes the extra junctional membrane and leaves the synaptic junction intact. Treatment with *N*-lauroyl sarcosinate (*NLS*), a stronger detergent, leaves only the postsynaptic density. [From ref. 116]

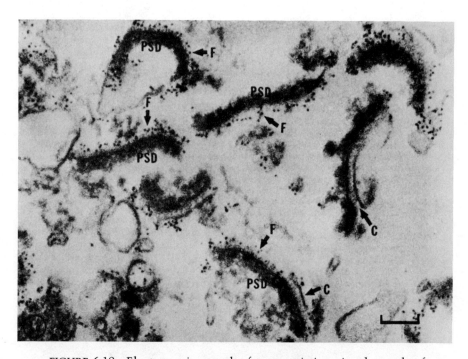

FIGURE 6.19   Electron micrograph of a synaptic junctional complex fraction isolated by treatment of a synaptic plasma membrane fraction with Triton X-100. The preparation has been incubated with concanavalin A–ferritin conjugates. In those complexes where the cleft is open, exposing the outer surface of the postsynaptic membrane, the ferritin molecules are seen localized on the external surface of the postsynaptic membrane (*F*) overlying the postsynaptic density (*PSD*). Little ferritin is found on the PSD. Ferritin is excluded from the cleft where the presynaptic membrane remains attached to the postsynaptic structure (*C*). Bar, 0.1 μm. [From ref. 99]

This complex of structures, called the *synaptic junctional complex* (*SJC*), has been isolated by treating the synaptic plasma membrane preparations with Triton X-100 or related nonionic detergents[99-102] (Figures 6.18 and 6.19). These synaptic membrane preparations are obtained by osmotic lysis of synaptosome preparations followed by their density-gradient separation. It is clear from electron micrographs of the isolated synaptic junctional complexes that the postsynaptic plasma membrane can survive these procedures. Detergent would normally be expected to dissolve this membrane, but the surrounding structures in the synapse may provide it with some protection (Figure 6.18). The extent to which the various components of the synaptic junctional complex are dissolved during preparation depends upon the nature and concentration of the detergent used.[103]

Contrary to expectation, these synaptosomal membrane preparations, even those derived from crude mitochondrial fractions, appear to carry little contamination from mitochondria or glial cells.[104] Contributions from these sources should therefore also be absent from the final preparation of synaptic complexes, and this appears to be so, as judged by the low content of cytochrome oxidase (a mitochondrial marker), 5'-nucleosidase (a plasma membrane marker), phospholipids, and nucleic acids.[102] Some procedures allow for removal of free mitochondria by a density gradient method before detergent treatment of synaptic plasma membranes.[99]

Postsynaptic Density and Lattice

Triton X-100 at the high concentration of 3 percent w/v, and other detergents that are ionic rather than neutral at concentrations of 3 percent w/v (e.g., sodium *N*-lauroyl sarcosinate, or sodium deoxycholate), dissolve all visible postsynaptic plasma membrane and produce fractions comprising 3–4 percent of the protein of the original synaptosomal plasma membranes. These fractions appear to consist entirely of the postsynaptic densities,[103] bereft of *presynaptic* dense projection and solubilized to varying degrees depending on the detergent employed[114,116] (see Figures 6.18 and 6.20). The nonionic detergent produces relatively electron-dense structures, identified as the postsynaptic densities proper. They are characteristically cup- or disk-shaped particles (200–500 nm × 40 nm) composed of a planar array of 13–30 nm subunits, sometimes with 10 μm filaments extending from the center of the density to beyond its extremities forming a subsynaptic web,[102] and sometimes containing a central "hole" (compare Figures 6.20 and 1.28).

Deoxycholate, in contrast, produces much less densely stained structures, containing 25 percent of the protein and composed essentially of a polygonal framework of short branching fibers or filaments 5 nm in diameter (Figure 6.21). This is most likely the bare substructure forming the cytoskeletal framework of the density. It is called the *postsynaptic*

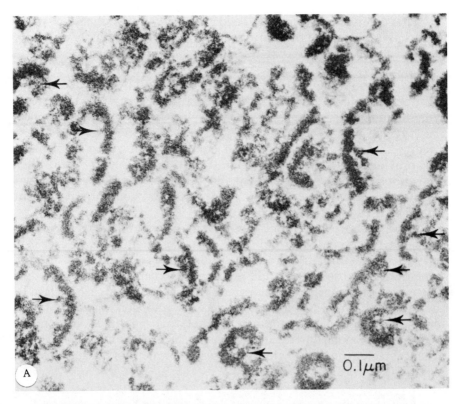

FIGURE 6.20   (A) Postsynaptic densities preparation isolated by treatment of synaptic membranes with Triton X-100 and positively stained
for electron microscopy. Arrows indicate subsynaptic bodies or subsynaptic web material. (B) High-powered selected PSD subsynaptic bodies (*sp*)
about 30 nm in diameter are attached by filamentous material. (C) Replica of PSD prepared by coating with platinum and carbon. Note disk-
shaped appearance with an apparent hole in center, and the filaments
emanating from this structure (*arrows*). [From ref. 102]

lattice. The hollow interior of the lattice polygons may be hydrophobic
and provide the hydrophobic anchorage sites for the proteinacious 10–20
nm globular structures seen in the whole structure. These globular structures are conceived as being functional units, such as receptors or enzymes, rather than as serving a structural role in the lattice.[104,117]

Protein Constituents of Synaptic Junctional Structures

Isolated synaptic junctional complexes (SJC) consist mainly of protein
with small amounts of lipids and carbohydrate. In general, there is con-

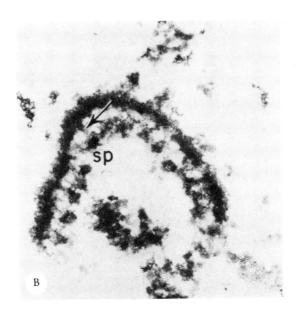

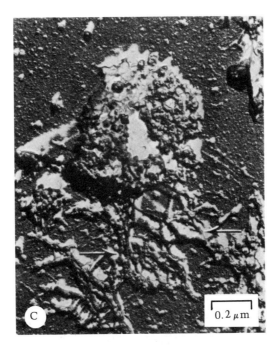

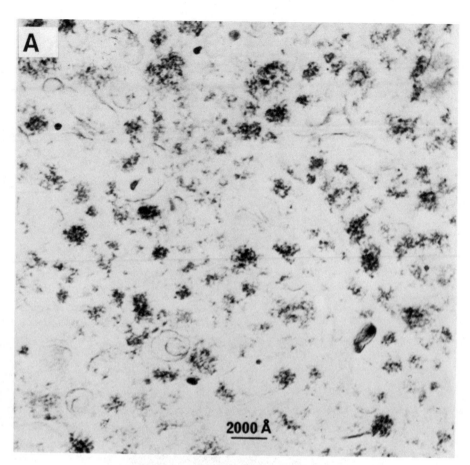

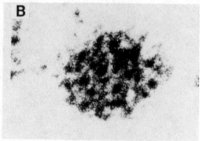

FIGURE 6.21 (A) Isolated synaptic junctional lattices in an enriched fraction obtained by digesting a synaptic plasma-membrane-enriched fraction with sodium deoxycholate. (B) The isolated junctional lattices appear as discrete flat rigid disks composed of uranyl acetate–stained subunits. [From ref. 104]

siderable agreement in reports from various laboratories on the nature of the major protein constituents of SJCs and postsynaptic density (PSD) fractions.[103,113,116,118] Both preparations contain polypeptides ranging in molecular weight from 8000 to 300,000. The PSD fraction characteristically contains far fewer major proteins than the SJC, with one protein of 50,000–52,000 molecular weight comprising more than 45 percent of the total present.[102,103,116,118] This protein appears to be unique to postsynaptic junctional structures, for which it is now a good candidate as a biochemical marker. There are also significant quantities of six other proteins with molecular weights in the range of 26,000 to 180,000,[1] and evidence exists for the presence of calmodulin, a protein kinase, and phosphoprotein, as well as high-affinity binding sites for GABA and other neuroactive amino acids.[118]

The whole densities, and the derived lattices, show differences in their protein components, which confirms that they represent somewhat different structures. What appears to be actin (mol wt 47,500) is a major protein of the lattice, and microtubular subunit proteins (tubulins) may also be lattice components on a small scale.[103,115,117]

Isolated synaptic junctions also contain several glycoproteins that have been characterized by their affinity for the lectin concanavalin A. These glycoproteins appear to be synapse-specific, and they are barely detectable in the synaptic plasma membranes from which the SJC is prepared. The major components have molecular weights of 160,000, 110,000, and 95,000.[118] Cytochemical studies have shown that these glycoproteins are localized within the synaptic cleft on the external surface of the postsynaptic plasma membrane[114] (Figure 6.19). A great deal of work remains to be done to properly characterize these various proteins and to clarify whether they are proteolytic degradation products of molecules of larger size.[103,117]

Whatever the molecular composition of the various synaptic junctional components turns out to be, at present there is no satisfactory description of the function of these structural elements in synaptic transmission. The intimate contact seen to exist morphologically between microtubules and postsynaptic densities under certain conditions of fixation may be a strong lead,[105] and it has been proposed that the postsynaptic density immobilizes receptor glycoproteins by anchoring them and preventing their tendency to diffuse laterally out of the area of synaptic contact. Also, interaction between presynaptic densities and microtubules has been observed morphologically and may have relevance in vesicle translocation during the process of transmitter release[106] (see Chapter 7). But these observations are at present simply the basis for speculation,[117,118] and a great deal of research effort needs to be applied to this topic in order to establish definitive functions for these elements.

REFERENCES

General Texts and Review Articles

A   Jones, D. G. (1976) *Synapses and Synaptosomes*. Chapman and Hall, London.

B   Bradford, H. F. (1975) "Isolated Nerve Terminals As an *In Vitro* Preparation for the Study of Dynamic Aspects of Transmitter Metabolism and Release," in *Handbook of Psychopharmacology*, vol. 1 (eds. L. L. Iversen, S. W. Iversen, S. H. Snyder), pp. 191–252. Plenum Press, New York.

C   Whittaker, V. P. (1984) "The Synaptosome," in *Handbook of Neurochemistry*, 2d ed., vol. 7 (ed. A. Lajtha). Plenum Press, New York.

D   Rodriguez de Lorez Arnaiz, G. R., and E. de Robertis. (1972) "Properties of Isolated Nerve-Endings," in *Current Topics in Membranes and Transport*, vol. 3 (eds. F. Bronner and A. Kleinzeller), pp. 237–272. Academic Press, New York and London.

E   de Belleroche, J. J., and H. F. Bradford. (1973) "The Synaptosome: An Isolated Working Neuronal Compartment," in *Progress in Neurobiology*, vol. 1, pp. 275–298. Pergamon Press, Oxford.

Additional Citations

1   Cragg, B. G. (1975) *J. Comp. Neurol.* 160: 147–166.

2   Nafstad, P. H. J., and T. W. Blackstad. (1966) *Z. Zellforsch. Mikrosk. Anat.* 73: 234–245.

3   Gray, E. G., and V. P. Whittaker. (1962) *J. Anat.* 96: 79–88.

4   De Robertis, E., A. P. De Iraldi, G. R. De Lorez Arnaiz, and L. Salganicoff. (1962) *J. Neurochem.* 9: 23–35.

5   Bradford, H. F. (1969) *J. Neurochem.* 16: 675–684.

6   Whittaker, V. P. (1969) In *Structure and Function of Nervous Tissue*, 2d ed., vol. 3 (ed. G. A. Bourne), pp. 1–24. Academic Press, London.

7   Csillag, A., and F. Hajos. (1980) *J. Neurochem.* 34: 495–503.

8   Jones, D. G., and H. F. Bradford. (1971) *Brain Res.* 28: 491–499.

9   Keesey, J. C., H. Wallgren, and H. McIlwain. (1965) *Biochem. J.* 95: 289–300.

10   Verity, M. A. (1972) *J. Neurochem.* 19: 1305–1318.

11   Bradford, H. F., D. G. Jones, H. K. Ward, and J. Booher. (1975) *Brain Res.* 90: 245–259.

12   Kini, M. M., and J. H. Quastel. (1959) *Nature* (Lond.) 184: 252–256.

13   Bradford, H. F., and A. J. Thomas. (1969) *J. Neurochem.* 16: 1495–1504.

14   Bradford, H. F. (1974) *Biochem. Soc. Trans.* 2: 680–682.

15   Balfour, D. J. K., and J. C. Gilbert. (1970) *Biochem. Pharmacol.* 20: 1151–1156.

16   McIlwain, H., and H. S. Bachelard. (1971) *Biochemistry and the Central Nervous System*, 4th ed., p. 73. Churchill Livingstone, Edinburgh and London.

17   Ling, C. M., and A. A. Abdel-Latif. (1968) *J. Neurochem.* 15: 721–729.

18   Hosie, R. J. A. (1965) *Biochem. J.* 96: 404–410.

19   Bradford, H. F., E. K. Brownlow, and D. B. Gammack. (1969) *J. Neurochem.* 13: 1283–1297.

20   Diamond, I., and R. A. Fishman. (1973) *Nature* (Lond.) 243: 519–520.

21 Marchbanks, R. M., and C. W. B. Campbell. (1976) *J. Neurochem.* 26: 973–980.

22 Archibald, J. T., and T. D. White. (1947) *Nature* (Lond.) 252: 595–596.

23 Whittaker, V. P., and H. Stadler. (1980) In *Proteins of the Nervous System*, 2d ed. (eds. R. A. Bradshaw and D. M. Schneider), pp. 231–255. Raven Press, New York.

24 Keen, P., and T. D. White. (1970) *J. Neurochem.* 17: 565–571.

25 Blaustein, M. P., and J. M. Goldring. (1975) *J. Physiol.* (Lond.) 247: 589–615.

26 Blaustein, M. P. (1975) *J. Physiol.* (Lond.) 247: 617–655.

27 Davila, H. V., B. M. Salzberg, L. B. Cohen, and A. S. Waggoner. (1973) *Nature (New Biol.)* 241: 159–160.

28 Bradford, H. F. (1971) In *Cellular Organelles and Membranes in Mental Retardation* (ed. P. F. Benson), pp. 1–14. Churchill Livingstone, London.

29 Marchbanks, R. M. (1976) *Exp. Brain Res.* 24, 16

30 Nicholls, D. G., and K. E. O. Akerman. (1981) *Philos. Trans. R. Soc.* (Lond.), Ser. B, 296: 115–122.

31 Akerman, K. E. O., and D. G. Nicholls. (1981) *Biochem. J.* 117: 491–497.

32 Bradford, H. F. (1970) *Brain Res.* 19: 239–247.

33 Osborne, R. H., H. F. Bradford, and D. G. Jones. (1973) *J. Neurochem.* 21: 407–419.

34 de Belleroche, J. S., H. F. Bradford, and D. G. Jones. (1976) *J. Neurochem.* 26: 561–571.

35 Bradford, H. F., G. W. Bennett, and A. J. Thomas. (1973) *J. Neurochem.* 21: 495–505.

36 de Belleroche, J. S., and H. F. Bradford. (1972) *J. Neurochem.* 19: 1817–1819.

37 Wonnacott, S., and R. M. Marchbanks. (1976) *Biochem. J.* 156: 701–712.

38 Reichhardt, L. F., and R. B. Kelly. (1983) *Ann. Rev. Biochem.* 52: 871–926.

39 de Belleroche, J. S., and H. F. Bradford. (1972) *J. Neurochem.* 19: 585–602.

40 Fried, R. C., M. P. Blaustein. (1978) *J. Cell Biol.* 78: 685–700.

41 Fried, R. C., and M. P. Blaustein. (1976) *Nature* (Lond.) 261: 255–256.

42 Edwardson, J. A., G. W. Bennett, and H. F. Bradford. (1972) *Nature* (Lond.) 250: 554–556.

43 Edwardson, J. A., and G. W. Bennett. (1974) *Nature* (Lond.) 251: 425–427.

44 Wofsey, A. R., M. J. Kuhar, and S. H. Snyder. (1971) *Proc. Natl. Acad. Sci.* (USA) 68: 1102–1106.

45 Faivre-Bauman, A., J. Rossier, and P. Benda. (1974) *Brain Res.* 76: 371–375.

46 Hokfelt, T., and A. Ljungdahl. (1972) *Exp. Brain Res.* 14: 354–362.

47 Hokfelt, T., and A. Ljungdahl. (1972) *Adv. Biochem. Psychopharmacol.* 6: 1–36.

48 Raiteri, M., R. Frederico, A. Coletti, and G. Levi. (1975) *J. Neurochem.* 24: 1243–1250.

49 Levi, G., and M. Raiteri. (1976) *Int. Rev. Neurobiol.* 19: 51–74.

50 Cox, D. W. G., and H. F. Bradford. (1978) In *Kainic Acid* (eds. E. McGeer, J. Olney, and P. McGeer), pp. 71–93. Raven Press, New York.

51 de Belleroche, J. S., and H. F. Bradford. (1976) In *Transport Phenomena in the Nervous System: Physiological and Pathological Aspects* (eds. G. Levi, L. Battistin, and A. Lajtha), pp. 395–404. Plenum Press, New York.

52 Garcia-Segura, L. M., R. Martinez-Rodriguez, and R. Martinez-Murillo. (1978) *Experimentia* 34: 1598.

53　Dvorak, D. J., E. Gipps, and C. Kidson. (1978) *Nature* (Lond.) 271: 564–566.

54　Richardson, P. J., K. Siddle, and J. P. Luzio. (1983) *Biochem. J.* 219: 647–654.

55　Docherty, M., H. F. Bradford, J.-Y. Wu, T. Joh, and D. Reis. (1985) *Brain Res.*, in press.

56　Snyder, S. H., A. B. Young, J. P. Bennett, and A. H. Mulder. (1973) *Fed. Proc.* 32: 2039–2047.

57　Wofsey, A. R., M. J. Kuhar, and S. H. Snyder. (1971) *Proc. Natl. Acad. Sci.* (USA) 68: 1102–1106.

58　Starke, K. (1981) *Ann. Rev. Pharmacol. Toxicol.* 21:' 7–30.

59　de Belleroche, J. S., and H. F. Bradford. (1978) *Brain Res.* 142: 53–68.

60　de Belleroche, J. S., and H. F. Bradford. (1979) *Neurosci. Lett.* 11: 209–213.

61　Yagihara, Y., J. E. Bleasdale, and J. N. Hawthorne. (1972) *J. Neurochem.* 19: 355–367.

62　Yagihara, Y., J. E. Bleasdale, and J. N. Hawthorne. (1973) *J. Neurochem.* 21: 173–179.

63　Schact, J., and B. W. Agranoff. (1974) *J. Biol. Chem.* 247: 771–777.

64　Schact, J., and B. W. Agranoff. (1974) *J. Biol. Chem.* 249: 1551–1557.

65　Bleasdale, J. E., and J. N. Hawthorne. (1975) *J. Neurochem.* 24: 373–379.

66　Pickard, M. R., and J. N. Hawthorne. (1978) *J. Neurochem.* 30: 145–155.

67　Hawthorne, J. N., and M. R. Pickard. (1979) *J. Neurochem.* 32: 5–14.

68　Michell, R. H., S. S. Jafferfi, and L. M. Jones. (1977) *Adv. Exp. Med. Biol.* 83: 447–464.

69　Jones, L. M., and R. H. Michell. (1978) *Biochem. Soc. Trans.* 6: 673–688.

70　Griffen, H. D., and J. N. Hawthorne. (1978) *Biochem. J.* 176: 541–552.

71　Griffen, H. D., J. N. Hawthorne, and M. Sykes. (1979) *Biochem. Pharmacol.* 28: 1143–1147.

72　Gambetti, P., L. A. Autilio-Gambetti, N. K. Gonatas, and B. Shafer. (1972) *J. Cell Biol.* 52: 526–535.

73　Wedege, E., Y. Luqmani, and H. F. Bradford. (1977) *J. Neurochem.* 29: 527–537.

74　Verity, M. A., W. J. Brown, and M. Cheung. (1980) *J. Neurosci. Res.* 5: 143–153.

75　Cotman, C. W., and D. A. Taylor. (1971) *Brain Res.* 29: 366–372.

76　Gambetti, P., L. A. Autilio-Gambetti, N. K. Gonatas, and B. Shafer. (1972) *J. Biol. Chem.* 52: 526–535.

77　Cupello, A., and J. Hyden. (1975) *J. Neurochem.* 25: 399–406.

78　Ramirez, G., I. B. Levitan, and W. E. Mushynski. (1972) *J. Biol. Chem.* 247: 5382–5390.

79　Henn, F. A., D. J. Anderson, and D. G. Rustard. (1976) *Brain Res.* 101: 341–344.

80　Henn, S. W., and F. A. Henn. (1983) In *Handbook of Neurochemistry*, 2d ed., vol. 2 (ed. A. Lajtha), pp. 147–161. Plenum Press, New York.

81　Cotman, C., H. Herschman, and D. Taylor. (1971) *J. Neurobiol.* 2: 169–180.

82　Delaunay, J.-P., F. Hog, F. V. de Feudis, and P. Mandel. (1979) *J. Neurochem.* 33: 611–612.

83　Wheeler, G. H. T., H. F. Bradford, A. N. Davison, and E. Thompson. (1979) *J. Neurochem.* 33: 331–337.

84　Redburn, D. A. (1978) *J. Neurochem.* 31: 939–946.

85　Ward, H. K., and H. F. Bradford. (1979) *J. Neurochem.* 33: 339–342.

86　Matus, A. I., D. H. Jones, and S. Mughal. (1976) *Brain Res.* 103: 171–175.

87 Wilson, W. S., R. A. Schulz, and J. R. Cooper. (1973) *J. Neurochem.* 20: 659–668.
88 Briggs, C. A., and J. R. Cooper. (1981) *J. Neurochem.* 36: 1097–1108.
89 Hajos, F., G. Willain, J. Wilson, and R. Balazs. (1975) *J. Neurochem.* 24: 1273–1278.
90 Dowdall, M. J., and V. P. Whittaker. (1973) *J. Neurochem.* 20: 921–935.
91 Dowdall, M. J., and E. J. Simon. (1973) *J. Neurochem.* 21: 969–982.
92 Jones, D. G. (1967) *J. Cell Sci.* 2: 573–586.
93 Morel, N., M. Israel, R. Manarache, and P. Mastour-Fracon. (1977) *J. Cell Biol.* 75: 43–55.
94 Morel, N., M. Israel, and R. Manarache. (1978) *J. Neurochem.* 30: 1553–1557.
95 Israel, M., and B. Lesbats. (1981) *J. Neurochem.* 37: 1475–1483.
96 Whittaker, V. P., I. A. Michaelson, and R. J. A. Kirkland. (1964) *Biochem. J.* 90: 296–303.
97 De Lorez Arnaiz, R., M. Alberici, and E. de Robertis. (1967) *J. Neurochem.* 14: 215–312.
98 Kornguth, S. E., A. Flangas, R. Siegel, R. Geison, J. O'Brien, J. Lamar, and G. Scott. (1972) *J. Biol. Chem.* 246: 1177–1184.
99 Cotman, C. W., and D. Taylor. (1972) *J. Cell Biol.* 55: 696–711.
100 Davis, G., and F. E. Bloom. (1973) *Brain Res.* 62: 135–153.
101 Matus, A. I., B. B. Walters, and D. H. Jones. (1975) *J. Neurocytol.* 4: 357–367.
102 Cohen, R., F. Blomberg, K. Berzins, and P. Siekevitz. (1977) *J. Cell Biol.* 74: 181–203.
103 Matus, A. I., and D. H. Taff-Jones. (1978) *Proc. R. Soc.* (Lond.) *Ser. B.* 203: 135–151.
104 Matus, A. I. (1978) *Methods Membrane Biol.* 9: 203–236.
105 Westrum, L. E., and E. G. Gray. (1977) *J. Neurocytol.* 6: 505–518.
106 Gray, E. G., P. R. Gordon-Weekes, and R. D. Burgoyne. (1982) In *Neurotransmitter Interaction and Compartmentation* (ed. H. F. Bradford) NATO Advanced Study Series, vol. A48, pp. 1–13. Plenum Press, New York.
107 Arregui, A., W. J. Logan, J. P. Bennett, and S. H. Snyder. (1972) *Proc. Natl. Acad. Sci.* 69: 3485–3489.
108 Balazs, R., F. Hajos, A. L. Johnson, R. Tapia, and G. Wilkin. (1974) *Brain Res.* 70: 285–299.
109 Whittaker, V. P., M. J. Dowdall, and A. F. Boyne. (1972) In *Neurotransmitters and Metabolic Regulation* (ed. R. M. S. Smellie), Biochemical Society Symposium 36, The Biochemical Society, London.
110 Whittaker, V. P., and M. N. Sheridan. (1969) *J. Neurochem.* 12: 363–372.
111 Peters, A., and I. A. Kaiserman-Abramof. (1969) *Z. Zellforsch. Mikrosk. Anat.* 100: 487–506.
112 Aghajanian, G. K., and F. E. Bloom. (1967) *Brain Res.* 6: 716–723.
113 Blomberg, F., R. S. Cohen, and P. Siekevitz. (1977) *J. Cell Biol.* 74: 204–255.
114 Cotman, C. W., and P. T. Kelly. (1980) *Cell Surf. Rev.* 6: 505–533.
115 Kelly, P. T., and C. W. Cotman. (1978) *J. Cell Biol.* 79: 173–183.
116 Kelly, P. T., and C. W. Cotman. (1977) *J. Biol. Chem.* 252: 786–793.
117 Matus, A. I. (1981) *Trends Neurosci.* 4: 51–53.
118 Cotman, C. W., E. E. Mena, and M. N. Sampredo. (1982) In *Neurotransmitter Interaction and Compartmentation* (ed. H. F. Bradford) NATO Advanced Study Series, vol. A48, pp. 17–38. Plenum Press, New York.

119 Michell, R. H., and C. J. Kirk. (1981) *Trends Pharmacol. Sci.* 2: 86–89.

120 Csillag, A., and F. Hajos. (1980) *J. Neurochem.* 34: 495–503.

121 Hajos, F. (1975) *Brain Res.* 93: 485–489.

122 Umbach, J. A., C. B. Gundersen, and P. F. Baker. (1984) *Nature* (Lond.) 311: 474–477.

123 Carvalho, A. P. (1983) In *Handbook of Neurochemistry* (ed. A. Lajtha), vol. 1, pp. 69–116. Plenum Press, New York.

124 Schellenberg, G. D., L. Anderson, and P. D. Swanson. (1983) *Mol. Pharmacol.* 24: 251–258.

125 Blaustein, M. P., R. W. Ratzlaff, and E. S. Schweitzer. (1980) *Fed. Proc.* 39: 2790–2785.

126 Berridge, M. J., and R. F. Irvine. (1984) *Nature* (Lond.) 312: 315–321.

127 Whittaker, V. P. (1984) *Biochem. Soc. Trans.* 12: 561–576.

128 Sihra, T. S., I. G. Scott, and D. G. Nicholls. (1984) *J. Neurochem.* 43: 1624–1630.

7

Synapses:
Their Development
and Dynamics

The neurons of the adult mammalian nervous system make intercon-
necting patterns that are a marvel of complexity and sophistication. Typical pyramidal neurons of the cortex may have up to 100,000 synaptic
connections with other cells, and these connections are not random—
they are specifically placed on the cell body or on the dendrites. This is
but a hint of the precision involved in establishing the wiring of the
nervous system. We are far from understanding how this elaborate network of connections is achieved.

NERVE FIBER GROWTH
AND SYNAPSE FORMATION

We now know a great deal about the cellular migrations that lay down
the basic pattern of the nervous system. We know much less about the
events that guide the axons to their target sites and the cells to their
"final addresses." The possible role of radial glial cells in forming a radially organized scaffolding—often right across the thickness of the neural
tube—and guiding young neurons to their final positions[131] (Figure 7.1)
was referred to in Chapter 2. How neurons such as the pyramidal cells of
the cerebral cortex become perfectly aligned, with their axons directed
towards the underlying white matter and their prominent apical dendrites directed towards the cortical surface, is still a mystery. Biochem-

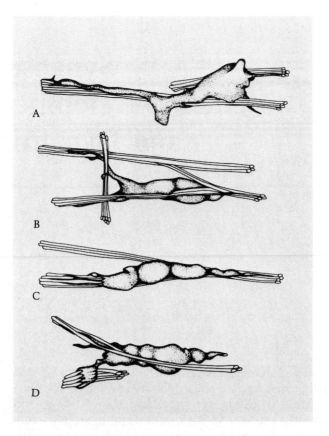

FIGURE 7.1   Reconstructions from serial electron micrographs of migrating neurons in monkey cortex, showing the complex patterns of leading processes associated with radial glia. Neurons are migrating from right to left in each case. [From ref. 130]

ical influences are the most likely determinants. Different classes of cell surface molecules specifically concerned with cell–cell orientation might be responsible, or surface macromolecules initially responsible for the cells aggregation might undergo a selective redistribution.[B,C]

Pathfinding by Neuronal Fibers During Development

After reaching their final address, neurons begin to generate dendritic and axonal processes to enable them to make and receive contacts from other cells. They usually generate far more contacts than survive to form part of the mature and differentiated adult neuron. Axonal processes are directionally guided by both mechanical and chemical factors.[A,168] Observations of growing axons in vivo agree in most respects with observations gleaned from studies of the growth of processes in tissue culture.

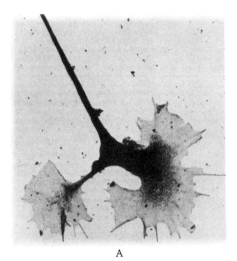
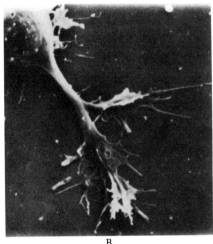

A   B

FIGURE 7.2   Growth cones of neurons in tissue culture. (A) Transmission electron micrograph shows a pair of growth cones at the end of an axon of a sympathetic ganglion cell. The fine, fingerlike extensions are filopodia; the flattened veillike sheets between them are lamellipodia. (B) Scanning electron micrograph shows a growing dendrite of a neuron from hippocampus. [From ref. 132; Courtesy of J. M. Cochran, M. B. Bunge, and S. R. Rothman]

In tissue culture (e.g., of the dorsal root ganglion), the axon emerges from the cell body of the neuron bearing a specialized varicose tip or growth cone, from which undulating fingerlike or sheetlike membranes, called filopodia, extend in all directions. Time-lapse cinematography shows the constant exploratory and contractile movements of these membranes in the immediate environment of the growth cone[1,2,154] (see Figure 7.2). These structures guide the direction of the growing axonal tip and determine where it will make its connections transitory or permanent. (Growth cones exist in vivo but are not so readily visible.[1]) The growing axons contain the usual subcellular organelles such as neurotubules, neurofilaments, mitochondria, coated and smooth vesicles, and endoplasmic reticulum, together with some ribosome clusters.[1] Microtubules appear to be necessary to form the structural scaffolding that maintains the stability of elongated fibers. Treatment with colchicine, which depolymerizes neurotubules (see Chapter 1), causes complete retraction of processes in spinal ganglion cells in culture.[2,154] It is likely that neurotubules also play a role in axoplasmic transport (see Chapter 1), and therefore their disruption could also prevent transport from the cell body of materials required for assembly of membranes and other structures at the growth cone where axonal growth occurs. Actin-like proteins of the type that constitutes microfilaments probably undergo contraction in a fashion analogous to skeletal muscle, since myosin is also present.[2] Such contractile activity of microfilaments could well underlie the movements of the filopodia of the growth cone and of the axon itself. For instance, cytochalasin-B, a drug that prevents the assembly of microfilaments, reversibly inhibits the movement of growth cones and axons of dorsal root ganglion cells.[2,154] (See Chapter 1, pages 16, 20.)

Current evidence strongly favors the view that individual axons are "guided" in their directional growth to their points of synapse formation. The alternative concept is that many outgrowing axons project towards the target but only one is successfully "recognized" and stabilized after synapse formation.[3,8,9,45,168] In the latter theory, the supernumerary outgrowths that did not stabilize would be retracted or would degenerate.[A,3,9] Additional evidence for "guidance" of growing axons to their targets comes from experiments in which axons are displaced some distance from their normal course and are able to adjust their route to achieve their target.[A,B,8,45] Eyes transplanted to the middle of the forehead in toad embryos can still grow to their target and make functionally successful connections to the optic tectum in the brain.[4] And motoneurons in embryonic chick spinal cord will grow to the correct muscles even when segments of the cord are reversed.[8,45,168,184]

Exactly what the guiding influences are is still in the realm of speculation.[142,143] Apart from mechanical restriction of direction by glial cells, specific components on their cell surfaces might provide environmental clues. Intrinsic differences between axons must enable them to orient differently in response to the same set of cues. Once a fiber reaches its destination, other fibers growing in the same tract would be able to follow the leader by contact guidance. Secondary fibers have followed in close surface contact after successful connection to a single "leader" fiber, for instance, in the growth of axons from the individual visual detection units (ommatidia) of the compound eye of *Daphnia* to the eye's optic cartridges.[5] And where a leading axon has been experimentally diverted from its path, other axons have followed the incorrect path; an example is the regeneration of fibers from the sensory bristles of the assassin bug *Rhodnius* to their connections in the central nervous system. Here, the fibers appear to grow in random directions until they encounter other fibers that have already established their central nervous system connections. At this point, the regenerating fibers grow along the surfaces of the connected fibers and reestablish their own synaptic contact.[6]

Chemical Influences on Nerve Fiber Growth

The second major guiding influence on nerve fibers is likely to involve chemospecificity of another kind. The role of diffusible chemicals in guiding axonal growth has not been established, but it is clearly an attractive possibility. The chemospecificity theory of Sperry proposes that each nerve cell acquires a particular genetically determined chemical identification due to its composition. Specific macromolecules present on its surface, or released to its immediate environment by diffusion, would enable the cell to be identified by an exploring axon and either accepted or rejected, thereby indicating whether or not it was following the correct pathway. In this way an axon would seek out a three-dimensional route characterized by a sequence of neurons with precise chemical "flavors" determined originally by specific gene expression. A few biochemical substances that are known to influence the direction of growth of fibers, or cell recognition and aggregation, have been identified and characterized. For instance, after disaggregation of embryonic brain tissue by mild mechanical or chemical treatment to form a mixture of cell types from different regions, the cells will reaggregate preferentially with other cells from the same brain region.[11,12] They show a selective adhesiveness, probably because of the appearance on their surfaces of macromolecules that serve to both "recognize" cells of the same kind and bind them together.[10] These macromolecules appear to be specific to each cell type and to change in number and distribution as development proceeds.[148,149,151,169] In addition, factors promoting this specific aggregation are released into the medium of retinal or cerebral tissue cultures. These factors have been purified.[11,12]

As well as showing a selective aggregation per se, randomly mixed neural cells in culture will migrate to be close to cells that were their neighbors in the original tissue,[13,14] reflecting the presence of organizing factors that are no doubt key agents in embryonic development. One such substance, purified from the surface of neural retinal cells of the chick embryo, was found to be a glycoprotein of molecular weight 50,000. It induces cell aggregation in a tissue-specific fashion, and there is good evidence that it may be a component of the retinal cell surface mechanisms involved in morphogenetic cell recognition and cell association into organized tissues. The name *cognin* was coined for cell-surface molecules having this function. Cognins and related proteins have been detected in a range of embryological tissues.[144,145] Immunolabeling of purified chick retinal cognin in conjunction with scanning electron microscopy has shown the localization of this substance on the surface of embryonic retinal cells,[11] but only during the period of retinal histogenesis. Complement-mediated chick retinal cell immunolysis confirmed the surface location of retinal cognin.[146] Although nonneural embryonic tissue did not interact with this antiserum, tests with other neural tissues—such as the optic tectum—showed the presence of cognin, or a closely related molecule, on the cell surface of a high proportion of cells.[11]

Another group of cell-surface glycoproteins isolated from chick neural retina or liver includes the *neural cell adhesion molecule* (N-CAM)[145] and the *liver cell adhesion molecule* (L-CAM,[169] originally isolated from liver). N-CAMs have a basic molecular weight of 160,000, and the additional presence of carbohydrate moieties—large amounts of sialic acid, linked as polysialic acid—gives a wide range of higher molecular weights.

The N-CAM molecule appears to function as a ligand in cell-to-cell recognition and adhesion.[145,146,147,150,169] The amino-terminal domain of the N-CAM on one cell is involved in binding to a similar domain of N-CAM on an apposing cell (homophilic binding), and the process is Ca^{2+}-independent. The polysialic acid, situated in the middle domain of the molecule, does not appear to participate directly in the binding process, but changes in its structure do have a dramatic effect on its binding properties—probably because of changes in the total negative charge.[173] During development, the sialic acid content of N-CAM reduces to one-third at a time when various subclasses of N-CAM are appearing. The carboxyl-terminal of the molecule embeds the N-CAM in the outer surface of the cell membrane, though it remains freely diffusible in the plane of the membrane.[169] The L-CAM glycoprotein from liver, with a molecular weight of 124,000, does not possess sialic acid residues, and the cell-to-cell adhesion it mediates is dependent on Ca^{2+}. In the absence of Ca^{2+}, L-CAM undergoes rapid proteolytic destruction.

The two CAMs undergo substantial changes in quantity at the cell surfaces of organ rudiments during development and cell migration. A strong steering influence might be provided in early morphogenesis by the differential expression of several CAM genes and by alterations in the transport of the different CAMs to the cell surface.[169] Various subclasses of N-CAM appear during development. They differ in their sialic acid content and some possess greater capacities for interaction. In chick embryos, at the stage just before the differentiation of glia, a secondary category of CAM appears on neuronal surfaces. Designated Ng-CAM, it is thought to mediate neuron–glia adhesion by Ca^{2+}-independent binding.[174] Ng-CAM is not found on glial cell surfaces, and another, unidentified, CAM is presumed to appear on glial cells and participate in a heterophilic interaction. Such a G-CAM could participate in the guidance of nerve processes by glia during early development. Glial cell neuronal guidance has been observed to occur in cerebral and cerebellar cortex and in other systems (see below). It is possible that families of a dozen or so structurally related CAMs exist, but it seems unlikely that we will find the hundreds or even thousands of CAMs that would be required to code for unique neural addresses of developing neural processes as originally conceived in the chemoaffinity hypothesis of Sperry. Instead, CAMs might provide for control of the patterns of neural development by means of genetic control of dynamic changes in the populations of relatively few CAM-like surface determinants, and by control of the agents that interact with them.[169]

Nerve growth factor (NGF) is another substance with intriguing actions on neuronal growth and development.[15,16,17] A protein existing as a zinc-containing complex (mol wt about 140,000), its biological activity resides in its beta subunit, which consists of two identical 118-residue chains each of mol wt 13,259. There is also a regulatory alpha subunit and a gamma subunit, the latter a potent peptidase.

In the form of a large molecular weight complex, the NGF is protected from proteolysis. It seems that NGF is synthesized as a precursor peptide (mol wt 22,000), two molecules of which are cleaved by the peptidase of the gamma subunit to form the active β-NGF dimer.[15,16,17] This last substance has profound and potent neuronotrophic actions on catecholamine-containing neurons. Thus, NGF (1) increases the number of neuroblasts if applied at an early stage; (2) increases neuronal size and axonal growth of peripheral sympathetic and sensory ganglia both in vivo and in vitro; and (3) increases neuronal size and neurotransmitter production in these ganglia when applied after synapses have been formed and process elongation has ceased (Figure 7.3). When antibodies to NGF are injected into newborn mice or rats, the sympathetic ganglia become shriveled and degenerated, implying an involvement of NGF in the events ordering normal development (Figure 7.4). Adrenal chromaffin cells show degeneration if NGF antibody is administered before birth. When NGF antiserum is given to adult rats or mice, only a transitory decrease in catecholamine biosynthetic enzymes is seen, which further reveals the importance of NGF in the developing nervous system rather than in the adult system. In culture, dissociated embryonic sympathetic and sensory neurons deteriorate and die unless the medium contains NGF.[18] One action that does seem to be independent of age is the NGF-mediated selective induction of tyrosine hydroxylase and of dopamine β-hydroxylase in sympathetic neurons and adrenal medullary cells. This inductive effect occurs at the posttranscriptional level involving RNA synthesis[19] and is probably unrelated to its growth-promoting properties.

There is also evidence of an action of NGF on cholinergic cells, at least in sympathetic ganglion cells in culture. For instance, acetylcholine content per neuron increases as the NGF concentration is raised, and this occurs in parallel with a rise in norepinephrine levels. In fact, NGF stimulates production of both transmitters to the same extent.[166]

Here, then, is a protein with a profound influence on neuronal growth and development, especially of the adrenergic system. It also has clearcut chemotactic influences on innervation and reinnervation patterns in vivo and in vitro. For instance, if mouse connective-tissue tumor (sarcoma), which is rich in NGF, is grafted onto chick chorioallantoic membrane, it stimulates the appearance of multiple outgrowths of the chick sensory and sympathetic ganglia, which invade the tumor tissue. Such outgrowths will also invade spinal cord and follow the trail of NGF injected into the cord.[20,21] Similar tropic, chemotactic, or "direction-determining" properties can be demonstrated in vitro for fibers growing from sympathetic ganglia[22] (see Figure 7.3).

Nerve growth factor, apparently so crucially involved in the growth and formation of synaptic connections for sympathetic neurons, is probably only one of a series of such compounds; others must exert similar influences over the growth of neuronal cell types other than the adrenergic variety.[155] This series of compounds, together with the mechanical

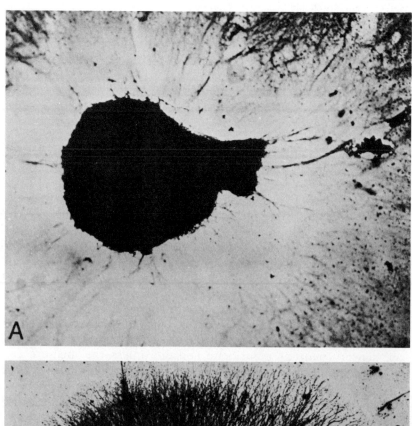

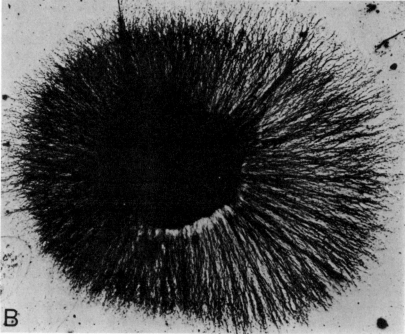

FIGURE 7.3 Effect of nerve growth factor on neurite growth. Photomicrographs of sensory ganglia of a 7-day chick embryo cultured for 24 hours in (A) a control medium, and (B) in the same medium supplemented with nerve growth factor (0.01 μg/ml). [From ref. 152]

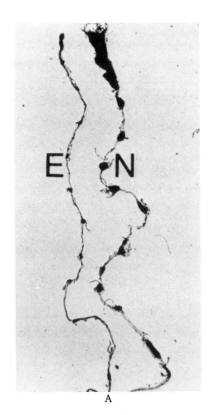
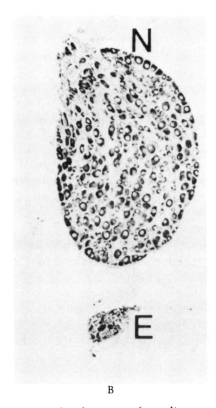

A   B

FIGURE 7.4   Effect of nerve growth factor on development of ganglia.
(A) Normal sympathetic ganglion chain (N) of 1-month-old mouse compared with another chain (E) treated with antiserum to nerve growth
factor for 5 days following birth. (B) Transverse section through the superior cervical ganglion of a 9-day-old mouse (N). The same ganglion (E)
after treatment with antiserum to nerve growth factor for 3 days following birth. [From refs. 152, 153]

cues and other surface biochemical cues described above, are probably
responsible for bringing axons to their correct site for synapse formation
in the developing nervous system.

Synapse Formation

Once growing axons are close to their target cells, they can make their
synaptic connections. The process can be viewed as separate from path-
finding by the axons, and it is under the influence of different trigger and
control mechanisms.[A,B,142,143] At this stage, axonal elongation must cease

and synapse formation must begin. The signal to stop does not seem to be programmed into the cell in terms of distance covered; when the target moves farther away, elongation resumes. For instance, if one superior colliculus is removed in newborn hamsters, optic fibers, which normally synapse at this point, grow across the lesioned site and make contact with the superior colliculus on the opposite side.[23]

How the specificity of synaptic connections is determined is uncertain.[141,142] Possibly, an initial excessive formation of connections is followed by a degeneration of all but the "correct" ones (the cell death hypothesis). Evidence also favors the idea that fibers grow directly to the correct site and that the mature synapse forms at the site after an initial successful contact. Of course, the axonal tips (growth cones) must "sample" and reject many cell surfaces before concentrating on the specified site, and in some cases synapses do form at "temporary" sites. Examples of such unstable transitory synapses come from studies of developing cerebellum and from the formation of retinotectal synapses in the brains of developing toads of the genus *Xenopus*. In the developing cerebellum, transitory synapses are formed between climbing fibers and Purkinje cell bodies; in the mature animal, climbing fibers are found only on the proximal regions of the Purkinje cell dendrites.[24] However, the well-established and widespread phenomenon of extensive neuronal cell death in ganglia and other central nervous system regions during the period of synaptic formation probably also plays its part in eliminating fibers making inappropriate contacts and in matching the numbers of pre- and postsynaptic cell populations.[168]

It is likely that several additional factors act in unison to direct a given axon to a given synaptic site. Apart from chemical specificity between axon and target, the *timing* of differentiation is important. In terms of priority, the primary target must be present or synapses will be formed at secondary sites, since the central drive of neurons is to form connections. In cerebellar development, for example, granule cells normally migrate downward below the Purkinje cell layer to meet ascending mossy fibers and form synapses. If the granule cells are destroyed by virus infection, or their downward migration is delayed by X-irradiation, the ascending mossy fibers are unable to form their normal synapses with these cells. Instead, they grow through the Purkinje cell layer (see Figure 2.7), a region they do not normally enter. Here they may form synapses with Purkinje cells, basket cells, and other local cells[A,25,26]—a "second choice" for the elongating axons seeking to make synaptic connections.

The cardinal question is how a growing axon recognizes the site at which it will form a synapse. It has been proposed that macromolecules on the axonal surface provide recognition sites complementary to those on the postsynaptic membrane.

The nature, or identity, of any surface macromolecules that would allow such specificity in synapse formation remains unknown, but surface proteins like the cognins and N-CAMs described above for retinal cells are possible candidates.[148,149] One category with the required properties could be a target cell with surface glycoprotein that could interact with specific lectinlike substances on the surface of the target-seeking axon and at least allow the initial adhesion to occur. In support of this concept is the finding of concanavalin A (lectin) binding sites on the postsynaptic membrane.[A,29] Some researchers have suggested that an enzyme present on one membrane surface (e.g., ganglioside GM_1 synthetase) might couple with its substrate on the other membrane surface (e.g., ganglioside GM_2). Indirect evidence exists for just such a system in the developing optic tectum.[142]

Once contact is secured, transsynaptic influences come into play to induce the formation of the postsynaptic density in the postsynaptic cell as well as the associated pre- and postsynaptic structures and cleft material that together form the elaborate structure of the mature synapse (see Figure 6.16). This stage is followed by the first appearance, or the enhanced synthesis, of specific enzymes for the biosynthesis, degradation, and transport of the appropriate neurotransmitter system on the presynaptic side and the positioning of specific neurotransmitter receptor proteins on the postsynaptic cell surface. Evidence from the formation of the neuromuscular junction indicates that these receptors are initially present but are dispersed over the entire cell surface, becoming localized into patches during synapse formation. It is very likely that synaptic activity in the form of an interaction between these neurotransmitter receptors and the released neurotransmitter is necessary to "stabilize" the synapse, or make it semipermanent. Such "functional stabilization" has been described for neuromuscular junctions as well as for synaptic connections in the optic tectum.[27,C]

The factors determining *which* neurotransmitter the presynaptic terminal will develop are probably both genetic and environmental. Studies with developing sympathetic neurons in tissue culture have revealed that substances produced by nonneural cells can directly influence whether a neuron will develop as an adrenergic or cholinergic cell. One such substance, apparently a glycoprotein of mol wt 50,000, induces the production of choline acetylase activity and can increase the ratio of acetylcholine production to catecholamine production a thousandfold without affecting neuronal growth or survival.[166] Equivalent factors are likely to be controlling the development of other neurotransmitter systems. As synapses become active, they too can condition the way in which the neurotransmitter system will develop in the postsynaptic cell. Evidence for this comes from observations of the ability of sympathetic ganglion neurons in culture to change from being adrenergic to cholinergic under the influence of conditioning factors. Stimulation or depolarization of these cells makes them refractory to such influence, probably because of the Ca^{2+} influx during such treatment.[166]

Regeneration of Synapses in Adult Brains

It was long believed that damage to the adult brain in which neuronal processes are severed leads to irreversible denervation. The past decade has shown that neurons are able to "sprout" new axonal branches from existing fibers and form new synapses to replace those that have been lost.[D,29,143] The usual sequence of events involves two main steps: (1) The degenerating terminal is ingested by local astrocytic glial cells that move in to occupy the space vacated by the terminal. (2) A newly sprouted axon from an adjacent axon makes its way, possibly guided by the glial cells, to the vacated site where the postsynaptic density remains in position. It has been suggested that the ingested terminals provide the local glial astrocyte with a chemical means of promoting the sprouting and growth of the new terminals[30,143] (Figure 7.5). In some cases the nature of the neurotransmitter in the regrowing axon may determine its target site.[28]

Sprouting of this kind, with regrowth to vacated postsynaptic sites, has been demonstrated in various regions of the central nervous system, including the hippocampus,[167,31,D] olfactory bulb,[32] septal nuclei,[30,33] and the optic tract projection to the superior colliculus.[34,C] The severed ends of axons can regrow to occupy these vacated synaptic sites or even invade other related regions.[28,33,34,167,C,D]

The *functional* value of these new contacts remains to be established. Obviously, where synapses using one neurotransmitter grow to a site previously occupied by the terminal for another neurotransmitter system, anomalous effects will result.[28] Equally, reinnervation from an inappropriate pathway with the "right" neurotransmitter would produce similar anomalous results. Thus, sprouting and fiber regeneration in the central nervous system may not lead to functional recovery as it often does in the peripheral nervous system, though such recovery has been demonstrated in some cases.[35,36,D]

Brain Tissue Grafts

Recent times have seen a surge of interest in the survival of tissue brain grafts or implants.[28,30,176–179] Provided embryonic or neonatal tissue is used, and certain other conditions are met, considerable success is achieved.[176] It is important, for instance, to select tissue of the appropriate developmental age and to ensure that the grafted tissue is placed in a position that allows its rapid revascularization and connection to the blood and cerebrospinal fluid circulation of the host. The grafts usually establish extensive networks of reciprocal axonal connections with the host brain. The newly formed connections may be transmitter-specific, reflecting the pattern of innervation of the intact brain. Such connections are enhanced by removing the input of those host neurons that form the target sites for the ingrowing axons (i.e., deafferentiation).[176] These observations indicate that in certain circumstances grafts may be able to rees-

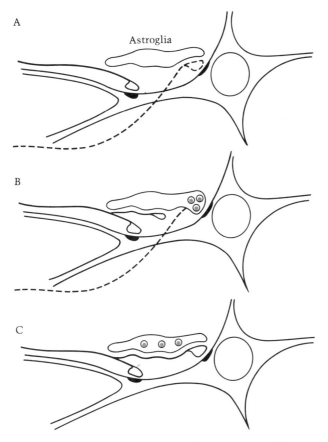

FIGURE 7.5   Diagram showing degeneration and regeneration (sprouting) of nerve terminals. (A) Incoming nerve fiber degenerates. (B) Degenerating nerve terminal is ingested by a local glial cell. (C) Chemical signal due to ingested material causes another nerve fiber to sprout and occupy the vacant synaptic site.

tablish neuronal circuitries damaged by various forms of lesion, permitting the return of disrupted physiological and behavioral functions.[176]

For example, grafts of rat embryonic ventral mesencephalon, which contains the dopaminergic cells of the substantia nigra, have been placed in the cerebral cortex of adult rats previously treated unilaterally with 6-hydroxydopamine to destroy the dopaminergic innervation of the striatum in the injected hemisphere.[177] The graft was placed in direct contact with the dorsal or lateral surface of the neostriatum. Over a period of 2 to 6 months, bundles of axons entered the host neostriatum to form a dense plexus extending to one-third of the caudate-putamen. Dopamine levels and turnover rates were restored to normal in the reinnervated regions.[178] There was an associated functional recovery in the animals, which correlated with the extent of striatal reinnervation. The spontaneous or amphetamine-induced ipsilateral rotation, and the contralateral apomorphine-induced rotations, were absent or much reduced (see Chapter 8). Thus, the grafts partially or completely compensated for the dopaminergic locomotor functions lost from one side of the brain. Similar results were found when suspensions of disaggregated embryonic nigral tissue were injected into the neostriatum, but in this case the responses developed more rapidly, within 3 to 6 weeks.[179,180]

A similar pattern of functional recovery followed implantation of a suspension of disaggregated embryonic striatal tissue into the caudate-putamen of rats previously receiving striatal injections of ibotenic acid, which (like kainic acid) destroys the intrinsic neurons of the striatum.[181] These grafts almost completely prevented the nocturnal locomotor hyperactivity that followed ibotenic acid injection into striatum. At the same time, the grafts normalized the reduction in glucose utilization that occurs in various extrapyramidal brain regions in treated animals. This metabolic recovery was judged by studies of the rates of $[^{14}C]$-2-deoxy-D-glucose utilization (see Chapter 3).

These various nigral or striatal tissue brain grafts exert their effects both via the observed nerve fiber connections which they establish with host neurons and via a neurohumeral action due to a diffuse release of the active compounds (neurotransmitters, neuromodulators) into the host tissue, or into the cerebrospinal fluid.

These findings are potentially important for the treatment of neurological and psychiatric disorders.[176] There is already considerable interest in their relevance for the treatment of Parkinson's disease (grafts of nigral or other dopamine-producing tissue) and Huntington's chorea (grafts of striatal tissue) (see Chapter 8).

Maintenance of Synaptic Regions

Retrograde axoplasmic transport of growth-regulating substances from axonal tips to cell bodies is also likely to serve a critical role in both establishing and maintaining synaptic connections. The presence of retrograde transport is readily demonstrated by the movement of horseradish peroxidase from the axonal tips to neuronal cell bodies, which provides a technique for allowing pathways to be traced between one brain region and another. The same route is used by growth-regulating factors, which are able to provide information to cell bodies on the state of organization and preservation of mature nerve terminals. Nerve growth factor, for instance, carries such information from peripheral target organs to adrenergic and sensory ganglion cells.[15] The uptake of NGF at nerve terminals appears to involve an interaction with specific receptor sites before entry and transport. Horseradish peroxidase and many other substances are taken up nonspecifically by pinocytosis (endocytosis), though high concentrations may have to be used to provoke this uptake.

Indirect evidence for this messenger role of NGF comes from observations following disruption of the connection between the neuronal cell body and the target organ. The connection is severed by surgical or chemical methods, for example, by using 6-hydroxydopamine. Or it may be interrupted by blocking axoplasmic transport, for example, by applying the drugs vinblastine or colchicine. Blockade causes a degeneration of the cell bodies in the newborn and temporary functional impairment in adults. Reduced function may be detected as lowered levels of tyrosine hydroxylase, or dopamine β-hydroxylase, or by aberrant ganglionic transmission. All of these effects can be prevented by local or systemic application of NGF,[15] suggesting that NGF or a similar agent is responsible for maintaining the functional integrity in the intact system.

SYNAPTIC EVENTS AND NEUROTRANSMITTER RELEASE MECHANISMS

Calcium Entry and Neurotransmitter Release

Action potential invasion of the nerve terminal involves a wave of depolarization of the terminal membrane. The principal neurochemical event initiated by this invasion is the influx of Ca^{2+} into the terminal cytoplasm. It is this Ca^{2+} signal that triggers a pulse of neurotransmitter release. For example, injection of Ca^{2+} into the giant nerve terminal of the squid in the absence of depolarization leads to neurotransmitter release.[44] The bioluminescent protein aequorin reacts with small quantities of calcium $(10^{-8}\ M)$ to produce light, and when injected into the presynapse it allows the internal Ca^{2+} concentration to be monitored. Depolarization of nerve terminals containing aequorin, such as at the squid giant synapse,[46,47] leads to neurotransmitter release accompanied by a brief increase in luminescence, indicating an increase in free Ca^{2+} levels.[46,47]

When Na^+ and K^+ fluxes across the membrane are blocked (e.g., with tetrodotoxin and tetraethylammonium chloride respectively), stepwise depolarization of the terminal membrane by applied electrical currents leads to correlated increases in both neurotransmitter release and bioluminescence of the aequorin. This continues until the equilibrium potential for Ca^{2+} is reached, at which point no further Ca^{2+} influx can occur (this is the "suppression potential"). At this point both neurotransmitter release and light emission cease simultaneously.[46,47] Thus, the crucial part played by Ca^{2+} in the depolarization-coupled release of neurotransmitter is revealed.

The time-course of events between the onset of nerve terminal depolarization and the activation of the postsynaptic membrane can give information on the nature of the intervening events. First is the period

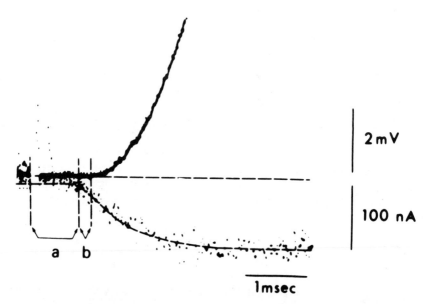

FIGURE 7.6   Calcium influx and neurotransmitter release in presynaptic terminal of squid stellate ganglion. Shown are details of the inward current (due to calcium entry) and the postsynaptic response. The upper curve is the postsynaptic response, the lower curve indicates the time course of the inward Ca^{2+} current. The first broken vertical line marks the onset of the depolarizing potential; the second, the onset of the inward current; the third, the onset of the postsynaptic response. Segment *a* represents the time to the initiation of the calcium conductance change (Ca^{2+} entry). Segment *b* represents the time between the onset of Ca^{2+} entry and the onset of postsynaptic response (first sign of neurotransmitter release). The scale shows *b* to be about 200 microseconds. The total synaptic delay is composed of segments *a* and *b*. [From ref. 156]

between application of the depolarizing potential and the first appearance of the inward Ca^{2+} current (component *a*). This is usually the lengthiest component, for example, lasting a millisecond in the squid giant synapse.[37] The period between the first influx of Ca^{2+} and the onsent of the postsynaptic potential (component *b*) is a measure of the period required for Ca^{2+} to initiate neurotransmitter release[156] (Figure 7.6). In the squid giant synapse, this has been found to be a very short time, 200 microseconds,[37] and it is likely to be much shorter in the mammalian neuromuscular junction at 37°C, for example, 50 microseconds in the rat diaphragm.[54] Since only a small fraction of component *b* is required for diffusion of the neurotransmitter molecules across the synaptic cleft (e.g., less than 1 microsecond for a gap width of 20 nm and a diffusion coefficient of 10^{-5} cm^2 per second),[37] most of the 200 microseconds must be necessary for the interaction of the influxing Ca^{2+} with the neurotransmitter releasing system and its subsequent mobilization. If vesicular stores are the source of the neurotransmitter released into the cleft, then the appropriate vesicles must be "in position" at the presynaptic membrane ready to discharge their content the moment Ca^{2+} activation of the system occurs. Movement of vesicles from deeper regions of the viscous terminal cytoplasm up to the presynaptic membrane would take too long and must be discounted. Equally, the channels for Ca^{2+} entry would be best positioned close to the membrane because of the short time available for its diffusion and interaction. The high selectivity of Ca^{2+} in promoting neurotransmitter release strongly suggests that specific binding of Ca^{2+}, probably to a protein, is involved in the process.

So far little progress has been made in describing in molecular detail the events leading to neurotransmitter release after the voltage-dependent increase in Ca^{2+} conductance[44] and Ca^{2+} entry. It is now widely accepted that the actions of intracellular Ca^{2+} are mediated by a four-domain Ca^{2+}-binding protein called *calmodulin*.[38,42,182] This protein is present in all nucleated cells and seems to regulate many cellular activities by acting as an intracellular intermediary for Ca^{2+}.[39,40] It binds one calcium ion to each domain as the latter's intracellular concentration increases in response to a stimulus, and it undergoes conformational changes as a result. Brain tissue is especially rich in calmodulin (24 μmol/ kg brain),[182] but only a fraction of this is involved in Ca^{2+}-mediated enzyme activation. The rest are proteins that undergo Ca^{2+}-dependent associations with calmodulin (calmodulin-binding proteins), and they also bind Ca^{2+} itself (e.g., calcineurin, calspectin). Their precise functions are not yet established but they include inhibition of enzymes activated by Ca^{2+} and calmodulin. Calcineurin inhibits cyclic nucleotide phosphodiesterase.[181] These proteins may also be involved in controlling free calcium levels in cells. Calmodulin from bovine brain has a molecular weight of 16,680 and binds four calcium ions, one to each of its four domains. Calcineurin, and other calcium-binding or calmodulin-binding proteins, have larger molecular weights (e.g., calcineurin, mol wt 70,000). The calcium–calmodulin complex is capable of binding to and activating several enzymes, including 3',5'-nucleotide phosphodiesterase (which breaks down cyclic AMP and cyclic GMP) and adenylate cyclase—suggesting a central involvement of calmodulin in regulating cellular levels of cyclic AMP[40] (Figure 7.7). Calmodulin also appears to mediate the action of Ca^{2+} in promoting actin and myosin interaction to produce contraction in smooth muscle. This again involves the activation of a phosphorylating enzyme (myosin light-chain kinase) by the calcium–calmodulin complex. Interestingly, calmodulin itself does not appear to serve this purpose in skeletal muscle. But troponin C, the Ca^{2+} receptor protein of this tissue, is structurally very similar to calmodulin. In addition, phosphorylase kinase of skeletal muscle, which is activated by Ca^{2+}, contains four polypeptide chains, one of which is calmodulin.[39,40] Phosphorylase kinase converts the enzyme phosphorylase into the active form in which it hydrolyzes glycogen to glucose.

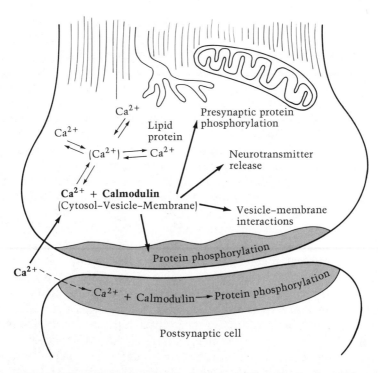

FIGURE 7.7   Schematic model of the possible role of calmodulin in mediating several of the effects of Ca^{2+} on synaptic function. Calmodulin
bound to synaptic vesicles serves as a Ca^{2+} receptor in close proximity to
the synaptic junction. Calmodulin may also be present in the synaptosome cytosol and synaptic membrane. Following the depolarization-
dependent entry of Ca^{2+} into the nerve terminal, the Ca^{2+} is immediately
bound to calmodulin near the membrane. The binding of Ca^{2+} to calmodulin in the presynaptic terminal could then initiate several processes:
presynaptic protein phosphorylation, neurotransmitter release, vesicle–
membrane interactions, and other Ca^{2+}-regulated synaptic functions.
[From ref. 38]

In nerve endings, a close correlation has been reported[38] between in
vitro transmitter release from isolated synaptic vesicles and vesicle phosphorylation. This phosphorylation is promoted by Ca^{2+} together with a
vesicle protein closely resembling calmodulin.

Thus, a calcium–calmodulin complex may be promoting phosphorylation of specific protein components of those synaptic vesicles that
take an active part in the neurotransmitter release process. In addition,
calmodulin promotes phosphorylation of the protein β-tubulin in synaptosomes,[41] and β-tubulin is the protein that polymerizes to form neurotubules, the cytoskeletal organelles (see Chapter 1) whose presence in nerve
terminals in large peripheral ring formations has created considerable
interest.[43] Neurotubules might themselves be involved in neurotransmit-

ter release mechanisms through interactions with the presynaptic membrane, synaptic vesicles, or both,[43] perhaps enhancing fusion or creating reactive attachment sites for colliding vesicles.

Vesicular Theory of Neurotransmitter Release

The traditional concept of events leading to neurotransmitter release embodies the central dogma of the local fusion of the terminal and vesicle membranes at critical contact zones followed by expulsion of discrete packets or "quanta" of many thousands of neurotransmitter molecules into the synaptic cleft from synaptic vesicles by a process of exocytosis (Figure 7.8). Alternatively, some form of "docking" of the vesicles close to the membrane on a gridlike structure (presynaptic grid; see Figure 1.25) leads to expulsion of neurotransmitter into the cleft from the vesicle through a fixed channel[165] (synaptopore; see Chapter 1, pages 42–43). In this model, the vesicle membrane does not fuse with the presynaptic membrane, nonetheless the released transmitter has a vesicular origin (Figure 7.9).

The "vesicular" theory was developed[E] to account for the observed pattern of release of acetylcholine at the neuromuscular junction in a fashion that generated short-lived, low-amplitude pulses of electrical discharge in the postsynaptic muscle fiber.[52,164] Called miniature end-plate potentials (m.e.p.p.), the pulses occur spontaneously when the nerve-muscle system is in a state of rest and there is neither impulse traffic nor visible contraction of muscle fiber.[E] These m.e.p.p.'s are about 0.5 mV in amplitude and all show the same characteristic asymmetric time course (Figure 7.10). When a nerve impulse arrives at the neuromuscular junction it produces a much larger electrical potential in the postsynaptic muscle fiber, and this potential is identical in overall form (shape, time course) to the m.e.p.p. In fact, this impulse-evoked end-plate potential (e.p.p.) is made up of hundreds of m.e.p.p.'s. The nerve impulse merely accelerates the ongoing process of spontaneous m.e.p.p. generation by depolarizing the presynaptic membrane and allowing the influx of Ca^{2+}.[E] Unlike the m.e.p.p., however, the e.p.p. is large enough to cause the muscle fiber to twitch. Miniature end-plate potentials appear to reflect the basic mode of neurotransmitter release at the neuromuscular junction: short, discrete bursts of large numbers of molecules acting simultaneously on the postsynaptic membrane. Since m.e.p.p.'s have been recorded at neuron-neuron synapses in both the central and peripheral systems,[48] one may conclude that mechanisms of neurotransmitter release are likely to be fundamentally similar throughout the nervous system.

The number of neurotransmitter molecules likely to be simultaneously expeled into the synaptic cleft to produce an m.e.p.p. has been estimated by three independent and ingenious approaches[49,50,51] to be in the range of 1000 to 10,000 molecules. Such a number of acetylcholine

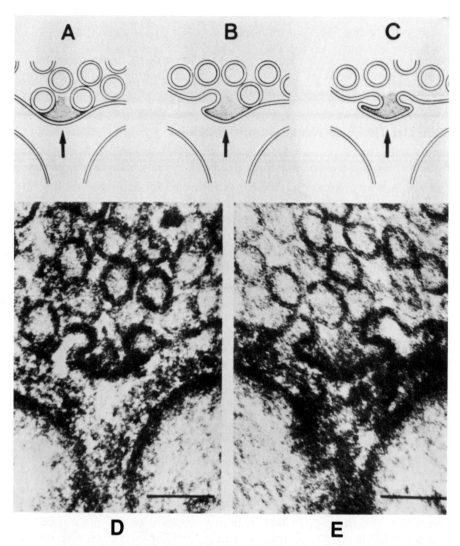

FIGURE 7.8 Active zones of frog neuromuscular junctions seen in sections parallel with the axis of an axon terminal branch and perpendicular to the double rows of synaptic vesicles attached to them. Each is located above a postsynaptic fold. Upper: Schematic drawings of active zones (see arrows). (A) Two synaptic vesicles, adjacent to the densly stained central region, are attached to the presynaptic membrane; (B and C) show one and two synaptic vesicles respectively opening into the synaptic cleft at two active zones. (D and E) Two active zones with one or two wide-open synaptic vesicles corresponding to B and C respectively. Bar, 0.1 μm. [From refs. 157, 158]

molecules could be accommodated in a single hollow synaptic vesicle of 40-nm internal diameter (50-nm external diameter), though it would give a hypertonic concentration variously estimated at 70 to 500 mM, according to the system studied and the extent of any intravesicular binding. Fusion of such a vesicle with the presynaptic membrane, and the subse-

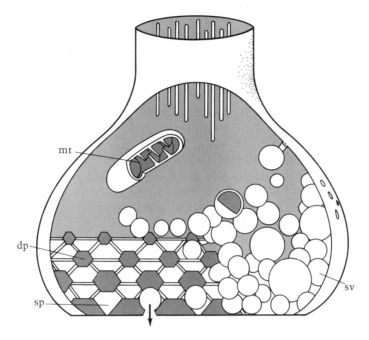

FIGURE 7.9   Organization of the presynaptic vesicular grid. Note hexagonal arrangement of presynaptic dense projections (*dp*) surrounding synap-
topores (*sp*) which provide "docking" points for synaptic vesicles (*sv*).
One vesicle is shown emptying its contents through the synaptopore into
the synaptic cleft (*arrow*). Vesicles of different sizes, mitochondria (*mt*),
and neurotubules are also represented. [From ref. 165]

quent exocytosis of its contents, could be expected to give rise to a briefly
lived m.e.p.p.—of the kind observed in muscle fiber—through rapid interaction with a patch of postsynaptically positioned receptor macromolecules. This interaction would be followed by their equally rapid inactivation by the hydrolytic action of acetylcholinesterase. It has been
estimated that each acetylcholine molecule would have time to interact
with a receptor molecule only once before its destruction.[49]

   We see, then, from the above outline that the vesicular theory of
neurotransmitter release accounts for many of the anatomical, physiological, and biochemical features of the organization and operation of synapses, and in particular of the neuromuscular junction. Nevertheless, a
number of experimental features of this process in cholinergic systems
have been difficult to encompass in the theory as it stands.[186]

Vesicles or Cytoplasm: The Source of
Released Neurotransmitter

Consideration of these experimental anomalies had led to a reappraisal of
the possibility that neurotransmitter is released into the cleft not from

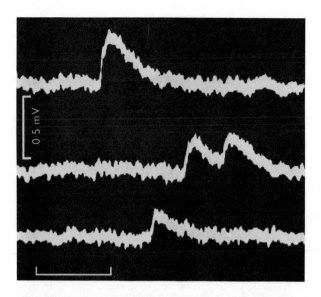

FIGURE 7.10   Spontaneous miniature end-plate potentials (m.e.p.p.'s) in frog sartorius muscle with resting membrane potential of 96 mV. These potentials are due to the random release of small amounts (quanta) of acetylcholine by the presynaptic nerve terminal at the neuromuscular junction. Note the fast rise time and slower decay time of the potentials. The potentials appear randomly, typically with a frequency of 0–4 per second. Time bar, 5.0 milliseconds. [From ref. 164]

vesicles but directly from the cytoplasm via a preset membrane "shuttle" or "carrier" mechanism.[5] Such a system might be conceived of as opening for a predetermined interval and thereby allowing the exit of a fixed number of neurotransmitter molecules. Such a system has been postulated to have a more definite physical counterpart and to exist as a specialized region of the presynaptic membrane or as part of the highly organized presynaptic grid structures described in Chapter 1. Terms such as "operator" or "vesigate" have been proposed for this hypothetical neurotransmitter carrier system. It is assumed to charge itself automatically and to remain saturated with neurotransmitter, which is conceived as binding to the carrier directly from the cytoplasm.

Evidence from Electron Microscopy   Evidence for a fusion of synaptic vesicles with nerve terminal membranes comes from conventional electron microscopy of many types of synapse (Figure 7.8), where "omega," or flask-shaped, profiles can be seen opening into the cleft region. These are not as commonly seen, however, as might be predicted from the vesicular theory. Omega profiles have been observed in freeze-fractured specimens (Figure 7.11), which demonstrates that they are not artifacts of fixation. In an elegant series of experiments,[F,55] rapid (<10 milliseconds) freezing

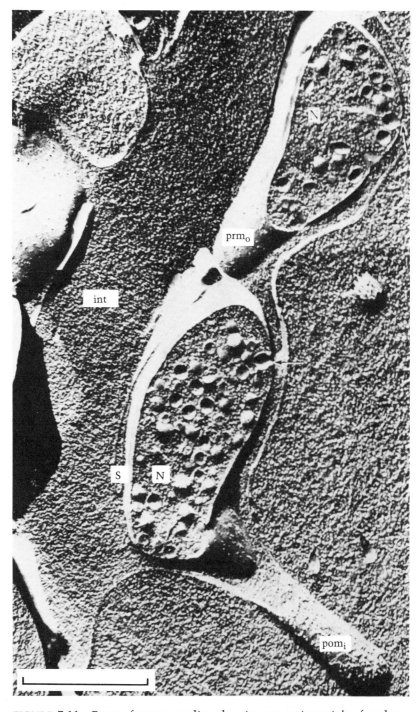

FIGURE 7.11   Freeze-fracture replica showing synaptic vesicles fused to the presynaptic membrane (prm) of a nerve terminal (N) and open to the synaptic cleft (*arrows*). Seen also are the postsynaptic membrane (pom), interstitial space (int), and Schwann cell process (S). The subscripts i and o indicate views of the membrane from the inner or outer sides respectively from *Torpedo* electric organ. Bar, 0.6 μm. [From ref. 163]

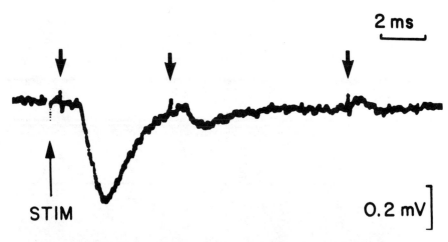

FIGURE 7.12   Extracellular recording from a curarized frog neuromuscular junction exposed to 1 m*M* 4-aminopyridine (an agent that prolongs
action potentials) in normal 2 m*M* calcium-Ringer's medium. At this
level of calcium the axon fires repetitively after a single electrical stimulus (*lower arrow*). The three small biphasic spikes (*upper arrows*) just before the larger downward deflections (postsynaptic end-plate current) represent three successive action potentials in the nerve. This repetitive
firing is prevented by raising calcium levels in the medium to 10 m*M*.
[From ref. 55]

of frog neuromuscular junctions allowed freeze-fracture pictures to be
taken at the very peak of the postsynaptic e.p.p., when the rates of transmitter release were at their maximum (Figures 7.12 and 7.13). In these
experiments, acetylcholine release was evoked by electrical pulses delivered to the neuromuscular junction while it was soaking in 1 m*M* 4-
aminopyridine and 10 m*M* Ca^{2+}. This agent increases the size of the e.p.p.
by blocking K^+ channels, which prolongs depolarization and therefore
allows entry of extralarge amounts of Ca^{2+}. Freeze-fracture pictures taken
5 milliseconds after stimulation at the maximum amplitude of the e.p.p.
(Figure 7.14) showed numerous pits along the line of the active zones,
where synaptic vesicles are known to align on the inner face (Figure 7.15).

Active zones are regions of the end plate from which acetylcholine is
most likely to be released; they are sited opposite that part of the postsynaptic muscle membrane which bears the highest concentration of acetylcholine receptors. At 3 milliseconds, however, no such pits were visible
(Figure 7.13). Counts showed that a positive correlation exists between
the number of pits formed and the quanta of acetylcholine released. This
correlation was measured at different levels of neurotransmitter release;
release increased as the concentration of 4-aminopyridine was raised and
was measured either as e.p.p. amplitude or in terms of the concentrations
of curare required to block the responses.

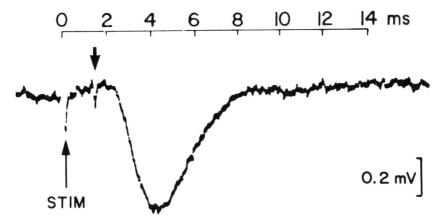

FIGURE 7.13 Extracellular recording from a curarized frog neuromuscular junction exposed to 1 m*M* 4-aminopyridine and 10 m*M* calcium and given a single electrical stimulus (*lower arrow*). The small biphasic spike at approximately 1.5 milliseconds after the stimulus (*upper arrow*) represents the nerve action potential. The large downward deflection beginning 1 millisecond later represents the postsynaptic end-plate current evoked by acetylcholine discharged from the nerve. This trace reveals that acetylcholine released from 4-aminopyridine-treated nerves is not only much greater than normal (cf. Figure 7.12), but also lasts longer. Normal postsynaptic end-plate currents are over by 5 milliseconds. This is presumed to occur because 4-aminopyridine prolongs the action potential and holds open the calcium channels in the presynaptic membrane longer than usual. [From ref. 55]

These experiments seem to provide clear evidence for fusion of synaptic vesicles with the correct region of the presynaptic end-plate membrane in a fashion that correlates well with end-plate synaptic activity. It is tempting, therefore, to assume that acetylcholine is expelled into the cleft by these vesicles at the moment of their fusion. A principal consideration—which brings caution to this conclusion—is the absence of pits, at times earlier than 5 milliseconds from the onset of stimulation, for example, 3 milliseconds (Figures 7.12 and 7.13). At this stage, neurotransmitter release was clearly occurring, even if not maximally. Technical reasons have been advanced to explain this anomaly.[55]

Another correlation between vesicle fusion and synaptic activity comes from the finding that vesicle number falls with *high* levels of stimulation, because vesicle formation cannot keep up with the rate of vesicle disappearance.[58,160] That stimulation releases the "most recently synthesized" neurotransmitter in preference to that already present—a widespread finding—provides another area of difficulty. This neurotransmitter necessarily shows a high degree of labeling from radioactive precursors. The synaptosomal pool most closely matching the released transmitter in its level of specific radioactivity is the cytoplasmic frac-

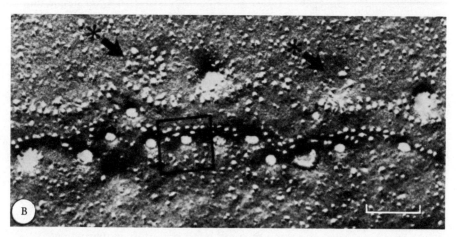

FIGURE 7.14   Freeze-fracture electron micrographs of frog neuromuscular junction. (A) View of protoplasmic fracture surface of active zone in nerve terminal soaked in 1 m*M* 4-aminopyridine (which prolongs action potentials) and 10 m*M* calcium for 30 minutes after one electrical stimulus applied 3 milliseconds before freezing. By 3 milliseconds the action potential is presumed to have arrived at the terminal but not to have altered the resting appearance of the active zone, which is recognized as a slight ridge bordered by parallel rows of large intramembrane particles. (B) View of protoplasmic fracture surface of an active zone from a nerve terminal after applying one electrical stimulus 5 milliseconds before freezing. In the interval between 3 and 5 milliseconds, just at the time when the nerve begins to discharge large numbers of acetylcholine quanta, many membrane perturbations appear along the edges of the active zone. These are exocytotic stomata (openings into the underlying synaptic vesicles). The two areas marked with arrows are vesicles that have collapsed flat after opening. These areas display a cluster of two or more large intramembranous particles similar to those found in intact synaptic vesicle membranes. The area in the square illustrates the characteristic domain of exocytosis. Bar, 0.1 μm. [From ref. 55]

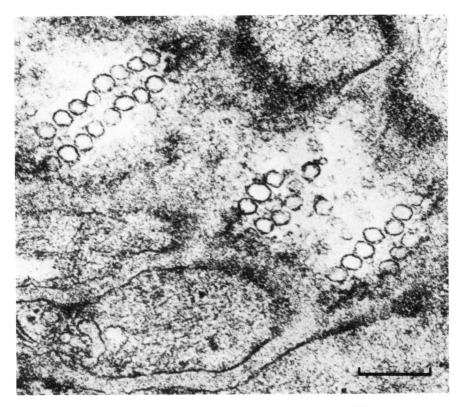

FIGURE 7.15   Tangential section of presynaptic region of frog neuromuscular junction, showing the double rows of synaptic vesicles attached to active zones. Three such rows of vesicles can be partly seen. Bar, 0.2 μm. [From ref. 159]

tion rather than the vesicle fraction. This finding is also widespread, and it applies to the synapses of many systems, including the cholinergic, catecholaminergic, and amino acid types.[F,56,57,58] The vesicle fraction usually shows a substantially lower level of labeling than the released transmitter.

Evidence from* Torpedo *Electric Organ   One proposed explanation is that only part of the vesicle population is actively involved in "exocytosing" transmitter at any one time, and this subpopulation is therefore predicted to have a specific radioactivity similar to that of the released transmitter. A vesicle subpopulation matching these requirements has been isolated from the electric organ of the ray *Torpedo marmorata* after stimulation.[58] It was found to be both smaller and heavier than the bulk vesicle fraction, and therefore descended to a lower level in sucrose density-gradients, probably owing to its smaller water content.[68] This same fraction was found to take up extracellular markers (e.g., dextran), which has been taken as evidence that the interior of these vesicles must be open to the extracellular space at some point. This could happen during exocytosis, or during a subsequent endocytosis. The denser vesicles in this fraction were associated with membranes and were smaller in diameter than those found in the lighter vesicle fraction.[58]

During continuous stimulation of the electric organ at low frequencies, when no fall in vesicle number was observable, the average content of acetylcholine and of ATP fell in the vesicles, indicating that refilling of the vesicles could not keep up with the rate of release of neurotransmitters. At the same time the number of smaller diameter vesicles increased, and each could be shown to take up high-molecular-weight extracellular markers such as dextran[58,59] (mol wt 40,000; Figure 7.16). False transmitters generated in the electric organ, or in brain synaptosomes, from choline analogues such as triethylcholine, pyrrolcholine, or homocholine

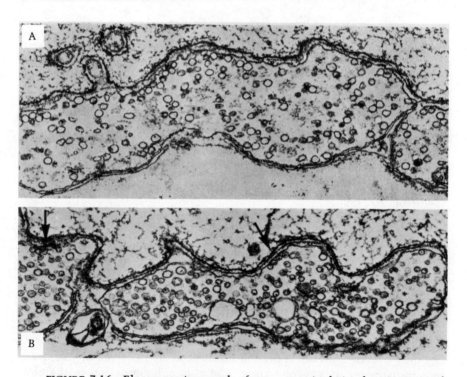

FIGURE 7.16   Electron micrograph of nerve terminals in electric organ of *Torpedo marmorata*. The terminals were stimulated while being perfused with dextran-containing medium. (A) Tissue after 5 hours of perfusion. The vesicle cores remain electron-lucent. (B) Tissue after activation of the nerve terminal by stimulating the nerve. Many vesicles are now of smaller diameter and contain dextran granules in their lumen. Only a few of the remaining large vesicles contain a dextran granule. Arrow indicates patches of small dextran-containing vesicles lined up along the presynaptic membrane. [From ref. 59]

become incorporated into synaptic vesicles. Stimulation releases mixtures of acetylcholine and false transmitter in a ratio closer to that found in vesicles than to that of the soluble cytoplasmic fraction or whole tissue.[60,61,65,66]

After depletion of vesicular acetylcholine by stimulation of the *Torpedo* electroplaque, the newly synthesized acetylcholine generated from radiolabeled choline was found mainly in the heavier small-vesicle fraction. Again, released transmitter was similar in specific radioactivity to that found in these small and dense vesicles, suggesting that release could well have been from this organelle—although in these experiments cytoplasmic levels were not measured for comparison.[62]

This denser subfraction of small-diameter vesicles is associated with many membrane profiles when first taken from the sucrose gradient, and it does not give the appearance of homogeneity typical of the lighter vesicle fraction. This membrane contamination can be removed without substantially changing the balance of the data, thus eliminating the contamination as the site of acetylcholine binding or occlusion.[67] In these experiments with *Torpedo* electric organs, the absence of cytoplasmic fractions for comparison with vesicle fractions has limited the conclusions that can be drawn in distinguishing between vesicle and cytoplasm as the source of released acetylcholine. More recent data from experiments in which cytoplasmic fractions were identified by their content of nucleotides and false cholinergic transmitters have supported the concept of a vesicular site of origin of released acetylcholine.[59,60,66]

Evidence Supporting Cytoplasmic Release The general and much wider debate on this question continues, however,[53] since several issues remain to be settled. For instance, as mentioned in Chapter 4 (page 166), choline acetyltransferase, the biosynthetic enzyme for acetylcholine, is present in soluble form in the cytoplasm of cholinergic terminals and is not bound to synaptic vesicles. Acetylcholine synthesis is therefore likely to take place in the cytoplasm rather than in vesicles, with transfer to vesicles taking place at a later stage. Thus, the "newly synthesized" neurotransmitter in cholinergic terminals, at least, would appear in the soluble cytoplasm before any subsequent move to the vesicles.[186]

Use has been made of the hydrolytic enzyme acetylcholinesterase in an attempt to settle the question of the origin of synaptically released acetylcholine in neurons of the sea snail *Aplysia*.[58,63] About 2 hours after injection of the enzyme into the cell body of a cholinergic neuron in the buccal ganglion of *Aplysia*, the postsynaptic potential decreased and eventually disappeared. This period was required for diffusion to the nerve terminals, a distance of 300–500 μm. Acetylcholinesterase does not attack acetylcholine contained in synaptic vesicles of vertebrate cholinergic neurons, and vesicular acetylcholine is assumed to be protected in the *Aplysia* neuron. Electron microscopy showed that no change occurred in vesicle number, and vesicles did not contain any acetylcholinesterase. It was therefore assumed that vesicles remained filled with neurotransmitter while cytoplasmic transmitter was being destroyed, and that this depletion of the cytoplasmic pool caused the disappearance of synaptic potentials. In these experiments, the cytoplasm is indicated as the site of origin of the released transmitter. In addition, other investigators[56,64] have reported no change in content and no uptake of "newly synthesized" acetylcholine into the total vesicle population of *Torpedo* electroplaque during periods of stimulation, while the cytoplasmic transmitter pool was found to decrease in size and increase in specific radioactivity. In these studies no subfractions of the bulk vesicle fractions were separated for study.

Studies of acetylcholine release from synaptosomes isolated from *Torpedo* electroplaque[69] show that it is the subfraction, or compartment, that is liberated by freezing and thawing the synaptosomes which is the source of the acetylcholine released by depolarizing agents. This compartment is interpreted as being the cytoplasm rather than synaptic vesicles.[69]

Exocytosis   The mechanics of any vesicle fusion that occurs during neurotransmitter release are assumed to mirror those involved in the secretion of other cell products by exocytosis. These are perhaps most relevant in systems such as the secretion of epinephrine to the blood from the chromaffin granule ("vesicle") of the adrenal medulla, or the release of oxytocin or vasopressin from the terminals of neurons in the posterior pituitary gland[155] (neurohypophysis; Figure 7.17). The process involves contact and fusion followed by exocytosis, and then recovery of the vesicle membrane by endocytosis to avoid the otherwise inevitable growth in area of the terminal membrane. The latter process may result in the formation of "coated" (complex) vesicles (see Chapter 1). Because of the absence of certain plasma membrane components in the vesicle membrane, it has been suggested that the vesicle membrane maintains its identity after fusion and is recaptured as a new vesicle on endocytosis. Several proteins, including Na^+K^+-ATPase and several major protein antigens detectable by immunoelectrophoresis, are absent from vesicle fractions.[F] But morphological evidence suggests that proteins are "squeezed out" of the plasma membrane during the process of fusion with the vesicle. In this way, only phospholipids would be involved in the fusion of the two membranes. For instance, Figure 7.18 shows the apparent absence of intramembranous protein particles in the bulging regions of plasma membrane at the point of contact with vesicles of *Torpedo* electroplaque. These intramembranous particles are assumed to be points of exocytosis; this seems a reasonable assumption since a similar clearing of membrane proteins appears to accompany secretion by mast cells and fusion of chromaffin granules.[F]

The points of fusion of vesicles with the plasma membrane in the region of the active zones in at least three different kinds of synapse may

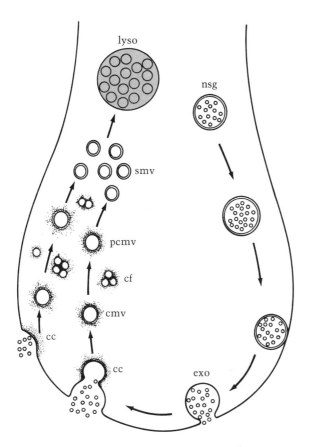

FIGURE 7.17   Diagram showing proposed exocytosis-vesiculation sequence in nerve terminals of posterior pituitary gland. Contents of neurosecretory granules (*nsg*) are emptied directly into extracellular space by exocytosis (*exo*). The granule membrane is retrieved from the terminal surface by micropinocytosis (vesiculation), producing coated caveolae (*cc*), or "coated pits," that "pinch-off" to form coated microvesicles (*cmv*) and then shed coat fragments (*cf*) to become partly coated microvesicles (*pcmv*) and, finally, smooth (synaptic) microvesicles (*smv*). These, in turn, may be incorporated into lysosomal bodies (*lyso*). [From ref. 161]

be identifiable by the appearance of large intramembranous protein particles after any transient fusion pits have vanished. These characteristics clusters of two to four large membrane particles are seen in the inner leaflet of the plasma membrane at the points of fusion; they at least double in number during stimulation of neurotransmitter release by a variety of methods[70,55] (see Figure 7.14B). Since similar clusters of large particles can be seen in the synaptic vesicles of these synapses, the large particles are assumed to be added to the plasma membrane during the exocytotic process.[F,70,55]

FIGURE 7.18   Freeze-fracture view of bulging vesicle contacts seen in *Torpedo* electric-organ nerve terminals fixed in high concentrations of calcium (50 m*M*). A fracture passes along the presynaptic membrane and into the nerve interior to reveal synaptic vesicles bulging into the membrane from the axoplasm. [From ref. F; micrograph by J. Heuser]

Another characteristic feature of vesicles is the formation of electron-dense calcium spots (only one per vesicle) on their inner membrane walls when exposed to high concentrations of calcium (e.g., 50 m*M*) during fixation.[F] The spots can be formed using other divalent cations (e.g., La^{2+}); they probably identify a region of high negative charge on the vesicle wall—which is of possible importance either in fusion processes or in the storage of organic cations.

A critical question bearing on the relevance of all these observations for the process of neurotransmitter release is whether exocytosis, as seen in neural and other secretory processes, can be fast enough to match the short time-period involved from entry of calcium to neurotransmitter release (about 200 microseconds in squid giant axon). Current evaluation shows that 200 microseconds is much faster than the rate of secretions from glandular structures.

Cotransmission: The Release of More than One Neurotransmitter

The concept that each nerve cell makes and releases only one neurotransmitter at all its branches and terminals is known as *Dale's principle.*[O] In recent years this concept has been reexamined in the light of evidence to

TABLE 7.1   Properties of neurotransmitters, neuromodulators, and cotransmitters

Neurotransmitter	Neuromodulator	Cotransmitter
Acts on postsynaptic receptor to produce a direct postsynaptic effect (an ionotropic or metabotropic action).	Acts on postsynaptic receptor to influence the action of a neurotransmitter without having a direct postsynaptic action itself.	Coexists with a neurotransmitter in a given neuron population and coreleased with the neurotransmitter. May be a neuromodulator or a second neurotransmitter.
May act on presynaptic receptor to control its own release through "autoreceptors."	Acts on presynaptic receptor to influence release or synthesis of a neurotransmitter.	Acts on synaptic sites on the same or different cells.

the contrary.[71,72,73,O,P] Certainly, substances other than neurotransmitters are known to be released from certain kinds of nerves. For instance, adrenergic nerves may release ATP, dopamine β-hydroxylase, a protein called chromogranin A, and, in addition, norepinephrine—all from the same vesicles. Recently, the concept has been advanced that the substances called neuromodulators may be released together with neurotransmitters—that is, coreleased—and thereby affect the overall action of the latter on the postsynaptic membrane potential. The precise difference between a substance acting as a neuromodulator or as a cotransmitter has not yet been defined in internationally accepted terms (see Table 7.1). Where postsynaptic receptors for a released substance can be demonstrably linked to ionotropic or metabotropic responses, it should be regarded as a neurotransmitter rather than as a neuromodulator *at that synapse*, since it activates its own sequence of postsynaptic events. Neuromodulators are conceived of as acting at some point on the postsynaptic receptor complex to cause a change in the *efficiency* of the neurotransmitter in its ion gating or other postsynaptic action. Its specific binding sites will have all the usual properties of a receptor. Neuromodulators may also exert their action by altering the *rate of release or synthesis* of the neurotransmitter by interaction at a presynaptic receptor site. It now seems to be clear that a given substance may act as a neurotransmitter, neuromodulator, or even as a hormone at different synaptic or other cellular sites that bear its receptors—and that classification should be in terms of the action rather than the specific substances.

One interesting example of the modulation of postsynaptic neurotransmitter action is the interaction of the benzodiazepine system with postsynaptic GABA receptors[137] (see Chapter 4). It is postulated that an endogenous benzodiazepinelike ligand is either coreleased with GABA from a subclass of GABAergic receptors or that it is released from a more distant site and interacts with the supramolecular GABA receptor complex at the benzodiazepine recognition site (Figure 7.19). When it occu-

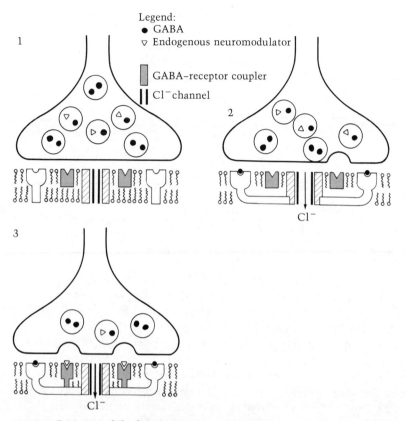

FIGURE 7.19   Model of GABAergic synapse functioning with a double signal generation, normally GABA and a neuromodulator whose actions are mimicked by benzodiazepine drugs. The coupler for the GABA recognition site and chloride gate is shown as a postsynaptic membrane protein. [From ref. 140]

pies this site it is able to enhance or diminish the ion gating (Cl^-) action initiated by GABA when it binds to its own separate GABA recognition site. In experimental terms, the neuromodulator action of benzodiazepines on the action of GABA is seen in their enhancing effect (at micromolar concentrations) on the high-affinity binding of GABA to synaptic membranes. They also facilitate the inhibitory action of GABA at certain synapses and increase the frequency of opening of Cl^- channels elicited by GABA.[136] Added alone, benzodiazepines have no action in this respect.

Other substances that may be coreleased and may influence primary neurotransmitter action postsynaptically include the "microamines" phenylethylamine and paratyramine in adrenergic nerves, and ATP or adenosine in a wide range of nerves including autonomic nerve terminals, adrenal medulla, and phrenic nerves.[73] Adenosine may well exist in its own right as a neurotransmitter (the so-called purinergic system; see Chapter 4). ATP can be coreleased with norepinephrine or acetylcholine

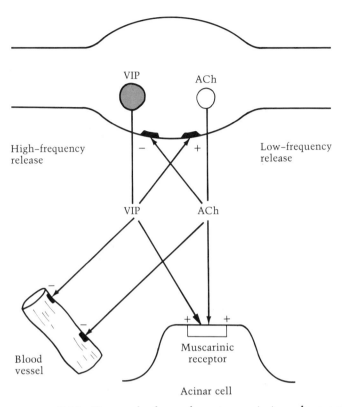

FIGURE 7.20 Proposed scheme for cotransmission where a classical neurotransmitter, ACh, coexists with VIP in parasympathetic nerves supplying the cat salivary gland. Note that ACh and VIP are stored in separate vesicles; they can be released differentially at different stimulation frequencies to act on acinar cells and glandular blood vessels. Cooperation is achieved by selective release of ACh at low-impulse frequencies and of VIP at high-impulse frequencies. Pre- and postsynaptic modulation is indicated. [From ref. 141]

(probably from the same vesicles), and it influences their rate of release through presynaptic P_1 receptors as well as acting postsynaptically in the same synapse on its own separate P_2 postsynaptic receptors.[73] (See Figure 7.21; see also Figure 4.47.)

Some sympathetic neurons have the potential to synthesize both norepinephrine and acetylcholine at early stages in development. These cells require NGF to survive, and they respond to its action with the production of both tyrosine hydroxylase and choline acetyltransferase. They also possess a high-affinity uptake system for catecholamines. Subsequently the cells become either cholinergic or adrenergic, with a gradual loss of one or the other enzyme as appropriate. The cholinergic cells become unresponsive to NGF and lose their catecholamine transport system. A substantial body of evidence suggests, however, that some of these cells retain the ability to synthesize and release both norepinephrine and acetylcholine.[72]

Evidence also exists for the coexistence of certain neuroactive peptides in monoaminergic neurons[M,O,P] (see Chapter 5). For instance, somatostatinlike immunoreactivity has been observed in 60–70 percent of all principal adrenergic ganglion cells of the inferior mesenteric ganglion and in the celiac superior mesenteric ganglion complex.[74] Enkephalinlike immunoreactivity has been demonstrated in the cervical ganglion of the rat in norepinephrine-containing ganglion cells.[75] Substance P has been detected in certain 5-HT-containing neurons in the lower brain stem,[76] and vasoactive intestinal peptide (VIP) appears to be concentrated in certain peripheral and central cholinergic neurons.[73,O,P,129] It seems that peptides are present in separate large granular vesicles. Whether both neuroactive substances are always released during nerve stimulation remains to be clarified, but in certain cases there is strong evidence of such corelease. In the salivary gland, acetylcholine released from sympathetic nerves at low frequencies of stimulation causes salivary secretion from acinar cells and some dilation of blood vessels in the gland. At higher frequencies of stimulation, VIP is released and produces marked vasodilation. It has no direct effect on acinar cell secretion but substantially enhances the action of acetylcholine on acinar cell secretion, and it also enhances the release of acetylcholine from the nerve endings by an action at presynaptic receptors[162] (Figure 7.20). At the same time the vasodilation caused by VIP acting directly on blood vessels would provide for the increased metabolic needs of the tissue during demanding periods when neural activity is high.

If it becomes established that two different neurotransmitters are released into the same synaptic cleft, then new vistas are opened for the manipulation and control of the neural signal as it is transduced from its electrical to its chemical phase en route through the neural network. Mechanisms can be envisaged that affect both the size and nature of the signal, each pathway being characterized by certain qualitative features. Where cotransmitters exist in a common vesicle they cannot be differentially released, and one must rapidly facilitate the actions of the other. Cotransmitters stored in separate vesicles can be differentially released at different stimulation frequencies and then exert their actions on two different postsynaptic target sites (Figure 7.20).

PRESYNAPTIC RECEPTORS: THE CONTROL OF NEUROTRANSMITTER RELEASE

Neurotransmitter release is very much under the control of Ca^{2+}, whose influx triggers the mechanism that leads to the secretion. This cardinal influence of Ca^{2+} is itself controlled through the active removal of cytoplasmic Ca^{2+} into mitochondria[77] and smooth endoplasmic reticulum.

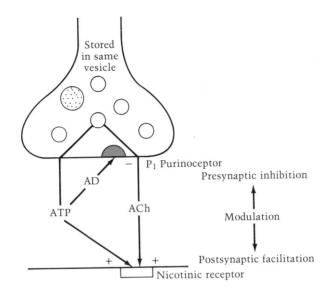

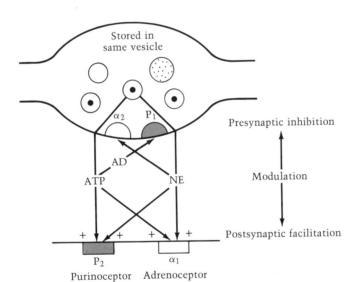

FIGURE 7.21   Proposed scheme for cotransmission where classical neuro-transmitters coexist with ATP. (A) The skeletal neuromuscular junction. ACh and ATP are stored in the same vesicle. Upon release, ACh occupies receptors that lead to excitation of the postsynaptic cell. Coopera-tion is achieved initially by rapid facilitation of postsynaptic actions of ACh by ATP and later by inhibition of ACh release by adenosine (arising from extracellular hydrolysis of ATP). (B) Sympathetic nerves to the vas deferens and some blood vessels. Upon release from a common vesicular store, both norepinephrine and ATP occupy postsynaptic receptors that lead to synergistic excitation of the effector cell. Cooperation is achieved initially by rapid mutual facilitation of postsynaptic excitation and sub-sequently by mutual inhibition of transmitter release. [From ref. 141]

Although such controlling action of mitochondria and smooth endoplasmic reticulum within nerve terminals clearly depends on the size, number, and surface area of these organelles, it must be assumed that the Ca^{2+}-sequestering capacity available in the nerve terminal is always in excess of requirement, since studies with aequorin show only brief rises in Ca^{2+} levels (half-life 1–2 milliseconds). Not surprisingly, the last decade has seen the discovery of additional mechanisms controlling neurotransmitter release, some of which allow manipulation of Ca^{2+} levels. Prominent among these are *presynaptic* receptor systems.[G,H,I,J] Concepts of their organization and precise modes of action are still largely in a state of formulation, though all appear to involve interaction between a neurotransmitter and a specific receptor system positioned on the nerve terminal or on terminal varicosities (peripheral nervous system) leading to a reduction, and sometimes an increase, in neurotransmitter release from that terminal or varicosity.

Categories of Presynaptic Receptor

Presynaptic receptors appear to be of several kinds and they occur at a variety of synaptic junctions in the central and peripheral nervous systems, including specialized axoaxonic (serial) synapses in the central nervous system (see Figures 1.30 and 1.33) and on varicosities in the peripheral nervous system. Some populations of presynaptic receptors are specific for a neurotransmitter other than that released by the presynaptic terminal (Figure 7.22D).

A second category of presynaptic receptor includes those which are sensitive to a neurotransmitter other than secreted from their terminal but that do not form part of an organized and morphologically recognizable synaptic structure (Figure 7.2D). A third category encompasses receptors specific to the neurotransmitter released by the terminal in question. They are sometimes called "autoreceptors"[78] that mediate "autoinhibition." These last two categories of presynaptic receptor often appear on the terminal varicosities of peripheral catecholaminergic systems (Figure 7.22). Apart from controlling the extent of transmitter release by feedback or feed-forward mechanisms, presynaptic receptors provide the principal means by which neurotransmitter systems may interact with one another, allowing subtle modulation of neural activity and exquisitely fine control of the integration and balance of neural activity.

The precise mechanisms by which presynaptic receptors achieve their influence on neurotransmitter release are not clear. In some cases it is likely to be a question of controlling the extent of transmitter biosynthesis presynaptically, and for dopaminergic systems this can be achieved by a direct influence on tyrosine hydroxylase, the rate-limiting enzyme of synthesis (see Chapter 4, pages 179–182). This enzyme may form a physical part of the presynaptic receptor unit. Other aspects of presynaptic

Single neurotransmitters

A Norepinephrine (NE)

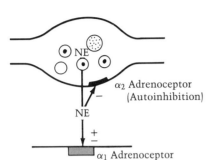

B Acetylcholine (ACh)

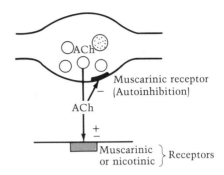

C Adenosine–5′–triphosphate (ATP)

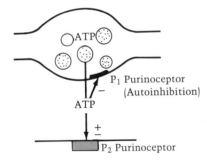

D Cross–talk between closely apposed
   adrenergic & cholinergic terminals

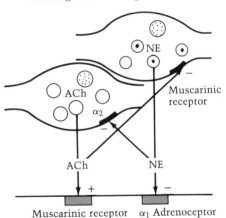

FIGURE 7.22 Proposed schemes·for operation of pre- and postsynaptic receptors in neurotransmission involving single neurotransmitters. Schematic representation of transmission at (A) adrenergic, (B) cholinergic, and (C) purinergic neuroeffector junctions. Note the presence of autoinhibition of all three synapses, with negative feedback of the neurotransmitter on specific presynaptic receptors. (D) A schematic representation of the interactions between norepinephrine (noradrenaline, *NA*) and ACh released from closely apposed adrenergic and cholinergic varicosities. The neurotransmitters have opposite postsynaptic actions but have mutual inhibitory actions on their release. [From ref. 141]

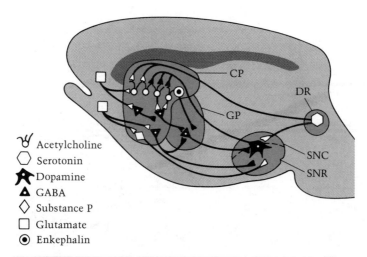

𝒱 Acetylcholine
◯ Serotonin
★ Dopamine
▲ GABA
◇ Substance P
□ Glutamate
◉ Enkephalin

FIGURE 7.23   Summary of the principal known pathways between the basal ganglia, indicating the neurotransmitter attributed to each pathway. *CP*, caudate-putamen; *DR*, dorsal raphe; *GP*, globus pallidus; *SNC*, zona compacta of the substantia nigra; *SNR*, zona reticulata of substantia nigra. [From ref. 121]

receptor action may include ion gating mechanisms leading to hyperpolarization or depolarization, depending on whether the resulting influence on neurotransmitter release is inhibitory or facilitatory. Calcium gating mechanisms could exert a similar range of control through enhancement or attenuation of rates of Ca^{2+} entry through receptor-linked, voltage-sensitive Ca^{2+} channels. Little clear-cut information is available on these processes, though interactions among a wide range of neurotransmitters have been firmly established.[79,G,H,I,J,M]

The strategies for recognizing presynaptic receptors include both direct and indirect approaches. Direct approaches include the use of synaptosome preparations devoid of *functional* postsynaptic structures and the use of brain slices containing only the nerve endings for a given neurotransmitter system and not the cell bodies or dendrites (e.g., dopamine in substantia nigra and norepinephrine in cerebellum or cerebral cortex). Indirect approaches employ a wide range of relatively intact preparations such as brain slices, superfused brain, heart, vas deferens, spleen, pineal gland, or gut. Neurotransmitter agonist or antagonist drugs with a well-defined specificity of action for a given neurotransmitter system when applied to these preparations can be shown to enhance or depress neurotransmitter release evoked by electrical pulses or by depolarizing chemicals. Such responses indicate the presence of a transmitter "receptor-mediated" control of transmitter from active functional pools, and a presynaptic location on nerve terminals for such receptors is the most likely explanation of these phenomena. At present there is no known basis for a retrograde transsynaptic action from postsynaptic receptors influencing neurotransmitter release from the corresponding presynaptic structures. Feedback effects from postsynaptic to presynaptic neurons via recurrent collateral axons do occur, but any action due to them can usually be eliminated by certain strategies: (1) Removal of the relevant nerve terminals by degradation following pathway ablation (axotomy, 6-hydroxydopamine) removes the effect. (2) The preparations contain only the nerve endings, and not the cell bodies or dendrites, for the neuronal transmitter system in question (e.g., brain slices prepared from corpus striatum contain only dopamine terminals)—in this case, the possibility of increased release being due to a *changed firing-rate* in the cell body of the neuron is eliminated. (3) The properties of the presynaptic receptors mediating the transmitter-release effect are different in drug sensitivity from those which are postsynaptic in the synapse under study, indicating the presence of quite different categories of receptor (e.g., where α- and β-noradrenergic receptors are present specifically at pre- or postsynaptic sites; see below). (4) The response—this agent prevents action-potential generation and would prevent feedback from postsynaptic to presynaptic structures via nerve fibers without preventing field-effect stimulation.

Peripheral Presynaptic Receptors

The early observations on presynaptic receptor function were mostly obtained with preparations from the peripheral nervous system. Alpha-adrenergic receptor blockers (e.g., phenoxybenzamine) were found to increase the rate of outflow of norepinephrine stimulated by nerve activity (e.g., in cat spleen[80]). In preparations such as guinea pig atria, cat heart, and rabbit heart, the postsynaptic receptors are of the beta adrenoreceptor type and would therefore not interact with the alpha type of receptor antagonist,[81,82] thus eliminating the possibility that postsynaptic structures are responsible for the effect. Moreover, alpha adrenoreceptor agonists inhibit norepinephrine release during nerve stimulation independently of the alpha or beta nature of the postsynaptic adrenoreceptor that mediates the response of the effector organ.[5,6] The frequency of stimulation is, however, an important factor here, with low frequencies allowing the larger effects to be displayed. This phenomenon is probably due to an involvement of Ca^{2+} fluxes in the response (see below).

When endogenous pools of norepinephrine are reduced with synthesis blockers (e.g., α-methyltyrosine), the amounts of transmitter that are released and that subsequently accumulate in the synaptic cleft are also reduced. As a result, any negative feedback effect due to the transmitter accumulating in the cleft would be reduced and would even disappear if a "threshold" concentration was required. Such events do take place, as is evidenced by the greatly reduced effectiveness of alpha adrenoreceptor blockers in enhancing release of norepinephrine after synthesis blockade.[81,82]

Differences in the drug sensitivity of the pre- and postsynaptic receptors has led to the concept that two receptor types exist, alpha 1 (postsynaptic) and alpha 2 (presynaptic). Clonidine (agonist) and yohimbine (antagonist) are more effective on presynaptic alpha 2 receptors than on postsynaptic alpha 1 receptors,[83,84] and the agonists phenylephrine and methoxamine have a high affinity for the alpha 1 adrenoreceptors and little or no affinity for the alpha 2 adrenoreceptors.[81]

Differential blockade of pre- and postsynaptic receptors has been used to illustrate the physiological importance of presynaptic receptors in vivo. Where presynaptic negative-feedback receptors are of the alpha type and the postsynaptic receptors are of the beta type, then exposure to alpha blockers should enhance the response of the effector organ by causing enhancement of neurotransmitter release. This is found to be true of guinea-pig atria and several other preparations.[81,85]

Further demonstration of the existence of presynaptic receptors comes from the use of neurotransmitter-uptake transport blockers. These would be expected to allow the concentration of the transmitter in the cleft to rise rapidly to the threshold level for activating presynaptic receptors during nerve stimulation and initiating modulatory effects on release. Such responses were found to follow when, for instance, cocaine was used as an uptake blocker for norepinephrine.[81,82]

Excitatory Peripheral Presynaptic Receptors

So far, we have described only inhibitory presynaptic receptors that mediate a reduction in transmitter release by negative feedback cycles. Positive feedback loops also occur, and they result in enhanced neurotransmitter release following presynaptic receptor activation. Inhibitory and facilitatory receptors may sometimes occur on the same terminals, thereby creating the opportunity for very fine tuning of the control mechanisms.

These facilitatory adrenoreceptors on peripheral nerves appear to be of the beta type. Low concentrations (e.g., 10^{-8} M) of isoprenaline (a beta adrenoreceptor agonist) enhances norepinephrine release during low-frequency stimulation of several organs with a nonradrenergic innervation, including guinea pig atria, cat thoracic aorta, and perfused spleen or pineal glands.[21,82,86] This response is often prevented by propranolol, a specific beta antagonist.

How the actions of inhibitory and facilitatory presynaptic receptors on the same terminals are integrated appears to be as follows: When low levels of transmitter are present in the synaptic cleft, the facilitatory beta receptors are activated, leading to increased release of neurotransmitter. When higher concentrations are reached in the cleft, the presynaptic alpha receptors come into play, resulting in a reduction in the level of release. Thus, differential sensitivity to neurotransmitter concentration would allow a delicate balance to be maintained between the opposing actions of the two types of presynaptic receptor.

Dopaminergic and Cholinergic Presynaptic
Interactions in the Peripheral Nervous System

In addition to the noradrenergic type, peripheral dopaminergic and cholinergic presynaptic receptors influencing norepinephine release have also been detected. The presence of the dopaminergic systems is revealed by blocking the noradrenergic presynaptic alpha receptors with an alpha antagonist (e.g., phentolamine) and demonstrating the inhibition of norepinephrine release mediated by dopamine itself or by its agonists (e.g., apomorphine). The responses due to dopaminergic agents are, in turn, suppressed by dopamine antagonists (e.g., pimozide), thus confirming the involvement of stereospecific receptors in the process.[81,82,88] Dopaminergic presynaptic receptors are not as widespread as the adrenergic type and have been reported to be present in only a few organs (e.g., cat heart and spleen, dog heart). Since blockade of these receptors with dopaminergic antagonists does *not* lead to increased norepinephrine release during nerve stimulation, in contrast to the effects of α-adrenergic blocking agents (see above), their in vivo significance remains uncertain.

Both nicotinic and muscarinic cholinergic presynaptic systems, which influence norepinephrine release due to nerve stimulation, have been shown to occur in many saline-perfused organs and other in vitro preparations.[87,89,90,94,95] Well-studied are the heart, vascular preparations, and spleen, where added acetylcholine or carbechol reduces norepinephrine release in the presence of nicotinic antagonists and enhances its release when muscarinic antagonists are present. This modulation reveals the activities of inhibitory muscarinic and excitatory nicotinic presynaptic receptors sited on noradrenergic nerve terminals in these preparations. The cholinergic systems add their effects to those of other presynaptic receptors controlling norepinephrine release. Thus, as would be expected, atropine itself allows enhanced norepinephrine release during peripheral nerve stimulation by blocking the inhibitory actions of acetylcholine originating in local, often structually linked, cholinergic nerve endings. Muscarine and nicotine themselves are, of course, effective in respectively reducing or enhancing norepinephrine release from various perfused peripheral organs during nerve stimulation.[87,89,90] In all of these presynaptic control mechanisms, however, it is usually the *inhibitory* influence that predominates over the positive-feedback countersystem, as in most aspects of neural organization.

Enhanced norepinephrine release evoked by "releasing agents" such as amphetamine or tyramine shows quite different characteristics from release produced during nerve stimulation by the various receptor-targeted adrenergic or cholinergic compounds described above. Apart from the much larger amounts released by these agents, and the lack of requirement for calcium, their actions are not necessarily altered by nerve stimulation and are not prevented by receptor antagonists. The events mediated via presynaptic receptors use sophisticated mechanisms of a quite distinct and separate nature and may well involve at least partially separate neurotransmitter pools. Though such features emphasize the likely physiological importance of presynaptic receptors, clear demonstration of the extent and precise nature of this physiological role is still awaited for several of these systems,[88,90] particularly the cholinergic and dopaminergic category.

Presynaptic Receptors Operated by Other Neurotransmitters and Modulators

A bewilderingly large number of receptor-mediated interactions influence the extent of nonrepinephrine release from the terminal varicosities of peripheral noradrenergic neurons they include actions at the relevant low concentrations for adenosine,[91] prostaglandins,[92] peptides such as enkephalin[93] (see Chapter 5), and cholinergic agents.[89,94,95] Since these are all naturally occurring compounds, the possibility that they have an influence on norepinephrine release in vivo must be considered. A picture is emerging of a curiously complex array of presynaptic receptors on the terminal varicosities of noradrenergic neurons, all working in concert to effect the optimal degree of neurotransmitter release for a given situation.

Presynaptic Receptors in the Central Nervous System

Presynaptic receptor mechanisms similar to those described for the peripheral nervous system have now been demonstrated in various central nervous system preparations.[185] Direct approaches employing synaptosome preparations or brain slices containing nerve endings (but not the cell bodies) for a given neurotransmitter allow fairly straightforward tests to be made for presynaptic systems. For instance, presynaptic receptors controlling dopamine release from slices and synaptosomes from corpus striatum, which is enriched with dopaminergic terminals, have been widely reported. These have been detected by the actions of various drug agonists and antagonists on both resting and depolarization-induced dopamine release to incubation media. Thus dopamine agonists (e.g., apomorphine) can reduce Ca^{2+}-dependent dopamine release from caudate slices induced by low-frequency electrical stimulation, and antagonists (e.g., haloperidol) increase this release.[93,96,97,98] Such effects reveal the presence of inhibitory feedback control of dopamine transmitter release; modification of the experimental design allows detection of a facilitatory presynaptic receptor for dopamine. The conclusion that these responses are presynaptic in origin is based on a number of assumptions, for example, the absence of functional postsynaptic elements in the case of synaptosomes, or the absence of any known mechanism through which a postsynaptic receptor could locally control presynaptic release. It remains possible that feedback loops from the postsynaptic cell to the presynaptic neurons, or even from small intercalated local neurons, could mediate control of release in whole tissue slices, but there are often grounds for ruling this out—for instance, if tetrodotoxin, which blocks action potentials, is without effect on a response induced by field stimulation.

Using similar strategies, presynaptic receptor systems have been described that allow additional control of the release of most of the identified neurotransmitters in the brain.[9,M]

Presynaptic Receptors Controlling Central Dopamine Release Dopamine terminals are present in high density in the nigrostriatal system. As its anatomy is well known, the majority of release studies have employed preparations from this system. Inhibitory and facilitatory presynaptic receptors in caudate slices,[96,97] control both the rates of synthesis of dopamine and the extent of its release.[98,99,185] The effects on synthesis are most likely due to tyrosine hydroxylase being either structurally or functionally linked to the surface receptor.[100]

Just as norepinephrine release from peripheral nerves shows multiple control by various transmitter systems, so does dopamine release centrally. For instance, when acetylcholine and cholinergic agonists (e.g., carbechol) are added to caudate slices or synaptosomes, they stimulate dopamine release, whereas cholinergic antagonists block the responses. These effects are observed in the absence as well as the presence of depolarizing agents, and they involve the "newly synthesized" fraction of neurotransmitter, normally the source for transmitter released by nerve activity. There is evidence that a muscarinic type of presynaptic receptor is mediating inhibition of dopamine release and that a nicotinic category mediates facilitation of its release.[101,102,104,105,106,185] In addition, studies of ligand binding indicate the presence of presynaptically located muscarinic and nicotinic receptors.[103] For example, a reduction in the binding of a specific receptor-binding agent (e.g., α-bungarotoxin) after the destruction of terminals by axotomy or specific chemical lesion (e.g., 6-hydroxydopamine) points to the presence of the relevant receptors (e.g., nicotinic cholinergic) on presynaptic sites.[103]

Cholinergic receptors within the corpus striatum must be located on the endings of dopaminergic neurons since the cell bodies are in the substantia nigra, which is well removed from the corpus striatum to which the terminals are projected. The acetylcholine operating these receptors probably originates in the intrinsic cholinergic interneurons of the striatum itself; as deduced by histochemical and lesion studies, these interneurons contain most of the striatal acetylcholine.[107] Since the nigrostriatal dopaminergic neurons cause tonic inhibition of some striatal cholinergic neurons, these presynaptic receptors could be involved in a local feedback regulation of dopamine release at their synapses (Fig. 7.21).

Other neurotransmitter systems influencing dopamine release in the striatum include L-glutamate (stimulation),[108] met-enkephalin (inhibition),[109] GABA (stimulation),[110] and serotonin (stimulation).[111,185]

Agents modulating the release of dopamine and other transmitters usually have equivalent effects on the rates of synthesis and catabolism of these compounds, and this offers another parameter to measure when estimating the size and extent of these responses.[99,101,102,120]

In vivo monitoring of dopamine release with superfusion salines introduced through push-pull cannulae implanted in the caudate nucleus has shown that many of these *in vitro* responses probably reflect important physiological mechanisms.[112] It has been shown that nonisotopic dopamine releases newly synthesized [³H]dopamine from corpus striatum by a stereospecific receptor-mediated mechanism whose drug sensitivity strongly suggests the involvement of presynaptic structures.[113]

Presynaptic Receptors Controlling Central Norepinephrine Release

Since α-adrenergic agonists reduce the electrically and potassium-evoked release of norepinephrine from cerebral cortex slices and antagonists enhance this release, a presynaptically sited inhibitory noradrenergic mechanism is clearly indicated for this brain region.[87,93,185] Equivalent findings have been reported for other brain regions (e.g., hypothalamus, cerebellum).[93] Here, as for the other systems described in this chapter, the possibility can be ruled out that the increased transmitter levels detected in incubation media could be due to a block of reuptake by the drugs employed. Drugs showing no uptake-inhibition properties (e.g., phentolamine) produce equivalent responses.[87,93]

Mechanisms of Presynaptic Receptor Action

Several kinds of mechanism appear to be involved in presynaptic receptor actions. In each case receptor activation must lead to either an increase or a decrease in the normal rate of neurotransmitter release from the nerve terminal during action potential invasion. This implies an increase or decrease in the amount of neurotransmitter released per impulse. Such modulation could be achieved by control of one or more of the key factors involved in the usual release mechanism, such as calcium entry (see Chapter 4, presynaptic inhibition). Equally, manipulation of the extent of the prevailing membrane depolarization induced by invading action potentials influences the extent of release, for instance, by changing the rate of voltage-dependent calcium entry. Altering the size of the "releasable" neurotransmitter pool by modifying the activity of a rate-limiting biosynthetic enzyme can also change the rate of release.

Restriction of calcium entry does, indeed, appear to be involved in the operation of peripheral α-adrenergic presynaptic receptors. Evidence for calcium involvement comes from various sources. For example, the inhibition of neurotransmitter release caused by exposure to norepinephrine was much greater when normal calcium levels were reduced by 75 percent.[81,114]

The dependence of the size of presynaptic receptor responses on the frequency of nerve stimulation can also be explained in terms of calcium sensitivity. Neurotransmitter release evoked by *high-frequency* stimulation of nerves and other neural preparations is not easily modulated by transmitter agonists or antagonists, unlike release evoked by medium and low frequencies. During high-frequency stimulation, large amounts of calcium are likely to enter neurons and their nerve terminals, where cytoplasmic accumulation could lead to saturation of calcium-sensitive sites. Under these conditions agents working through a capacity to modulate calcium entry would become less effective.[115]

Peripheral presynaptic beta adrenoreceptors whose activation enhances nonrepinephrine release may well operate through a mechanism that includes the formation of cyclic AMP. Evidence for cyclic AMP involvement comes from the actions of dibutyryl cyclic AMP and other cyclic nucleotide analogues as well as from those of theophylline (a drug that reduces cyclic AMP breakdown) and several other phosphodiesterase inhibitors. All of these drugs increase the release of norepinephrine and dopamine β-hydroxylase from vas deferens, spleen, and other organ preparations during nerve stimulation[81,116] and during depolarizing treatments applied in vitro.[117]

Just how the intrasynaptic cyclic AMP is producing its facilitatory actions on neurotransmitter release from these terminals remains to be established. One possibility is that it accelerates tyrosine hydroxylase activity, the rate-limiting enzyme of catecholamine synthesis, thereby increasing the amount of norepinephrine available for release. A parallel effect is the increase in tyrosine hydroxylase activity, which occurs during sympathetic nerve stimulation. This, too, is associated with an increase in cyclic AMP levels in the nerve terminals and clearly involves a "switching-on" of adenylate cyclase, the enzyme that forms cyclic AMP.[1,118] Both phenomena may be operating through a similar chain of events, and the adenylate cyclase must be functionally and even physically linked to the presynaptic receptor in the nerve terminal membrane.

Another mode of action of intraterminal cyclic AMP generated by activation of these presynaptic beta adrenoreceptors could be through phosphorylation of a membrane protein, leading to changes in "ionic" fluxes across the terminal membrane that result in adjustments in membrane potential. In this case, a shift towards depolarization would enhance Ca^{2+} entry and so increase transmitter release. Phosphorylation of key membrane proteins leading to membrane-potential changes of this kind are envisaged in the mechanism of action of second messengers operating in the postsynaptic component of the synapses[119] following postsynaptic receptor activation (see Figure 4.4).

NEUROTRANSMITTER INTERACTIONS

Presynaptic receptors provide a means by which a given neurotransmitter in one neuronal pathway is able to influence the effects mediated by another neurotransmitter in a separate pathway.[K,L]

Interaction at the level of neuronal interconnections in the "wiring" patterns of the nervous system often involves presynaptic receptor mechanisms (e.g., presynaptic inhibition; see Figure 4.43). The presence of neurons using different neurotransmitters impinging on a third category of cell is a common feature of neural organization and indicates how a given group of neurons can receive inputs mediated by a range of neurotransmitters.

Neurotransmitter Interactions in the Basal Ganglia

One well-studied example of such a system in the mammalian brain is the interaction between pathways of the basal ganglia[121,127,175,K,L] (see Chapter 4, page 205). The basal ganglia are several closely grouped neuronal nuclei situated below and on each side of the cerebral cortex; and they are concerned with movement control, particularly at the finer level (the so-called extrapyramidal pathway). They include the *neostriatum*, consisting of the *caudate nucleus* and *putamen*, the two shells of the *globus pallidus*, and the *subthalamic nucleus*. The substantia nigra, although more caudal in position and situated in the region of the brain stem, is included as a basal ganglion structure because of its close functional and anatomical relationships with the neostriatum. The *pars compacta* of the inner substantia nigra consists largely of dopamine-containing cell bodies whose dendrites penetrate into the outer, *pars reticulata*, layer of this nucleus. These dopaminergic neurons project fibers through the brain to the caudate-putamen complex of the same side (ipsilateral), where they appear to synapse with various neuron types, including cholinergic interneurons[121,175] (Figure 7.23; see also Figure 9.5), and produce an inhibitory effect. This nigrostriatal pathway reduces the activity of these excitatory cholinergic interneurons. The subsequent connections that lead to the output of this caudate-putamen complex are not yet known, but several proposed wiring diagrams show the cholinergic interneurons making connections with GABA containing cells within the striatum[175] (Figure 7.23). It is likely that GABA-containing neurons in the globus pallidus project back to the dendrites (pars reticulata) of the substantia nigra to form part of the *striatonigral* pathway. These GABAergic terminals exert inhibitory actions on the dopaminergic neurons of the substantia nigra and thereby complete the circuit of the polysynaptic polytransmitter pathway. A reduction in the inhibitory action of these GABA terminals in the substantia nigra would allow an increase in the activity of the dopaminergic nigrostriatal pathway and vice versa. Here, then, are three neurotransmitters systems interacting in a serial pathway forming a loop. Several other transmitters are also known to be present and also interact in this basal ganglia circuitry. For instance, serotonin-containing terminals project from the dorsal raphe nucleus in the upper region of the brain stem to the caudate-putamen, where they make inhibitory synapses with the cholinergic interneurons. Similar fibers also project to the substantia nigra and interact with dopaminergic neurons. The evidence suggests that these serotoninergic terminals exert an inhibitory action in both cases.

Peptide transmitters also appear to be involved; excitatory substance P–containing neurons (see Chapter 5) with cell bodies in the neostriatum project to the substantia nigra. And enkephalin-containing neurons in the caudate appear to project to the globus pallidus, where they are likely to inhibit the output to the substantia nigra. They also project to the subthalamic nuclei, which exchange fiber pathways with the globus pallidus[121,122,127] (Figure 7.23). Moreover, angiotensin-containing interneurons have been detected in the striatum, although how they participate remains unclear.[121]

Another very large input to the caudate nucleus from the cerebral cortex is the corticostriatal pathway, which appears to arise principally from the frontal cortex. Although not yet firmly established, current evidence[123] strongly indicates that glutamate is one of the transmitters in this excitatory input to the basal ganglia complex which represents about 30 percent of the total input to the caudate-putamen (See Figure 4.38).

There are other inputs to the basal ganglia system from thalamus, midbrain, and elsewhere, and each will add the influence of another transmitter type to this complex cascade of interactions. For instance a minor noradenergic input extends from the locus ceruleus to the striatum.[121]

Dendritic Release of Neurotransmitter

An unusual feature of dopaminergic neurons of the substantia nigra is the Ca^{2+}-dependent release of neurotransmitter from their dendrites. This has been concluded from both in vivo and in vitro experiments in which dopamine release was evoked from substantia nigra with depolarizing agents.[124,125] Since only the cell bodies and dendrites of dopaminergic neurons are present in the substantia nigra, one or both of these must have been the site of origin of the released dopamine. In vitro experiments with separated pars reticulata and pars compacta regions show a more efficient release from pars reticulata, which contains the dendrites.[125,K] These dendrites not only have dopamine (in vesicles) and its synthetic enzymes but also take dopamine up by high-affinity transport processes.

The present hypothesis is that dopamine released from dendrites during activity of the dopaminergic neurons of the pars compacta interacts with receptors on the external surface of these dendrites, leading to "self-inhibition" of these neurons through what must be regarded as dopaminergic "autoreceptors" (see above, page 390). Dopamine released from these sites is likely to modify the activity of other neurons in the vicinity, including the GABAergic, cholinergic,[126] and serotonergic variety.

There is also evidence that acetylcholine is synthesized and stored in the dendrites of cholinergic neurons in the neostriatum, from which it is released to act at nicotinic or muscarinic receptors on dopaminergic nerve endings.[102,K]

The Output of the Basal Ganglia

The overall picture,[K,L] far from being finally accepted, describes the convergence of largely excitatory inputs from cerebral cortex and elsewhere into the caudate-putamen (striatum), where they interact with the excitatory "pacemaker" neurons and with the inhibitory command and local-circuit neurons.[127,175] The chief switching mechanism for turning on a sequentially patterned, smooth, and coordinated output to the globus pallidus, and on to the thalamus and other regions, probably comes from the pathway projecting from the substantia nigra. After feedback through the cerebral cortex, these neural activities ultimately initiate and guide the desired bodily movement via appropriate discharges down the pyramidal tract and on to the motoneurons of the spinal cord, smoothed by preprogrammed inputs from the cerebellum (Figure 7.24).

Why several different kinds of excitatory and inhibitory transmitters are required to achieve this output from the basal ganglia remains unclear but is presumably owing to the subtle differences in the nature of the interactions of these transmitters with their receptors—for instance, their different time courses of action, and, in some cases, an ability to act at a distance by diffusion.

Failure at the biochemical level in the basal ganglia leads to failure in the coordination of movement. This is expressed in diseases such as Parkinson's disease, characterized by tremor and a relative poverty of movement (see Chapter 9). The dyskinesias, disorders characterized by abnormal and sometimes dramatically energized or bizarre involuntary movements, include Huntingdon's chorea, dystonia, torticollis, and myoclonus. These conditions mostly include behavioral as well as movement and locomotory abnormalities, reflecting the likely additional role of the basal ganglia in the behavioral output of the brain.[128] This point is discussed in more detail in Chapters 8 and 9.

IS NEUROTRANSMITTER RELEASED FROM ALL REGIONS OF THE NEURONAL CELL SURFACE?

Neurotransmitter release has been traditionally regarded as occurring from nerve terminals existing in close apposition to their postsynaptic junctional membrane. But transmitter release from dendrites can also be demonstrated experimentally, at least for the dendrites of the dopaminergic neurons in the substantia nigra, as described above. In fact, a

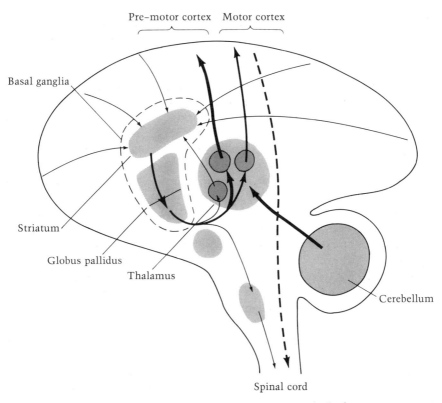

FIGURE 7.24   Extrapyramidal pathways in the control of movement. Thin arrows represent pathways from various parts of the cortex. Reprocessed signals from the basal ganglia and cerebellum (*solid line*) pass through the thalamus to the motor cortex, which then relays them to spinal cord (*dotted line*).

body of evidence has now accumulated indicating that in a variety of neuronal systems, dendrites may act as sites for the storage and release of neurotransmitters.[133,134] These include the GABAergic axonless granule cell dendrites of the olfactory bulb and substance P–releasing dendritic projections of peptide-containing primary sensory neurons. In the olfactory system, the mitral cell dendrite (the input stage from the olfactory nerve) is excitatory to the granule cell dendrite, and the granule cell dendrite is inhibitory to the mitral cell dendrite. This led to the concept of reciprocal dendrodendritic synapses, which was reinforced by ultrastructural studies showing unusual features for dendritic profiles and contacts.[133] Such dendrodendritic synapses have been demonstrated for a number of neural systems (see Figure 1.32) and synaptic vesicles that might represent the transmitter storage sites, have been reported to occur in dendrites of substantia nigra neurons.[135]

Apart from dendrites, both axons and cell bodies have been shown to release neurotransmitters (e.g., acetylcholine, norepinephrine, amino acids) by calcium-dependent mechanisms when stimulated.[134,136,137,170, 171,172] They include sympathetic preganglionic axons, the hypogastric nerve, and postganglionic cell bodies. In fact, neurotransmitters may be released from most, if not all, of the neuron surface during passage of the action potential, though it may not show the same pattern of modulation by drugs as synaptically released neurotransmitter.[170]

Both catecholamine and serotonin axons, particularly in the autonomic system, have extensive regions bearing vesicle-filled varicosities (1–2 μM diameter) that can release neurotransmitter during passage of an action potential. Very often they show few or none of the ultrastructural pre- or postsynaptic specializations typical of conventional synapses[73,138,139] (see Figure 1.29). Varicosities are also seen in monoaminergic fiber systems of the central nervous system.[138] Transmitters or other neuroactive substances released from these nonterminal neuronal regions may have to diffuse some distance in extracellular fluid, perhaps several microns, before interacting with the appropriate postsynaptic receptors. As might be expected, they have a localized hormonelike action, now termed a *paracrine action* to distinguish it from the *endocrine action* of hormones released to the blood and acting over much greater distances. Paracrine release mechanisms are themselves likely to be subject to modulatory influences from other locally released or circulating substances.

Thus, monoamine and peptide neurotransmitters released from the central and peripheral regions of the nervous system from nonsynaptic structures must be viewed as exerting a slow tonic influence over a relatively large distance, or volume, of tissue during extended periods of release,[73,134,O] rather in the fashion of a hormone.

REFERENCES

General Texts and Review Articles

A Cotman, C. W., and G. A. Banker. (1974) *The Making of a Synapse* (eds. S. Ehrenpreis, and I. J. Kopin). Reviews of Neuroscience, vol. 1, Raven Press, New York.

B Varon, S. S., and G. G. Somjen, eds. (1979) *Neuron–Glia Interactions. Neurosci. Res. Prog. Bull.* 17 (1): 1–239.

C Edds, M. V., R. M. Gaze, G. E. Schneider, and L. N. Irwin, eds. (1979) *Specificity and Plasticity of Retinotectal Connections. Neurosci. Res. Prog. Bull.* 17 (2): 243–375.

D E. Eidelberg, and D. G. Stein, eds. (1974) *Functional Recovery after Lesions of the Nervous System. Neurosci. Res. Program Bull.* 12 (2). 191–303.

E Katz, B. (1969) *The Release of Neural Transmitter Substances.* Liverpool University Press, Liverpool.

F Llinás, R. R., and J. E. Heuser, eds. (1977) *Depolarization-Release Coupling Systems in Neurons. Neurosci. Res. Prog. Bull.* 15 (4): 557–687.

G Paton, D. M., ed. (1979) *The Release of Catecholamines from Adrenergic Neurons.* Pergamon Press, Oxford.

H Langer, S. Z., K. Starke, and M. L. Kubocovich, eds. (1979) *Presynaptic Receptors.* Pergamon Press, Oxford and New York.

I Usdin, E., and W. E. Bunney, eds. (1975) *Pre- and Post-Synaptic Receptors.* Marcel Dekker, New York.

J de Belleroche, J. S., ed. (1982) *Presynaptic Receptors.* Ellis-Horwood, Chichester, U.K.

K Garattiac, S., J. F. Pujol, and R. Samanin, eds. (1978) *Interaction Between Putative Neurotransmitters in the Brain.* Raven Press, New York.

L Butcher, L. L., ed. (1978) *Cholinergic–Monoaminergic Interactions in the Brain.* Academic Press, New York.

M Bradford, H. F., ed. (1982) *Neurotransmitter Interaction and Compartmentation.* Plenum Press, London and New York.

N Jacobson, M. (1978) *Developmental Neurobiology,* 2d ed. Plenum Press, New York and London.

O Osborne, N. N., ed. (1983) *Dale's Principle and Communication Between Neurons.* Pergamon Press, Oxford.

P Cuello, A. C., ed. (1982) *Cotransmission.* Macmillan Press, London.

Q Kretsinger, R. H., ed. (1981) *Mechanisms of Selective Signaling by Calcium. Neurosci. Res. Prog. Bull.* 19 (3): 213–328.

R Spitzer, N. C., ed. (1982) *Neuronal Development.* Plenum Press, New York and London.

S Hopkins, W. G., and M. C. Brown (1984) *Development of Nerve Cells and Their Connections.* Cambridge University Press, Cambridge.

Additional Citations

1 Tennyson, V. M. (1970) *J. Cell Biol.* 44: 62–79.

2 Yamada, K., B. S. Spooner, and N. K. Wassells. (1971) *J. Cell Biol.* 49: 614–635.

3 Goldberg, S., and A. J. Coulombre. (1972) *J. Comp. Neurol.* 146: 507–518.

4 Sharma, S. C. (1972) *Nature (New Biol.)* 238: 286–287.

5 Macagno, E. R. (1978) *Nature* (Lond.) 275: 318–320.

6 Wigglesworth, V. B. (1953) *Q. J. Microsc. Sci.* 94: 93–100.

7 Sperry, R. W. (1963) *Proc. Natl. Acad. Sci.* (USA) 50: 703–710.

8 Landmesser, L. (1980) *Ann. Rev. Neurosci.* 3: 279–302.

9 Pettrigrew, A. G., R. Lindeman, and M. R. Bennett. (1979) *J. Embryol. Exp. Morphol.* 49: 115–137.

10 Merrell, R., M. W. Pulliam, L. Randono, L. Boyd, R. Bradshaw, and L. Glaser. (1975) *J. Biol. Chem.* 250: 5055–5059.

11 Ben-Shaul, Y., R. E. Hausman, and A. A. Moscona. (1979) *Dev. Biol.* 72: 89–101.

12 Garber, B. B., and C. T. Moscona. (1972) *Dev. Biol.* 27: 235–243.

13 Gottlieb, D. I., R. Merrell, and L. Glaser. (1974) *Proc. Natl. Acad. Sci.* (USA) 71: 1800–1802.

14 Gottlieb, D. I., and L. Glaser. (1975) *Biochem. Biophys. Res. Comm.* 63: 815–821.

15 Harper, G. P., and H. Thoenen. (1980). *J. Neurochem.* 34: 5–16.

16 Bradshaw, R. A. (1978) *Ann. Rev. Biochem.* 47: 191–216.
17 Server, A. C., and E. M. Shooter. (1977) *Adv. Protein Chem.* 31: 339–409.
18 Varon, S. (1975) *Exp. Neurol.* 48: 75–92.
19 Rohver, H., U. Otten, and H. Thoenen. (1978) *Brain Res.* 159: 436–439.
20 Levi-Montalcini, R., and P. U. Angeletti. (1968) *Physiol. Rev.* 48: 534–569.
21 Levi-Montalcini, R. (1976) *Prog. Brain Res.* 45: 235–258.
22 Letourneau, P. C. (1978) *Dev. Biol.* 66: 183–196.
23 Schneider, G. E. (1971) *Anat. Rec.* 169: 420.
24 Kornguth, S. E., and G. Scott. (1973) *J. Comp. Neurol.* 146: 61–82.
25 Altman, J. (1973) *J. Comp. Neurol.* 149: 153–160.
26 Llinás, R., D. Hillman, and W. Precht. (1973) *J. Neurobiol.* 4: 69–94.
27 Changeux, J.-P., and A. Dauchin. (1976) *Nature* (Lond.) 264: 705–711.
28 Lewis, E. R., and C. W. Cotman. (1983) *Neuroscience* 8: 57–66.
29 Cotman, C. W. (1976) *J. Supramol. Struct.* 4: 319–327.
30 Raisman, G., and P. M. Field. (1973) *Brain Res.* 50: 241–264.
31 McWilliams, J. R., and G. S. Lynch. (1978) *J. Comp. Neurol.* 180: 581–615.
32 Barber, P. C. (1981) *Brain Res.* 216: 239–251.
33 Field, P. M., D. E. Coldham, and G. Raisman. (1980) *Brain Res.* 189: 103–113.
34 Schneider, G. E. (1973) *Brain Behav. Evol.* 8: 73–109.
35 Tsukahara, N., F. Murakami, and H. Hultborn. (1975) *J. Neurophysiol.* 38: 1359–1372.
36 Steward, O., C. W. Cotman, and G. Lynch. (1973) *Exp. Brain Res.* 18: 396–414.
37 Llinás, R., I. Z. Steinberg, and K. Walton. (1976) *Proc. Natl. Acad. Sci.* (USA) 73: 2918–2922.
38 De Lorenzo, R. J. (1980) *Ann. N.Y. Acad. Sci.* (USA) 356: 92–109. *Proc. Natl. Acad. Sci.* (USA) 76: 1838–1842.
39 Marx, J. L. (1980) *Science* 208: 274–276.
40 Cheung, W. T. (1980) *Science* 207: 19–27.
41 Burke, B., and R. J. De Lorenzo. (1981) *Proc. Natl. Acad. Sci.* (USA) 78: 991–995.
42 Sobue, K., K. Kanda, K. Yamagami, and S. Kakiuchi. (1982) *Biomed. Res.* 3: 561–570.
43 Gordon-Weeks, P. R., R. D. Burgoyne, and E. G. Gray. (1981) *Neuroscience* 6: 1793–1811.
44 Miledi, R. (1973) *Proc. R. Soc.* (Lond.) *Ser. B.* 183: 421–425.
45 Lance-Jones, C., and L. Landmesser. (1980) *J. Physiol.* (Lond.) 302: 581–602.
46 Llinás, R., J. R. Blinks, and C. Nicholson. (1972) *Science* 176: 1127–1129.
47 Llinás, R., and C. Nicholson. (1975) *Proc. Natl. Acad. Sci.* (USA) 72: 187–190.
48 Weakly, J. N. (1969) *J. Physiol.* (Lond.) 204: 63–77.
49 Katz, B., and R. Miledi. (1979) *Proc. R. Soc.* (Lond.) *Ser. B.* 205: 369–378.
50 Kuffler, S. W., and D. Yoshikami. (1975) *J. Physiol.* (Lond.) 251: 465–482.
51 Fletcher, P., and T. Forrester. (1975) *J. Physiol.* (Lond.) 251: 131–144.
52 Fatt, P., and B. Katz. (1952) *J. Physiol.* (Lond.) 117: 109–128.
53 Tauc, L. (1979) *Biochem. Pharmacol.* 28: 3493–3498.
54 Hubbard, J. I. (1973) *Physiol. Rev.* 53: 674–723.

55 Heuser, J. E., T. S. Reese, M. J. Dennis, L. Jan, and L. Evans. (1979) *J. Cell Biol.* 81: 275–300.

56 Dunant, T., M. Israel, B. Lesbats, and R. Manarache. (1972) *J. Neurochem.* 19: 1987–2002.

57 De Belleroche, J. S., and H. F. Bradford. (1977) *J. Neurochem.* 29: 335–343.

58 Zimmerman, H. (1979) *Neuroscience* 4: 1773–1803.

59 Zimmerman, H., and C. R. Denston. (1977) *Neuroscience* 2: 695–714.

60 Luqmani, Y. A., G. Sudlow, and V. P. Whittaker. (1980) *Neuroscience* 5: 153–160.

61 Schwarzenfeld, I. von. (1979) *Neuroscience* 4: 477–493.

62 Suszkin, J. B., H. Zimmerman, and V. P. Whittaker. (1978) *J. Neurochem.* 20: 1269–1280.

63 Tauc, L., A. Hoffmann, S. Tsuji, D. H. Hinzen, and L. Faille. (1974) *Nature* (Lond.) 250: 496–498.

64 Dunant, Y., J. Gautron, M. Israel, B. Lesfats, and R. Manaranche. (1974) *J. Neurochem.* 23: 635–643.

65 Schwarzenfeld, I. von, G. Sudlow, and V. P. Whittaker. (1979) *Prog. Brain Res.* 49: 163–174.

66 Whittaker, V. P., and Y. A. Luqmani. (1980) *Gen. Pharmacol.* 11: 7–14.

67 Giompres, P. E., H. Zimmerman, and V. P. Whittaker. (1981) *Neuroscience* 6: 765–774.

68 Giompres, P. E., H. Zimmerman, and V. P. Whittaker. (1981) *Neuroscience* 6: 775–785.

69 Israel, M., and B. Lesbats. (1981) *J. Neurochem.* 37: 1475–1483.

70 Heuser, J. E. (1979) In *Neurobiology of Chemical Transmission* (eds. M. Otsuka and Z. W. Hall), pp. 3–11. Wiley, New York.

71 Burnstock, G. (1976) *Neuroscience* 1: 239–248.

72 Burnstock, G. (1978) *Progr. Neurobiol.* 11: 205–222.

73 Burnstock, G. (1981) *J. Physiol.* (Lond.) 313: 1–35.

74 Hökfelt, T., L. G. Elfvin, R. Elde, M. Schultzberg, M. Goldstein, and R. Luft. (1977) *Proc. Natl. Acad. Sci.* (USA) 74: 3587–3591.

75 Schultzberg, M., T. Hökfelt, L. Terenius, L. G. Elfvin. J. M. Lundberg, J. Brandt, R. P. Elde, and M. Goldstein. (1979) *Neuroscience* 4: 249–270.

76 Chan-Palay, V., G. Jonsson, and S. L. Palay. (1978) *Proc. Natl. Acad. Sci.* (USA) 75: 1582–1586.

77 Blaustein, M. P., C. F. McGraw, A. V. Somlyo, and E. S. Schweitzer. (1980) *J. Physiol.* (Paris) 76: 459–470.

78 Carlsson, A. (1975) In *Pre and Post Synaptic Receptors* (eds. E. Usdin and W. E. Bunney), pp. 49–66. Marcel Dekker, New York.

79 Mestikawy, S. E., J. Glowinski, and M. Hamon. (1983) *Nature* (Lond.) 302: 830–832.

80 Brown, G. L., and J. S. Gillespie. (1957) *J. Physiol.* (Lond.) 138: 81–102.

81 Langer, S. Z. (1979) In *The Release of Catecholamines from Adrenergic Neurons* (ed. D. M. Paton), pp. 59–85. Pergamon Press, Oxford.

82 Langer, S. Z. (1977). *Brit. J. Pharmacol.* 60: 481–497.

83 Starke, K. E. Borowski, and T. Endo. (1975) *Eur. J. Pharmacol.* 34: 385–388.

84 Starke, K., H. Montel, K. W. Gay, and R. Merker. (1974) *Naunyn-Schmiedeberg's Arch. Pharmacol.* 285: 113–150.

85 Langer, S. Z., E. Adler-Graschinsky, and O. Giorgi. (1977) *Nature* (Lond.) 265: 648–650.

86   Langer, S. Z. (1976) *Clin. Sci. Mol. Med.* 51: 423–426.
87   Starke, K. (1977) *Rev. Physiol. Biochem. Pharmacol.* 77: 1–124.
88   Enero, M. A., and S. Z. Langer. (1975) *Naunyn-Schmiedeberg's Arch. Pharmacol.* 289: 179–203.
89   Muscholl, E. (1979) In *The Release of Catecholamines from Adrenergic Neurons* (ed. D. M. Paton), pp. 87–110. Pergamon Press, Oxford.
90   Löffelholz, K. (1979) In *The Release of Catecholamines from Adrenergic Neurons* (ed. D. M. Paton), pp. 275–302. Pergamon Press, Oxford.
91   Clanachan, A. S. (1979) In *The Release of Catecholamines from Adrenergic Neurons* (ed. D. M. Paton), pp. 263–274. Pergamon Press, Oxford.
92   Stjärne, L. (1979) In *The Release of Catecholamines from Adrenergic Neurons* (ed. D. M. Paton), pp. 111–142. Pergamon Press, Oxford.
93   Starke, K. (1979) In *The Release of Catecholamines from Adrenergic Neurons* (ed. D. M. Paton), pp. 143–184. Pergamon Press, Oxford.
94   Lindmar, R., K. Loffelholz, and E. A. Muscholl. (1968) *Brit. J. Pharmacol.* 32: 280–294.
95   Westfall, T. C. (1974) *Life Sci.* 14: 1641–1652.
96   Arbella, S., J. Z. Nowak, and S. Z. Langer. (1982) In *Presynaptic Receptors* (ed. J. S. de Belleroche), pp. 30–45. Ellis-Horwood, Chichester.
97   Farnebo, L. O., and B. Hamberger. (1971) *Acta Physiol. Scand. Suppl.* 271: 35–44.
98   Christiansen, J., and R. F. Squires. (1974) *J. Pharm. Pharmacol.* 26: 367–369.
99   Westfall, T. C., M. J. Besson, M. F. Giorguieff, and J. Glowinski. (1976) *Naunyn-Schmiedeberg's Archiv. Pharmacol.* 292: 279–287.
100 Horn, A. S., J. Korf, and B. H. C. Westerink. (1979) *The Neurobiology of Dopamine.* Academic Press, London.
101 de Belleroche, J. S., and H. F. Bradford. (1978) *Brain Res.* 142: 53–68.
102 de Belleroche, J. S., and H. F. Bradford. (1978) *Adv. Biochem. Psychopharmacol.* 19: 57–73.
103 de Belleroche, J. S., Y. Luqmani, and H. F. Bradford. (1979) *Neurosci. Lett.* 11: 209–213.
104 Giorguieff, M. F., M. L. Le Floc'h, T. C. Westfall, J. Glowinski, and M. J. Besson. (1976) *Brain Res.* 106: 117–131.
105 Giorguieff, M. F., M. L. Le Floc'h, J. Glowinski, and M. J. Besson. (1977) *J. Pharmacol. Exp. Ther.* 200: 535–540.
106 Westfall, T. C. (1974) *Fed. Proc.* 33: 524.
107 Butcher, S. G., and L. L. Butcher. (1974) *Brain Res.* 71: 167–171.
108 Giorguieff, M. F., M. L. Kemel, and J. Glowinski. (1977) *Neurosci. Lett.* 6: 73–77.
109 Subramanian, V., P. Mitzivegg, W. Sprügel, W. Domschke, S. Domschke, E. Wünsch, and L. Demling. (1977) *Naunyn Schmiedeberg's Arch. Pharmacol.* 299: 163–166.
110 Giorguieff, M. F., M. L. Kemel, J. Glowinski, and M. J. Besson. (1978) *Brain Res.* 139: 115–130.
111 de Belleroche, J. S., and H. F. Bradford. (1980) *J. Neurochem.* 35: 1227–1234.
112 Glowinski, J., A. Cheramy, and M. F. Giorguieff. (1979) In *The Neurobiology of Dopamine* (eds. A. S. Horn, J. Korf, and B. H. C. Westerink), pp. 199–216. Academic Press, London.
113 Cheramy, A., A. Nieoullon, R. Michelot, J. Glowinski. (1977) *Neurosci. Lett.* 4: 105–109.
114 Langer, S. Z. (1980) *Pharmacol. Rev.* 32: 337–362.

115 Bennett, M. R., and T. Florin. (1975) *Brit. J. Pharmacol.* 55: 97–104.

116 Cubeddu, L. X., E. Barnes, and N. Weiner. (1975) *J. Pharmacol. Exp. Ther.* 193: 105–127.

117 Pelayo, F., M. L. Dubocovich, S. Z. Langer. (1978) *Nature* (Lond.) 274: 76–78.

118 Roth, R. D., J. R. Walters, L. C. Muriu, and V. H. Morgenroth. (1975) In *Pre- and Post-Synaptic Receptors* (eds. E. Usden and W. E. Bunney), pp. 5–46. Marcel Dekker, New York.

119 Greengard, P. (1976) *Nature* (Lond.) 260: 101–108.

120 Patrick and Barchas. (1974) *J. Neurochem.* 23: 7–15.

121 McGeer P. L., E. G. McGeer, and T. Hattori. (1979) *Prog. Brain Res.* 51: 286–301.

122 Watson, S. J., H. Akil, S. Sullivan, and J. D. Barchas. (1977) *Life Sci.* 21: 733–738.

123 McGeer, P. L., E. G. McGeer, U. Scherer, and K. Singh. (1977) *Brain Res.* 128: 369–373.

124 Nieoullon, A., A. Cheramy, and J. Glowinski. (1977) *Nature* (Lond.) 266: 375–377.

125 Geffen, L. B., T. M. Jessell, A. C. Cuello, and L. L. Iversen. (1976) *Nature* (Lond.) 260: 258–260.

126 Korf, J., M. Zieleman, and B. H. C. Westerink. (1976) *Nature* (Lond.) 260: 257–258.

127 Dray, A. (1979) *Neuroscience* 4: 1407–1439.

128 Marsden, C. D. (1980) *Trends Neurosci.* 3: 284–287.

129 Eckenstein, F., and R. W. Baughman. (1984) *Nature* (Lond.) 309: 153–155.

130 Rakic, P., L. J. Stensaas, P. E. Sayre, and R. L. Sidman. (1974) *Nature* (Lond.) 250: 31–34.

131 Rakic, P. (1972) *J. Comp. Neurol.* 145: 61–84.

132 Cowan, W. M. (1979) *Sci. Am.* 241(3): 113–133.

133 Cuello, A. C. (1983) In *Dale's Principle and Communication Between Neurons* (ed. N. N. Osborne), pp. 69–82. Pergamon Press, Oxford.

134 Vizi, E. S. (1983) In *Dale's Principle and Communication Between Neurons* (ed. N. N. Osborne), pp. 83–112. Pergamon Press, Oxford.

135 Hajdu, F., R. Hassler, and J. Bak. (1973) *Z. Zellforsch. Mikrosk. Anat.* 146: 207–221.

136 Esquerro, E., V. Cena, P. Sandez-Garcia, S. M. Kirpekar, and A. G. Garcia. (1980) *Eur. J. Pharmacol.* 61: 183–186.

137 Johnson, D. A., and G. Pilar. (1980) *J. Physiol.* (Lond.) 299: 605–619.

138 Beaudet, A., and L. Descarries. (1978) *Neuroscience* 3: 851–860.

139 Leger, L., and L. Descarries. (1978) *Brain Res.* 145: 1–13.

140 Costa, E., C. M. Forchetti, A. Guidotti, and B. C. Wise. (1983) In *Dale's Principle and Communication Between Neurons* (ed. N. N. Osborne), pp. 161–177. Pergamon Press, Oxford.

141 Burnstock, G. (1983) In *Dale's Principle and Communication Between Neurons* (ed. N. N. Osborne), pp. 7–35. Pergamon Press, Oxford.

142 Feldman, J. (1978) In *Intercellular Junctions and Synapses* (eds. J. Feldman, N. B. Gilula, and J. D. Pitts), pp. 181–214. Receptors and Recognition, Series B, vol. 2. Chapman and Hall, London.

143 Slater, C. R. (1978) In *Intercellular Junctions and Synapses* (eds. J. Feldman, N. B. Gilula, and J. D. Pitts), pp. 217–239. Receptors and Recognition, Series B, vol. 2. Chapman and Hall, London.

144 Merrell, R., D. I. Gottlieb, and L. Glaser. (1975) *J. Biol. Chem.* 250: 5655–5659.
145 Rutishauser, U., J.-P. Thiery, R. Brackenbury, B. A. Sela, and G. M. Edelman. (1976) *Proc. Natl. Acad. Sci.* (USA) 73: 577–581.
146 Hausman, R. E., and A. A. Moscona. (1979) *Exp. Cell Res.* 119: 191–204.
147 Rutishauser, U., S. Hoffman, and G. M. Edelman. (1982) *Proc. Natl. Acad. Sci.* (USA) 79: 685–689.
148 Barondes, S. H. (1982) *Trends Neurosci.* 5: 280–282.
149 Trisler, D. (1982) *Trends Neurosci.* 5: 306–310.
150 Cunningham, B. A., S. Hoffman, U. Rutishauser, J. J. Hemperly, and G. M. Edelman. (1983) *Proc. Natl. Acad. Sci.* (USA) 80: 3116–3120.
151 Hausman, R. E., and A. A. Moscona. (1975) *Proc. Natl. Acad. Sci.* (USA) 72: 916–920.
152 Levi-Montalcini, R. (1964) *Science* 143: 105–110.
153 Levi-Montalcini, R., and S. Cohen. (1960) *Ann. N.Y. Acad. Sci.* (USA) 85: 324–341.
154 Letourneau, P. C. (1981) *Dev. Biol.* 85: 113–122.
155 Adler, R., and S. Varon. (1981) *Dev. Biol.* 81: 1–11.
156 Llinás, R. R. (1977) In *Society for Neuroscience Symposia*, vol. 2 (eds. W. M. Cowan and J. A. Ferrendelli), pp. 139–160. Bethesda, Maryland.
157 Couteaux, R., and M. Pecót-Dechavassine. (1973) *Arch. Ital. Biol.* 3: 231–262.
158 Couteaux, R., and M. Pecót-Dechavassine. (1970) *Can. R. Acad. Sci. (D)* 271: 2346–2349.
159 Couteaux, R., and M. Pecót-Dechavassine. (1974) *Can. R. Acad. Sci. (D)* 278: 291–293.
160 Fried, R. C., and M. P. Blaustein. (1978) *J. Cell Biol.* 78: 685–700.
161 Douglas, W. W. (1974) In *The Adrenal Gland* (ed. R. O. Greep, et al.). *Handbook of Physiology*, Section 7, Endocrinology, vol. 6, pp. 367–388. American Physiology Society, Washington, D.C.
162 Lundberg, J. M. (1981) *Acta Physiol. Scand.* 112 (Suppl. 496): 1–57.
163 Nickel, E., and L. T. Potter. (1970) *Brain Res.* 23: 95–100.
164 Katz, B., and R. Miledi. (1969) *J. Physiol.* (Lond.) 203: 689–706.
165 Akert, K., K. Pfenninger, C. Sandri, and H. Moore. (1972) In *Structure and Function of Synapses* (eds. G. D. Pappas, and D. P. Purpura), pp. 67–86. Raven Press, New York.
166 Patterson, P. H. (1979) *Prog. Brain Res.* 51: 75–82.
167 Cotman, C. W. (1975) *Prog. Brain Res.* 51: 203–215.
168 Hopkins, W. G., and M. C. Brown. (1984) *Development of Nerve Cells and Their Connections*. Cambridge University Press, Cambridge.
169 Edelman, G. M. (1984) *Trends Neurosci.* 7: 78–84.
170 Vizi, E. S., K. Gyives, G. T. Somogyi, and G. Ungváry. (1983) *Neuroscience* 10: 967–972.
171 Johnson, D. A., and G. Pilas. (1980) *J. Physiol.* (Lond.) 299: 605–619.
172 Wheeler, D. D., L. L. Boyarsky, and W. H. Brooks. (1966) *J. Cell Physiol.* 67: 141–147.
173 Hoffman, S., and G. M. Edelman. (1983) *Proc. Natl. Acad. Sci.* (USA) 80: 5762–5766.
174 Grumet, M., U. Rutihauser, and G. M. Edelman. (1984) *Proc. Natl. Acad. Sci.* (USA) 81: 267–271.

175 Lehmann, J., and S. Z. Langer. (1983) *Neuroscience* 10: 1105–1120.

176 Dunnett, S. B., A. Bjorkland, and U. Stenevi. (1983) *Trends Neurosci.* 6: 266–270.

177 Dunnett, S. B., A. Bjorkland, U. Stenevi, and S. D. Iversen. (1981) *Brain Res.* 229: 457–470.

178 Schmidt, R. H., M. Ingvar, O. Lindvall, U. Stenevi, and A. Bjorkland. (1982) *J. Neurochem.* 38: 737–748.

179 Schmidt, R. H., A. Bjorkland, and U. Stenevi. (1981) *Brain Res.* 218: 347–356.

180 Dunnett, S. B., R. H. Schmidt, A. Bjorkland, U. Stenevi, and S. D. Iveresen. (1981) *Neurosci. Lett. Suppl.* 7: 176.

181 Isacson, O., P. Brundin, P. A. T. Kelly, F. H. Gage, and A. Bjorkland. (1984) *Nature* (Lond.) 311: 458–460.

182 Kakiuchi, S. (1983) *Neurochem. Int.* 5: 159–169.

183 Watterson, D. M., F. Sharief, and T. C. Vanaman. (1980) *J. Biol. Chem.* 225: 962–975.

184 Landmesser, L. (1984) *Trends Neurosci.* 7: 336–339.

185 Chesselet, M.-F. (1984) *Neurosci.* 12: 347–375.

186 Dunant, Y., and M. Israel. (1985) *Sci. Am.* 252(4):40–48.

8

Central
Neurotransmitter
Systems and
Behavior

The goal of defining mammalian behavior in terms of neurotransmitter systems and neural pathways is still a very long way from being achieved. Of course, all behavior is initiated in the brain and may be manifested via the peripheral nervous system as complex voluntary or involuntary responses or via the hypothalamus as changes in endocrine secretion. The involuntary manifestations of central activity, expressed through the autonomic nervous system, may involve modulation of neural control of smooth muscle in the viscera or blood vessels, or changes in movement, posture, or glandular secretion. In common language this means such behavior as crying, trembling, blanching of the face, sweating, turning of the stomach, or pounding of the heart. But many brain states or moods may not be so clearly expressed through bodily responses and are not therefore readily accessible for study and measurement. And the neurotransmitters involved in producing bodily responses at the autonomic and peripheral level may well be different from those involved in the central initiating processes within the brain. We consider in this chapter only those neurotransmitter systems which are involved in central processes occurring within the brain and spinal cord.

The brain structures that appear to be primarily involved in neural mechanisms generating emotion are known collectively as the *limbic system*. This system includes the cingulate gyrus of the cerebral cortex, the hippocampus, amygdala, septum, hypothalamus, olfactory bulb, thalamus, and fornix. Fear, anger, satisfaction, and pleasure are typical behavior aspects influenced by the limbic system.

RESEARCH STRATEGIES

A particular neurotransmitter system within the brain may play an initiating or dominant role in triggering or carrying through specific behavioral programs, but it is never the only system involved. Hierarchies of neurotransmitter interactions are a feature of every behavioral pattern being investigated.

A wide range of well-defined body movements, postures, and attitudes occurring in experimental animals as a result of chemical or surgical intervention in the brain may be classified as patterns of behavior. Pharmacological manipulation of the level of activity of a particular neurotransmitter system may be achieved by the application of specific agonists, antagonists, depleters, or neurotoxins. These manipulations can produce well-defined and sometimes simple responses. Relatively well defined forms of evokable behavior may be quantified to some extent or at least rated on a severity scale; a list of these behaviors for rodents and other laboratory animals would include compulsive stereotypic movements involving repetitive acts such as sniffing, licking, and gnawing. In addition, there is withdrawal, sleep, arousal, rage, aggression, sexual behavior, and exploration of the environment. Unusual patterns of eating or drinking, and characteristic head or limb movements, or curious locomotory movements such as circling behavior (locomotion in a tight circle often in one direction)[1] can all be induced in laboratory rodents by administering specific drugs. Sometimes responses are seen as patterns of generalized hyperactivity that can be measured as whole-body movements by "activity meters" and are often resolved into various subcomponents, such as forepaw-treading, head-weaving, or myoclonic limb-jerks.[2,3,27]

Other whole-body movements that can be evoked in certain laboratory animals include such unusual behavior as "wet-dog shakes" and "backwards-walking." These relatively simple patterns of behavior obviously have immediate links with basic neurological processes. Certain kinds of more complex and subtly motivated behavior can also be analyzed—to some extent by pharmacological manipulation—in terms of the neurotransmitter systems involved. One experimental approach, operant conditioning, involves training animals to push a lever that results in the delivery of food or that leads to other rewards, such as electrical stimulation of the "pleasure center" in the hypothalamus through implanted electrodes (self-stimulation). Equally, animals can learn not to push a lever that delivers a pain or punishment stimulus. Maze-running and other learning paradigms, as well as exploring or performing balancing or climbing routines, can all be studied under controlled conditions and after drug or surgical treatments.

Again, the influence of various compounds on the *extinction* of this conditioned behavior can give leads on which neurotransmitter systems are involved in the neural mechanisms underlying learning and memory. Extinction is the enforced cessation of particular conditioned behavior either by disruption of the association between conditioned stimulus and conditioned response or by removal of the reinforcement for that particular behavior.

Research linking even simple behavior in, say, mammals to specific experimental intervention is complicated by the fact that categories of behavior linked to particular neurotransmitter systems, or groups of systems, will often be manifested somewhat differently at the motor level in different species. Added to this difficulty are the problems associated with variation in behavioral response according to the site or route of drug administration (e.g., intraventricularly as opposed to injection into a specific brain region) as well as variations in dosage. And different methods of attempting lesions in specific pathways (e.g., mechanical versus chemical, cutting versus electrolytic lesion) can lead to different results in the same animal due to nonspecific[4] lesion of neighboring neural pathways.

CATECHOLAMINES AND BEHAVIOR

Dopaminergic Systems

Dopamine has been known for some time now to be a neurotransmitter system of major importance within the brain. It participates in a range of apparently qualitatively different brain mechanisms. These include the central control of such key functions as locomotion and of the neural processes generating motivational behavior, mood, and emotion.

Motor Behavior Intrahypothalamic and intrastriatal injections of dopamine induce motor hyperactivity. Injection of either dopamine or norepinephrine into the striatum of one hemisphere causes turning in a circle directed away from the side receiving the injection, that is, contralateral turning.[1,4] These and many other observations suggest a function for central catecholaminergic neurons in controlling motor activity. This appears to be exerted through the basal ganglia system and the extrapyramidal system, whose pathways descend into the spinal cord to control the activity of body musculature.

Destruction of brain catecholamine pathways by intraventricular administration of 6-hydroxydopamine results in a reduction of general motor activity. Specific destruction of dopaminergic or noradrenergic pathways can be achieved according to the site of intracerebral injection. Within the ascending noradrenergic dorsal bundle of the forebrain, 6-hydroxydopamine selectively depletes norepinephrine. In contrast, injection into the nigrostriatal dopaminergic bundle or the substantia nigra after pretreatment with desimipramine, the specific norepinephrine uptake blocker—to block 6-hydroxydopamine uptake to noradrenergic cells—results in selective loss of dopaminergic cells.[K] Similar dopamine losses occurring in Parkinson's disease victims result in the characteristic reduction in motor activity (ataxia) (see Chapter 9).

The precise role of dopamine versus norepinephrine in this behavior remains unclear, but the present hypothesis attributes alertness and behavioral reinforcement to the noradrenergic system and motor activity and drive to the dopaminergic system. Complete lesion of all dopamine-containing neurons in the zona compacta (cell body region) of the substantia nigra (the A9 group; see Figure 4.24) and the ventral tegmental area deprives rats of all overt behavior:[5] they are akinetic, aphagic, and adipsic (i.e., do not move, eat, or drink). Depression of the postsynaptic activity of dopaminergic neurons by application of specific antagonists such as chlorpromazine or haloperidol (members of the various classes of neuroleptic drugs) also depresses motor activity to varying degrees. The actions of the drugs may not exclusively reflect the response of dopaminergic receptors.[4] For instance, α-adrenergic blockers such as phentolamine will also reduce motor activity. In contrast, moderate doses of dopamine agonists such as apomorphine *stimulate* motor activity, as does L-dopa, the dopamine precursor—though once again, augmented actions on noradrenergic receptors owing to increased norepinephrine synthesis cannot be ruled out. Curiously, very low doses of dopamine agonists cause *sedation*, probably because an action on *presynaptic* dopamine receptors inhibits dopamine release.[K]

The amine-depleting agent, amphetamine, releases monoamines from their storage sites in nerve terminals, and at low doses (0.5–1.5 mg/kg) causes an increase in their extracellular concentrations. It also blocks uptake of these amines, which further increases their extracellular concentrations. Administration of amphetamine at low doses to experimental animals results in a considerable increase in motor activity. At higher doses (>5 mg/kg) it leads to stereotyped behavior patterns involving repetitive licking, sniffing, chewing, and assuming various purposeless postural attitudes. This latter behavioral phenomenon has features in common with amphetamine-induced psychosis in humans, which itself has much in common with paranoid schizophrenia[4,6] and often involves similar stereotyped posturing.

Whether dopamine alone is involved in generating stereotyped behavior remains uncertain since the influence of noradrenergic pathways has not yet been excluded.[4] The fact that inhibitors of dopamine β-hydroxylase, the synthetic enzyme for norepinephrine do not reduce the effect[15] suggests that norepinephrine is not involved. Also supporting the key role for dopamine in this behavioral response is that d-amphetamine is some tenfold more effective than the l-isomer in precipitating the condition. The d-isomer is known to be 7–10 times more effective on dopaminergic systems than the l-isomer, whereas the two are equipotent on noradrenergic systems.[K]

At the other end of the spectrum of activity is catalepsy, a state of transfixation and minimal behavior. This condition can be induced by dopaminergic antagonists such as the neuroleptic groups of drugs[7] (see Table 4.7).

Circling Behavior   Turning or circling behavior can be induced in experimental animals by administering amphetamine after making a unilateral lesion to allow degeneration of the nigrostriatal pathway, with parallel loss of dopamine-containing terminals. The dopamine released by the drug in the intact striatum on the unlesioned side of the brain is believed to stimulate activity locally, causing an imbalance in motor commands and a consequent turning towards the lesioned side, which has lower striatal dopaminergic activity.[1] The intensity of amphetamine-induced rotation is proportional to the degree of loss of dopamine function in the striatum on the lesioned side. Apomorphine, a dopamine agonist, produces turning in the *opposite* direction by stimulating the supersensitive postsynaptic dopamine receptors that develop on the lesioned side.

Noradrenergic pathways seem able to modulate amphetamine and apomorphine-induced circling behavior while not being primarily involved in producing the effect. Lesioning studies show that the pathway involved arises in the locus ceruleus and projects through the ventral noradrenergic bundle.[K]

Dopamine Pathways   A major question is the identity of the dopaminergic pathways involved in these various behavioral responses. Some of the answers are now emerging. The two main dopamine pathways ascending from the ventral regions are (1) the nigrostriatal projection from substantia nigra to corpus striatum (extrapyramidal area), (2) the mesolimbic system projecting from the ventral tegumental area to the nucleus accumbens, amygdala, septum, olfactory tubercle, and frontal (cingulate) cortex of the limbic system (see Chapter 4 and Figure 4.24).

Specific lesions applied to these two pathways should expose any specific influence each may exert. Bilateral infusion of 6-hydroxydopamine into the nucleus accumbens destroys the catecholaminergic projection to this structure and also selectively abolishes the amphetamine-induced hyperactivity, but it does *not* prevent the appearance of the stereotyped behavior. Lesion of the ventral caudate nucleus, on the other hand, destroys local terminals and reduces the level of dopamine in the striatum but does not effect dopamine levels in the more distant limbic cortex.[4] After such treatment, stereotyped behavior diminishes but stimulation of locomotor behavior is maintained.

The nigrostriatal pathways and the mesolimbic pathways are involved in generating somewhat different forms of behavior, with the former being directly involved in locomotion and the latter more concerned with controlling stereotyped and postural behavior. Bilateral destruction of catecholamine neurons in substantia nigra with 6-hydroxydopamine abolishes both locomotor and stereotyped behaviors, showing that both are to some extent dependent on the functional integrity of the nigrostriatal pathway.[9]

Other features of dopaminergic control of behavior may be revealed by lesioning the nigrostriatal system. A lesion on one side leads to a behavioral condition called sensorimotor neglect, in which there is absence of response to sensory inputs from the contralateral side of the body.[10,11,19,55] There is no overt response to stimulation of the body surface on that side, and the limbs appear uncoordinated and adopt unusual postures. The animals may not respond to objects in the visual field on the contralateral side. These observations indicate that in the reception and analysis of sensory information, as well as in the execution of the appropriate motor responses, an essential part is played by the dopamine projection to the striatum. This conclusion is reenforced by in vivo studies showing complex modulation of dopamine release and uptake in the nigrostriatal system of the cat after unilateral delivery of mild electric shocks to the forelimb.[12]

The capacity of the dopaminergic extrapyramidal system for receiving and integrating sensory information indicates a role in the cognitive processes that enable an animal to perceive and act appropriately on sensory information received from its external environment.

Motivational Behavior   Dopaminergic systems are also likely to be involved in motivational behavior.[58] One valuable experimental model illustrating this is intracranial self-stimulation,[C,13] which allows study of the neural pathways in the brain that mediate the mechanisms of reward. In this model, electrodes implanted in an animal deliver electrical stimuli to specific brain regions such as the hypothalamic area or the limbic system structures.[C] The animal is able to deliver the stimulus to itself by performing some action that allows the passage of electric current: bar pressing, breaking a circuit involving activation of a photocell, etc. The stimulation motivates the animal because of the apparent pleasurable sensation it produces[D] and also reinforces the behavior the animal must perform to cause its delivery. At some sites the animal will only self-stimulate at a high rate if it is hungry or thirsty, suggesting that the pathways being activated could be involved in the normal behavior patterns mediating hunger and thirst. Many other normal patterns of behavior mediated by the hypothalamus can be evoked by stimulation, including grooming and sexual behavior.[C,D] The pathways activated in the hypothalamus connect with other regions, including the parts of the limbic system and the brain stem that incorporate several catecholamine systems (see Figures 4.23 and 4.24).

Catecholaminergic Systems and Self-Stimulation

There is ample evidence of participation of both dopamine and norepinephrine in the pathways mediating self-stimulation.[C,17,87] Though the relative importance of each is a matter for continuing debate, opinion seems to have settled on dopamine as the more important catechol-amine.[C,K] Compounds that selectively block dopamine receptors, such as neuroleptic drugs, decrease the effect. For example, haloperidol and pimo-zide (see Table 4.7) suppress intracranial self-stimulation of the median forebrain bundle,[C] and spiroperidol suppresses the effect in nucleus accumbens, hippocampus, and a range of other brain regions. Dopamine-receptor stimulants such as apomorphine consistently increase intracranial self-stimulation or restore it after administration of reserpine, a catecholamine-depleting drug. Destruction of both dopamine and nor-epinephrine neurons by 6-hydroxydopamine treatment (see Tables 4.4 and 4.7) at high doses drastically reduces self-stimulation of the median forebrain bundle. Local injection of this drug into the substantia nigra greatly depletes the dopamine content of the caudate-putamen and re-duces self-stimulation of the region by 80 percent[86] (cf. Figures 8.1 and 8.2).

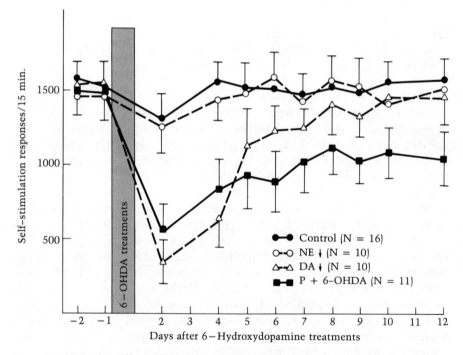

FIGURE 8.1   Effect of norepinephrine, dopamine, or catecholamine deple-tion on self-stimulation, obtained from electrodes placed in the lateral hypothalamus. Prior to treatment with 6-hydroxydopamine (6-OHDA), rats were assigned to groups to match the mean self-stimulation rate for each. *NE* ↓ refers to rats treated with 3 × 25 μg of 6-OHDA to deplete brain norepinephrine. *DA* ↓ refers to rats that received desipramine (25 mg/kg) 1 hour before receiving 200 μg of 6-OHDA to deplete dopamine. *P + 6-OHDA* refers to rats that were treated with pargyline (50 mg/kg) 30 minutes prior to intracisternal injection of 200 μg of 6-OHDA to re-duce both catecholamines in the brain. *N*, number of rats. [From ref. 16]

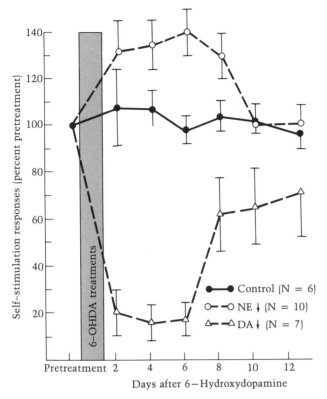

FIGURE 8.2 Effect on locus ceruleus self-stimulation of different 6-hydroxydopamine (6-OHDA) treatments that deplete norepinephrine (*NE* ↓) or dopamine (*DA* ↓). Animals selected for the various groups had similar self-stimulation rates prior to treatment with 6-OHDA. Treatment regimens are the same as those described in Figure 8.1. Significant depression of response on days 2, 4, and 6 (P < 0.001) was seen in the dopamine-depleted group (*DA*). Treatment that reduced norepinephrine levels (*NE*) elevated responses significantly on days 4, 6, and 8 after treatment (P < 0.05). *N*, number of rats. [From ref. 16]

Self-stimulation can be obtained with electrodes positioned in the locus ceruleus, a purely noradrenergic nucleus, though not at brain stem sites lateral, ventral, or medial to it and not in the cerebellum.[17] Effective self-stimulation at the locus ceruleus requires dopamine neurons to be intact in other brain regions.[C] Treatments lowering only norepinephrine levels (e.g., 6-hydroxydopamine at low doses or dopamine β-hydroxylase inhibitors; see Table 4.4) do not prevent self-stimulation of the locus ceruleus or the lateral hypothalamus[C,16,87] (Figures 8.1 and 8.2). Although there is continuing debate on the precise specificity of these treatments, the conclusion is that noradrenergic pathways are not necessary for effective self-stimulation.[C,K] Instead, the effect centers on dopamine neurotransmitters.[C,K] When the electrodes are located in the ventral mesen-

TABLE 8.1   The effects of lesioning and electrically stimulating ascending catecholamine systems

CA pathway	Cell body groups	Effects of lesions	Electrical self-stimulation (e.s.s.)	Rate of acquisition of response to e.s.s.	Behavior associated with e.s.s.
Dopamine neurons	A8,A9,A10	Bilateral: akinesia aphagia adipsia	Yes	Fast	Motor excitation: elements of the sniffing, licking, gnawing syndrome
		Unilateral: contralateral sensory neglect			
Dorsal bundle of norepinephrine neurons	A6,A4	Learning deficits	Yes	Slow	Motor excitation absent
Ventral bundle of norepinephrine neurons	A1,A2	Increased food intake and weight gain	No	–	–

Note: e.s.s. = electrical self-stimulation. Data are from ref. 15.

cephalic group of dopamine neurons, marked activation of motor behavior is seen. Rats appear excited, and they will sniff, lick, and gnaw at the lever in an energetic fashion. They are also easily trained. Stimulation of the noradrenergic locus ceruleus causes no such activation of behavior and the animals usually take much longer to train (Tables 8.1 and 8.2).

Self-Administration Experiments

Another experimental approach providing a different line of evidence for norepinephrine and dopamine participation in neural mechanisms that direct central reward and learning processes comes from the self-administration paradigm. This involves the voluntary self-administration of drug solutions through a chronic indwelling intravenous cannula by pressing a lever. The experimental animal is usually a rat, dog, or monkey. Self-administered infusions of dopaminergic drugs are thus seen as reinforcing learning behavior. Amphetamine is administered by lever-pressing, which is altered in rate to maintain a constant brain level as the drug concentration in the infusate is altered. Dopamine blockers such as pimozide increase the rate of lever-pressing, indicating that the amphetamine is acting through dopaminergic systems.[K] Dopamine receptor stimulants such as apomorphine or piribedil are also very effectively

TABLE 8.2 Monoamines and cholinergic pathways in behavior

Neurotransmitter	Cell body groups	Pathway	Nerve terminal location	Behavior
Norepinephrine	A4,A6	Dorsal bundle	Cerebral and cerebellar cortex	Reward; reinforcement in learning
	A1,A2,A5,A7	Ventral bundle	Hypothalamus	Satiety mechanisms
Dopamine	A8,A9,A10	Nigrostriatal pathway	Corpus striatum	Control of motor activity; circling behavior; hyperactivity syndrome; general motor activation; reception and analysis of sensory information
	A8,A10	Mesolimbic pathway	Nucleus accumbens; frontal cortex; limbic system	Stereotyped behavior; incentive and motivation mechanisms; postural behavior; emotional states
Serotonin	B4 to B9	Ascending pathways	Cerebral cortex; hypothalamus; corpus striatum	Inhibitory action on motor social, and sexual behavior; self-stimulation; circling behavior
	B1,B2,B3	Descending pathway	Ventral and intermediate horns of spinal cord	Stimulation of motor activity; hyperactivity syndrome (particulary forepaw-treading, head-weaving, head tremor, and hindlimb abduction)
Acetylcholine	Nucleus basalis	Extrinsic and intrinsic neurons of extrapyramidal system; limbic system	Intrinsic	Tremor; involuntary control of movement; postural and circling behavior; emotional states; aggression; self-stimulation
	Nucleus basalis	Ascending from brain stem	Cerebral cortex	Cortical arousal, states of attention; sleep

self-administered in rats. The administration rate is not changed by depletion of presynaptic amine stores (e.g., with α-amino paratyrosine), emphasizing the direct action of these agonists on dopamine receptors. As expected, self-administration of amphetamine is increased when presynaptic amine stores are reduced—an attempt to maintain the level of stimulation by the released dopamine.[K] Self-administration of other drugs such as cocaine or morphine is also susceptible to dopamine antagonists or other interference with dopaminergic pathways (e.g., from 6-hydroxydopamine lesions). Thus, self-administration of a number of substances may involve these pathways in reinforcement and reward.

Relevant to a possible role of norepinephrine in reward mechanisms is the observation that rats will effectively self-administer clonidine, a noradrenergic alpha-2 agonist. Because the response is blocked by the alpha blocker phenoxybenzamine, direct stimulation of norepinephrine receptors is likely to be a component of this reward and reinforcement behavior.[90,M] Other treatments, however, provide evidence against any wide participation of noradrenergic pathways. For instance, depleting norepinephrine with synthesis inhibitors that block dopamine β-hydroxylase does not affect self-administration of morphine or clonidine. Alpha and beta noradrenergic blockers are without action on amphetamine or cocaine self-administration.[91]

The Catecholamine Hypothesis Although not yet resolved, these curious and sometimes conflicting findings have been embodied in a "general catecholamine hypothesis" in order to explain the different contributions dopaminergic and noradrenergic pathways may make to the neural mechanisms of reinforcement[C] and, indeed, to other patterns of behavior[17,C] (Table 8.2; Figure 8.3). The dopamine pathways are seen as governing the incentive, drive-induction, and motor-activation mechanisms that lead the animal to approach the reinforcing stimulus (in the rat, this principally means food-seeking) or to activate the self-stimulation current.[57,58,K] Norepinephrine pathways (locus ceruleus system) may be more concerned with facilitating learning, possibly by strengthening and consolidating the response, and are therefore involved with the actual process of reinforcement,[C] perhaps directly by consolidating memory processes[17] (Table 8.2). Alternatively, norepinephrine may be acting indirectly through a role in the neural processes underlying selective "attentional" behavior.[K] In such case it would be promoting motivational intensity through concentration on a single stimulus dimension, irrelevant stimuli being ignored. There is now much evidence to implicate norepinephrine in such "attentional" functioning.[K,88,89]

Destruction of the noradrenergic locus ceruleus, or the linked noradrenergic dorsal bundle, causes no significant change in motor activity or in food and water intake.[17] Lesion of the ventral noradrenergic bundles (arising from a subceruleus region in the brain stem; see Figure 4.23) does lead to increased food intake and weight gain. In fact, this ventral-bundle

Sensorimotor integration

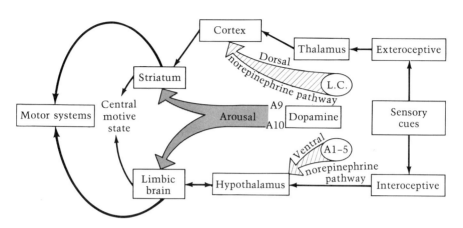

Sensorimotor integration

FIGURE 8.3   Diagram summarizing the possible involvement of catecholamine pathways in behavior of laboratory rodents. Environmental (exteroceptive) or internal (interoceptive) sensory cues are neurally processed by various brain regions in the sequence indicated by arrows. The norepinephrine or dopamine pathways are postulated to influence this neural processing at the points indicated by the projection of the pathways. The dorsal noradrenergic pathway is postulated to be involved in the neural mechanisms of reward and reinforcement; the ventral noradrenergic pathway is involved in satiety mechanisms controlling food intake. The ascending dopaminergic pathways participate in the neural mechanisms for arousal and incentive, and they initiate the motor activity shaping the animal's response to the environmental stimuli (cf. Figures 4.22 and 4.23). L. C., locus ceruleus. A1–A5, A9, and A10 refer to groups of neurons. [From S. D. Iversen, unpublished figure]

pathway may function as a component of the satiety mechanism, though it is not a site for effective self-stimulation (Table 8.1; Figure 8.3).

Serotonin in Self-Stimulation   There is also evidence of serotonin mechanisms forming part of the self-stimulation process, since *p*-chlorophenylalanine, an inhibitor of serotonin synthesis (see Table 4.8), reduces intracranial self-stimulation in a pattern correlating with the fall and recovery of brain serotonin levels.[18] Thus, several other neurotransmitter systems may be involved as equal or junior partners to the catecholamines.[K] Whichever pathways are involved, the phenomenon of intracranial self-stimulation with its components of reward and reinforcement of behavior is a valuable model for identifying the basic neural elements and pathways underlying arousal, drive, and motivation, and their linkage to learning and memory (Table 8.2).

Dopamine Receptor Subpopulations

Four kinds of dopamine receptor have been identified: D_1, D_2, D_3, and D_4 (see Chapter 4). The D_1 type, linked to cyclic AMP formation (metabotropic), and the D_2 type, linked only to ionic fluxes (ionotropic; see Table 4.6). Stereotyped behavior in rats appears to be mediated through selective stimulation of D_2 receptors by the appropriate drugs (e.g., bromocriptine, lisuride, lergotrile, pergolide).[20]

SEROTONIN SYSTEMS AND BEHAVIOR

Serotonin pathways arise from a series of some nine nuclei situated in the midline of the brain stem, the raphe nuclei (see Figure 4.25). As well as descending to the ventral and intermediate horns of the spinal cord (medullary raphe nuclei), the projections from these nuclei (midbrain raphe nuclei) are diffusely distributed to the corpus striatum along with the dopamine projection and to the cerebral and cerebellar cortex along with the norepinephrine projections. There are two major ascending pathways, a medial route innervating mainly hypothalamic and preoptic structures, and a lateral projection to the cortical areas (see Figure 4.25). A much smaller and more lateral projection innervates the corpus striatum. A descending pathway arises from the pontine and medullary raphe nuclei. These facts have mainly been established for the brains of rodents and other experimental animals. However, studies of the human fetus show that an essentially similar pattern of distribution of serotonin pathways occurs in humans.[17]

Inhibitory Actions

It is possible to deplete the brain of serotonin over a period of days by treatment with p-chlorophenylalanine, which selectively inhibits tryptophan hydroxylase and therefore serotonin synthesis (see Table 4.8). Responses to depletion are similar in a range of laboratory animals and include increased motor activity and often insomnia. (These responses can be reversed by administering 5-hydroxytryptophan, the immediate precursor of serotonin.) The animals show increased responsiveness to most kinds of sensory stimulation and to pain stimuli. These behavioral changes are accompanied by increased aggressiveness and irritability as well as increased sexual activity and enhanced social interaction in general.[17] Other methods of reducing serotonin levels, such as the use of a selective neurotoxin (dihydroxytryptamine, Table 4.8) for serotonin neurons or electrolytic lesions of midbrain raphe nuclei, also lead to a general increase in activity. These findings show that a *reduction* in the levels of serotonin are associated with arousal, increased activity, and sensitivity; therefore, the overall behavioral action of serotonin in the normal brain is likely to be *inhibitory* irrespective of the local synaptic action. In fact, the close anatomical distribution of catecholamine fiber projections with ascending serotonin projections and the somewhat different effects they mediate suggest that the two kinds of system may work in opposition, the catecholamine systems being *excitatory* in action and the serotonin system *inhibitory*.

More direct evidence of the inhibitory influence of serotonin in behavior comes from experiments in which the tissue levels of serotonin are manipulated upwards. For instance, administration of the precursors 5-hydroxytryptophan and tryptophan leads to decreased motor activity.[21,22] Direct infusion of serotonin into brain ventricles also leads to reduced spontaneous locomotor activity.[23,25]

Excitatory Actions

There are also clear-cut examples in which serotonin and some related indoles *activate* rather than inhibit motor behavior. Autonomic excitatory effects, tremor, myoclonus, and even generalized seizures can be induced by precursors,[21,24] agonists, and releasing agents for serotonin. In fact, any manipulation resulting in increases in the levels of synaptic serotonin or stimulation of postsynaptic serotonin receptors produces the same effects. Considerable hyperactivity can be induced in rats and mice by tryptophan or by 5-hydroxytryptophan provided an MAO inhibitor (e.g., tranylcypromine, pargyline; see Table 4.8) is given as a pretreatment, or some equivalent treatment is given to enhance and maintain serotonin levels.[26] These examples of stimulated behavior due to serotonergic systems are specialized and are qualitatively different from simple arousal and locomotor activation; they involve complex and repetitive limb, trunk, or head movements.

This "serotonin hyperactivity syndrome"[27] is accompanied by raised body temperature (hyperpyrexia). It may be that only the activity of serotonin neurons contributing to the pathways *descending* to the spinal cord are involved, since sectioning just above the caudal raphe, which isolates the *ascending* serotonin pathways, does not prevent the response (see Figure 4.25). The overall serotonin syndrome has been analyzed into its subcomponent behaviors, which include hyperactivity, locomotor activation, hyperreactivity, resting tremor, rigidity, hindlimb abduction, lateral head-weaving, head twitching, myoclonic jerks, reciprocal forepaw treading, and the "Straub tail phenomenon"[27,28,56] in which the tail is held in a rigid upward curving position.

The complete syndrome, with the exception of hindlimb abduction and locomotor activation, appears to be mediated by neural mechanisms present in the pons, medulla, and spinal cord and is largely due to the activity of the descending pontine and medullary serotonergic raphe nuclei.[28,56] Behavioral syndromes of this level of complexity, however, necessarily involve the activity of several neurotransmitter systems. In this serotonin syndrome, the locomotor excitation component is almost certainly due to enhancement of dopaminergic activity. Other results suggest that GABAergic mechanisms also participate.[30] But the major involvement of serotonin can be seen through the inhibitory action of serotonin antagonists, such as metergoline, which prevent the appearance of the syndrome.[27] Moreover, blockade of serotonin uptake by chlorimipramine potentiates the syndrome (see Table 4.8).

In summary, this paradoxical shift in behavior from decreased locomotion to the hyperactivity syndrome, both behaviors apparently being mainly mediated by serotonin, seems to be associated with raised extracellular levels of serotonin and enhanced postsynaptic serotonergic receptor activation. It is likely that exaggerated serotonergic activity in the *descending* pathways, which are probably excitatory, predominates over the actions of inhibitory *ascending* forebrain pathways,[56] but it is clear that serotonin can have *direct* excitatory actions in some of its pathways. A somewhat similar hyperactivity syndrome follows treatment of rats with L-dopa plus MAO inhibitors, but this is qualitatively different in nature, showing a predominance of locomotor activation with squeaking, jumping hyperreactivity and piloerection. Neuroleptic drugs such as haloperidol and chlorpromazine as well as α-methylparatyrosine, a tyrosine hydroxylase inhibitor, will block the effect, showing that dopaminergic systems are predominantly responsible.[29]

Some features of the serotonin behavioral syndrome appear to be exclusively mediated by activation of serotonin receptors. These features include the simultaneous appearance of side-to-side head-weaving, tremor, forepaw-padding, and splaying of hindlimbs, since these are modulated by drugs or treatments affecting only serotonergic systems.[31]

Other Serotonin-Mediated Behaviors

The unusual whole-body movement called wet-dog shakes, involving periodic, repetitive shaking movements, is seen in rats after their immersion in cold water and during morphine withdrawal. It can be induced by systemic administration of 5-hydroxytryptophan, serotonin-releasing drugs (e.g., fenfluramine), or various serotonin agonists.[31,32] It is also frequently observed after intraventricular treatment with serotonin or with L-dopa plus MAO inhibitors).[31] This behavioral phenomenon clearly involves both serotonergic and dopaminergic pathways.

Indeed, the joint application of a catecholamine releaser (amphetamine) and a serotonin releaser (p-chloroamphetamine or fenfluramine) produces wet-dog shakes and hindlimb abduction as well as characteristic stereotyped behavior, though at a greatly reduced level. In addition, a markedly different pattern of behavior occurs, including the backwards-walking and circling movements.[33] It is concluded from the results of various drug treatments that this novel behavior pattern is due to simultaneous actions of serotonin and dopamine and requires the release of both neurotransmitters for its appearance.[33]

Backwards walking, circling, or pivoting may represent a model for observing hallucinatory behavior in rodents and other mammals.[33,34] Amphetamine and many other hallucinogens induce this abnormal locomotory behavior. The hallucinogenic actions of the agonist compounds lysergic acid diethylamide (LSD) and dimethyltryptamine (DMT) are thought to be mediated via presynaptic serotonin receptors. These hallucinations are usually visual and tactile and can be distinguished from the frequently auditory hallucinations of psychotic states. In view of the localization of the raphe nuclei close to the brain stem's consciousness-alerting system, called the ascending reticular activating system (ARAS), it is suggested that LSD could influence the gating function of this system for afferent sensory information. Thus a reduction in serotonin release mediated by inhibitory presynaptic serotonin receptors would be likely to reduce the tonic inhibition in the ARAS due to serotonin and thus allow abnormal stimulation of the visual and other relevant areas of the brain,[E] causing hallucination.

Other interactions between serotonin and various neurotransmitters have been reported. For instance, the cataleptic states characterized by a "frozen" posture and inability to respond to stimuli, which are induced in rats by certain neuroleptic drugs, can be reduced by lesions to the medial and dorsal raphe nuclei[7] or by depletion of serotonin through the loss of serotonergic neurons (accomplished with the specific neurotoxin 5,7-dihydroxytryptamine).[35] The serotonergic projection to the nucleus accumbens seems to be of special importance in producing catalepsy and its associated abnormal behavioral states. In fact, an imbalance in the dopamine and serotonin interactions in certain brain regions, including mesolimbic and extrapyramidal areas, has been implicated in the etiology of some forms of schizophrenia (see Chapter 9), and these findings could have special relevance in this respect.[35]

A dopamine–serotonin interaction can be seen in dopamine-dependent stereotyped behavior in rats. Modulation of serotonin levels, or serotonin activity, greatly affects the manifestation of this behavioral syndrome,[36] leading to attenuation of the stereotyped features. In the nucleus accumbens and nigrostriatal system, dopamine and serotonin exert mutually antagonistic effects on stereotyped behavior, and it is likely that the selective suppression of various components of stereotyped behavior is mediated by these systems.[36]

Circling behavior, too, can be modified by serotonin,[1] since it is decreased by administration of serotonin agonists. Conversely, blockade of serotonin receptors with antagonists such as methysergide, or block of its synthesis with *p*-chlorophenylalanine, leads to an increase in circling behavior.[1]

Serotonin and Sleep

There has been a long-standing interest in the possible involvement of serotonin in the neural mechanisms causing sleep states.[37] Serotonin given to animals with a permeable blood-brain barrier (e.g., chickens) induces cortical synchronization and sleep. Intracerebral application of serotonin or its precursor 5-hydroxytryptophan to a wide range of mammals induces signs of, or actual, slow-wave sleep. Equally, administration of *p*-chlorophenylalanine (the inhibitor of tryptophan hydroxylase), which leads to depletion of serotonin, produces a prolonged state of EEG wakefulness, which is reversed by administration of 5-hydroxytryptophan. Slow-wave sleep and the rapid eye movements (REM characteristic of deep sleep) can be prevented by monoamine releasers such as reserpine and are restored by 5-hydroxytryptophan. This sleep is reduced by serotonin antagonists (e.g., methysergide) and by lesions in the raphe nuclei. However, some of these raphe lesions could have involved destruction of the ARAS at sites involving other neurotransmitters. And other (conflicting) drug studies have failed to support the conclusion that serotonin plays the central role in sleep mechanisms.[37]

Serotonin Receptor Subpopulations

As with so many other neurotransmitter systems,[50] subpopulations of serotonin receptors, S_1 and S_2, have been described as a result of studies at both the ligand-binding and behavioral levels[51] (see Chapter 4, page 200). The current view is that 5-HT$_1$ receptors mediate inhibition and 5-HT$_2$ receptors mediate excitation.[50,51]

At the behavioral level, the two receptor types can be distinguished by the profile of action of the various serotonergic drugs[27,51,52] (see Table 4.8). The hyperactivity syndrome discussed above is blocked by serotonin antagonists (e.g., cyproheptadine) acting preferentially on S_2 receptors with potencies correlating with their binding to S_2 sites.[51]

More recently, evidence has been advanced for the existence of separate central tryptamine receptors that mediate other aspects of behavior under the control of indoleamine systems[53] and impinge directly on the expression of behavior involving activity of the serotonergic pathways[54] (Table 8.2).

MONOAMINERGIC NEURAL ORGANIZATION AND BEHAVIOR: A SUMMARY

Each of the monoamine pathways projects from discrete nuclei, or groups of nuclei, to provide a diffuse innervation of various brain structures (Table 8.2). The terminals of these systems often consist of strings of varicosities, usually without the conventional features of organized synapses that typically involve closely opposed postsynaptic structures and receptor fields (see Figure 6.16). Some researchers have suggested, therefore, that monoamines may be released at one site and act at another, distant, site in a quasi-hormonal fashion (paracrine action). This mechanism would create a "field effect" rather than localized action at well-defined points and would perhaps favor neurons in a particular functional state.[15] Where this results in the formation or activation of new circuits, it could be the basis of the noradrenergic reinforcement system described above, perhaps providing the neural changes that underlie learning and also the tonic inhibition produced by serotonergic systems.[15]

In summary, the locus ceruleus, and the dorsal (ascending) noradrenergic bundle to which it gives rise, appear to act by modulating the actions of dopamine and several other neurotransmitters (e.g., opioids, other peptides) that may be more directly involved in the neural mechanisms underlying behavior. There may be some role in the "reward" and "reinforcement" mechanisms associated with intracranial self-stimulation, but much of the earlier work supporting this postulate has been reevaluated and found wanting.[C,K] This pathway appears to be of cardinal importance in extinction behavior, possibly through a primary role for norepinephrine in selective attentional mechanisms.[88,89,K] It does not appear to be directing any other well-defined behavior of the kind that can be revealed by sectioning of the pathway. Its fibers project mainly to cortical areas. The ventral noradrenergic bundle, arising in nuclei of the caudal medulla and projecting mainly to hypothalamus, appears to be part of a satiety mechanism controlling food intake. It can terminate a positively rewarded behavioral sequence. Its activation does not have rewarding effects.

The ascending dopaminergic pathways are likely to be directing "arousal" and "incentive" mechanisms and translating them into motor action through the striatum and limbic brain regions[57,58,C,K] (Figure 8.3). Rats will self-stimulate these dopaminergic pathways and show marked activation of motor and other behavior. Such responses do not follow self-stimulation of the dorsal noradrenergic system.

The serotonin pathways, which all arise from the midbrain raphe nuclei, exert an inhibitory or restraining influence on a variety of social and sexual behaviors and are probably involved in mediating the effects of punishing stimuli on "inappropriate" behavior performed in test situations. Lesions of the median raphe nuclei suggest that the projections of this system exert an inhibitory influence on motor behavior (Table 8.2).

ACETYLCHOLINE SYSTEMS AND BEHAVIOR

Cholinergic neurons play an important part in the neural mechanisms of behavior at all levels of complexity.[F] Not least among these is the central cholinergic participation in motor systems. In the first place, the motoneurons of the spinal cord are cholinergic, which highlights the role of these central acetylcholine neurons in the reflex regulation of movement and in postural mechanisms. But the command neural mechanisms are complex and involve many other neurotransmitter systems in the brain.

Tremor

At the level of the brain itself, cholinergic pathways are involved with the involuntary control of movement and with posture. Centrally active cholinergic agonists (e.g., carbachol; see Table 4.3) cause parkinsonianlike tremors, other dyskinetic movements, catalepsy, and circling behavior. The nuclei comprising the extrapyramidal system—globus pallidus, caudate nucleus, putamen, and substantia nigra—are known to be rich in cholinergic nerve endings, the majority of which seem to belong to intrinsic local-circuit neurons.[39] Infusion of acetylcholine, carbachol (a cholinomimetic agent), or anticholinesterase drugs (e.g., eserine, diisopropylfluorophosphate) into the globus pallidus or caudate nucleus causes a resting tremor which is prevented by hemicholinium and other anticholinergic agents (see Table 4.3).

Cholinergic receptor *antagonists*, such as atropine, can be effective in relieving the tremor of parkinsonism, indicating the normal participation of these acetylcholine pathways together with the dopaminergic system in the control of tremor and posture (see Figure 9.5).

The muscarinic agonist agents tremorine and oxotremorine (Figure 8.4) increases cerebral acetylcholine content and precipitate a tremoring and shuddering condition in a wide range of experimental animals by an action at various sites within the brain.[G] Tremorine must be metabolized to oxotremorine, which is the active agent. This compound is sufficiently hydrophobic to readily enter the central nervous system, where it probably acts by enhancing muscarinic cholinergic transmission in the extrapyramidal system in the brain stem and in the dorsal mesencephalic tegmentum.[G] The tremor produced is enhanced during voluntary movement, which is characteristic of malfunction of the *extrapyramidal system* (basal ganglia) and there is also a resting tremor that is characteristic of extra*pyramidal malfunction.*[G] Because the drug causes profound hypothermia, part of the tremor has been attributed to a shivering response, since these too are tremors of the shuddering and postural type. But the tremors are not abolished by placing affected animals in a warm atmosphere. The tremors due to oxotremorine and those due to shivering are probably similar because both are mediated by the same neural mechanisms in the mesencephalic reticular formation.

Arecoline, a muscarinic cholinergic agonist extracted from betel nut (Figure 8.5) has similar tremor-inducing properties.[G] It acts both centrally and peripherally, producing both postural and resting tremor. The central

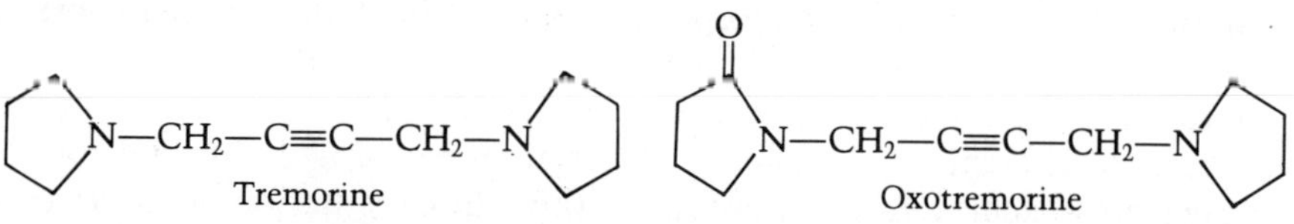

FIGURE 8.4   The structures of tremorine and oxotremorine.

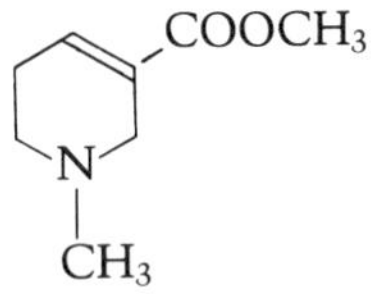

FIGURE 8.5   The structure of arecoline.

origin of the tremor is confirmed by the potent blocking action of muscarinic antagonists. Nicotinic cholinergic antagonists are without effect.

Circling Behavior

Participation of cholinergic neural mechanisms in circling behavior was
inferred in early work from the effects of infusing diisopropylfluorophosphate, the irreversible acetylcholinesterase inhibitor, into one caudate
nucleus of the rabbit. Circling movements towards the noninjected side
resulted.[1,40,41] This *contralateral* movement correlated with the degree of
inhibition of acetylcholinesterase; a 40 percent inhibition had to occur
before the circling behavior was seen.[41] In contrast, infusion of acetylcholine into one common carotid artery resulted in ipsilateral rotation, that
is, rotation in the direction opposite to that seen after injection of diisopropylfluorophosphate.[42]

Further evidence for a cholinergic involvement in circling comes
from the systemic administration of anticholinergic drugs such as scopolamine, benztropine, or atropine to animals with a lesion in one side of the
nigrostriatal system. In this case too, there is circling towards the side of
the lesion and, in addition, there is enhancement of the actions of dopamine agonists. In contrast, cholinergic *agonists* in this model reduce
the intensity of circling due to dopamine agonists. Such amelioration is
in keeping with all the other evidence for a balanced antagonism between
cholinergic and dopaminergic neurotransmitter systems in movement
control (see Chapters 7 and 9). Indeed, the circling produced by anticholinergic drugs alone clearly requires intact dopaminergic function for its
appearance, since neuroleptic and other antidopaminergic drugs block
the acetycholine-mediated effect.[1]

Arousal

The participation of cholinergic mechanisms of the brain stem in arousal
and in states of consciousness has long been suspected.

The "arousal" or "ascending reticular activating system" (ARAS) of
the brain emanates from the reticular formation of the brain stem. It has
subcomponents from ascending spinal tracts and peripheral inputs from
spinal and cranial nerves. These inputs are channeled into the final common pathway of the ARAS, which projects via the nonspecific thalamic

nuclei and hypothalamus to the cerebral cortex, hippocampus, and other structures to cause activation of the cerebral cortex and limbic system. Stimulation of the ARAS causes a release of acetylcholine from many cortical areas, including those receiving the ARAS input.[F] Low-frequency stimulation of the mesial thalamic nuclei produces an equivalent effect, and injection of carbachol, or of acetylcholine itself, produces a state of arousal in rats that markedly increases responsiveness to sensory stimulation. Adrenergic agents produce quite opposite effects.[43] Transection of the ARAS leads to a decreased cortical release of acetylcholine and release of cortical GABA.[44]

A considerable volume of research has demonstrated a correlation between states of behavioral arousal and raised tissue levels of acetylcholine and enhanced rates of its release from central cholinergic systems. Sleep and diminished activity, on the other hand, correlate well with low levels of these parameters.[44,F]

There is little doubt that ascending acetylcholine pathways represent only one of several systems controlling states of arousal and "sleep-wakefulness" or states of consciousness. Other neurotransmitters that seem to be involved include GABA, glutamate, catecholamines, and serotonin, as shown through pharmacological manipulation of their function and also by studies of their patterns of cortical release during various states of arousal.[44,45,F]

Cholinergic pathways have been implicated in the control of emotional states, aggression, avoidance behavior, and self-stimulation.[F] For instance, during intracranial self-stimulation of the median forebrain bundle, the anticholinesterase agent diisofluorophosphate facilitates the response, probably via facilitatory synapses on the catecholaminergic neurons involved.[C,45] A role for cholinergic pathways in intracranial self-stimulation would, of course, also imply a participation in motivated behavior. Acetylcholine and its agonists stimulate drinking, illustrating cholinergic involvement in a quite different form of behavior.

The precise interactions between acetylcholine and the other neurotransmitters involved in these various behaviors are yet to be established and remain an important topic for future research (see Table 8.2).

PEPTIDES AND BEHAVIOR

The discovery that peptides can act in the central nervous system to profoundly influence physiological functions and elicit striking and often complex behavioral responses has stimulated a great deal of neuroscience research.[H,I,L] Behavioral responses range from highly specific actions on drinking, mating, or locomotion to less specific effects on arousal, learning, and memory. These substances also appear to participate in neural mechanisms generating euphoria and stress, and they may be involved in certain mental illnesses.

Evidence for peptide involvement in these phenomena has come from direct administration of only femtogram or picogram quantities of the peptides or their more stable substituted analogues, usually by intra-cerebral microinjection into a specific brain region. Evidence has also come from studies involving pharmacological manipulation of the level of the appropriate peptide or blockade of its specific receptors (see Chapter 5).

Angiotensin and Drinking Behavior

A specific example of peptide-elicited behavior that is a clear-cut, readily reproducible response is angiotensin-induced drinking (see Chapter 5, page 288). This has a short latency, usually less than a minute, and results in copious drinking compared with the usual rates of water in-take. This is not, however, a frenetic response but consists of relaxed drinking over extended periods.[1,46] The evidence suggests that endoge-nous angiotensin II contributes to thirst and drinking behavior in the normal animal when there is a deficit of body fluid. Specific angiotensin antagonists such as saralasin [Sar^1,Ala^8]-angiotensin II block the drinking response evoked by angiotensin II or by substances giving rise to it (i.e., renin, synthetic renin substrate, or angiotensin I, see Figure 5.12). Each of these precursors is less potent than the angiotensin II octapeptide, and is active only due to formation of angiotensin II. Thus, preventing the con-version of angiotensin I to angiotensin II by inhibiting the converting enzyme blocks the drinking response to all components except angioten-sin II itself.[1]

The most effective intracerebral injection sites are the subfornical organ and the *organum vasculosum* of the lamina terminalis, two highly vascularized structures lying outside the blood-brain barrier (see Figure 2.22). It is possible that angiotensin II causes local vascular contraction (smooth-muscle action), which leads to the emptying of blood from the vessels of these structures. Such an apparent fall in blood volume would be an important component of the central mechanisms regulating thirst and would act as a stimulus to initiate drinking behavior.[1]

Luteinizing Hormone–Releasing Hormone and Sexual Behavior

Luteinizing hormone–releasing hormone (see Chapter 5, page 287) evokes specific sexual behavioral responses. Systemic injection stimu-lates lordosis in rats.[47] This receptive posture in female animals allows males to mount and copulate. LHRH and other hypothalamic peptides are synthesized in special neurons in the preoptic hypothalamic nucleus and released into the hypothalamo-neurohypophyseal portal blood system, by which they reach the anterior pituitary cells and modulate the release of trophic hormones (see Figure 5.9). However, extrahypothalamic pathways are likely to mediate the sexual behavior influenced by LHRH.[47,1] No doubt these neural mechanisms are multifactorial and involve more than one other neurotransmitter, including the catecholamines.

Thyrotropin-Releasing Hormone, Somatostatin, and Physiological Response

Thyrotropin-releasing hormone (TRH) and somatostatin (Chapter 5, pages 282 and 284) have counterbalancing behavioral and physiological effects that reflect their opposite endocrine actions. TRH stimulates locomotor activity, reduces eating behavior, and antagonizes the actions of depressant drugs. Somatostatin has the opposite effects[48] (see Tables 5.5, and 5.6). When given intracisternally in mice, both somatostatin and neurotensin (Chapter 5, pages 284 and 291) enhance pentobarbital-induced sleeping time and mortality. Neurotensin produces a similar action by lowering body temperature, but this is not so for somatostatin.[48,1]

Many of the behavioral effects of neuropeptides found in the hypothalamus can be attributed to neural mechanisms other than those mediated by the pituitary gland via trophic or target hormones, since removal of the hypothalamus does not prevent or reduce these effects and they cannot be mimicked by administering appropriate trophic or target hormones (e.g., TSH or thyroid hormones).

Interaction between peptidergic systems and other neurotransmitters is likely to be the common pattern underlying the organization of most physiological functions or behaviors that involve peptides as neurotransmitters or endocrine agents. For instance, the dopamine-mediated behavioral excitement seen after administering L-dopa and an MAO inhibitor is substantially enhanced by TRH, somatostatin, melanophore-inhibiting factor, and by LHRH.[49,1,H,L]

Some researchers have proposed that TRH may be involved in the neural mechanisms subserving arousal. It appears to have many affects in common with amphetamine, well known to cause behavioral arousal. Both agents increase locomotor activity, enhance the L-dopa and 5-HTP–induced hyperactivity syndrome (see above), cause an alert EEG pattern, and antagonize the sedative actions of barbiturates[59] (see Table 5.6). Both appear to have acute antidepressant actions in humans. Unlike amphetamine, however, TRH does not increase self-stimulation or induce stereotypic behavior. Grooming, wakefulness, and rearing behavior in animals are all enhanced by TRH. Because of its putative role in conditioning arousal states, there has been considerable interest in its possible use as a clinical anticonvulsant.[59,60] Not all studies have confirmed its potency in the treatment of depression,[61,62] and its usefulness in this respect remains equivocal.

Neurotensin and Dopaminergic Systems

Neurotensin may be anatomically and functionally linked with catechol-aminergic systems[63] (see Chapter 5). It has many actions in common with the neuroleptic drugs, which have a dopamine-blocking action,[64] and when injected intracerebrally it prevents the stereotyped behavior and hyperactivity responses normally evoked by *d*-amphetamine. Neuro-tensin microinjected into the A10 dopamine cell-body group (see Figure 4.24) causes an *increase* in exploratory behavior,[65] presumably by activating the dopaminergic pathway originating in this cell group.

Cholecystokinin and Feeding Behavior

It is now well established that cholecystokinin (CCK) and the structurally related peptide cerulin (see Figure 5.15) inhibit feeding behavior in hungry laboratory animals when microinjected into the brain or given parenterally.[66,67] Feeding by sheep is reduced by intraventricular injection of small doses of CCK,[68] and it has been shown to reduce food intake in humans.[69] There are reports that the brains of fasting mice contain significantly more CCK receptors in olfactory bulb and hypothalamus but not in other brain regions.[70]

The levels of CCK in the blood rise towards the end of a meal, and this is thought to trigger neural mechanisms producing satiety, therefore causing the animal to cease feeding.[71,73] Investigators have concluded that mechanisms operating in the peripheral nervous system are primarily responsible for the satiety effect, rather than a direct action of circulating CCK on the brain, since the actions of CCK on food intake when given systemically are not observed after sectioning the vagus nerve.[71,72] It is proposed that CCK is released from endocrine cells in the small intestine to the blood due to "food stimuli," and that it subsequently interacts directly with the intestinal or gastric smooth muscle. Feedback through visceral afferents to the central nervous system then results in satiety and cessation of feeding.[71] However, the potent anorexic actions of CCK administered intracerebrally, and the distribution and the changes in receptor populations in the brain during fasting, suggest that central mechanisms involving CCK interactions with the hypothalamus are also cardinally involved in the control of feeding behavior.[70,73,74,75]

Vasopressin and Memory Processes

Vasopressin and oxytocin are present in many nerve fibers projecting from hypothalamic nuclei to many regions of the brain and spinal cord (see Chapter 5). The suggestion that vasopressin could facilitate memory consolidation and retrieval has come from experiments with rats. The extinction of active shock-avoidance behavior in shuttle-boxes was found to be facilitated by removal of the posterior pituitary gland, which normally secretes oxytocin and vasopressin. The effect was prevented by systemic injections of pitressin or lysine-8-vasopressin, which are not readily broken down in the blood.[76] When vasopressin is administered either intracerebrally or systemically in normal rats, the extinction of learned avoidance behavior is delayed.[75,L] Since intracerebral injections are effective at much lower doses than when given systemically, it has been proposed that the vasopressin acts at a site within the central nervous system to produce these actions on learning and memory. Vasopressin antiserum injected intracerebrally reduces retention of passive avoidance behavior.[76] The Brattleboro rat, which is homozygous for the diabetes insipidus trait, produces almost no vasopressin and has been reported to display a severe memory impairment, which is alleviated by administration of vasopressin, though this has not been consistently confirmed.[77] When a peptide antagonist of vasopressin, which does not penetrate into the brain was given systemically, it abolished the effects of systemically injected arginine-vasopressin, which prolonged extinction of a pole-jump avoidance behavior for as long as 6 hours.[78] But the antagonist also abolished the increase in blood pressure caused by vasopressin, raising the possibility that vasopressin produces its actions on memory and learning indirectly, through influences on the autonomic nervous system.[77] But vasopressin derivatives lacking vasopressor activity have been reported to be more potent than vasopressin itself in these behavioral effects.[79,80] It is possible that vasopressin has direct actions in the brain that increase arousal[81] and, as a result, indirectly improves the performance of the animal in the test situation by altering its motivation.

Endogenous Opioid Peptides

The potent naloxone-reversible actions of intracerebrally administered β-endorphin in inducing profound immobility was referred to in Chapter 5. These actions occur at doses lower than those required to produce analgesia and are produced by delta receptor ligands (Chapter 5). The enkephalins, with typical mu receptor ligands, such as morphine, are much less effective.[82] Some of the observed immobility is likely to be due to limbic seizures induced by the peptides.

Opiate receptors and innervation rich in endogenous enkephalins are often found in association with dopaminergic neurons (see Chapter 5). Injection of morphine, enkephalin, or the stabilized enkephalin analogues D-Ala^2, Met^5-enkephalinamide or D-Ala^2,D-Leu^5-enkephalinamide into brain regions containing dopaminergic neuronal cell bodies or receiving dopaminergic as well as enkephalinergic innervation (e.g., ventral tegmental area, globus pallidus, or nucleus accumbens) causes enhanced locomotor activity (at low doses) or stereotyped behavior (at higher doses). The effects are inhibited by naloxone, indicating the involvement of opioid receptors. In the *ventral tegmental area* the responses are also prevented by dopamine blockers (e.g., α-flupenthixol, haloperidol) or by lesioning the dopamine neurons (e.g., with 6-hydroxydopamine). These findings indicate that opioid peptides stimulate dopaminergic neurons when injected into certain brain regions such as the ventral tegmental area, which contains the cell bodies of the A9 and A10 dopaminergic neurons (see Figure 4.24). Since enkephalin normally produces inhibitory actions when applied locally (Chapter 5), it probably interacts with dopaminergic neurons indirectly via local circuits that themselves involve inhibitory neurons. However, the hypermotility and behavioral arousal seen to follow injection of enkephalin into *globus pallidus or nucleus accumbens* is not antagonized by antidopamine drugs but is sensitive to naloxone,[83,84] showing that opioid peptide receptors in these brain regions are not coupled to behavioral arousal and locomotor responses via dopaminergic interactions.

There have been many studies of the influence of endogenous opioid peptides on learning and memory in experimental animals.[85,94,L,M] For example, met-enkephalins and α- and β-endorphins, like vasopressin, have been shown to delay extinction of pole-jumping avoidance behavior in rats. But since these effects cannot be opposed by specific opiate antagonists, a rather different mode of action is indicated, involving neurolepticlike and amphetaminelike activity.[85,M] Curiously, γ-endorphin facilitated the extinction of pole-jumping avoidance, while Des-Tyr-β-endorphin appeared to have a cataleptogenic action (an akinetic-rigid state) similar to that induced by haloperidol, the neuroleptic drug.[84]

Various studies on the influence of opioid peptides on learning and memory have shown that they reverse amnesia and delay extinction of aversively motivated tasks learned by rats. Gamma-endorphin again produced effects opposite to those of other endorphins or enkephalins. In learned behavior motivated by food, endorphins have the opposite action and improve performance or prolong extinction. None of these behavioral influences of the opioid peptides are naloxone-sensitive, and their significance in terms of functional roles awaits clarification.[95]

Naloxone has been administered to simulate the effect of reduced levels of endorphins in the brain in order to investigate endorphin participation in various forms of behavior.[M] Drinking and food intake are reduced by naloxone, indicating that the drug is modifying their behavioral reinforcing value. Naloxone also enhances sexual performance in male rats. These actions indicate that endorphins promote feeding and inhibit sexual behavior.

Endorphins may have a special role to play in adaptation to stress. Both β-endorphin and ACTH are released from the pituitary gland in response to stress, and the former is thought to be responsible for the well-known phenomenon of stress-induced analgesia.[92]

In summary, the neural pathways employing endogenous opioid peptides apparently influence food and water intake, anxiety, behavioral arousal, aspects of learning and memory, and sexual drive. They may well be neuromediators of euphoria and drive-reduction reward.[93] Seen in another light, they may mediate the global response to starvation.[M] In this respect, endogenous endorphins can be seen as promoting foraging behavior by stimulating feeding, reducing sexual activity, and minimizing anxiety and painful stress.[M]

REFERENCES

General Texts and Review Articles

A Horn, A. S., J. Korf, and B. H. C. Westerink, eds. (1979) *The Neurobiology of Dopamine.* Academic Press, London.

B Cotman, C. W., and J. L. McGaugh. (1980) *Behavioral Science: An Introduction.* Academic Press, New York and London.

C Hall, R. D., F. E. Bloom, and J. Olds, eds. (1977) *Neuronal and Neurochemical Substrates of Reinforcement. Neurosci. Res. Program Bull.* 15: 133–314.

D Campbell, H. J. (1973) *The Pleasure Areas.* Eyre Methuen, London.

E Kruk, Z. L., and C. J. Pycock. (1979) *Neurotransmitters and Drugs.* Croom-Helm, London.

F De Feudis, F. V. (1974) *Central Cholinergic Systems and Behavior.* Academic Press, London.

G Brimblecombe, R. W., and R. M. Pinder. (1972) *Tremors and Tremorgenic Agents.* Scientechnica, Bristol.

H Liebeskind, J. C., R. K. Dismukes, J. L. Barker, P. A. Berger, I. Creese, A. J. Dunn, D. S. Segal, L. Stein, and W. W. Vale, eds. (1978) *Peptides and Behavior: A Critical Analysis of Research Strategies. Neurosci. Res. Program Bull.* 16: 489–635.

I Iversen, L. L., R. A. Nicoll, and W. W. Vale, eds. (1978) *Neurobiology of Peptides. Neurosci. Res. Program Bull.* 16: 211–370.

J Iversen, S. D., and L. L. Iversen. (1975) *Behavior Pharmacology.* Oxford University Press, New York.

K Mason, S. T. (1984) *Catecholamines and Behavior.* Cambridge University Press, London.

L De Wied, D., and P. A. Van Keep. (1980) *Hormones and the Brain.* MTP Press, Lancaster.

M Smith, J. E., and J. D. Lane, eds. (1983) *The Neurobiology of Opiate Reward Process.* Elsevier Biomedical Press, Amsterdam.

N Trimble, M. R., and E. Zarifian, eds. (1984) *Psychopharmacology of the Limbic System.* Oxford University Press, Oxford.

Additional Citations

1 Pycock, C. J. (1980) *Neuroscience* 5: 461–514.

2 Green, A. R., and D. G. Grahame-Smith. (1976) *Nature* (Lond.) 260: 487–491.

3 Jacobs, B. C. (1976) *Life Sci.* 19: 777–786.

4 Laverty, R. (1973) *Prog. Neurobiol.* 3: 33–70.

5 Ungerstedt, U. (1971) *Acta Physiol. Scand. Suppl.* 367: 95–122.

6 Hornykiewicz, O. (1978) *Neuroscience* 3: 773–783.

7   Costall, B., D. H. Fortune, R. J. Naylor, C. D. Marsden, and C. Pylock. (1975) *Neuropharmacology* 14: 859–868.

8   Fog, R. (1972) *Acta Neurol. Scand.* 48 (Suppl. 50): 3–66.

9   Creese, I., and S. D. Iversen. (1975) *Brain Res.* 83: 419–436.

10 Mason, S. T., and S. D. Iversen. (1980) *Brain Res. Rev.* 1: 107–137.

11 Iversen, S. D., and P. J. Fray. (1982) In *The Neural Basis of Behavior* (ed. A. Beckman), pp. 229–264. Spectrum, New York.

12 Leviel, V., M. F. Chesselet, J. Glowinski, and A. Chéramy. (1981) *Brain Res.* 223: 257–272.

13 Fibiger, H. C., and A. G. Phillips. (1976) *Brain Res.* 116: 23–33.

14 Fibiger, H. C. (1978) *Ann. Rev. Pharmacol. Toxicol.* 18: 37–56.

15 Crow, T. J. (1979) In *Biochemical Correlates of Brain Structure and Functions* (ed. A. N. Davison), pp. 137–174. Academic Press, London.

16 Cooper, B. R., J. M. Cott, and G. R. Breese. (1974) *Psychopharmacology* 37: 235–248.

17 Crow, T. J., and S. Wendlandt. (1976) *Nature* (Lond.) 259: 42–44.

18 Phillips, A. G., D. A. Carter, and H. C. Fibiger. (1976) *Psychopharmacol. Bull.* 49: 23–27.

19 Ungerstedt, U., and T. Ljungberg. (1975) In *Catecholamines and Schizophrenia* (eds. S. Matthysse, and S. S. Kety), pp. 149–150. Pergamon Press, Oxford.

20 Costall, B., and R. J. Naylor. (1979) In *The Neurobiology of Dopamine* (eds. A. S. Horn, J. Korf, and B. H. C. Westerink), pp. 555–576. Academic Press, London.

21 Stewart, R. M., J. H. Grondon, D. Concion, and R. J. Baldessarini. (1976) *Neuropharmacology* 15: 449–455.

22 Everett, G. M. (1974) *Biochem. Pharmacol.* 10: 261–262.

23 Green, R. A., J. C. Gillin, and R. J. Wyatt. (1975) *Psychopharmacology* 51: 81–84.

24 Klawans, H. L., C. Goetz, and W. J. Weiner. (1973) *Neurology* 23: 1234–1240.

25 Warbritton, J. D., R. M. Stewart, and R. J. Baldessarini. (1978) *Brain Res.* 143: 373–382.

26 Grahame-Smith, D. G. (1971) *J. Neurochem.* 18: 1053–1066.

27 Green, A. R., J. E. Hall, and A. R. Rees. (1981) *Brit. J. Pharmacol.* 73: 703–719.

28 Jacobs, B. L., and H. Kleinfuss. (1975) *Brain Res.* 100: 450–457.

29 Heal, D. J., A. R. Green, D. J. Boullin, and D. G. Grahame-Smith. (1976) *Psychopharmacol.* 49: 287–300.

30 Green, A. R., A. F. C. Tordoff, and M. R. Bloomfield. (1976) *J. Neural. Transm.* 39: 103–112.

31 Sloviter, R. S., E. G. Drust, and J. D. Connor. (1978) *J. Pharmacol. Exp. Ther.* 206: 339–347.

32 Bedard, P., and C. J. Pycock. (1977) *Neuropharmacology* 16: 663–670.

33 Curzon, G., J. C. R. Fernando, and A. J. Lees. (1979) *Brit. J. Pharmacol.* 66: 573–579.

34 Davis, W. M., W. A. Bedford, J. C. Buelke, M. M. Guinn, H. T. Hatoum, I. W. Waters, M. C. Wilson, and M. C. Brande. (1978) *Tox. Appl. Pharmac.* 45: 49–62.

35 Carter, C. J., and C. J. Pycock. (1978) *Naunyn-Schmiedeberg's Arch. Pharmacol.* 304: 135–139.

36 Carter, C. J., and C. J. Pycock. (1981) *Neuropharmacology* 20: 261–265.

37　Drucker-Colin, R. R. (1976) *Prog. Neurobiol.* 6: 1–22.

38　Beaudet, A., and L. Descaires. (1978) *Neuroscience* 3: 851–860.

39　Butcher, S. G., and L. L. Butcher. (1974) *Brain Res.* 71: 167–171.

40　Freedman, A. M., and H. E. Himwich. (1949) *Am. J. Physiol.* 156: 125–128.

41　White, R. P. (1956) *Proc. Soc. Exp. Biol. Med.* 93: 113–116.

42　Aprison, M. H., P. Nathan, and H. E. Himwich. (1956) *Science* 119: 158–159.

43　Grossman, S. P. (1968) *Physiol. Behav.* 3: 777–786.

44　Jasper, H. H., and I. Koyama. (1972) *Canad. J. Physiol. Pharmacol.* 47: 889–905.

45　Coons, E. E., N. Schupf, and L. G. Ungerleider. (1976) *J. Comp. Physiol. Psychol.* 90: 317–342.

46　Fitzsimons, J. T. (1980) *Proc. R. Soc.* (Lond.) *Ser. B* 210: 165–182.

47　Moss, R. L., and M. M. Foreman. (1976) *Neuroendocrinology* 20: 176–181.

48　Moss, R. L., S. M. McCann, and C. A. Dudley. (1975) *Prog. Brain Res.* 42: 37–46.

49　Prange, A. J., I. C. Wilson, G. R. Breese, and M. A. Lipton. (1975) *Prog. Brain Res.* 42: 1–9.

50　Snyder, S. H., and R. R. Goodman. (1980) *J. Neurochem.* 35: 5–15.

51　Peroutka, S. J., R. M. Lebovitz, and S. H. Snyder. (1981) *Science* 212: 827–829.

52　Leysen, J. E. (1981) *J. Physiol.* (Paris) 77: 351.

53　Luscombe, G., P. Jenner, and C. D. Marsden. (1982) *Neuropharmacology* 21: 1257–1265.

54　Cox, B., T. F. Lee, and D. Martin. (1981) *Brit. J. Pharmacol.* 72: 477–482.

55　Marshall, J. F., N. Berrios, and S. Sawyer. (1980) *J. Comp. Physiol. Psychol.* 94: 833–846.

56　Warbritton, J. D., R. M. Stewart, and R. J. Baldessarini. (1980) *Brain Res.* 183: 355–366.

57　Iversen, S. D. (1980) *Psychol. Med.* 10: 522–539.

58　Beninger, R. J. (1983) *Brain Res.* 287: 173–196.

59　Prange, A. J., C. B. Nemeroff, M. A. Lipton, G. R. Breese, and I. C. Wilson. (1978) In *Handbook of Psychopharmacology*, vol. 13, (eds. L. L. Iversen, S. D. Iversen, and S. H. Snyder), pp. 1–117. Plenum Press, New York.

60　Huey, L. Y., D. S. Janowsky, A. J. Mandell, L. L. Judd, and M. Pendery. (1975) *Psychopharmacol. Bull.* 11: 24–27.

61　Morley, J. E. (1979) *Life Sci.* 25: 1539–1550.

62　Yarborough, G. G. (1979) *Prog. Neurobiol.* 12: 291–312.

63　Nemeroff, C. B., and A. J. Prange. (1982) In *Handbook of Psychopharmacology* (eds. L. L. Iversen, S. D. Iversen, and S. H. Snyder), pp. 363–466. Plenum Press, New York.

64　Nemeroff, C. B. (1980) *Biol. Psychiatr.* 15: 283–302.

65　Kalivas, P. W., C. B. Nemeroff, and A. J. Prange. (1982) *Eur. J. Pharmacol.* 78: 471–474.

66　Smith, G. P., and J. Gibbs. (1979) *Prog. Psychobiol. Physiol. Psychol.* 8: 179–242.

67　Kulkosy, P. J. (1980) *Behav. Neurol.* 29: 111–116.

68　Delta-Fera, M. A., and C. A. Baile. (1979) *Science* 206: 471–473.

69　Kissileff, H. R., F. X. Pi-Sunyer, J. Thornton, and G. P. Smith. (1981) *Am. J. Clin. Nutr.* 34: 154–160.

70 Saito, A., J. A. Williams, and I. D. Goldfine. (1981) *Nature* (Lond.) 289: 599–600.

71 Smith, G. P., C. Jerome, B. J. Cushin, R. Etemo, and K. J. Simansky. (1981) *Science* 213: 1036–1037.

72 Morley, J. E., A. S. Levine, J. Kneip, and M. Grace. (1982) *Life Sci.* 30: 1943–1947.

73 Simpson, A., and T. Shiraishi. (1980) *Brain Res. Bull.* 5: 153–158.

74 Saito, A., H. Sankaran, I. D. Goldfine, and J. A. Williams. (1980) *Science* 208: 1155–1156.

75 McCaleb, M. L., and R. D. Myers. (1980) *Peptides* 1: 233–241.

76 De Wied, D. (1980) *Proc. R. Soc.* (Lond.) Ser. B. 210: 183–195.

77 Gash, D. M., and G. J. Thomas. (1983) *Trends Neurosci.* 6: 196–197.

78 Le Moal, M., G. F. Koob, L. Y. Koda, F. E. Bloom, M. Manning, W. H. Sawyer, and J. Rivier. (1981) *Nature* (Lond.) 291: 491–493.

79 Kovacs, G. L., and D. De Wied. (1983) *Neuroendocrinology* 37: 258–261.

80 Mannings, M., and W. H. Sawyer. (1984) *Trends Neurosci.* 7: 6–9.

81 Sahagal, A., A. B. Keith, C. Wright, and J. A. Edwardson. (1982) *Neurosci. Lett.* 28: 87–92.

82 Berger, P. A., H. Akil, S. J. Watson, and J. Barchas. (1982) *Ann. Rev. Med.* 33: 397–415.

83 Joyce, E. M., G. F. Koob, R. Strecker, S. D. Iversen, and F. E. Bloom. (1981) *Brain Res.* 221: 359–370.

84 Pert, C. B., and C. Sivit. (1977) *Nature* (Lond.) 265: 645–647.

85 De Wied, D., G. L. Kovacs, B. Hohus, J. M. Van Ree, and H. M. Greven. (1978) *Eur. J. Pharmacol.* 49: 427–430.

86 Phillips, A. G., D. A. Carter, and H. C. Fibiger. (1976) *Brain Res.* 104: 221–232.

87 Cooper, B. R., R. J. Konkol, and C. R. Breese. (1978) *J. Pharmacol. Exp. Ther.* 204: 592–605.

88 Mason, S. T. (1980) *Life Sci.* 27: 617–631.

89 Mason, S. T. (1981) *Prog. Neurobiol.* 16: 263–303.

90 Davis, W. M., and S. G. Smith. (1977) *Life Sci.* 20: 483–492.

91 Roberts, D. C. S., M. E. Corcoran, and H. C. Fibiger. (1977) *Pharmacol. Biochem. Behav.* 6: 615–620.

92 Bodnar, R. J., D. D. Kelly, M. Brutus, and M. Glusman. (1980) *Neurosci. Biobehav. Rev.* 4: 87–100.

93 Belluzzi, J. D., and L. Stein. (1977) *Nature* (Lond.) 266: 556–558.

94 Le Moal, M., G. F. Koob, and F. E. Bloom. (1979) *Life Sci.* 24: 1631–1639.

95 Koob, G. F., and F. E. Bloom. (1983) in *Brain Peptides* (eds. Kreiger, D. T., M. J. Brownstein, and J. B. Martin), pp. 369–388. Wiley, New York.

9

The Failure of Neurotransmitter Systems in Disorders of the Brain and Nervous System

In view of the highly sophisticated and controlled nature of the feed-forward and feedback traffic of communication between populations of neurons as they project from center to center within the brain and nervous system, it is to be expected that malfunction of synaptic mechanisms will lead to distortions of neural performance and therefore to neurological and psychiatric diseases. Of course, degeneration of other regions of neurons or changes in glial cells are often the primary event in such diseases, but these changes too are usually expressed as distorted patterns of neural communication, whether from nerve to muscle, or nerve to gland, or between neuron and neuron. In addition, complex multiple interactions between neurotransmitter systems impinging on one neuronal center may be totally distorted by the failure of just one of the component systems. Moreover, the delicate balance of pre- and post-synaptic mechanisms within defined fiber projections can be upset by a failure or reduction in the formation of a neurotransmitter or a malfunction of its receptors. Finally, exaggerated release of transmitter or changed sensitivity of its receptors will lead to similar derangement of function.

IMBALANCE OF SYNAPTIC ACTION

Such changes have two kinds of effects. First, the loss of one system will cause an alteration in the functional state of one or another of the interacting transmitter systems, leading to a shift in the functional homeostasis in that brain region. Such "chain reactions" mean that malfunction of several transmitter systems is a feature of the disease state, some being primary and others secondary in the condition.

The second and related effect is the shift in the prevailing balance between excitation and inhibition in the neuronal systems involved. This shift is the inevitable result of the functional changes referred to above, and though simple in concept, it is often devastating in consequence.

Both neurological and psychiatric disorders can result from these changes. The distinction between these two classes of disease is now largely artificial and reflects the mode of treatment rather than any intrinsic differences. The term *neuropsychiatric* probably provides a better description of the whole spectrum of nervous-system disorders featuring emotional (affective) disorders, dementia, or psychotic states to varying extents or not at all. In both classes, there is often detectable malfunction or pathology at the anatomical, neurophysiological, or biochemical level. For instance, progressive degenerative diseases such as senile dementia and Huntington's chorea are characterized by both encephalopathy and deterioration of mental function, and changes at the biochemical level are detectable in postmortem brains of schizophrenic patients. However, some of these changes may be induced by the drugs used in treatment, and care is needed in appraising their true significance for the etiology of the disease (see below).

MOVEMENT DISORDERS

Movement disorders are characterized by certain symptoms involving poor control of body musculature, but changes in intellectual and emotional performance may also be features of this type of disease. The loss of control of muscles may be manifested as hypokinesia, a considerable reduction, or paucity, of voluntary movement with limited ability to move the limbs, as in Parkinson's disease. Other conditions, the dyskinesias, can involve excessive, exaggerated, or bizarre involuntary movement, as in Huntington's chorea, tardive dyskinesia, dystonia, and torticollis.[B,H] These various movement disorders all seem to result from disorders of the basal ganglia, often involving not only structural degeneration but also changes in one or more neurotransmitter systems.

Parkinson's Disease

The disease state known as Parkinson's disease was first described in detail by a London physician, Dr. James Parkinson, in 1817. It is a relatively common disease of the second half of life and has an incidence in most populations examined of about 1 in 1000, rising exponentially after age fifty.

The disease state is primarily due to a loss of the dopaminergic substantia nigra neurons and the consequent deficiency in the dopamine systems or other systems affected by this loss. The cause of the disease is unknown, but it can commence many years after an encephalitic infection or exposure to manganese. There is some evidence of an infective element and the involvement of latent, or "slow" viruses.[G] The salient features of parkinsonism, known collectively as the akinetic-rigid syndrome, are tremor, muscular rigidity, relatively little limb movement, and a pattern of walking with small slow steps. The outstanding structural change in the disease is the degeneration of the substantia nigra[101] (Figure 9.1). This involves loss of the neuronal cell bodies that form the *zona compacta* of this region and is therefore accompanied by a large reduction in dopamine content both in the substantia nigra and in the corpus striatum to which these cell bodies project (Figure 9.2). The pathological processes are not strictly localized; degeneration is also seen in other pigmented nuclei, such as the locus ceruleus, as well as in the raphe nuclei, the hypothalamus, the pons, and the sympathetic ganglia. Characteristic neuronal inclusions called *Lewy bodies* appear in substantia nigra and locus ceruleus in the disease state and have attracted a great deal of attention (Figure 9.3). Lewy bodies consist of an electron-dense amorphous core surrounded by radiating filamentous material.[1,2] Their precise nature and role in the disease process remains unclear. Another cell in-

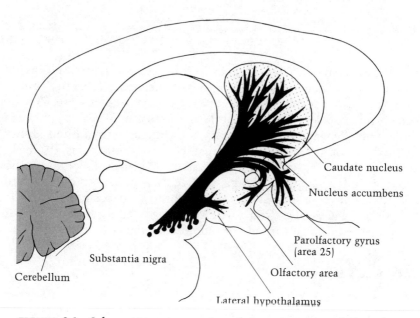

Caudate nucleus

Nucleus accumbens

Parolfactory gyrus (area 25)

Olfactory area

Substantia nigra

Cerebellum

Lateral hypothalamus

FIGURE 9.1   Schematic representation of the mesotelencephalitic dopamine system in the human brain, projected into a paramedian sagittal plane. The term *substantia nigra* includes all dopamine containing mesencephalic cell groups; the term *caudate nucleus* includes the corpus striatum (caudate plus putamen) and the globus pallidus. [From ref. 3]

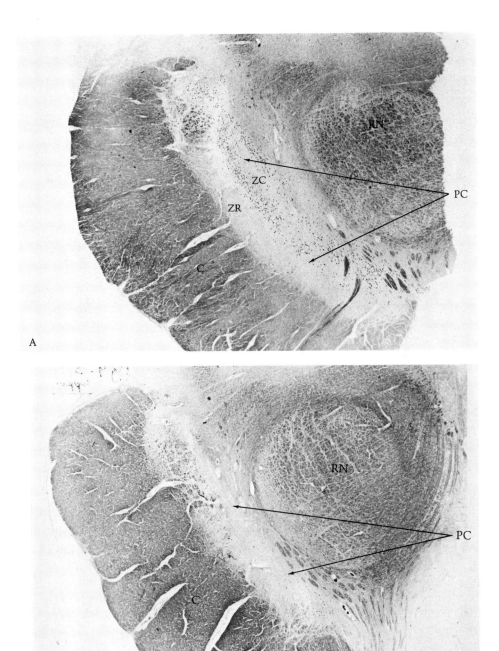

FIGURE 9.2   Characteristic lesions of the substantia nigra seen in Parkinson's disease. (A) A normal population of pigmented cells (*PC*) in the zona compacta region in the substantia nigra of a 69-year-old man. (B) The loss of these cells in a 69-year-old woman with Parkinson's disease. *RN*, red nucleus; *C*, crus cerebri; *ZC*, zona compacta; *ZR*, zona reticulata. Kluver-Barrera stain. Bar, 2 mm. [From ref. 101]

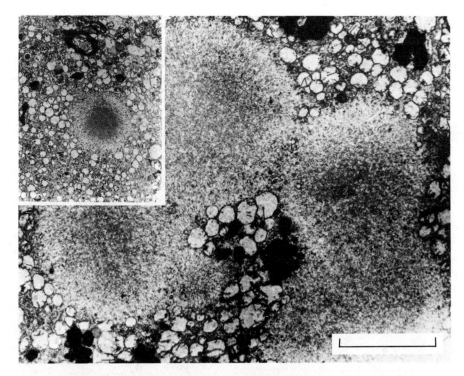

FIGURE 9.3   Multiple confluent Lewy bodies in a pigmented neuron in the substantia nigra, and (*inset*) a small Lewy body in the locus ceruleus, in patients with Parkinson's disease. Bar, 5 μm. [From ref. 1]

clusion, the neurofibrillary tangle[1] (Figure 9.4), is characteristic of some forms of Parkinson's disease, particularly that which develops after encephalitis, a form that constitutes about a quarter of the cases. Neurofibrillary tangles are seen in the substantia nigra. They consist of large aggregates of neurofilaments, contrasting with the normal parallel arrays (see Chapter 1). These tangles are better known as pathological features of senile dementia, or Alzheimer's disease (see below), but Lewy bodies are also seen in Alzheimer's disease. Such observations indicate possible common factors in the two diseases,[82] and may have some bearing on the question of mental deterioration in Parkinson's disease.

In terms of its pathophysiology, Parkinson's disease may be seen as resulting from overactivity of the caudate-putamen and globus pallidus owing to loss of the fine control normally exercised over them by the substantia nigra. This would partly explain why, in some instances, improvement may follow destruction of the globus pallidus or its efferent pathway.[109]

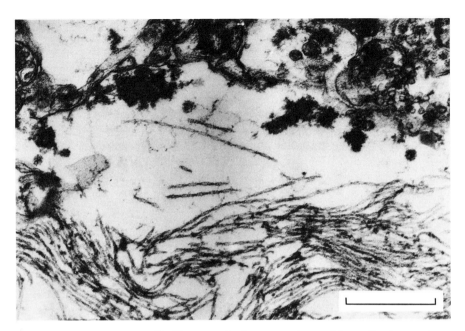

FIGURE 9.4 Neurofibrillary tangle from a patient with senile dementia of the Alzheimer type. The tangles consist of proliferation of abnormal intracellular fibrillary material occupying most of the neuronal cytoplasm; they are thought to be pairs of helically twisted neurofilaments. Bar, 0.25 μm. [From ref. 34]

Neurotransmitter Changes in Parkinson's Disease Studies of postmortem brains have shown that the most prominent change in neurotransmitter profile in parkinsonism is the large reduction in the dopamine content of the corpus striatum due to loss of the fiber projections from the dopaminergic cell bodies present in the substantia nigra and adjoining areas[3] (Figure 9.2). Experimentally induced lesions of this pathway, or induced striatal dopamine deficiency in animals or humans, leads to symptoms of Parkinson's disease: akinesia, rigidity, and tremor. Loss of dopamine is accompanied by the loss of its main metabolite, homovanillic acid (see Figure 4.19). These dopamine losses are not restricted to the zona compacta of the substantia nigra and its projection to the caudate nucleus, the globus pallidus, and the putamen. Very substantial depletion also occurs in the nucleus accumbens, the lateral hypothalamus, and the cortical region of the parolfactory gyrus (Table 9.1). These are limbic forebrain regions and are concerned with the control of mood and emotion.

TABLE 9.1   Dopamine and tyrosine hydroxylase levels in postmortem human brain with parkinsonism

Brain region	Dopamine (μg/g tissue)	Tyrosine hydroxylase (nmol CO_2/30 min/100 mg protein)
Putamen		
Control	5.06 ± 0.39 (17)	17.4 ± 2.4 (3)
Parkinsonism	0.14 ± 0.13 (3)	3.1 ± 1.2 (3)
Caudate nucleus		
Control	4.06 ± 0.47 (18)	18.7 ± 2.0 (3)
Parkinsonism	0.20 ± 0.19 (3)	3.2 ± 0.5 (3)
Globus pallidus		
Control	0.5 (6)	3.5 (1)
Parkinsonism	0.2 (4)	1.7 ± 0.2 (3)
Substantia nigra		
Control	0.46 (13)	17.4 (1)
Parkinsonism	0.07 (10)	6.1 ± 1.5 (3)

Note: Values are mean ± SEM for number of samples in parentheses. Data are from various sources quoted in ref. 3.

The enzymes involved in the synthesis of dopamine, tyrosine hydroxylase and dopa decarboxylase (see Figure 4.12), show very large reductions in basal ganglia and in substantia nigra[3] (Table 9.1); like dopamine itself, they are located in the degenerating dopaminergic neurons.

This large decrease in the population of dopaminergic neurons leads to certain functional shifts with associated functional distortions. For instance, there appears to be a heightened response to dopamine and its agonists by neurons that are postsynaptic to the dopaminergic innervation. This "supersensitivity" is manifested by the much enhanced responsiveness of parkinsonian patients to dopaminomimetic drugs and to the dopamine precursor L-dopa. The supersensitivity appears to be the result of an increase in the numbers of specific postsynaptic dopamine receptors (D2 type) located in the caudate nucleus and putamen,[4] and it may represent a homeostatic compensatory response aimed at maximizing the effectiveness of the remaining dopamine neurons and minimizing the functional deficit. These D2 receptors, which are detected by the binding of radiolabeled neuroleptics, are of the ionotropic category (see Chapter 4). In contrast, the adenylate cyclase–coupled dopamine receptors (D1, metabotropic type) appear to be reduced in population in the striatum.[4] These two receptor types could be mediating the inhibitory and excitatory actions of dopamine in the striatum,[5] though which receptor subtype is responsible for each action is unknown at this time. A reduction in the metabotropic D1 type of receptor could be another feature of a functional compensation in parkinsonism. Certainly all the antiparkinsonian activity of dopamine-substituting agents is generated via the D2 ionotropic dopamine receptors. Identical shifts in dopamine receptor populations follow experimental destruction of the nigrostriatal pathway in animals and are accompanied by a supersensitivity in behavioral response to drugs acting on dopaminergic systems (see Chapter 8).

Involvement of Other Neurotransmitters The multiple neurotransmitter systems present in the basal ganglia, and their complex interactions (see Chapter 7), make inevitable a functional shift in the balance of excitatory and inhibitory action with the progressive failure of the dopamine system. The observable neurochemical features of such secondary effects include decreased concentrations of striatal and nigral serotonin, parallel losses of the GABA-synthesizing enzyme glutamate decarboxylase, and losses of GABA receptors, as well as a fall in the concentration of norepinephrine in the substantia nigra and nucleus accumbens[3] (Table 9.2). Increases in striatal enkephalin receptors and decreases in choline acetylase, the enzyme responsible for acetylcholine formation, have also been reported.[3,6] Substance P, enkephalins, CCK, and neurotensin all show localized decreases in basal ganglia regions in this disease.[110,119,128]

Some changes in parkinsonism result from parallel degeneration of nondopaminergic neurons. For instance, norepinephrine loss is largely due to degeneration of the noradrenergic neurons of the locus ceruleus. Other changes are a direct result of degeneration of dopaminergic neurons. There is loss of GABA receptors in the substantia nigra, where a large proportion are located on nigral dopamine neurons.[3,6-8]

TABLE 9.2   Transmitter and enzyme changes in postmortem human brain with parkinsonism

Brain region	Norepinephrine (μg/g tissue)	Serotonin (μg/g tissue)	GAD (nmol CO_2/hour per mg protein)	ChAT (pmol/minute per mg protein)
Putamen				
Control	0.10 ± 0.01 (12)	0.32 (6)	622 ± 110 (3)	1656 (3)
Parkinsonism	0.05 ± (6)	0.14 (5)	292 (3)	780 (3)
Caudate nucleus				
Control	0.09 ± 0.01 (11)	0.33 (6)	659 ± 93 (13)	1460 (3)
Parkinsonism	0.04 (6)	0.12 (5)	321 (2)	694 (3)
Globus pallidus				
Control	0.06 ± 0.02 (13)	0.23 (6)	553 ± 15 (13)	231 (2)
Parkinsonism	0.05 ± 0.03 (5)	0.13 (5)	388 (2)	39 (2)
Substantia nigra				
Control	0.23 ± 0.05 (9)	0.55 (6)	637 ± 149 (13)	63 (6)
Parkinsonism	0.11 ± 0.03 (3)	0.26 (5)	263 (2)	12 (3)

Note: Values are mean ± SEM for number of samples in parentheses. Data are from various sources quoted in ref. 3.

Dopamine–Acetylcholine Interactions The precise relationships between the various neurotransmitter systems remain unclear. One clear
feature is that the dopaminergic input to the striatum inhibits the activity of the resident cholinergic neurons and creates an imbalance in the
striatal dopaminergic interaction with a shift towards cholinergic dominance, which is a special characteristic of parkinsonism (Figures 9.5 and
9.6). This imbalance occurs perhaps due to the relative lack of degeneration of cholinergic neurons. Cholinergic overactivity seems to be directly
involved in producing the characteristic tremor and rigidity of the condition, since anticholinergic agents relieve these symptoms. In addition,
administration to animals of cholinergic agonists to simulate the overactivity produces an akinetic–rigid syndrome, catalepsy.[3,6,8,H] The loss of
part of the population of noradrenergic neurons in the locus coerulus
results in a similar functional shift since these neurons, which diffusely
innervate many brain regions (see Chapter 4), exert a facilitatory action
on the dopamine systems controlling motor output.[2,3] In contrast, non-
ceruleus noradrenergic neurons inhibit these dopaminergic forebrain

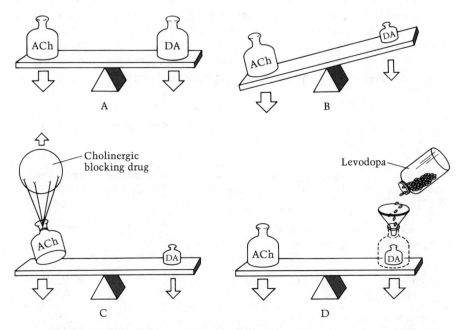

FIGURE 9.5 Diagrammatic representation of dopamine (*DA*) and acetylcholine (*ACh*) acting as antagonistic neurotransmitters in the corpus
striatum. (A) Normally there is a balance (B) In parkinsonism, dopaminergic function is reduced, so that the balance is disturbed in the
direction of cholinergic dominance. (C) This disturbance may be corrected by reducing the effect of acetylcholine with cholinergic (muscarinic) blocking drugs such as atropine or benzhexol. (D) Alternatively,
the balance may be restored by administering the immediate precursor of
dopamine, levodopa (L-dopa). [From ref. H]

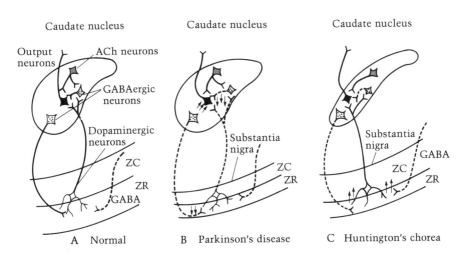

FIGURE 9.6 (A) Normal function of neurotransmitters. (B) Malfunction of neurotransmitters in the pathways between the corpus striatum and substantia nigra in Parkinson's disease. The inhibitory nigrostriatal dopaminergic pathway is lost with degeneration of the substantia nigra. The inhibitory GABAergic terminals that act on striatal neurons, and the striatonigral GABAergic terminals that normally mediate inhibition of dopaminergic neurons in the substantia nigra are also lost. There is an associated decrease in the GABA receptor population. The result is hyperactivity of cholinergic neurons and modification of the output neurons of the striatum. (C) Malfunction of neurotransmitters in the pathways between the corpus striatum and substantia nigra in Huntington's chorea. The cholinergic and GABAergic interneurons, and the inhibitory GABAergic neurons projecting to the zona reticulata of the substantia nigra, are lost with degeneration of the corpus striatum and the zona reticulata. The result is hyperactivity of dopaminergic neurons projecting to the corpus striatum. There is also a supersensitivity of caudate neurons to dopamine. These changes are likely to cause a dopaminergic overdrive of the striatal neurons and modification of their output.

mechanisms. Therefore, loss of ceruleus neurons could be expected to result in a shift towards diminished motor performance emphasized by the inhibitory norepinephrine systems (Figure 9.6A).

Changes in the GABA System The diminished content in parkinsonism of the GABA-producing enzyme glutamate decarboxylase (GAD) in all basal ganglia (Table 9.2) indicates a loss of inhibitory GABA terminals, which could be expected to result in further dominance of competing cholinergic systems in the striatum (Figures 9.5 and 9.6). The fall in nigral GABA-receptor numbers referred to above also reflects malfunction of the GABA system per se in this disorder.[3,6,8]

Changes in Opioid Peptide Systems Decreased binding of [³H]naloxone (mu receptor) in caudate nucleus and increased binding of [³H]enkephalins (delta receptors) in putamen, limbic cortex, and hippocampus of patients with Parkinson's disease[6] point to primary or secondary involvement of opioid systems.[110] Opiate binding sites are reduced by destruction of the dopaminergic nigrostriatal tract, suggesting that part of this receptor population is located on the striatal dopamine nerve terminals.[6,9] Opiate receptors should therefore diminish, in parallel with the degeneration of dopamine terminals, and they do. Since mu and delta receptors mediate different functions (see Chapter 4) and change differentially in the disease, the precise role of opioid systems in the pathophysiology of parkinsonism remains obscure.[109]

Replacement Therapy and Other Treatments   The administration of L-dopa is the most widely acclaimed therapy for parkinsonism. This agent, unlike dopamine itself, readily passes the blood-brain barrier and enters the brain. It is the direct natural precursor of dopamine, and its effectiveness is almost certainly due to the increase in striatal dopamine concentration it produces, presumably improving the performance of the remaining dopaminergic nerve terminals. At the same time part of the lost GAD activity is restored, possibly owing to readjustment of compensatory homeostatic mechanisms. Although L-dopa therapy improves the symptoms of the disease, the underlying progressive loss of neurons in the substantia nigra continues.[6] In advanced cases L-dopa treatment becomes less effective, and direct stimulation of striatal postsynaptic dopamine receptors with agonist drugs such as bromocriptine[H,10] and related compounds (ergot derivatives) may then have a beneficial effect.

Anticholinergic drugs are used to reduce the cholinergic dominance in parkinsonism (Figure 9.5), and they diminish the tremor and rigidity characteristic of the disease.[H] In addition, many of them inhibit reuptake of dopamine from the synaptic cleft, thereby prolonging the action of the dopamine whose release is diminished by the disease process.

Dyskinesias

Dyskinetic conditions are characterized by excessive, exaggerated, bizarre, and sometimes explosive muscle or limb movements or postures. These movements may include quick, ticlike choreiform (dance-like) movements or slow, writhing (athetotic) movements of hands, feet, and digits. Essentially, these are abnormal involuntary movements occurring either spontaneously at rest or superimposed on voluntary movements causing distortion and apparent lack of control. The various forms, from tremor to chorea, give rise to the names of the various dyskinesias. Each type of dyskinesia may be caused by specific neurological disorders or may be induced by drugs. Malfunction of the basal ganglia appears to be involved in most forms, though no clear correlations with damage to particular regions of the basal ganglia have been established. The mobile, hyperkinetic state of dyskinesia can be induced in patients with the akinetic-rigid state of Parkinson's disease by treatment with L-dopa. It seems likely, therefore, that the parkinsonian condition reflects certain basal ganglia mechanisms operating at a *low* level of activity and the dyskinesias are due to *overactivity* of these same mechanisms. Further evidence of this intimate connection between the two states is that dopaminergic antagonist drugs which effectively control dyskinesia often induce the akinetic-rigid parkinsonian state. Thus, dyskinesias may be the result of excessive striatal dopaminergic stimulation, whereas parkinsonism is due to a much reduced and insufficient dopaminergic activity (Figure 9.6). Of course, these states can be exacerbated by failure of other neurotransmitter systems.

Choreic muscle spasms consist of quick, ticlike, disorganized involuntary muscular contractions that cause jerky movements of trunk or limbs, and distort normal movement or posture. The muscles of the face, mouth, tongue, and hands may be involved in a range of dyskinetic movements from involuntary tongue protrusion to progressive involuntary grimaces and grotesque distortions of the mouth and face,[B,11] constituting the slow, writhing movements called athetosis.

Huntington's Chorea

Huntington's chorea is an inherited progressive degenerative disease that is transmitted as an autosomal dominant trait with full penetrance—half of the children of parents with the disease will become sufferers. The disease was first fully described in 1872 by George Huntington, a physician who, over a long period, observed cases in a family living near his home in Long Island, New York. His description emphasized the combination of progressive dementia with bizarre involuntary movements and odd bodily postures. It affects about one person in about 20,000. There is usually a delayed onset, with symptoms not appearing until middle life,[B,11] about 35 to 45 years of age. Both mental deterioration leading to dementia and progressively severe chorea are now well established as the usual features of the disease. Death usually occurs some 15 years from onset. Juvenile cases, about 10 percent of those affected, run a more rapid course.

The most prominent pathological feature of Huntington's chorea is marked atrophy and degenerative changes in the large basal ganglia structures that result in a much shrunken brain and enlarged ventricles. There is severe loss of the small and intermediate-sized neurons, with relative sparing of the larger neurons. The caudate and putamen are particularly affected, often being reduced to half their normal size[12] (Figure 9.7), with loss of the striopallidal fiber bundles. Substantial neuronal loss is also found in the globus pallidus. The substantia nigra, however, in contrast to the pathology seen in Parkinson's disease, is much less affected, the zona reticulata (dendrite region) showing some atrophy[12] while the zona compacta (cell body region) shows little change, though the substantia nigra as a whole is more densely pigmented than normal. Degeneration of

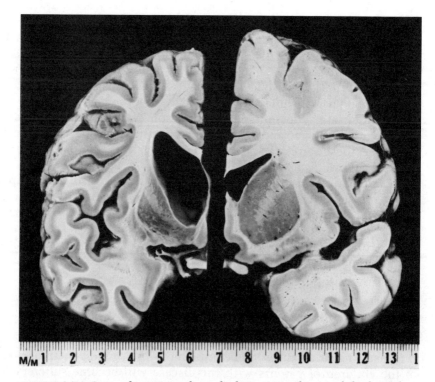

FIGURE 9.7   Coronal sections through the anterior horns of the lateral
ventricles show changes in brain structure seen in Huntington's chorea.
The hemisphere at left is from a 49-year-old woman with Huntington's
chorea (brain weight 1065 g). The hemisphere at right is from a normal
69-year-old woman (brain weight 1415 g). Note the considerable atrophy
of the corpus striatum. The head of the caudate nucleus, which normally
bulges into the floor of the ventricle, has shrunk to a narrow concave
ribbon. [From ref. 12]

the cerebral cortex is evident, with widespread loss of neurons, particularly in the 3d, 5th, and 6th layers, and diffuse loss of nerve fibers.[12] Thus,
the major degeneration occurs first in the striatum, which leads to the
chorea, and second in the cerebral cortex, which causes the dementia.

Neurotransmitter Changes in Huntington's Chorea   Studies of the
neurotransmitter-system profiles of postmortem brains have provided
the key information on the perturbations occurring in this degenerative
disease process.[13,111] The levels of both GAD and GABA in the striatum
and substantia nigra are much depleted. The level of GABA receptors
have been reported to be either unchanged or reduced in the striatum,[14,112] and greatly increased levels of GABA receptors are seen in the
substantia nigra[16] (Table 9.3). These findings have been interpreted as
caused by a denervation supersensitivity towards GABA in cells (probably dopaminergic cells) that are postsynaptic to GABAergic neurons in

TABLE 9.3 Transmitter and enzyme changes in postmortem human brain with Huntingdon's chorea

Brain region	GAD (μmol/$^{14}CO_2$ evolved/hour per g tissue)	GABA (μmol/g tissue)	ChAT (μmol/hour per g tissue)	Muscarinic acetylcholine receptors (pmol/mg protein)	GABA receptors (pmol/mg protein)
Putamen					
Control	4.4 ± 0.4 (45)	4.45 ± 0.33 (13)	21.8 ± 1.6 (41)	18.9 ± 2.1 (7)	0.36 ± 0.01*
Huntington's chorea	0.9 ± 0.2 (41)	1.76 ± 0.24 (11)	9.1 ± 1.3 (42)	8.7 ± 0.8 (6)	0.39 ± 0.03*
Caudate nucleus					
Control	5.1 ± 0.4 (68)	2.87 ± 0.22 (13)	11.9 ± 0.8 (60)	13.9 ± 0.5 (14)	0.15 ± 0.01
Huntington's chorea	1.5 ± 0.2 (66)	1.57 ± 0.26 (11)	5.4 ± 0.7 (64)	8.8 ± 0.8 (11)	0.11 ± 0.01
Globus pallidus					
Control	7.1 ± 1.3 (24)	—	0.9 ± 0.2 (9)		
Huntington's chorea	1.9 ± 0.5 (14)	—	1.5 ± 0.3 (8)		
Substantia nigra					
Control	6.5 ± 1.0 (38)	5.58 ± 0.28 (13)	1.2 ± 0.4 (12)		0.037 ± 0.004 (5)
Huntington's chorea	2.2 ± 0.3 (40)	2.18 ± 0.22 (11)	1.0 ± 0.5 (14)		0.087 ± 0.006 (4)

Note: Values are mean ± SEM for number of samples in parentheses. Data from refs. 14, 16, and 17.
* Data are for frontal cortex.

the substantia nigra. The relevant GABAergic neurons appear to form part of the striatonigral tract and project their terminals to the substantia nigra, where they inhibit the activity of dopaminergic cell bodies. In addition, small GABAergic interneurons in the striatum degenerate. The decreased levels of GAD and GABA in the corpus striatum and substantia nigra reflect this loss of GABAergic neurons.[13,14,15] (Table 9.3).

Cholinergic Changes The cholinergic systems of the choreic brain also show changes. The activities of both choline acetylase, the enzyme that synthesizes acetylcholine (Chapter 4), and muscarinic cholinergic receptors show large and selective reductions in the basal ganglia (Table 9.3), presumably reflecting loss of cholinergic neurons in these structures.[17] This loss appears later than the degeneration of the GABAergic neurons,[13,14,15] and it may be these GABAergic cells that receive the cholinergic input (i.e., cholinoceptive neurons), thereby explaining the reduction in muscarinic receptors.

Serotonin in Huntington's Chorea Other neurotransmitter systems in Huntington's chorea reportedly showing aberration in tissue or cerebrospinal fluid content or in receptor density include serotonin and a variety of peptides. Receptors for serotonin were found to be substantially reduced in the globus pallidus (75 percent reduction) and caudate-putamen (50 percent reduction).[17] Postmortem tissue levels of serotonin in the brains of patients with this disease do not show such consistent changes, though elevations in putamen and central gray matter have been detected[17,111,112] and could represent a response to the decline in the number of receptors present.

Peptide Neurotransmitters Decreases in the levels of substance P, enkephalins, angiotensin, and cholecystokinin have been detected in the postmortem brains of Huntington's patients. Cholecystokinin is decreased only in globus pallidus and substantial nigra. Receptors for this peptide are reduced by 70 percent in basal ganglia and 40 percent in cerebral cortex.[111] The substantia nigra and pallidum show considerable reductions in met-enkephalin, while angiotensin and angiotensin-converting enzyme (its biosynthetic enzyme) are depleted in corpus striatum and substantia nigra.[18,112] Substance P is much reduced in level in the medial pallidum and in the pars reticulata of the substantia nigra, indicating losses in the striatopallidal and striatonigral projections respectively.[111,112] Less marked reductions have been observed in the caudate and the putamen, and in the lateral pallidum and pars compacta of the substantia nigra. Both somatostatin and neuropeptide Y are increased in the basal ganglia in this disease state, though this could reflect a relative sparing of neurons containing these peptides during the shrinkage of the basal ganglia that is characteristic of Huntington's chorea.[106,111] The elevated levels, however, could represent a real increase in the number of fibers containing these peptides as well as an artifactual elevation due to tissue shrinkage. Thus, the levels of somatostatin in the nucleus accumbens are also raised two- or threefold, even though this brain region shows much less atrophy.[111] Some researchers have hypothesized that somatostatin is functionally involved in producing the symptoms of Huntington's chorea, possibly by contributing to the excess of dopaminergic activity in the disease by enhancing the release and the action of dopamine.[111] It is notable that somatostatin is decreased in the cerebral cortex, while cholecystokinin is unchanged.[119]

Other neuropeptides showing changed levels in the disease are thyrotropin-releasing hormone (TRH), which is increased in the caudate and amygdala, and neurotensin, which shows elevated levels in the caudate nucleus.[111,112]

Dopamine Systems in Huntington's Chorea   The dopaminergic transmission of the basal ganglia in Huntington's disease is being distorted by the failure of systems impinging upon it, and dopaminergic neurons also show intrinsic changes. The activity of tyrosine hydroxylase is increased more than twofold in the substantia nigra, and homovanillic acid has been reported to be high in cerebrospinal fluid. Dopamine content of the caudate nucleus may be slightly lowered. In the substantia nigra, it may be raised.[14,15,112] Moreover, the dopamine receptor population of the substantia nigra appears reduced, probably due to its localization on degenerating cholinergic neurons.[8]

In spite of these conditions, there is a hypersensitivity, or exaggerated response, to dopamine. For instance, administration of L-dopa substantially increases existing choreiform movements in Huntington's chorea patients (and induces them in parkinsonian patients).[14,15] This hypersensitivity is no doubt partly due to the loss of the GABA inhibitory action, which normally impinges on these neurons both in the corpus striatum and the substantia nigra (Figure 9.6B). Drugs that are active dopamine antagonists greatly reduce choreic syndromes, and this finding supports the proposition that overactivity of dopamine is at the root of the pathophysiology. The reduced activity of the cholinergic neurons will further shift the imbalance between the dopamine and cholinergic systems in favor of a dopaminergic inhibitory overdrive.

Animal experiments in which kainic acid was injected intracerebrally to destroy cell bodies of the corpus striatum caused the appearance of many of the symptoms of Huntington's chorea.[16,20,115] Since GABAergic neurons are highly sensitive to kainic acid, this adds strong support to the possibility that there is loss of small neuronal cell bodies with associated decreases of GABA-mediated inhibition, as well as losses of cholinergic and peptidergic neurotransmission, in patients with Huntington's chorea. Others have argued that glutamic acid, an excitatory neurotransmitter in the corticostriate pathway, is partly responsible for the neuronal cell death in the striatum. Like kainic acid (but much less potently), it causes neuronal degeneration when injected into the striatum, and the drug baclofen (β-p-chlorophenyl-GABA), which inhibits glutamate release from nerve terminals, has been found effective in the treatment of the disease.[111,112] But an unidentified endogenous "kainate-like" substance, distinct from glutamate and acting at kainate receptors, has also been proposed as the agent causing the neurodegeneration.[116]

Treatment of Huntington's Chorea   The nature of the primary genetic defect causing the degenerative changes in the choreic brain remains unknown although researchers, using recombinant DNA techniques, have made some progress toward detecting the faulty genetic material associated with the disease.[96] There are no effective treatments to halt the progression of the disorder. Unsuccessful attempts at therapy have included the use of GABAergic drugs to overcome the deficiency in the GABA systems, and cholinomimetic drugs (including choline itself) to overcome the loss of cholinergic neurons.[14,15] Large doses of tryptophan, the precursor of serotonin, have been administered because of the reported serotonin receptor deficiency,[16,17] but also without effect. The severity of the chorea can be decreased with agents that reduce dopaminergic transmission[H] (e.g., tetrabenazine, pimozide). Drugs that raise the level of striatal acetylcholine (e.g., physostigmine) have been reported to decrease the choreic movements; those that antagonize acetylcholine action, such as benztropine, appear to increase these movements.[H] Thus, the hypothesis that a *hyper*dopaminergic-*hypo*cholinergic-*hypo* GABAergic imbalance prevails in the disease is supported by the nature of the drugs that provide an effective therapy.[14,15,H]

Drug-Induced Dyskinesias

The advent of the modern tranquilizing drugs has brought the appearance of drug-induced dyskinesia in patients who use them for several months or more. Phenothiazines, butyrophenones, and reserpine, which constitute the major groups of these drugs, may produce the *tardive dyskinesias*, movement disorders of late onset, after long-term therapy.[H,15,21,22] These conditions are characterized by involuntary choreoathetoid movements in the tongue, mouth, face, neck, trunk, and limbs, particularly in older patients. They persist after drug withdrawal, and may even appear for the first time after drug withdrawal.

    Brain amines are principally implicated in the etiology of drug-induced dyskinesias. Although the precise pathophysiology remains obscure, tardive dyskinesias appear to result from dopaminergic hyperactivity, possibly involving supersensitivity of excitatory dopamine receptors,[21] an increase in dopamine synthesis and release, and a decrease in dopamine reuptake. Therapy with ʟ-dopa might be expected to both cause and exacerbate the condition. Phenothiazines and butyrophenones are usually employed to treat the severe anxiety of psychosis and stress. They block catecholamine receptors, and their prolonged use leads to a fall in brain catecholamine content.[H] These agents may cause tardive dyskinesia during long-term medication by establishing a chemical "denervation supersensitivity" to dopamine. Under these conditions, normal levels of dopamine in the synapse could evoke movements similar to those induced by L-dopa after long-term treatment of parkinsonism. Supporting the "dopamine-hyperactivity" hypothesis is the finding that tardive dyskinesias can be treated by reblockade of dopamine receptors with haloperidol and also by amine depletors such as tetrabenazine.[H,15] Moreover, amine depletors such as reserpine can induce the akinetic-rigid tremor symptoms of Parkinson's disease by causing a loss of dopamine and norepinephrine, and the condition can be reversed by L-dopa and serotonin.[15]

PSYCHOLOGICAL DISORDERS

The biogenic amines are thought to have a special role in the etiology of a number of mental disorders, and these particular malfunctions are also associated with disturbances in the neurotransmitter systems of the basal ganglia and limbic systems. The apparent involvement of the amines of the basal ganglia in both psychological and movement disorders emphasizes the close association in these brain structures of movement control and the control of mechanisms generating mood and emotional states.[23]

Mental illness in its various forms presents a major social problem. In the United Kingdom, for example, as many as 50 percent of all hospital beds during the 1960s were occupied by patients of this category. This figure has now fallen considerably owing to the effectiveness of modern drug therapy, which has allowed many patients to leave the hospital and live in the community in special hostels. However, these drugs do not cure but rather relieve the symptoms of the disorders; many patients return to the hospital, and most are unable to take regular employment. For these reasons the social problem has not disappeared but has been shifted from the mental hospital into the community at large. Identifying the causes of these disorders and providing effective chemotherapies for them are therefore a major objective of clinical neurochemists and neuropharmacologists.

Schizophrenia

Most current theories of the cause of schizophrenia, a widespread psychotic disorder characterized by withdrawal, confused thinking, delusions, paranoia, and hallucinations, once again implicate a failure of dopamine neurotransmission.[1] There are probably a whole group of disorders, the "schizophrenias," within the boundary of the definable clinical syndrome, each differing somewhat in symptomatology. The cause is unknown, but there may be a hereditary susceptibility that allows the condition to develop where psychologically stressful environments persist.

Neurochemical Leads: Dopaminergic Systems   The effectiveness of certain neuroleptic drugs in ameliorating the symptoms of schizophrenia indicates the involvement of dopaminergic mechanisms in the disorder. Agents such as chlorpromazine are likely to produce their antipsychotic effects by blocking dopamine receptors.[15,23,24,F,I] There is a good correlation between the ability of neuroleptics to block dopamine receptors and their clinical effectiveness as antipsychotic drugs.[15] But there is an unexpected delay of several weeks between effective dopamine receptor blockade—as assessed by the rise in prolactin secretion in schizophrenic patients, a response controlled by inhibitory dopamine receptors—and the onset of the clinical response. This delay suggests a dimension of complexity in any dopamine malfunction in the disorder. Perhaps receptor blockade itself is necessary in order to allow changes in other related neurotransmitter systems.[23,24]

It was first reported that neuroleptic drugs potently antagonized the stimulation of dopamine-sensitive adenyl cyclase but that correlation with clinical effectiveness was best with phenothiazines (e.g., chlorpromazine) and thioxanthenes (e.g., flupenthixol)—the butyrophenones (e.g., haloperidol) not fitting in very well.[25] Subsequent studies of the blockade of receptor-binding, employing radioactive butyrophenone ligands, showed a good correlation for these drugs too. Thus, different categories of dopamine receptors having differing interactions with neuroleptic drugs appear to be involved in the disorder. These variable responses could be related to different "agonist" and "antagonist" conformations of the receptors or to their multiple cellular locations.[15,27]

Other considerations also point to a dopamine-mediated malfunction in schizophrenia. Amphetamine induces a psychosis with symptoms similar to those of acute paranoid schizophrenia, the effects most likely being due to amphetamine's potent catecholamine-releasing actions. The amphetamine psychosis is prevented by chemical lesioning of dopamine neurons with 6-hydroxydopamine. Neuroleptic drugs also prevent the behavioral effects of amphetamine. Significantly, amphetamine exacerbates the symptoms of schizophrenic patients, and this, too, is controlled by neuroleptics.

These observations with a catecholamine-releasing agent suggest that increased dopamine release may be at the root of schizophrenia. But measurements of dopamine metabolites in cerebrospinal fluid do not reveal the consistent rise that would be expected to follow a chronic increase in dopamine release. And the decrease in prolactin secretion that should follow increased dopamine release by the tuberoinfundibular system is not seen in unmedicated schizophrenic patients. Moreover, since parkinsonism, a *hypodopaminergic* state, can coexist with schizophrenia,[24] increased dopamine release cannot be essential for schizophrenia to appear. However, reports have appeared of increases in postmortem schizophrenic brains of both the dopamine[107] and norepinephrine[108,113] content of striatal and limbic brain regions as well as the accumbens area and the olfactory tubercle.

Much interest has focused on dopamine receptors. Development of postsynaptic supersensitivity of dopamine receptors would give responses similar to those of an increase in dopamine release. Some evidence for this hypersensitive condition now exists, with the finding that specific [³H]haloperidol and [³H]spiroperidol binding in the corpus striatum and nucleus accumbens was substantially increased in a large group of schizophrenic patients.[28,29] It is possible, however, that part or all of this receptor density (or affinity) increase is due to the neuroleptic medication received by many of the patients.[33] This point remains to be fully clarified, particularly as small increases in dopamine-receptor density have been shown to follow neuroleptic administration to animals.[24]

Dopaminergic pathways in two closely related brain regions, the extrapyramidal pathway and limbic systems, appear to be fundamentally involved in two different brain functions: generating emotional (or psychotic) states and controlling general motor activity of the body. These two functions are intimately linked in the execution of various patterns of behavior, such as goal-seeking activity or the expression of emotions[12] (see Chapter 8).

Transmethylation Products in Schizophrenia   Some researchers have postulated that methylated catecholamines or indoleamines, which have hallucinogenic properties, might be formed in exaggerated amounts in patients with schizophrenia. Such products were reported to be present in the urine of schizophrenics, and some methylated catecholamines (e.g., 3,4-dimethoxyphenylethylamine; DMPEA) have structural similarities with mescaline and can produce catatonia in animals. Hallucinogenic methylated indoleamines (e.g., *N,N*-dimethyltryptamine) *can* be formed in the body, and administration of methionine to enhance methylation has been reported by many research groups as exacerbating the symptoms of schizophrenia. But some of the findings may be complicated by the drug and dietary influences to which institutionalized schizophrenic patients are often exposed, and the search for toxic hallucinogenic methylated monoamine metabolites of this kind has made little progress.[F]

Other Neurotransmitter Systems in Schizophrenia: Serotonin and Norepinephrine   Apart from deficits in the dopamine system, malfunction of other biogenic amine–mediated systems may underlie the etiology of the schizophrenias. An imbalance between the dopamine, serotonin, and noradrenergic systems, for instance, could occur either because of primary lesions in these other systems or as secondary effects of the dopaminergic malfunction. It would not be reasonable to imagine that a disturbance in only one system could underlie a specific psychiatric disorder, particularly since experimental manipulation of any one of the biogenic amine systems leads to abnormal behavior. Reports are accumulating that show lowered densities of serotonin receptors in brains from schizophrenic patients, and some neuroleptics block serotonin receptors as well as dopamine receptors. Amphetamine treatment has been shown to lower brain serotonin levels as well as catecholamine levels. Propranolol, a β-adrenergic blocker, is reported to have beneficial effects in chronic schizophrenic patients, yet it appears to act as a central serotonin antagonist without blocking dopamine receptors.[15] All in all, there are grounds for careful assessment of the possible involvement of other biogenic amines, and even of some of the new peptide transmitters with behavior-modifying properties, in the neurochemical causes or effects of the schizophrenias.[103]

Peptide Neurotransmitter Systems in Schizophrenia Recently, measurements of five neuropeptides in various regions of brains from schizophrenics have revealed that certain peptidergic systems also show some aberration in this disease state, particularly in the limbic lobes.[117,118] Thus, the hippocampus of the schizophrenic brains showed significant losses of cholecystokinin (CCK) and somatostatin and increases in its content of substance P. In the amygdala, CCK was reduced in level while vasoactive intestinal peptide (VIP) was significantly elevated. In the temporal lobe, CCK was substantially depleted. In contrast, neurotensin showed no changes in tissue content in any of the various brain regions examined. Many of these reported alterations in peptide content in these limbic brain regions are large, being in the 30–40 percent range. The corpus striatum showed no significant changes in neuropeptide levels. The restriction of these changes within the limbic lobes contrasts with the neuropeptide changes seen in Huntington's chorea and Parkinson's disease, which are limited to nigrostriatal projection areas.[119] The limbic system is very much involved in the organization and expression of emotional behavior in addition to its more widespread influence on the circuitries of the brain, and disorders of the limbic system are likely to result in disorders of the emotions and perceptions, including those typical of schizophrenia. However, some,[118,120] but not all, reports agree on the presence, range, or extent of these aberrations in neuropeptide levels in postmortem schizophrenic brains, and more studies are needed before malfunction of peptidergic neurons can be firmly implicated in schizophrenia. Differences in the diagnosis and subclassification of the schizophrenic patients whose brains have been studied and the precise brain regions taken for the analyses are no doubt all adding complexity to the interpretation of these findings.[18] Failure of peptidergic systems could be of special significance in those subclasses of schizophrenia that are resistant to treatment with neuroleptic drugs and are therefore less likely to involve dopamine malfunction.

Neurochemical Changes: Cause or Effect? Whether the observed neurochemical abnormalities in psychiatric disturbance are causative in the conditions, or whether they simply reflect abnormal performance of the brain centers concerned with generating and modulating emotion and mood, is a question of profound importance[l] and the subject of much public and professional debate. Perhaps the most significant facts bearing

on the question concern the actions of mood-changing and hallucinogenic drugs, including amphetamines and LSD. These drugs can clearly
initiate changes in psychological states that have many features in common with the psychiatric conditions of schizophrenia and clinical depression. The sites of action of the drugs are mostly known and usually
include interaction with the synthesis, storage, release, receptor interaction, or inactivation of biogenic amines and other neurotransmitter systems. Whether social factors affecting the nervous system through psychological stress could also act to disturb these parameters is not yet
clarified.

Depressive Illness

The group of conditions collectively known as depression is characterized by a severe and chronic change of emotional state with prominent
subjective feelings of sadness, withdrawal, apathy, lowered ego, and a
negative self-assessment. These states, also called affective disorders,
may be readily distinguished from the swings in mood of normal people
by their severity and chronic nature, and they can be assessed on a clinical rating scale. The two categories of the syndrome are *bipolar*, in which
both depression and mania occur, and *unipolar*, in which only depressive
episodes occur. The manic syndromes of bipolar depressive illness include elation and hostility of mood, a favorable self-image with grandiose
self-assessment, and physical overactivity.

Neurotransmitter Malfunction The cause of depression is unknown.
The illness might result from failure or malfunction of the neurotransmitter systems of the brain either as a primary effect, or it might be an
outcome of maladaptive behavioral responses to stress in the social environment. In the latter situation, overactivity of certain mood-condition-
ing neural mechanisms might lead to malfunction and chronic imbalance
of certain key neurotransmitter systems as a secondary response.[F,I]

There is a genetic element to the disorder;[F,I] it presumably involves
some kind of predisposition to developing depressive illness under the
social or physical conditions that lead to its onset.

The anxiety of depressive illness may be alleviated by drugs, though
amelioration is often slow and incomplete. Most of the relevant drugs act
on certain aspects of monoamine transmitter systems and their therapeutic actions have been linked with these effects.

Monoamines and Affective Disorders The two amines predominantly
associated with current views of the etiology of depression and mania are
serotonin and norepinephrine, and this has lead to the formulation of the
indoleamine and catecholamine hypotheses of depressive illness. Neither
type of monoamine has been demonstrated to be *primarily* involved in
causing these mental disorders, though, both are implicated by current
findings.

Serotonin and Affective Disorders The tricyclic antidepressant clomipramine (3-chloroimipramine) at low concentrations selectively reduces the uptake of serotonin at nerve endings and decreases its synthesis, thus implicating serotonin in the etiology of depressive illness[Q,I,30] (Table 9.4). Higher concentrations of clomipramine (and its main metabolite dimethylclomipramine) inhibit uptake of other monoamines. Measurements of the levels of the serotonin metabolite 5-hydroxyindoleacetic acid (5-HIAA) in cerebrospinal fluid can give an index of the metabolic turnover of serotonin in the central nervous system. In spite of conflicting reports, a large body of evidence points to a lowered serotonin level and turnover in depressive illness,[30] though this may not be the only monoaminergic system affected.[Q] This is matched by reports of lowered levels of serotonin or its metabolites in the brains of depressive suicides. Clomipramine and other serotonin uptake blockers, which prolong the presence of serotonin (as well as other monoamines) in the synaptic cleft, should help to counter effects due to lowered brain content of serotonin. The tryptophan hydroxylase inhibitor PCPA (Table 4.8) can reverse the antidepressant effects of imipramine; the catecholamine synthesis blocker α-methyl-p-tyrosine is ineffective.[30]

Agents raising the general level of serotonin should also have antidepressant properties. This seems to be so. In rats, when MAO type A inhibitors, which prevent serotonin (and catecholamine) oxidation, are given together with L-tryptophan, they cause a marked rise in brain serotonin levels and induce behavioral changes including hyperactivity (see Chapter 8). In humans this combination produces an antidepressant action,[15,30] though the effect could be due to tryptamine, the levels of which also rise under these conditions.

The fact that monoamine-depleting drugs such as amphetamine and reserpine cause depression in the longer term is further circumstantial evidence that lowered levels of brain monoamines could be involved in generating clinical depression.[95]

Biochemical findings in mania generally point to a rise in serotonin levels and an increase in its turnover in the brain.[15,30,F,J] Lithium treatment of mania has become widespread although its mode of action remains obscure. There are indications that serotonin turnover is increased by lithium, and this could lead to a fall in serotonin level in the "releasable" transmitter store. Existing data indicates that long-term therapy causes increased tryptophan uptake but decreased tryptophan hydroxylase activity.[30] Lithium is also used generally in the prophylaxis of manic-depressive illness, which presumably involves swings of neurotransmitter concentrations. This treatment may somehow lead to a new steady-state and a stabilization of the level in the releasable pool.[30]

Catecholamines and Affective Disorders The concentration of homovanillic acid, the acid metabolite of dopamine, has been widely reported to be below normal levels in the cerebrospinal fluid of patients

TABLE 9.4   Tricyclic antidepressant drugs as inhibitors of monoamine uptake

Tricyclics

Name	A	B	R	IC_{50} for inhibition of amine uptake-μM		
				5-HT	NA	DA
Amitriptyline	—CH_2—CH_2—	C	$=CH \cdot CH_2CH_2N(CH_3)_2$	0.49	0.05	4.0
Nortriptyline	—CH_2—CH_2—	C	$=CH \cdot CH_2 \cdot CH_2 \cdot N \overset{H}{\underset{CH_3}{\diagdown}}$	1.60	1.30	5.5
Imipramine	—CH_2—CH_2—	N	—$(CH_2)_3$—$N(CH_3)_2$	0.50	0.20	8.7
Desipramine	—CH_2—CH_2—	N	—$(CH_2)_3$—$N \overset{H}{\underset{CH_3}{\diagdown}}$	2.50	0.03	50.0
Clomipramine	—CH_2—CH_2— (3-CHLORO)	N	—$(CH_2)_3$—$N \overset{H}{\underset{CH_3}{\diagdown}}$	0.04	0.30	12.0

Note: IC_{50} refers to the drug concentrations required to cause 50 percent inhibition of the uptake of [³H]monoamines by brain homogenates in vitro, using monoamine concentrations less than 0.1 μM. From ref. 88.

suffering from clinical depression, suggesting a decrease in dopamine turnover in this condition.[30] No complementary changes have been detected in the postmortem brains of such patients. Other evidence for a catecholamine involvement in depressive illness comes from the antidepressant properties of certain tricyclic drugs (amitriptyline, protriptyline, nortriptyline) that selectively block uptake of norepinephrine[88] (Table 9.4). Lithium also affects norepinephrine levels in a fashion similar to its actions on serotonin, causing enhanced turnover.

The extent to which catecholamine and indoleamine neurotransmitter systems are involved in affective disorders will be better evaluated with the use of new drugs with greater specificity of action.[F,J]

Electroconvulsive Therapy and Neurotransmitters   Electroconvulsive therapy for the treatment of psychiatric illness was introduced in the 1930s on the rationale of an apparent mutual antagonism between epilepsy and schizophrenia, since epilepsy was claimed to be rarely seen in depression or schizophrenia. This hypothesis is probably incorrect but electroconvulsive therapy continues to be used to treat depressive illness. (Usually, six to eight treatments are given over a period of 2 to 3 weeks.) Several trials have shown it to be at least as effective as tricyclic antidepressants and possibly better, producing a more rapid response.[31] The mechanism of action of electroconvulsive therapy remains unclear. Animal experiments show that a course of daily electroconvulsive shocks can lead to increased locomotor activity (mice) several days after the last shock.[31] After reserpine pretreatment, there is enhanced locomotor response to postsynaptic agonists for dopamine and norepinephrine (e.g., apomorphine, clonidine) and increased norepinephrine turnover, suggesting increased sensitivity of catecholamine receptors after therapy. In addition, daily electroconvulsive shock of rats produces augmentation of the hyperactivity caused by administration of tranylcypromine (an MAO inhibitor) together with L-tryptophan or L-dopa.[F,32] These findings suggest that repeat electroshock causes a postsynaptic increase in sensitivity to serotonin and dopamine, since no changes in rates of neurotransmitter synthesis, a presynaptic response, were detected. The response to postsynaptic agonists for serotonin and dopamine were also increased.[32] Thus, the therapeutic actions of electroshock could well be due to changes in biogenic amine transmitter function, particularly at the level of postsynaptic response.

Not surprisingly, changes in other neurotransmitter systems have also been noted to follow electroconvulsive therapy, including decreased GABA turnover and regional changes in met-enkephalin concentrations.[32] The significance of these effects awaits interpretation.

Senile Dementia

At least one-third of the elderly patient population of mental hospitals suffers from senile dementia. The clinical symptoms include, in addition to dementia, an excess of senile degenerative changes such as brain atrophy, senile plaques, and neurofibrillary and granulovacuolar degeneration in the cerebral cortex and hippocampus. When this condition develops in patients before the age of 65, it is known as presenile dementia or Alzheimer's disease.[12,34] In practice, all dementias with these symptoms and pathology tend to be called Alzheimer's disease.[K] Two different subgroups may well exist,[39] the second termed senile dementia of the Alzheimer type. This type is the dementia developing in old age with all the pathological features of Alzheimer's disease.

A related dementia is Pick's disease, in which there is extreme atrophy of relatively circumscribed parts of the cerebral hemispheres. This disease involves severe neuronal loss, astrocytic proliferation, and fibrous gliosis.[12] Some characteristically swollen neuronal cell bodies may be seen, and these have become known as "Pick cells." In Creutzfeldt-Jakob disease, which is associated with different clinical symptoms and mostly occurs in middle age, a more generalized degeneration occurs in the cerebral cortex; it becomes characteristically spongy and with extensive astrogliosis. This disease is of special interest because it is transmissible, although the infective agent has not been isolated.[P,80] Huntington's chorea also displays the features of atrophy and cortical neuronal loss seen in these other dementias, but characteristic is the considerable atrophy of the corpus striatum.

The clinical features of advanced senile dementia include profound deterioration of most aspects of mental function, including memory, intellect, and judgment. There may be speech disorders (dysphasia), inability to perform intentional movements (apraxia), and impairment of recognition (agnosia).[K,12]

Apart from brain atrophy, which can reduce brain weight by 20 to 30 percent, senile plaques develop (15–200 μm in diameter; Figure 9.8), particularly in the cerebral cortex and hippocampus. These plaques consist of degenerating nerve endings, reactive nonneural cells, and a central core of amyloid fibrils. Such plaques are seen in normal senile brains but relatively rarely. The numbers of plaques may be related to the intellectual impairment of the patient.[34]

Other interesting morphological features are the appearance of *neurofibrillary tangles* (Figure 9.4) in the cortical and hippocampal neurons and the presence of Hirano bodies[34] in the hippocampus (Figure 9.9) together with vacuolation. *Hirano bodies* consist of oval or rodlike eosinophilic structures 30 μm long and varying from 8 to 15 μm across.[127]

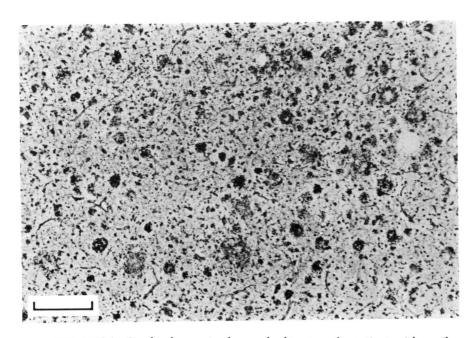

FIGURE 9.8 Senile plaques in the cerebral cortex of a patient with senile dementia of the Alzheimer type. Von Braunmuhl stain. Bar, 250 μm. [From ref. 34]

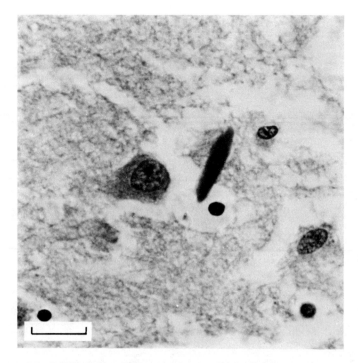

FIGURE 9.9   Large, elongated Hirano body lying between two pyramidal neurons in the hippocampus of a patient with senile dementia of the Alzheimer type. These Hirano bodies consist of layers of flat plates (20–30 nm thick) made of parallel rodlike structures (6–10 nm diam.). They are mainly found in hippocampus. Bar, 20 μm. [From ref. 34]

The neurofibrillary tangles are made up of arrays of paired 10-nm-diameter filaments twisted into helical configurations with a turn period of 65–80 nm and a diameter of about 22–34 nm. They have been mainly regarded as helically twisted neurofilaments, but this idea has now been challenged. Instead, they could be composed of unrelated fibrous proteins.[102]

Cholinergic Systems in Senile Dementia   The most marked, consistent, and selective change in Alzheimer's disease is a severe cholinergic lesion in the cerebral cortex and in the hippocampus.[38] This change seems to be relatively specific and is seen in the postmortem brain as a fall in content of choline acetylase and acetylcholinesterase in cerebral cortex. The pattern of the loss suggests damage to the diffuse ascending cholinergic (telencephalic) projections that probably arise in the nucleus basalis which lies within the substantia innominata, a large group of cells in the medial basal region of the forebrain below the basal ganglia, and, indeed, there are reports of losses and reductions in size of cholinergic neurons in the nucleus basalis of patients with Alzheimer's dis-

ease.[K,34,35,36,37,82,104,105,124,125] The extent of the cholinergic lesion correlates with the severity of the neuropathological damage and with the
severity of the dementia.[37] Scopolamine and other cholinergic antagonists can induce cognitive impairments such as short-term memory loss
in normal people, and these are reminiscent of the clinical features of
Alzheimer's disease. Such observations support what is now being called
the "cholinergic hypothesis" of Alzheimer's disease. The primary disease
process, however, could well be in another neurotransmitter system or
neuromodulator system whose activity bears on cholinergic function.[37,114]

GABA and Neuronin S6 in Senile Dementia   Other changes occurring in
the disease include loss of neuronin S6, a soluble acidic brain protein.[35]
There is also a decrease in the activity of glutamate decarboxylase in the
tissue. But these losses could be due to the hypoxia occurring in perimor-
tem states and the reduced cerebral blood flow of the condition. Ischemia
has been shown to cause loss of glutamate decarboxylase but not of choline acetylase. Thus, the terminal state of the patients, who often die of
bronchopneumonia, may be the cause of the postmortem levels of
neuronin S6 and glutamate decarboxylase[40] but is unlikely to underlie
the diminished levels of choline acetylase.

Other Neurotransmitters in Senile Dementia   Many research groups
have reported deficits in the noradrenergic system in postmortem
Alzheimer brains. This involves lowered levels of both norepinephrine
and dopamine-β-hydroxylase, particularly in the hypothalamus.[124] In addition, there is loss of neurons in the local ceruleus and the appearance of
numerous neurofibrillary tangles in this nucleus.

The serotonergic system too shows changes in senile dementia. The
raphe nuclei show both cell losses and neurofibrillary tangles. Serotonin
metabolites in brain and cerebrospinal fluid are reduced as are receptor
populations and uptake transport rates for this neurotransmitter.[124]

A wide range of neuropeptides have been measured in postmortem
brains from patients with this disease. A reduction in somatostatin levels
is the most consistent finding to emerge from these studies.[124]

Treatment of Senile Dementia   Drug treatment of the disease based on
cholinergic replacement therapy has largely failed so far. Orally administered choline and lecithin (its phospholipid form) have yielded variable
results, with reports of cognitive improvements in some cases.[K,37] Anticholinesterase agents such as physostigmine have not given clear indications of therapeutic success either.

The cause of senile dementia or Alzheimer's disease remains a mystery, with infective agents of the slow-virus type and some kind of complex autoimmune disease being hypotheses in vogue.[K,37]

EPILEPSY

The common neurological disorder known as epilepsy affects about one person in 200. The fits or convulsions that are the hallmark of this disease attest to its principal feature, spontaneous neuronal hyperactivity. This hyperactivity may be manifested in many different ways depending on its regional origin and localization, and it can involve psychological or emotional response as well as motor events. For example, temporal lobe psychomotor epilepsy involves automatic behavior beyond the control of the patient.

Categories of Fit

The wide range of symptoms of epilepsy defines the several different expressions of the disease that can affect people in any age group.[L]

Grand mal fits are generalized bilateral seizures involving hyperactivity of the whole cortex and usually resulting in loss of consciousness. The fits themselves are not painful, in spite of appearances, and the flailing limbs and tonic-clonic episodes simply indicate the progressive uncontrolled firing of many, or most, cortical and other neurons. In its worst form (status epilepticus), one fit will follow another in a continuous sequence with a short intervening recovery period.

Petit mal fits are also generalized seizures but of the mildest category, involving very brief losses of attention, or impairment of consciousness, most lasting only seconds. They appear to arise in brain stem structures. Petit mal, mostly affecting children, is accompanied by a transfixed stare, mumbling, or fumbling of the hands. The simple type of petit mal attack may occur very frequently, even as often as a hundred or more times a day.

Focal fits of the motor category involve myoclonic jerks and other explosive muscular excitations, which result from very localized (focal) hyperactivity in the cerebral cortex, brain stem, pons, or spinal cord.[51] They do not necessarily involve loss of consciousness. The *sensory* category of focal fit is associated with visual, auditory, olfactory, gustatory, or visceral sensations.

Each kind of epileptic condition produces a characteristic pattern in the electroencephalogram (EEG), which aids diagnosis and often allows localization of the hyperactive region. Since focal epileptic attacks may sometimes develop into generalized seizures, the boundaries between these various forms of epilepsy are often blurred, but for the purposes of drug treatment it is useful to classify patients according to the type of seizure they experience.

Etiology of Epilepsy

The primary hyperactive neuronal discharge marks the epileptic condition. This discharge, detected as an epileptiform EEG, is very often localized to a particular area called the *epileptic focus,* which may remain active continuously, only spreading out into the surrounding "normal" brain regions and precipitating a seizure state under certain conditions. Research directed at establishing the etiology of the disease concentrates on the events causing this hyperactivity in the epileptic focus. The underlying cause of the condition remains unknown, but it presumably involves a shift toward an imbalance between excitatory and inhibitory neurons in favor of excitation. This shift might result from loss of inhibitory interneurons, or it might be due to the greatly enhanced activity of excitatory neurons arising from internal nerve-cell changes. It seems unlikely that a single kind of pathophysiology and associated neurochemistry underlies all the forms of epilepsy, but the existence of a number of specific features in a wide range of human and experimental epileptic foci does indicate some degree of convergence in their mechanisms.

About 30 to 40 percent of human epilepsy is of the grand mal type and is associated with lesions in the temporal lobe and the underlying hippocampus.[C,L,41] Something of the order of 80 percent of these cases show a detectable histopathology, the structural abnormalities tending to occur in the medial region of the lobe and tending to be associated with the hippocampus.[42,43] This medial temporal sclerosis, as it is called, has an incidence of 50 to 60 percent in temporal-lobe epilepsy. In fact, the temporal lobes appear to be particularly susceptible to influences causing seizure, such as hypoxia and hypoglycemia. It may be that the organization of the blood supply to the hippocampus and neighboring structures, and their lateral anatomical position, renders them particularly vulnerable to hypoxia or mechanical damage.[45] Thus, birth trauma (i.e., localized ischemia of these structures due to bilateral skull deformation in the birth canal) or febrile convulsions occurring in the young during infective states such as scarlet fever (i.e., hypoxia due to enhanced, temperature-stimulated oxygen uptake) could result in histological lesions in the hippocampal region which, in later life, mature into epileptogenic foci.[42–46] The opposite case has also been strongly argued, that localized neuronal death and sclerosis may itself be *produced* by the seizures, or at least increased in extent by seizure activity,[43] possibly owing to the associated localized hypoxia.

Other forms of epilepsy are associated in a more clear-cut fashion with specific organic changes or traumas such as brain tumors, hematoma, cortical scarring through infection (e.g., meningitis), or head injury.[41] Brain scarring from missile wounds or other forms of head injury, such as depressed fractures, leads to an early onset or late onset epileptic condition in 5 to 30 percent of cases[47,48] depending on the features and site of the injury. For instance, acute intracranial hematoma carries a high risk of early epilepsy.[48]

The absence of a firm causal link between epileptic fits and specific patterns of neuropathology in many cases indicates the importance of seeking evidence of an underlying malfunction at the biochemical level.[89] Since only particular kinds of scarring lead to the development of spiking foci, it may be that the nature of the changed tissue organization in epileptogenic scars could give a clue to the factors involved. Various degrees of gliosis, for instance, with a change in ratio of neurons to glia, might critically alter neuronal–glial interrelationships at the metabolic or biochemical level. A change toward a higher permeability of the blood-brain barrier because of local vascular damage and vascular reorganization in active scars can provide a region of high permeability for local entry of excitant substances from the blood.

The Epileptic Focus

There are various features to be considered in the organization of an epileptic focus within a tissue lesion. There is the likely presence of neurons that survive the local tissue damage but remain functionally impaired. There is probably a gradient of decreasing impairment of this kind from the central necrotized inactive region of the focus to the periphery that is juxtaposed to normal tissue. Gradation implies the presence of an annulus of hyperactive cells at this boundary, and invasion of hyperactivity into the normal brain tissue is the point of initiation of the seizure.[42]

Thus, a useful distinction may be drawn between the processes initiating and maintaining the local hyperexcitability of the focus and the processes involved in the spread of the hyperactivity with recruitment of many normal neurons and the precipitation of the seizure.

Local hyperactivity could, in theory, result from changes in neuronal membrane properties that lead to abnormal patterns of ion conductance through the membrane or from the development of a localized supersensitivity to excitatory neurotransmitters. There may be losses of small inhibitory interneurons from the local circuitry in the focus, or a local accumulation of excitatory substances in the extracellular space because of leakage through membranes or reduced neuronal or glial high-affinity transport. However, there is either very little evidence or only circumstantial evidence in favor of these various possible causes of local hyperactivity in the epileptic focus.

The electrical properties of many acute and chronic experimentally induced foci show similarities. Apart from the epileptiform EEG, which is always present, these epileptic foci display a characteristic large depolarization shift called the *paroxysmal depolarization shift* (PDS).[89,92] These recurrent shifts of depolarization take the neuronal membrane potential through the critical level, causing increased spike discharge, but this is followed by a gross depolarization, with associated inactivity, owing to a condition called cathodal block. This condition is followed in turn by a period of hyperpolarization before onset of the next PDS. The cause of the PDS is unknown, but all evidence points to its being synaptically generated and graded in nature, having the properties of greatly enhanced excitatory postsynaptic potentials. Recordings from neurons in experimental foci *between* PDS events indicate the absence of any *intrinsic* hyperexcitability, suggesting that the membrane properties of the epileptic neurons themselves are not altered and that the PDS arises as the result of abnormal increases in *synaptic* activity.[92] Such enhanced excitatory synaptic activity could well result from increased excitatory neurotransmission, reduced inhibitory input, or direct stimulation of excitatory synaptic receptors by locally accumulating substances.

At present there is little understanding of the nature of the events causing a "quiescent" but active focus to interact with its surrounding normal tissue and initiate a detectable motor or sensory event such as a seizure or myoclonic jerk. Providing explanations of these events at the biochemical and physiological levels is critical to an understanding of epilepsy.

The factors imposing a temporal limitation on seizures and determining the length of the period between fits are not simple. Periods of seizure activity (ictal states) are always followed by quiescent periods during which seizures are absent and an inhibitory state prevails (interictal states). Whether quiescence commences because of neuronal exhaustion at the level of energy metabolism, or cathodal block of neuronal activity due to gross membrane depolarization, or the rise in activity of accessory inhibitory neuronal pathways remains a matter of debate. Hyperventilated animals experiencing seizure do not display large decrements in their reservoirs of energy substrates; ATP levels remain normal and phosphocreatine stabilizes at a reduced concentration[49,98] (see Chapter 3). But if hyoxia, hypoglycemia, or arterial hypotension occurs, ATP concentrations usually decrease. Therefore, inhibitory mechanisms at the level of synaptic interaction seem the more likely mechanisms operating to limit seizure length and produce inhibitory interictal states.

Animal Models of Epilepsy

A wide variety of animal models have been developed.[B,M] One category involves scarring localized brain tissue by applying small quantities of certain metals (e.g., cobalt) or metal oxide (e.g., aluminium oxide) to the brain surface or region. Localized rapid freezing of the cortex with low-temperature small-diameter metal probes, dry ice, liquid nitrogen, or an ethyl chloride spray also causes scarring.

The "kindling" model of epilepsy[M,49] is now very popular. It involves passing a small (200–300 μA) current into a brain region for one second each day for several days, or at shorter intervals. A considerable "afterdischarge" begins to appear after each stimulation, and this develops in the course of treatment into a full bilateral *grand mal* type of fit. Subsequently the animal's brain is permanently "kindled," and either stimulus-evoked or spontaneous seizures can be observed according to the animal species concerned. The electrical stimulation does not appear to cause tissue scarring in this model.

The "sensory reflex" models of epilepsy are usually genetic strains of laboratory animals having an exaggerated response to a certain form of sensory stimulation, and this response can be developed into a full grand mal convulsion. Examples are audiogenic seizures of a particular strain of mouse that responds with convulsions to certain auditory tones, and the photosensitive baboon (Papio papio) from Senegal in which seizures can be evoked by stimulating the retina with light.

These models, together with direct electroconvulsive cerebral shock of various laboratory animals and application of convulsant drugs or chemicals (e.g., pentylenetetrazole; penicillamine) have been employed by pharmaceutical companies in the detection, development, and testing of anticonvulsant drugs and by basic research workers studying the mechanisms underlying epilepsy.

Neurotransmitters Involved in Epilepsy

Most neurotransmitters and putative neurotransmitters have at some stage been implicated in the disease processes causing epilepsy.[91,98] The evidence for such involvement is usually the finding of a change in their level, or their action, recorded either in the brains of epileptic animal models or in epileptogenic tissue excised from human patients during brain surgery to remove temporal-lobe epileptic foci.[42,53,99,126]

The past few years have seen a concentration of interest on the excitatory and inhibitory amino acids as substances directly involved in causing focal epileptic hyperactivity and its spread.

GABA and Epilepsy   The inhibitory neurotransmitter GABA is widely employed throughout the brain and spinal cord (Chapter 4). Small GABAergic inhibitory interneurons in regions such as the cerebral cortex or hippocampus are commonly involved in the local control of neural activity by forming part of recurrent (feedback) pathways that normally prevent excessive synchronous and sustained firing of large groups of neurons. For this reason, diminished efficiency of GABAergic transmission would readily lead to epileptic discharges.

Failure of GABA synthesis or rates of release, or blockade of GABA receptors usually leads to seizure. Inhibition of glutamate decarboxylase (GAD), which leads to a reduction of GABA formation in nerve terminals, has allowed experimental investigation of the connection between GABA levels and epileptic fits. Antipyridoxal agents such as 4-deoxypyridoxine, thiosemicarbazide, and isoniazide inhibit GAD activity by an action on its coenzyme, pyridoxal phosphate. Because this coenzyme is an essential cofactor for all cerebral decarboxylases and transaminases, drugs acting via this mechanism (e.g., hydrazides) may have actions at multiple sites and tend to produce seizures that do not correlate with the degree of GAD inhibition or with the levels of GABA in cerebral tissues.[42,98,B] For instance, the GABA-catabolizing enzyme, GABA-transaminase, is also vulnerable to these drugs, and its impairment leads to an *accumulation* of GABA.

The correlation between GAD inhibition and seizure induction is more consistent with other agents, such as the simple competitive GAD inhibitor 3-mercaptopropionic acid or the irreversible catalytic inhibitor allylglycine. The former produces seizures after a very short latency in mice and baboons, with EEG features similar to those produced by picrotoxin, a specific GABA antagonist. In contrast, allylglycine and the pyridoxal phosphate antagonists produce brief repetitive seizures after a long latency.[98]

Seizures likely to be a result of reduced GABA synthesis are seen in children and animals as a result of a dietary deficiency of pyridoxine (vitamin B_6). Administration of pyridoxine has a rapid therapeutic effect.[98]

As might be expected, drugs that raise the levels of GABA in the nervous system tend to have anticonvulsant properties. Such agents include γ-vinyl and γ-acetylenic GABA, which irreversibly block GABA transaminase and thereby shut off the usual route of GABA breakdown[50] (Table 4.9). Other agents that block GABA-transaminase or succinic semialdehyde dehydrogenase raise GABA levels in the brain and cerebral spinal fluid and act as anticonvulsants (e.g., gabaculine, ethanolamine-O-sulfate, sodium N-dipropylacetate). Whether or not such compounds have clinical applications depends on whether they penetrate the blood-brain barrier and are sufficiently free of side effects. At present, only sodium dipropylacetate (Epilim, Depakene, Valproate) is in use to treat human epilepsy, but others are proving useful in clinical trials (e.g., γ-vinyl GABA).[123]

Benzodiazepine drugs such as clonazepam are widely used and effective anticonvulsants. How they work remains unclear, but their well-established enhancing action on GABA-mediated inhibition suggests that this is the mechanism of action.

It is currently postulated that endogenous "benzodiazepinelike" substances function as neuromodulators in certain GABAergic synapses by improving the efficiency of GABA in producing inhibition (see Chapter 4, Chapter 7, and Figure 7.19). Benzodiazepine drugs which possess anticonvulsant properties are likely to work by mimicking this action.

Glutamate, Aspartate, and Epilepsy It now seems certain that both glutamate and aspartate are employed by many pathways in the brain and spinal cord (see Chapter 4). Glutamate appears to be more widely employed than aspartate.[52] Both these excitatory dicarboxylic amino acids could be involved in either the generation or spread of epileptic hyperactivity. Significantly, excitatory receptors for these compounds, especially glutamate, are widely distributed throughout the brain. Both compounds exist in neural tissue at 3–10 mM concentrations. This is 100 to 1000 times higher than other established neurotransmitters (except GABA) and is due to their involvement in mainstream energy metabolism and protein biosynthesis. A potent high-affinity uptake system keeps their extracellular levels at a minimum (e.g., 5–10 μM in cerebrospinal fluid).

In the cerebral cortex, and in most other brain regions, the majority of neurons respond with firing to glutamate iontophoresed from micropipettes. Concentrations as low as 10–100 μM are usually sufficient to initiate action potential discharge. Any deficiency in the activity of the transport system affecting the reuptake of glutamate and leading to a rise in its extracellular concentration could precipitate local hyperactivity. Also, any exaggerated efflux of glutamate from neurons or glial cells due to changed membrane permeability would produce the same result. In the vicinity of receptors responsive to aspartate, the same would apply for this excitatory compound.[42,100]

A finding that points to the involvement of aspartate and glutamate in experimental and human epilepsy is their decreased concentrations in focal tissue excised during surgery to relieve epilepsy[N,53] and also in a range of experimental epileptic foci.[42,54,55,126] This depletion was not specific, GABA and taurine showing a similar pattern of loss whereas glycine and serine levels were elevated. Other studies with human tissue have detected only some of these changes.[99] Such reductions in pool size could reflect an increase in the rate of oxidation of substrates in the hyperactive tissue, whose oxygen uptake and mainstream and energy metabolism are known to be greatly augmented. However, several reports have appeared showing patterns of increased *release* of glutamate from animal cortical epileptic foci (and kindled amygdala) that correlate closely with the onset of epileptic hyperactivity.[55,56,60] Such release could itself be a result of the neuronal activity rather than its cause. More recently, researchers have shown that drugs which specifically block the excitatory actions of glutamate and aspartate, as well as blocking the actions of the synthetic excitatory dicarboxylic amino acids such as N-methyl-D-aspartic acid and quisqualic acid (see Chapter 4), suppress epileptic fits in cobalt implantation (rats),[57] audiogenic seizures (mice),[58] pentylenetetrazole (mice),[58] kindled amygdala[60] or hippocampus[97] (rats), and photosensitive reflex epilepsy (baboons).[59] The ω-phosphono-derivatives of α-carboxylic amino acids with chain lengths 4 to 8 are particularly effective (e.g, 2-amino-4-phosphonobutyrate, 2-amino-5-phosphonovalerate, and 2-amino-7-phosphonoheptanoate; see Chapter 4).

The observed suppression of seizures could be due to a curtailment of their *spread* via glutamate- or aspartate-operated synapses and does not prove a direct involvement of glutamate or aspartate in the initial processes leading to focal hyperactivity. There is also evidence that aspartate levels are substantially lowered by these and other anticonvulsants.[61]

It is unlikely that human epilepsy is caused by malfunction of only one neurotransmitter system. More likely, there are different forms of the disease, located in different brain regions, and each involving the failure of one or more systems in concert and leading to a net imbalance in favor of excitation.

Anticonvulsants: Mechanism of Action

The mechanisms of action of most anticonvulsant drugs are poorly understood.[89,90,93] Anticonvulsants can have their effect in two main ways: (1) direct action on the primary mechanisms, causing firing of the pathologically altered neurons in the epileptic focus, (2) prevention of the spread of hyperactivity from foci into normal brain tissue. Most anticonvulsants work at least partly by the second mechanism, since they all reduce seizures provoked by a range of different agents.[122]

Anticonvulsants have one or more of a number of actions, including the reduction of ion fluxes (Na^+, Ca^{2+}); the raising of seizure thresholds to convulsant stimuli; the potentiation of inhibition and reduction of posttetanic potentiation (PTT), which can act to facilitate action potential generation; and a general dampening of evoked electrical responses in neurons.[90] The actions of sodium N-dipropylacetate (Epilim, Depakene, Valproate) are thought to be due to the small rise in GABA levels it produces, and benzodiazepines (e.g., clonazepam) are likely to work by potentiating GABA-mediated inhibitory mechanisms. Carbamazepine (Tegretol), an iminostilbene, is chemically related to the tricyclic antidepressants and has actions similar to those of diphenylhydantoin (see below). The latter three drugs are the most recently introduced anticonvulsants, having come into use since 1965.

The long-established clinically effective anticonvulsants such as the barbiturates (e.g., phenobarbital) and hydantoins (e.g., diphenylhydantoin, phenytoin, Dilantin) continue to dominate the treatment (control) of epilepsy, although their modes of action are still a matter of speculation.[122] Where phenobarbital raises the threshold for precipitation of seizures by chemical convulsants, diphenylhydantoin does not, and other differences between their actions indicate different modes of action. To diphenylhydantoin has been attributed the property of "stabilizing" neuronal membranes by a dampening action on Na^+ and Ca^{2+} ionic current across the membrane.[122]

Small increases in the cerebral GABA content following systemic administration of hydantoins and barbiturates to animals have been reported, but these are not consistent and do not correlate with anticonvulsant action.[98]

Most anticonvulsants have been discovered empirically through their actions in various long-established test systems.[93,122] These involve the ability to antagonize convulsion induced chemically or electrically in laboratory animals.

HEPATIC COMA STATES

Chronic liver disease in humans, such as cirrhosis and liver failure, often involves the onset of comatose states progressing from confusion, slurred speech, drowsiness, and stupor to full coma. Experimental removal or bypass (portacaval anastomosis) of the liver in animal models produces comparable symptoms. Accompanying these states are raised levels of ammonia in the blood and the brain as well as disturbances in the composition of plasma that include altered levels of amino acids and fatty acids.

The nature of the processes leading to loss of consciousness are unknown, but an imbalance in the functional equilibrium of various neurotransmitter systems could well be involved.

Disturbances in Serotonin in Hepatic Coma

One line of evidence implicates changes in the monoamine neurotransmitters. For instance, if serotonin levels were to rise in brain, this might precipitate sleeplike or comatose conditions because of the possible involvement of this neurotransmitter in initiating and maintaining sleep states (see Chapter 8).

Because cerebral tryptophan hydroxylase is not normally saturated with tryptophan, it effectively controls the rate of serotonin synthesis. Increased transport of tryptophan from blood to brain could therefore theoretically lead to an increased production of serotonin in those neurons that initiate sleep processes. There is, indeed, good evidence from both clinical and animal studies that the turnover of brain serotonin increases during liver disease involving coma states and that brain and cerebrospinal fluid levels of tryptophan and 5-hydroxyindole acetic acid (5-HIAA) are also increased.[62,63] In the absence of coma, the changes were absent or much smaller—though not all reports agree on this point. An increase in the rate of tryptophan entry into the brain is the most attractive explanation of this augmentation in turnover, but whether this is the primary mechanism causing coma states in liver failure remains to be proved.

Standing against an involvement of monoamines is the failure of large oral or intravenous doses of tryptophan to induce coma and the lack of correlation between blood-free tryptophan levels and the incidence of coma in experimental animals or in patients with liver failure.[62]

Transport of tryptophan itself into the brain appears to be by two kinetically distinct process. (1) a low-capacity saturable system obeying Michaelis-Menten kinetics for which tryrosine, phenylalanine, valine, leucine, and isoleucine compete; (2), a larger capacity, unsaturable system that has no amino acid competitors and that accounts for 40 percent of normal tryptophan entry into the brain.[62,84]

Another important factor controlling tryptophan entry is the fraction of total plasma tryptophan in free solution, as opposed to that bound to proteins, especially albumin. Tending to accelerate entry into the brain are any influences releasing tryptophan from its binding sites on blood proteins, such as increased levels of plasma-unsaturated fatty acids. Since plasma albumin and unsaturated fatty acids are low in liver disease, this could at least partly account for the increased entry and turnover of tryptophan into the cerebrospinal fluid and brains of patients with liver disease.[62] In both patients with liver disease and animal models of liver failure, the levels of unbound tryptophan in the plasma are substantially raised.[62,85]

Another explanation for the enhanced tryptophan uptake to brain points to the lowering of plasma concentrations of certain branched-chain amino acids, which occurs in chronic liver failure in humans, dogs, and rats.[83,84,85,64] This change could decrease the degree of competition between this group of amino acids (e.g., leucine, valine, isoleucine) and plasma tryptophan for the saturable transport system at the blood-brain barrier.[63,64,83,84] The fall in plasma levels of these amino acids could also be linked to the increase in plasma insulin accompanying this disease state.[64] Increasing plasma insulin in experimental animals increases tryptophan entry to the brain.[65] Moreover administration of branched-chain neutral amino acids can reverse both experimental and human hepatic coma,[62] though controversy surrounds this claim.[64,66]

Other findings stand against the proposal that lowered competition between branched-chain amino acids and plasma tryptophan for entry to the brain is principally responsible for the increased levels of brain tryptophan observed in both human and experimental liver failure. For instance, in rats with acute liver failure caused by portocaval shunting of their liver and ligature of the hepatic artery, there are large *increases* in the plasma levels of these branched-chain amino acids, together with increases in most other amino acids, including tyrosine, phenylalanine, methionine, and threonine—which also compete with tryptophan transport. In spite of these changes, brain tryptophan levels are increased, indicating accelerated rates of tryptophan transport.[83,84] Also, the proportion of tryptophan entering via the unsaturable high-capacity transport system is raised to almost 75 percent of the total. Thus, although tryptophan entry to brain can be shown to be reduced by raising the plasma levels of competing amino acids,[84] evidence that a *lowering* of plasma levels has the opposite effect in experimental hepatic failure is not clear-cut. Instead, change in the levels of free versus bound plasma tryptophan, and changes in the kinetics of its transport into brain in these conditions, seem to offer equivalent or more likely explanations of the increased levels of brain tryptophan and turnover of serotonin.[62,83,84,85]

The high ammonia content of blood in these conditions is associated with a large rise, perhaps threefold, in the glutamine content of the brain. It has been argued that the efflux of excess glutamine from brain to blood via the large, aromatic neutral amino acid transport system enhances the countertransport of these compounds (i.e., tryptophan, tyrosine, phenylalanine).[64] In support of this suggestion, close correlation between the glutamine content and the tryptophan content of cerebrospinal has been reported.[64]

Changes in Other Monoamines in Hepatic Coma

The enhanced entry of tryptophan to the brain in these experimental conditions and in human liver disease is accompanied by a considerably increased transport of both tyrosine and phenylalanine into the brain,[84] perhaps fourfold, due to increased activity of the neutral amino acid transport system.[64] This, no doubt, leads to perturbances of the catecholamine neurotransmitter systems. For instance, the false neurotransmitters phenylethanolamine and octopamine are synthesized in considerably larger amounts[67] via the agency of aromatic amino acid decarboxylase, which is normally well below saturation with its amino acid substrate. The ratio of false neurotransmitter to catecholamine must rise substantially and could decrease noradrenergic and dopaminergic function at the relevant synaptic sites. There is also a marked fall in the norepinephrine content of brain in experimental and human liver failure.[67] Where this results in diminished activity of this mostly excitatory neurotransmitter there would be a shift in the balance in favor of inhibition and therefore towards comatose states.

In view of the wide spectrum of neurotransmitter disturbances seen in these conditions, it seems likely that no single change underlies the neurological symptoms but rather that the various maladjustments act in concert and perhaps greatly exacerbate a situation that is already moving towards coma for other reasons.

Amino Acid Neurotransmitters in Hepatic Coma   The disturbances of ammonia and of glutamine metabolism in the central nervous system that accompany experimental and human liver failure indicate that other categories of neurotransmitter could be involved in producing the neurological symptoms. The excitatory amino acid L-glutamate is produced from L-glutamine by the action of the enzyme glutaminase[68,69] (see Chapter 4), which, in turn, is inhibited by its products, ammonia and glutamate. The rising ammonia levels in blood, cerebrospinal fluid, and brain associated with liver failure could be inhibiting the production of tissue L-glutamate[70] such that, given the widespread distribution of excitatory glutamatergic pathways (Chapter 4), a substantial reduction in this neurotransmitter could lead to coma. Lowered levels of cerebral glutamate and aspartate have been reported for animals with experimental liver disease.[86,87]

The inhibition of brain glutaminase probably also adds to the large increase in brain and cerebrospinal fluid glutamine in liver failure and experimental hyperammonemia. The principal process leading to glutamine levels elevated two-to threefold has been assumed to be increased glutamine synthesis, but several studies suggest that this increase may not be so great as previously assessed.[71,72] In view of the high rate of glutamine turnover in brain,[73] a blockade of its catabolism via glutaminase could well account for most of its increased levels in hyperammonemic states.

As yet there has been no direct demonstration of any depletion of the glutamate *transmitter* pool when blood and brain ammonia are at high levels.[70,74] There is evidence, however, of a substantial (75 percent) decrease in the release of glutamate from perfused hippocampal slices, evoked by high potassium chloride (56 mM), when ammonium ion is present at 2–3 mM.[75] The time course and other features of this effect strongly suggest that an inhibition of glutaminase activity by ammonia is at the root of the depletion of glutamate in the relevant neurotransmitter pool.[75]

Others have argued that the inhibition of glutaminase by ammonium ions would lead to an increase in the metabolism of glutamine via the glutamine transaminase-ω-amidase pathway[76,77] (Figure 9.10). This pathway involves the formation of 2-oxoglutaramate, a compound that induces a large reduction in the locomotor activity of experimental animals when given intraventricularly at high doses. It is also elevated fourfold to tenfold in the cerebrospinal fluid of patients with hepatic encephalopathy, though it is not detectable in blood.[76,77] It has been suggested that the ring form of 2-oxoglutaramate (2-hydroxy-5-oxoproline) acts as an antagonist at glutamate receptor sites, thereby reducing glutamate-mediated neural activity in the central nervous system, and motor activity, eventually leading to coma.[76] Its congeners—proline, 5-oxoproline, and 1-hydroxy-3-amino pyrrolidone (HA966)—have been found to possess similar antiglutamate properties.[76]

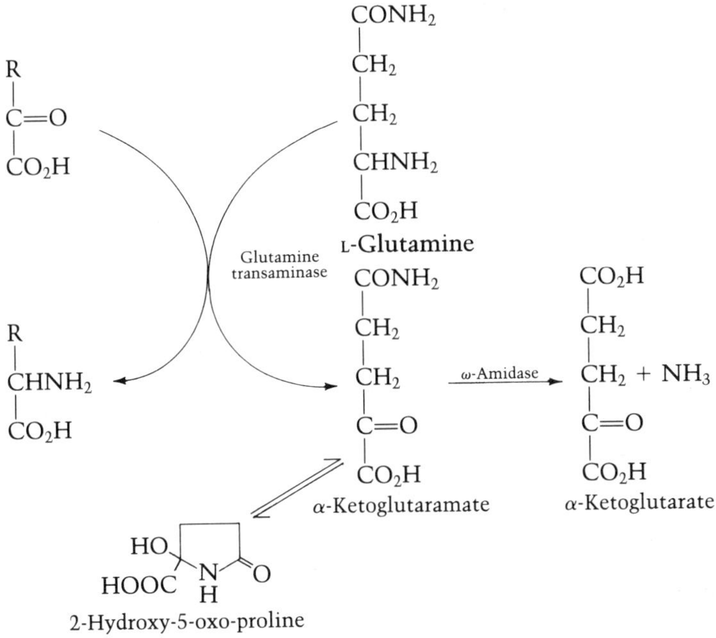

FIGURE 9.10   Transamination-deamidation pathway of glutamine metabolism.

MYASTHENIA GRAVIS

Myasthenia gravis seems to be due to a malfunction of neuromuscular transmission at the postsynaptic site. It involves failure of the nicotinic cholinergic receptor system, located on the muscle surface, which drives the voluntary musculature. There is an accompanying deterioration of the postsynaptic membrane. Myasthenia gravis is a progressive disease, involving severe muscular weakness and fatigability.[O] Its prevalence ranges from one to five persons in 50,000, depending on race. The condition seems to arise as a result of what is called an autoimmune reaction. Such reactions occur when the body's own immune system starts to produce antibodies to antigens present in its own organs. These antigens would perhaps not normally be accessible to the blood system, and the formation of antibodies to these "self-antigens" would normally be disallowed.

Myasthenia gravis can be mimicked by immunizing experimental mammals against acetylcholine receptor preparations purified from the electric ray *Torpedo marmorata* (see Chapter 4). Such immunized animals develop muscular weakness, and other features of the disease entity, because of a cross-reaction between the antibodies against the injected receptor preparation and the acetylcholine receptors of their own voluntary muscles. This animal model is known as experimental autoimmune myasthenia gravis[78,81] (Figure 9.11) and it indicates the considerable homology that must exist between the protein structure of the acetylcholine receptor of the *Torpedo* electric organ and that of the mammalian neuromuscular junction.

FIGURE 9.11 (A) Diagrammatic representation of a normal neuromuscular junction. (1) Acetylcholine-containing vesicles in the nerve ending. (2) The presynaptic membrane of the nerve. (3) The intersynaptic space across which acetylcholine must diffuse after release through specific regions on the presynaptic membrane. (4) Folds in the postsynaptic membrane of the muscle where AChR concentrated at the tips, closely presented to the sites of acetylcholine release. (B) The initial events in the immune assault on the postsynaptic membrane during experimental autoimmune myasthenia gravis (EAMG). (1) A small fraction of the AChR in the postsynaptic membrane is bound with antibodies. (2a) Complement is activated by some of the bound antibodies, resulting in focal destruction of the tips of postsynaptic folds where AChR is most concentrated. (2b) Specific proteolytic fragments of complement are released during the complement cascade, which induces phagocytic migration to the endplate. (C) The acute phase of EAMG and passively transferred EAMG. These events are seen within 1 day after passively transferring antibodies to AChR to a normal rat or 8 to 11 days after immunizing a rat with purified AChR. Phagocytes invade the endplate, recognize bound antibody, and strip off large areas of postsynaptic membrane. Thus binding of antibodies to a small number of AChR results in the destruction of large numbers of AChR. [From ref. 78]

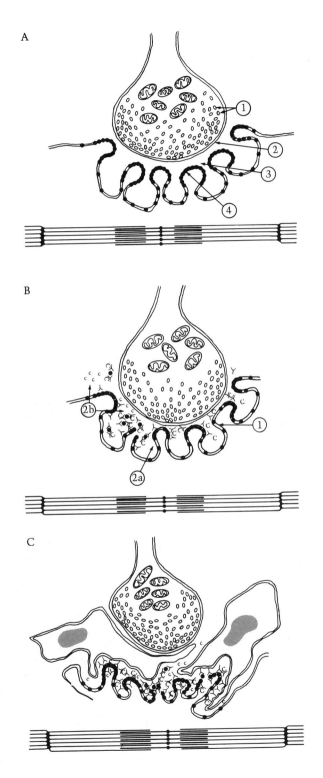

The loss of the postsynaptic membrane area and the consequent failure of neurotransmission appear to be due to a complement-mediated, antibody-dependent lysis, particularly at points on the folds where the receptors are concentrated. This loss is followed by macrophage invasion. Electrophysiological impairment follows junctional destruction. Miniature-end-plate potentials (see Chapter 7) are of decreased amplitude, and electromyograms show lowered potentials. Function on the presynaptic side seems to be normal, as judged by the number of quanta of acetylcholine released per impulse. Partial functioning of antibody-bound receptors may account for the precarious survival of some species of severely affected experimental mammals,[78,81] since the antibodies do not interact with the acetylcholine binding site.

Antibodies recognizing acetylcholine receptor protein, solubilized from human muscle, have been found in the serum of 90 percent of myasthenia gravis patients. The receptor content of the affected muscles was reduced by 60 percent, and half of the remaining content was bound to antibodies. These observations emphasize the likely similarity of the mechanisms operating in the experimental animal model and in the human disease. The greatly enlarged thymus gland in myasthenia gravis patients also highlights the presence of immunological factors in the disease.

What remains unclear is the nature of the factors in the human disease that allow the onset of the autoimmune response. Such possibilities as genetic predisposition,[79] release of the receptors from tissues during viral attack, or a chance similarity between antigenic sites on bacteria or viruses infecting the patient and those on the receptor protein surface[78] have all been considered.[81]

REFERENCES

General Texts and Review Articles

A　Davison, A. N., and R. H. S. Thompson, eds. (1981) *The Molecular Basis of Neuropathology.* Edward Arnold, London.

B　Bradford. H. F., and C. D. Marsden, eds. (1976) *Biochemistry and Neurology.* Academic Press, London.

C　Legg, N. J., ed. (1978) *Neurotransmitter Systems and Their Clinical Disorders.* Academic Press, London.

D　Taylor, A., and M. T. Jones, eds. (1978) *Chemical Communication Within the Nervous System and Its Disturbance in Disease.* Pergamon Press, Oxford.

E　Davison, A. N., ed. (1976) *Biochemistry and Neurological Disease.* Blackwell Scientific, Oxford.

F　Curzon, G., ed. (1980) *The Biochemistry of Psychiatric Disturbances.* Wiley, Chichester.

G　Marsden, C. D., and S. Fahn, eds. (1982) *Movement Disorders.* Butterworth Scientific, London.

H Calne, D. B. (1980) *Therapeutics in Neurology*, 2nd ed. Blackwells, London.

I Iversen, L. L., and S. P. R. Rose, eds. (1973) *Biochemistry and Mental Illness.* The Biochemical Society, London.

J Paykel, E. S., and A. Coppen, eds. (1979) *Psychopharmacology of Affective Disorders.* Oxford University Press, Oxford.

K Roberts, P. J., ed. (1980) *Biochemistry of Dementia.* Wiley, Chichester.

L Laidlaw, J., and A. Richens, eds. (1985) *A Textbook of Epilepsy*, 2d ed. Churchill Livingstone, London.

M Purpura, D. P., J. K. Penry, D. Tower, D. M. Woodbury, and R. D. Walter. (1972) *Experimental Models of Epilepsy.* Raven Press, New York.

N Tower, D. B. (1960) *Neurochemistry of Epilepsy.* Thomas, Springfield, Ill.

O Lunt, G. G., and R. M. Marchbanks. (1978) *The Biochemistry of Myasthenia Gravis and Muscular Dystrophy.* Academic Press, London.

P Illis, L. S., and D. C. Gajdusek. (1975) *Viral Diseases of the Central Nervous System.* Bailliere Tindall, London.

Q Van Praag, H. M. (1977) *Depression and Schizophrenia.* Spectrum, New York.

R Trimble, M. R., and E. Zarifian, eds. (1984) *Psychopharmacology of the Limbic System.* Oxford University Press, Oxford.

S Jobe, P. C., J. W. Dailey, and H. E. Laird, eds. (1984) "Epilepsy: Neurotransmitter Abnormalities As Determinants of Seizure Susceptibility and Severity," *Fed. Proc.* 43 (10): 2503–2531.

Additional Citations

1 Forno, L. S. (1982) In *Movement Disorders* (ed. C. D. Marsden and S. Fahn), pp. 25–40. Butterworth Scientific, London.

2 Duffy, P. E., and V. M. Tennyson. (1965) *J. Neuropath. Exp. Neurol.* 24: 398–414.

3 Hornykiewicz, O. (1982) In *Movement Disorders* (ed. C. D. Marsden and S. Fahn), pp. 41–58. Butterworth Scientific, London.

4 Lee, T., P. Seeman, A. Rajput, I. J. Farley, and O. Hornykiewicz. (1978) *Nature* (Lond.) 273: 59–61.

5 Cools, A. R., and J. M. Van Rossum. (1976) *Psychopharmacology* 45: 243–254.

6 Rinne, U. K. (1982) In *Movement Disorders* (ed. C. D. Marsden and S. Fahn), pp. 59–74. Butterworth Scientific, London.

7 Lloyd, K. G., L. Shemin, and O. Hornykiewicz. (1977) *Brain Res.* 127: 269–278.

8 Reisine, T. D., J. Z. Fields, H. I. Yamamura, E. D. Bird, E. Spokes, and P. S. Schreiner. (1977) *Life Sci.* 21: 335–344.

9 Reisine, T. D., M. Rossor, E. Spokes, L. L. Iversen, and H. I. Yamamura. (1979) *Brain Res.* 173: 378–382.

10 Liebermann, A. N., and M. Goldstein (1982) In *Movement Disorders* (ed. C. D. Marsden and S. Fahn), pp. 146–165. Butterworth Scientific, London.

11 Shoulson, I., and T. N. Chase. (1975) *Ann. Rev. Med* 26: 419–426.

12 Corsellis, J. A. N. (1976) In *Greenfield's Neuropathology*, 3d ed. (ed. W. Blackwood and J. A. N. Corsellis), pp. 796–848. Edward Arnold, London.

13 Bird, E. D., and L. L. Iversen. (1974) *Brain Res.* 97: 457–472.

14   Bird, E. D., and L. L. Iversen. (1977) In *Essays in Neurochemistry and Neuropharmacology*, vol. 1 (eds. M. B. H. Youdim, W. Lovenbury, D. F. Sharman, and J. Lagnardo), pp. 177–195. Wiley, Chichester.

15   Curzon, G. (1976) In *Biochemistry and Neurological Disease* (ed. A. N. Davison) pp. 168–227. Blackwell Scientific, London.

16   Cross, A. J., and J. L. Waddington. (1981) *J. Neurochem.* 37: 321–324.

17   Enna, S. J., J. P. Bennett, D. B. Byland, S. H. Synder, E. D. Bird, and L. L. Iversen. (1976) *Brain Res.* 116: 531–537.

18   Enna, S. J., L. Z. Starn, G. J. Wastek, and H. T. Yamamura. (1977) *Life Sci.* 20: 205–212.

19   Gale, J. S., E. D. Bird, E. G. Spokes, L. L. Iversen, and T. Jessell. (1978) *J. Neurochem.* 30: 633–634.

20   Coyle, J. T. (1978) *J. Biol Psych.* 14: 251–276.

21   Baldessarini, R. J. (1979) *Trends Neurosci.* 2: 133–135.

22   Barnes, T. R. E., and T. Kidger. (1979) *Trends Neurosci.* 2: 135–136.

23   Crow, T. J. (1979) *Trends Neurosci.* 2: 52–55.

24   Crow, T. J. (1980) In *The Biochemistry of Psychiatric Disturbances* (ed. G. Curzon) pp. 73–88. Wiley, Chichester.

25   Iversen, L. L. (1975) *Science* 188: 1084–1089.

26   Seeman, P., T. Lee, M. Chau-Wong, and K. Wong. (1976) *Nature* (Lond.) 261: 717–719.

27   Burt, D. R., I. Creese, and S. H. Snyder. (1976) *Mol. Pharm.* 12: 800–812.

28   Lee, T., P. Seeman, W. W. Tourtellotte, I. Farley, and O. Hornykiewicz. (1978) *Nature* (Lond.) 274: 897–900.

29   Owen, F., T. J. Crow, M. Poulter, A. J. Cross, A. Longden, and G. J. Riley. (1978) *Lancet* II: 223–225.

30   Green, A. R., and D. W. Costain. (1979) In *Psychopharmacology of Affective Disorders* (ed. E. S. Paykel and A. Coppen), pp. 14–40. Oxford University Press, Oxford.

31   Crow, T. J., and E. C. Johnstone. (1979) In *Psychopharmacology of Affective Disorders* (ed. E. S. Paykel and A. Coppen), pp. 108–122.

32   Green, A. R. (1980) In *The Biochemistry of Psychiatric Disturbances* (ed. G. Curzon) pp. 35–52. Wiley, Chichester.

33   Mackay, A. V. P., E. O. Bird, E. G. Spokes, M. Rossor, L. L. Iversen, I. Creese, and S. H. Snyder. (1980) *Lancet* II: 915–916.

34   Tomlinson, B. E. (1980) In *Biochemistry of Dementia* (ed. P. J. Roberts), pp. 15–52. Wiley, Chichester.

35   Bowen, D. M., C. B. Smith, P. A. White, and A. N. Davison. (1976) *Brain* 99: 459–496.

36   Rossor, M. N., N. J. Garrett, A. J. Johnson, C. Q. Mountjoy, M. Roth, and L. L. Iversen. (1982) *Brain* 105: 313–330.

37   Perry, E. K., and R. H. Perry. (1980) In *Biochemistry of Dementia* (ed. P. J. Roberts), pp. 135–184. Wiley, Chichester.

38   Rossor, M. N., C. Svendsen, S. P. Hunt, C. Q. Mountjoy, M. Roth, and L. L. Iversen. (1981) *Neurosci. Lett.* 28: 217–222.

39   Rossor, M. N., L. L. Iversen, A. J. Johnson, C. Q. Mountjoy, and M. Roth. (1981) *Lancet* II: 1422.

40   Bowen, D. M., C. B. Smith, P. A. White, M. J. Goodhardt, J. A. Spillane, R. H. A. Flack, and A. N. Davison. (1977) *Brain* 100: 397–426.

41 Currie, S., K. W. G. Heathfield, R. A. Henson, and D. F. Scott. (1971) *Brain* 94: 173–190.

42 Bradford, H. F., and P. R. Dodd. (1976) In *Biochemistry and Neurological Disease* (ed. A. N. Davison), pp. 114–167. Blackwell Scientific, Oxford.

43 Meldrum, B. S. (1975) *Modern Trends Neurol.* 10: 223–237.

44 Earle, K. M., M. Baldwin, and W. Penfield. (1953) *Arch. Neurol. Psychiat.* 69: 27–42.

45 Taylor, D. C., and C. Ounsted. (1971) *Epilepsia* 12: 33–46.

46 Falconer, M. A. (1971) *Epilepsia* 12: 13–31.

47 Newcombe, F. (1969) *Missile Wounds of the Brain*, p. 31. Oxford Neurological Monographs, Oxford University Press, Oxford.

48 Jennett, B. (1975) *Epilepsy After Non-Missile Head Injuries*, 2d ed. Heinemann Medical, London.

49 McNamara, J. O., M. C. Byrne, R. M. Dasheiff, and J. G. Fitz. (1980) *Prog. Neurobiol.* 15: 139–159.

50 Schechter, P. J., Y. Tranier, M. J. Jung, and A. Sjoerdsma. (1977) *J. Pharmacol. Exp. Ther.* 201: 606–612.

51 Swanson, P. D., C. N. Luttrell, and J. W. Magladery. (1962) *Medicine* 41: 339–356.

52 Fonnum, F., and D. Malthe-Sørenssen. (1981) In *Glutamate: Transmitter in the Central Nervous System* (eds. P. J. Roberts, J. Storm-Mathisen, and G. A. R. Johnston) pp. 205–222. John Wiley, Chichester.

53 Van Gelder, N. M., A. L. Sherwin, and T. Rasmussen. (1972) *Brain Res.* 40: 385–393.

54 Van Gelder, N. M., and A. Courtois. (1972) *Brain Res.* 43: 477–484.

55 Koyama, I. (1972) *Can. J. Physiol. Pharmacol.* 50: 740–752.

56 Dodd, P. R., and H. F. Bradford. (1976) *Brain Res.* 111: 377–388.

57 Coutinho-Netto, J. C., A. S. Abdul-Ghani, J. F. Collins, and H. F. Bradford. (1981) *Epilepsia* 22: 289–296.

58 Croucher, M. J., J. F. Collins, and B. S. Meldrum. (1982) *Science* 216: 899–901.

59 Meldrum, B. S., M. J. Croucher, G. Badman, and J. F. Collins. (1983) *Neurosci. Lett.* 39: 101–104.

60 Peterson, D., J. F. Collins, and H. F. Bradford. (1983) *Brain Res.* 275: 169–172.

61 Chapman, A. G., K. Riley, M. C. Evans, and B. S. Meldrum. (1982) *Neurochem. Res.* 7: 1089–1105.

62 Curzon, G. (1980) In *Biochemistry of Psychiatric Disturbances* (ed. G. Curzon) pp. 89–112. John Wiley, Chichester.

63 Munro, H. N., J. D. Fernstrom, R. J. Wurtman. (1975) *Lancet* I: 722–724.

64 James, J. H., V. Ziparo, B. Jeppsson, and J. E. Fischer. (1979) *Lancet* II: 772–775.

65 Daniel, P. M., E. R. Love, S. R. Moorehouse, and O. E. Pratt. (1979) *J. Physiol.* (Lond) 289: 87–88.

66 Eriksson, L. S., A. Persson, and A. Wahren. (1982) *Gut* 23: 801–806.

67 Dodsworth, J. M., J. H. James, M. G. Cummings, and J. E. Fischer. (1974) *Gastroenterology* 64: 881.

68 Bradford, H. F., H. K. Ward, and A. J. Thomas. (1978) *J. Neurochem.* 30: 1453–1460.

69 Hamberger, A., G. H. Chiang, E. S. Nylén, S. W. Scheff, and C. W. Cotman. (1979) *Brain Res.* 168: 513–530.
70 Bradford, H. F., and H. K. Ward. (1975) *Biochem. Soc. Trans.* 3: 1223–1226.
71 Cremer, J. E., D. F. Heath, H. M. Teal, M. S. Woods, and J. B. Cavanagh. (1975) *Neuropathol. Appl. Neurobiol.* 3: 293–311.
72 Matheson, D. F., and C. J. Van den Berg. (1975) *Biochem. Soc. Trans.* 3: 525–528.
73 Ward, H. K., C. M. Thanki, and H. F. Bradford. (1983) *J. Neurochem.* 40: 855–860.
74 Hawkins, R. A., A. L. Miller, R. C. Nielsen, and R. L. Veech. (1973) *Biochem. J.* 134: 1001–1008.
75 Hamberger, A., B. Hedquist, and B. Nystrom. (1979) *J. Neurochem.* 33: 1295–1302.
76 Duffy, T. E., F. Vergara, and F. Plum. (1974) In *Brain Dysfunction in Metabolic Disorders*, vol 53. (ed. F. Plum), pp. 39–51. Research Publications of the Association for Nervous and Mental Disorders, Raven Press, New York.
77 Vergara, F., F. Plum, and T. E. Duffy. (1974) *Science* 183: 81–83.
78 Lindstrom, J. (1978) In *The Biochemistry of Myasthenia Gravis and Muscular Dystrophy* (ed. G. G. Lunt and R. M. Marchbanks), pp. 135–156. Academic Press, London.
79 Fritze, D., C. Herrmann, F. Naeim, G. Smith, E. Zeller, and R. Walford. (1976) *Ann. N.Y. Acad. Sci.* 274: 440–450.
80 Matthews, W. B. (1975) In *Viral Diseases of the Central Nervous System* (ed. L. S. Ellis and D. C. Gajdusek) pp. 145–160. Bailliere Tindall, London.
81 Heilbron, E., and E. Stalberg. (1978) *J. Neurochem.* 31: 5–11.
82 Rossor, M. N. (1981) *Birt. Med. J.* 283: 1588–1590.
83 Mans, A. M., S. J. Saunders, R. E. Kirsch, and J. F. Biebuyck. (1979) *J. Neurochem.* 32: 285–292.
84 Mans, A. M., J. F. Biebuyck, S. J. Saunders, R. E. Kirsch, and R. A. Hawkins. (1979) *J. Neurochem.* 33: 409–418.
85 Record, C. O., B. Buxton, R. A. Chase, G. Curzon, I. M. Murray-Lyon, and R. Williams. (1976) *Eur. J. Clin. Invest.* 6: 387–394.
86 Biebuyck, J. F., J. Funovics, D. Dedrick, Y. D. Sherer, and J. E. Fischer. (1974) In *Artificial Liver Support* (ed. R. Williams and I. M. Murray-Lyon), pp. 51–52, Pitman Medical, London.
87 Hindfeld, B. (1972) *Scand. J. Clin. Lab. Invest.* 30: 245–255.
88 Iversen, L. L., and A. V. P. Mackay. (1979) In *Psychopharmacology of Affective Disorders* (ed. E. S. Paykel and A. Coppen), pp. 60–89. Oxford University Press, Oxford.
89 Pedly, T. A. (1978) *Ann. Neurol.* 3: 2–9.
90 Rall, T. W., and L. S. Schliefer. (1980) In *The Pharmacological Basis of Therapeutics*, 6th ed. (ed. A. Goodman-Gilman, L. S. Goodman, and A. Goodman), pp. 448–474. Macmillan, New York.
91 Snead, O. C. (1983) *Int. Rev. Neurobiol.* 24: 93–179.
92 Ajmone-Marsan, C. (1969) In *Basic Mechanisms of the Epilepsies* (ed. H. H. Jasper, A. A. Ward, and A. Pope), pp. 299–319. Churchill, London.
93 Krall, R. L., J. K. Penry, H. J. Kupferberg, and E. A. Swinyard. (1978) *Epilepsia* 19: 393–408.
94 Matthysse, S., and J. Lipinski. (1975) *Ann. Rev. Med.* 26: 551–565.
95 Willner, P. (1983) *Brain Res.* 287: 211–246.

96   Gusella, J. F., N. S. Wexler, P. M. Conneally, S. L. Naylor, M. A. Anderson, R. E. Tanzi, P. C. Watkins, K. Ottina, M. R. Wallace, A. Y. Sakaguchi, A. B. Young, I. Shoulson, E. Bonilla, and J. B. Martin. (1983) *Nature* (Lond.) 306: 234–238.

97   Peterson, D., J. Collins, and H. F. Bradford. (1984) *Brain Res.* 311: 176–181.

98   Meldrum, B. S. (1981) In *The Molecular Basis of Neuropathology* (eds. A. N. Davison and R. H. S. Thompson), pp. 265–301.

99   Perry, T. L., S. Hansen, J. Kennedy, J. A. Wajda, and G. B. Thompson. (1975) *Arch. Neurol.* 32: 752–754.

100 Van Gelder, N. M. (1981) *Rev. Pure Appl. Pharmacol. Sci.* 2: 293–376.

101 Oppenheimer, D. R. (1976) In *Greenfield's Neuropathology*, 3d ed. (eds. W. Blackwood and J. A. N. Corsellis), pp. 608–651. Edward Arnold, London.

102 Ihara, Y., C. Abraham, and D. S. Selkoe. (1983) *Nature* (Lond.) 304:727–730.

103 Berger, P. D. (1982) *Ann. Rev. Med.* 33: 397–415.

104 Whitehouse, P. J., D. L. Price, R. G. Struble, A. W. Clark, J. T. Coyle, and M. R. De Longi. (1982) *Science* 215: 1237–1239.

105 Nagai, T., T. Pearson, F. Peng, E. G. McGeer, and P. L. McGeer. (1983) *Brain Res.* 265: 300–306.

106 Copper, P. E., N. Aronin, E. D. Bird, S. E. Leeman, and J. B. Martin. (1981) *Neurology* 31: 64. Abstract.

107 Iversen, L. L., and A. V. P. Mackay. (1981) *Lancet* II: 149.

108 Farley, I. J., K. S. Price, E. McCullough, J. H. N. Deck, W. Hordyinski, and O. Hornykiewicz. (1978) *Science* 200: 456–458.

109 Schultz, W. (1984) *Life Sci.* 34: 2213–2223.

110 Pezzoli, G., A. E. Panerai, A. Di Gulio, A. Longo, D. Passerini, and A. Carenzi. (1984) *Neurology* 34: 516–519.

111 Martin, J. B. (1984) *Neurology* 34: 1059–1072.

112 Bird, E. D. (1980) *Ann. Rev. Pharmacol. Toxicol.* 20: 533–551.

113 Van Kammen, D. P. (1984) *Life Sci.* 34: 1403–1413.

114 Kaiya, H., T. Tanaka, K. Takenchi, K. Morita, S. Adachi, H. Shirakawa, H. Ueki, and M. Namba. (1983) *Life Sci.* 33: 1039–1043.

115 Schwarcz, R., A. C. Foster, E. D. French, W. O. Whetsall, and C. Kohler. (1984) *Life Sci.* 35: 19–32.

116 London, E. D., H. I. Yamamura, E. D. Bird, and J. T. Coyle. (1981) *Biol. Psychiatry.* 16: 155–162.

117 Roberts, G. W., I. N. Ferrier, Y. C. Lee, T. E. Adrian, D. T. O'Shaughnessey, T. J. Crow, J. M. Polak, and S. R. Bloom. (1982) *Regul. Peptides* 3: 81.

118 I. N. Ferrier, T. J. Crow, G. W. Roberts, E. C. Johnstone, D. G. C. Owens, Y. Lee, A. Baracese-Hamilton, G. McGregor, D. J. O'Shaughnessey, J. M. Polak, and S. R. Bloom. (1984) in *Psychopharmacology of the Limbic System* (eds. M. R. Trimble and E. Zarifian) pp. 244–254. Oxford University Press, Oxford.

119 Rossor, M. N., and P. C. Emson. (1982) *Trends Neurosci.* 5: 399–401.

120 Perry, R. H., G. J. Dockray, R. Dimaline, E. K. Perry, G. Blessed, and B. E. Tomlinson. (1981) *J. Neurol. Sci.* 51: 465–472.

121 Beal, M. F., E. D. Bird, P. J. Langlais, and J. B. Martin. (1984) *Neurology* 34: 663–666.

122 Spinks, A., and W. S. Waring. (1963) *Prog. Med. Chem.* 3: 261–331.

123 Schechter, P. J., N. F. J. Hanke, J. Grove, N. Huebert, and A. Sjoerdsma. (1984) *Neurology* 34: 182–186.

124 Hardy, J., R. Adolfsson, I. Alafuzoff, G. Bucht, J. Maraisson, P. Nyberg, E. Perdahl, P. Webster, and B. Winblad. (1985) *Neurochem. Internat.* in press.

125 McGeer, P. L., E. G. McGeer, J. Suzuki, C. E. Dolman, T. Nagai. (1984) *Neurology* 34: 741–745.

126 Lloyd, K. G., C. Munari, L. Bossi, C. Stoffels, J. Talairach, and P. L. Morselli. (1981) in *Neurotransmitters. Seizures and Epilepsy* (eds. P. L. Morselli, K. G. Lloyd, W. Löscher, B. S. Meldrum, and E. H. Reynolds), pp. 325–338. Raven Press, New York.

127 Schochet, S. S., and McCormick. (1972) *Acta Neuropath.* [Berl] 21: 50–60.

128 Agid, Y., and F. Javoy-Agid. (1985) *Trends Neurosci.* 8: 30–35.

Appendix: Sources on Mammalian Neuroanatomy

The books in the following list are clear, explanatory texts on gross anatomy of the mammalian brain and nervous system. They are of particular value to those with little or no acquaintance with the subject.

Students often have trouble visualizing the actual shapes and relationships of the brain structures. Very helpful in this respect are three-dimensional models of the human brain that can be disassembled into major subregions. Commercially available models of this kind are referred to below.

Most neuroanatomical textbooks describe human brain anatomy and are aimed at medical students. The brains of other mammalian species have the same general plan of organization but may show important differences in detail. I have therefore included texts on rat brain anatomy for comparison.

GROSS NEUROANATOMICAL TEXTS

Human Neuroanatomy

1 P. L. Williams and R. Warwick. (1975) *Functional Neuroanatomy of Man* (the neurology section from *Gray's Anatomy*, 35th ed.). Churchill Livingstone, Edinburgh, London, and New York.

2 M. B. Carpenter. (1978) *Core Text of Neuroanatomy*, 2d ed. Williams and Wilkins, Baltimore and London.

3 M. B. Carpenter and J. Sutin. (1983) *Human Neuroanatomy*, 8th ed. Williams and Wilkins, Baltimore.

4 J. B. Angevine, Jr. and C. W. Cotman. (1981) *Principles of Neuroanatomy.* Oxford University Press, New York and Oxford.

5 *The Brain.* (1979) A Scientific American Book. Freeman, San Francisco.

6 M. Liebman. (1983) *Neuroanatomy Made Easy and Understandable*, 2d ed. University Park Press, Baltimore.

7 P. F. A. Martinez. (1982) *Neuroanatomy.* Saunders, Philadelphia and London.

8 L. Heimer. (1983) *The Human Brain and Spinal Cord.* Springer Verlag, New York and Heidelberg.

9 J. G. Chusid. (1981) *Correlative Neuroanatomy and Functional Neurology*, 18th ed. Lange Medical Publications, Los Altos, California.

10 C. L. Noback. (1981) *The Human Nervous System: Basic Principles of Neurobiology*, 3d ed. McGraw-Hill, New York.

11 J. Nolte. (1981) *The Human Brain.* Mosby, St. Louis, Toronto.

12 M. L. Barr. *The Human Nervous System: An Anatomic Viewpoint*, 3d ed. Harper and Row, Hagerstown, Maryland.

13 W. J. H. Nauta and M. Feirtag. (1986) *Fundamental Neuroanatomy.* Freeman, New York.

Rat and Vertebrate Neuroanatomy

1 W. Zeman and J. R. Maitland Innes. (1963) *Craigie's Neuroanatomy of the Rat.* Academic Press, New York and London.

2 L. W. Hamilton. (1976) *Basic Limbic System Anatomy of the Rat.* Plenum Press, New York and London.

3 G. Paxinos and C. Watson. (1982) *The Rat Brain in Stereotaxic Coordinates.* Academic Press, New York and London.

4 R. Pearson and L. Pearson. (1976) *The Vertebrate Brain.* Academic Press, London and New York.

HUMAN BRAIN AND NERVOUS SYSTEM
MODELS AND LEARNING PROGRAMS

1 Adam Rouilly Plastic Models of Natural Size with key cards. These are three models of different complexity. The most complex model (MBZ) separates into nine parts: frontal and parietal lobes, temporal and occipital lobes, medulla, cerebellum, basilar artery with branches (17 × 16 × 17 cm; 1200 g). Also available is a model of the spinal cord with nerve branches (MB14). These can be obtained from Adam Rouilly London Limited, Crown Quay Lane, Sittingbourne, Kent ME10 3JG United Kingdom.

2   A wide range of brain and nervous system models is available from Marshall-Cordell, 7124 North Clark, Chicago, IL 60626 and from Carolina Biological Supplies, 2700 York Road, Burlington, NC 27215.

3   The Structure of the Nervous System (1981). Four slide/tape programs with self-assessment tests and answers assembled by Edward G. Jones for the (American) Society for Neuroscience. Each program consists of two 30 minute audio cassettes coupled to eighty 35 mm projector slides.

Part 1   Neuron and Synapses (No. 7201).

Part 2   Supporting Cells; Peripheral Nervous System and Aspects of Nervous Organization (No. 7202).

Part 3   An Introduction to the Development of the Nervous System (No. 7203).

Part 4   Basic Neuroanatomical Methods (No. 7204).

The programs are distributed by Audio Visual Medical Marketing Inc., 404 Park Avenue South, New York, NY 10016.

Index

Acetyl CoA:
    in acetylcholine synthesis, 165
    in citric acid cycle, 140
Acetylcholine:
    and aggression, avoidance behavior, and self-stimulation, 432
    and arousal, 431
    and the ascending reticular activating system, 432
    and behavior, 429–432
    and circling behavior, 431
    inactivation of, 167
    pharmacology of, 171, 173
    release of, 167
    storage of, 166
    synthesis of, 166
    and tremor, 430
Acetylcholine receptors:
    amino acid sequence of, 174
    molecular organization of, 175, 176
    postsynaptic, 174
    presynaptic, 172
Acetylcholinesterase, 168
    histochemical detection of, 171
γ-Acetylenic GABA, 235, 236

Actin, 20, 25
    in microfilaments, 20
    in nerve fibers, 355
Adenosine as a neurotransmitter or neuromodulator, 247
Adenylate cyclase, 163
    and histamine, 252
    and phosphorylation, 369
    and purinergic systems, 248
    in second messenger systems, 161
Adenylate energy charge, 134
ADP as a neurotransmitter, 246–250
Adrenergic receptors, 195
    alpha type, 195
    beta type, 195
    blockers for, 195, 196
    isolation of, 202, 203
    peripheral, 200
    physiological actions mediated by, 189
β-Alanine, 235, 243
Alanine aminotransferase, 144, 146
Alkyl guanidines, 177
Amino acids:
    formation of, from glucose, 141, 142
    as neurotransmitters, 210–246
Aminooxyacetic acid, 235, 236